Extensional Tectonics: Faulting and Related Processes

Part II

Geological Society Key Issues in Earth Sciences

KEY ISSUES IN EARTH SCIENCES

Extensional Tectonics: Faulting and Related Processes

Part II

COMPILED BY

R.E. HOLDSWORTH
University of Durham, UK

J.P. TURNER
University of Birmingham, UK

2002
Published by
The Geological Society
London

THE GEOLOGICAL SOCIETY

The Geological Society of London (GSL) was founded in 1807. It is the oldest national geological society in the world and the largest in Europe. It was incorporated under Royal Charter in 1825 and is Registered Charity 210161.

The Society is the UK national learned and professional society for geology with a worldwide Fellowship (FGS) of 9000. The Society has the power to confer Chartered status on suitably qualified Fellows, and about 2000 of the Fellowship carry the title (CGeol). Chartered Geologists may also obtain the equivalent European title, European Geologist (EurGeol). One fifth of the Society's fellowship resides outside the UK. To find out more about the Society, log on to *www.geolsoc.org.uk*.

The Geological Society Publishing House (Bath, UK) produces the Society's international journals and books, and acts as European distributor for selected publications of the American Association of Petroleum Geologists (AAPG), the American Geological Institute (AGI), the Indonesian Petroleum Association (IPA), the Geological Society of America (GSA), the Society for Sedimentary Geology (SEPM) and the Geologists' Association (GA). Joint marketing agreements ensure that GSL Fellows may purchase these societies' publications at a discount. The Society's online bookshop (accessible from *www.geolsoc.org.uk*) offers secure book purchasing with your credit or debit card.

To find out about joining the Society and benefiting from substantial discounts on publications of GSL and other societies worldwide, consult *www.geolsoc.org.uk,* or contact the Fellowship Department at: The Geological Society, Burlington House, Piccadilly, London W1J 0BG: Tel. +44 (0)20 7434 9944; Fax +44 (0)20 7439 8975; Email: *enquiries@geolsoc.org.uk.*

For information about the Society's meetings, consult *Events* on *www.geolsoc.org.uk.* To find out more about the Society's Corporate Affiliates Scheme, write to *enquiries@geolsoc.org.uk*.

Published by The Geological Society from:
The Geological Society Publishing House
Unit 7, Brassmill Enterprise Centre
Brassmill Lane
Bath BA1 3JN, UK
(*Orders:* Tel. +44 (0)1225 445046
Fax +44 (0)1225 442836)
Online bookshop: http://bookshop.geolsoc.org.uk

British Library Cataloguing in Publication Data
A catalogue record for this book is available from the British Library.

ISBN 1-86239-115-7

Typeset by Alden Bookset, UK
Printed by The Alden Press, Oxford, UK.

Distributors

USA
AAPG Bookstore
PO Box 979
Tulsa
OK 74101-0979
USA
Orders: Tel. +1 918 584-2555
Fax +1 918 560-2652
E-mail *bookstore@aapg.org*

India
Affiliated East-West Press PVT Ltd
G-1/16 Ansari Road, Daryaganj,
New Delhi 110 002
India
Orders: Tel. +91 11 327-9113
Fax +91 11 326-0538
E-mail *affiliat@nda.vsnl.net.in*

Japan
Kanda Book Trading Co.
Cityhouse Tama 204
Tsurumaki 1-3-10
Tama-shi
Tokyo 206-0034
Japan
Orders: Tel. +81 (0)423 57-7650
Fax +81 (0)423 57-7651

Contents

Extensional tectonics: faulting and related processes

J. P. TURNER[1] & R. E. HOLDSWORTH[2]
[1]*University of Birmingham, School of Earth Sciences, Birmingham B15 2TT, UK (e-mail: j.p.turner@bham.ac.uk)*
[2]*University of Durham, Reactivation Research Group, Dept of Geological Sciences, Durham DH1 3LE, UK (e-mail: r.e.holdsworth@durham.ac.uk)*

Introduction

Aims of this book and background

'Oil is found in the minds of men'. Except for the addition of '...and women', this aphorism is as true today as ever before. The present volume brings together a collection of key papers that exemplify some of the most exciting applications of structural geology to problems in petroleum geoscience over the last twenty years or so.

Petroleum geoscience is a highly creative process combining theoretical knowledge of the physics that governs how the Earth behaves with an intuitive understanding of the enormity of geological time (does anyone have a feel for a million years?), rates of geological processes and the sheer dimensions and magnitude of many geological phenomena. The wealth of research in applications of structural geology and tectonics to problems in petroleum geoscience has provided a rich vein of scientific publishing for the Geological Society of London. The research on which this prolific output was based has been facilitated by:

(i) the openness of the petroleum industry in sharing its commercially valuable subsurface datasets and ideas with the research community, and

(ii) the enthusiasm with which structural geologists embraced the shift from deeper crustal, 'ductile' deformation processes toward problems in brittle faulting and fracturing.

Scientific collaboration between the petroleum industry and the structural geological research community has been tremendously successful through the 1980s and 1990s. In the UK, a large number of collaborative research programmes and groups, funded jointly by industry and government agencies, have used structural geology and tectonics to tackle a number of major issues in petroleum geoscience, e.g. British Institutes Reflection Profiling Syndicate (BIRPS), UK DTI-LINK scheme, NERC Ocean Margins Programmes, Fault Analysis Group, Rock Deformation Research Group, Fault Dynamics Research Group.

The principal objective of the Geological Society's new *Key Issues in Earth Sciences* series is to provide a coherent overview of a current issue in geoscience drawing on a mixture of ground-breaking research papers and synoptic reviews that hitherto have been widely scattered among its many publications. As such, it aims to provide a readily accessible introduction to the leading papers for students, HE teachers and professionals. To convey an impression of the history of development of ideas contained herein, papers are mostly reproduced in the order in which they were published. This volume on faulting and fault-related processes is the second of a two-part contribution to the *Key Issues in Earth Sciences* series on the general theme of extensional tectonic processes and their implications for petroleum geoscience. The emphasis of the papers reproduced here is to present practical methods and conceptual models for improved fault interpretation from seismic data. The accompanying part on Regional Tectonics (Holdsworth & Turner 2002) addresses processes occurring on the scale of the sedimentary basin and greater.

This introductory paper summarises both the context of the present volume and the contents of each paper included here. Papers reproduced in the current volume are indicated in bold, whilst others referred to in the text are in the reference list as normal, including those from the companion part of this volume.

Role of structural geological research in petroleum geoscience

These have been exciting times for structural geologists working in petroleum geoscience, whether in academia, consultancy or oil companies. Fundamental advances in faulting and fault-related processes on which the Geological Society has published include:

- Elucidation of fault geometries in the crust using both geological and geophysical datasets (e.g. **Gibbs 1994**; **Jackson 1987**; **Walsh & Watterson 1991**; Yielding *et al.* 1991; **Roberts & Jackson 1991**; Morley 1995).

- Demonstration of the link between sedimentation and deformation and the development of models of interacting

From Holdsworth, R.E. & Turner, J.P. (compilers) *Extensional Tectonics: Faulting and Related Processes*. The Geological Society, Key Issues in Earth Sciences, **2 (2)**, 1–9.
1476-315X/02/$15.00

faulting and sedimentation (e.g. Leeder & Gawthorpe 1987; Leeder *et al.* 1991).

- Widespread recognition of fault reactivation in a variety of geodynamic settings and recognition of the concepts and underlying controls of sedimentary basin inversion (e.g. Williams *et al.* 1989; Hayward & Graham 1989; Sibson 1995; **Eisenstadt & Withjack 1995**; Holdsworth *et al.* 1997).

- Improved understanding of how faults grow and interact with adjacent faults on a variety of scales using geological and geophysical data (e.g. **Childs *et al.* 1995**; **Walsh *et al.* 1996**) and the results of analogue modelling experiments (e.g. **Vendeville *et al.* 1987**; **McClay 1995**).

- Development of quantitative methods to describe the spatial attributes and growth of natural fracture networks to better predict how these systems evolve and influence reservoir quality and fluid flow pathways (e.g. **Walsh & Watterson 1991**; **Manzocchi *et al.* 1998**; **Gibson 1998**; **Fisher & Knipe 1998**; **Ottesen Ellevset *et al.* 1998**; **Lonergan *et al.* 1998**).

A natural consequence of the sharing of subsurface data between the petroleum industry and research communities has been widespread access to state-of-the-art datasets and the use of leading edge interpretational software in higher education. Virtually every Earth Sciences degree programme includes courses that offer the opportunity to work with seismic images of producing oil and gas fields and many students have had first-hand experience of the sometimes spectacular capabilities of 3-D reflection seismic data. Whilst many of the papers reproduced in this volume are based at least partly on data collected from traditional field studies, seismic interpretation has become a new kind of virtual fieldwork. Students enjoy making geological interpretations of seismic data and they benefit from the constraints it allows them to place on the scale of geological processes in a way that outcrop geology rarely permits.

Has all this research led to the discovery and production of more oil and gas? Undoubtedly the answer is yes, particularly with the increased maturity of major petroleum provinces and prioritization of the search for the so-called subtle trap. However, the precise impact of geological research on exploration and production is difficult to quantify. Early in the exploration and production history of major petroliferous basins like the North Sea, the influence of research was probably minimal because most discoveries were made by drilling structural culminations, or 'highs', and more or less hoping for the best! However, the application of increasingly sophisticated geological modelling in combination with high-resolution reflection seismic data, especially 3-D seismic, has meant that subtle trapping configurations such as sub-salt structures, stratigraphic pinchouts, hanging wall closures and even fractured basement reservoirs are much more common drilling targets today than twenty years ago.

In addition to the tangible achievements of structural geological research in petroleum geoscience, it is worth remembering also its important contribution toward raising the general level of intellectual curiosity among petroleum geoscientists, with attendant benefits for professional efficacy and overall quality of life. For versatile scientists with a general interest in the structure and evolution of our planet, commerce and technology, geoscientific work in the petroleum industry is still the most stimulating of careers.

Basic terminology

This volume deals with geological faulting and fault-related processes. A fault is a shear fracture across which a marker horizon, such as a sedimentary layer, has undergone measurable displacement. Faults therefore differ from joints – brittle fractures that exhibit no measurable displacement, most of which form under conditions of zero resolved shear stress. In practice, faults of more than ten or so centimetres displacement are more accurately described as fault zones because their cumulative displacement is accommodated by an array of individual slip surfaces. Experimental rock mechanics and observations from active crustal-scale faults show that they classically develop at angles of 30° to the maximum principal stress axis.

Displacements can occur in any orientation within the fault surface, constrained by two end-member categories: dip-slip, where movement is parallel to the dip of the fault plane, and strike-slip, where movement is parallel to the strike of the fault. Dip-slip faults are further subdivided into normal and reverse according to whether they form in stress fields in which the greatest or least principal stress is vertical, respectively (Anderson 1951). Thus, normal faults are generally relatively steeply inclined (*c.* 60°) with their overlying, or hanging wall moving down relative to the opposing footwall. Normal faults generally lead to crustal extension, or stretching. Reverse faults, however, are usually gently inclined (*c.* 30°) in which case they are termed thrusts. They are characterized by upward movement of the hanging wall which is thrust up and over the footwall. Hence, thrusts generally lead to crustal shortening, or contraction. Steeply inclined reverse faults owe their steep inclination either to back-rotation and steepening of the fault after its formation, or they may have initiated as normal faults, being subsequently reactivated as reverse structures during compression (basin inversion). Strike-slip faults are subdivided into right-lateral (or 'dextral') and left-lateral (or 'sinistral') according to the sense of relative displacement across the fault of a reference marker when seen in plan view. In practice, many faults display oblique-slip kinematics in which there is a significant component of both strike-slip and dip-slip displacement.

Why so much petroleum-sponsored research on faulting?

Some 75% of petroleum fields are significantly fault-controlled. The basin- to lithosphere-scale aspects of this fault control are examined in the *Key Issues in Earth Sciences* volume that accompanies this book, (Holdsworth & Turner 2002). In the context of the present volume, fault control of petroleum traps encompasses two main classes of phenomena:

(1) Structural geometry and genesis. In order to conserve the cutoff angle between unfaulted sedimentary layering and the incipient fault plane, movement of a hanging wall sequence over a curviplanar fault profile causes it to deform systematically. With dip-slip faults, this results in the formation of secondary synthetic-antithetic faulting and passive folding of the hanging wall succession. In areas of normal faulting, 'rollover anticlines' are important petroleum traps in the major prospective Tertiary deltas such as those of Nigeria, Borneo and Egypt. Ramp anticlines form important traps in petroliferous thrust belts in Canada, Iran and Colombia.

(2) Fault seal. In so called juxtaposition traps like those associated with the Brent Province of the northern North Sea, faults provide critical permeability barriers inhibiting cross-fault migration of oil into finer grained, non-reservoir facies in their hanging walls.

These may be termed first order controls in that the integrity of the trap is entirely dependent on the configuration of the faulting. However, faults also exert a multitude of second order influences over the prospectivity of petroleum systems. For example:

(i) differential compaction can generate very large wavelength anticlines above basement horsts (e.g. the 250 km-long Gharwar Field, Saudi Arabia);

(ii) footwall uplift and consequent erosion is an important source of synrift reservoir rocks (e.g. Brae Field, South Viking Graben, North Sea);

(iii) sediment entry points to the depositional basin often coincide with the positions of relay ramps, or stepover zones between overlapping faults (e.g. Beryl Embayment, South Viking Graben);

(iv) flushing of overpressured aqueous fluid along faults and through their wall rocks may significantly alter reservoir quality for better or worse;

(v) an understanding of the compartmentalizing effect of low permeability, intra-reservoir fault rocks has direct implications for optimizing production strategies (e.g. Sleipner Vest Field, Norwegian North Sea);

(vi) secondary migration from source rock kitchen to reservoir is often strongly dependent on the presence of non-sealing faults juxtaposing permeable 'carrier' beds at critical points along the migration pathway (e.g. Wytch Farm Field, Wessex basin, southern England); and

(vii) in fields like Brent, where the Middle Jurassic reservoir is charged by Upper Jurassic source rock, fault throw (the vertical component of fault displacement) needs to be sufficiently large to place source rock physically beneath Middle Jurassic reservoir.

Context of the papers in the present volume

Geometry and spatial attributes of normal faults

For many geoscientists used to observing phenomena on the scale of small outcrops, the greatest challenge in making structural interpretations from seismic reflection data is one of scale. At typical reservoir depths, and assuming reasonable interval velocity and dominant frequency, minimum vertical and horizontal resolution is respectively 15–20 m and 40–50 m on migrated data, i.e. the scale of a sizeable outcrop. On the other hand, the ability to view high-quality 3-D seismic data in more or less any orientation has given structural geologists a superb insight into the 3-D architecture of mesoscopic fault systems. Much of the emphasis of recent research into the geometry and spatial attributes of faults has therefore focused on reconciling problems of scale and coherence with the new understanding of fault architecture brought about by 3-D seismic visualization.

For a variety of reasons – such as the natural filtering effect of the Earth, problems of resolution and the multitudinous sources of 'noise' – seismic data at reservoir depths and greater afford considerable latitude as to how geoscientists choose to make their structural interpretations. Because of this, seismic interpreters thrive on analogues and precedents as an illustration of the plausibility of an interpretation. As one of the first to attempt to apply ideas of 'thin-skin faulting' to extensional sedimentary basins, **Gibbs** provided seismic interpreters with a taste of an extensive new range of extensional fault geometries to test on their datasets.

Development of the concepts of thin-skin faulting in the 1960s and 1970s provided structural geologists with a powerful new model for understanding crustal-scale orogenic fault systems in terms of networks of curviplanar faults diverging from a common basal detachment (e.g. McClay & Price 1981). Balanced, or retrodeformable, cross-sections depict the geometry of faults and folds within the deforming thin skin, above the basal detachment. They take their name from the idea that the area and/or bed length of the deformed cross-section and its restored counterpart should balance. Thus, deformation is assumed to conserve rock volume throughout its progression. The balanced cross-section methodology allows for direct quantification of the magnitude of fault-related strain (i.e. shortening or lengthening) and it gives valuable insights to the geometric evolution of thin-skin fault systems. Gibbs took these well-established concepts from thrust belts and applied them for the first time to extensional systems.

Thrust systems evolve in response to the balance between the gravitational potential energy of the thickened wedge of shortened crust (promoting collapse of the thrust wedge) and the strength of the wedge and its basal detachment (serving to support the wedge). The gravitational potential energy of extensional fault systems is overall fairly low and most crustal-scale extensional faults appear to form thick-skin systems in which individual faults are rooted in the mid-crustal brittle-ductile transition. However, at the margins of extensional basins, where the combination of footwall uplift and hanging wall subsidence elevates rift shoulders high above the basin floor, gravity-driven faulting during extension is important. Consequently, it is at the basin margins where Gibbs' ideas on thin-skin extensional systems have been most valuable. The development of flat-lying extensional detachment faults is thought to require the presence of weak layers such as salt or overpressured shale. As a result, the extensional thin-skin terminology has also been particularly successfully applied to halokinetic fault systems (e.g. Central North Sea Chalk, Gulf of Mexico) and to the gravitational collapse of Tertiary deltaic sedimentary sequences (e.g. Niger, Nile, Barram deltas).

Most petroleum geoscientists view normal faults as imaged on time-migrated seismic reflection profiles. However, it can be difficult to gain an accurate impression of fault geometry because:

(1) The progressive increase in acoustic velocity with depth means that seismic time sections tend to progressively 'pull-up' inclined planar surfaces with depth, thereby creating artificial concave-up (listric) fault geometries.

(2) The typical 1:25,000/10 cm/sec aspect ratio with which most seismic data are displayed is strongly vertically exaggerated.

Consequently, seismic interpreters tend to rely on empirical rock mechanics data to infer the inclinations of faults. These may not be valid in the top 3 km of the crust, from where 90% of petroleum is produced. For many years it has been assumed that brittle failure in the upper crust is governed by the Mohr-Coulomb failure criterion. Using a representative angle of internal friction (an empirical constant per rock type) of 30°, this predicts that faults and fractures will develop at 30° to the maximum principal stress which, in extensional basins, is vertical (Anderson 1951). Presenting fault dip data from several thousand faults measured from coalmine plans mainly from the East Pennines Coalfield, **Walsh & Watterson** demonstrate that 70° rather than 60° is a better approximation of the inclination of upper crustal (<4 km depth) faults.

Rock mechanics theory shows that the chief control over the angle between fracture surfaces and the principal stress axes is not the absolute magnitude of the stress tensors, but the ratio between differential stress – the difference between the maximum and minimum principal stress – and the tensile strength of the rock undergoing failure. As a rule of thumb, if this ratio is more than 8, shear failure will occur according to the Mohr-Coulomb criterion (i.e. fractures develop at 30° to maximum principal stress), and if it is less than 4, tensile fractures will form at 90° to the minimum principal stress and parallel to the maximum principal stress. Walsh & Watterson conclude that failure in the upper crust takes place according to the so called modified Griffith criterion in which the ratio lies between 4–8. Thus, faults and fractures develop at angles of between zero and 30° to maximum principal stress. This theory also provides an elegant explanation for listric fault geometry without needing to invoke Gibbs' thin-skin basal detachments because:

- At very shallow depths, or in overpressured rocks where effective stress (the difference between applied normal stress and pore fluid pressure) is small, failure will be tensile and the resulting faults will be sub-vertical.
- At greater depths, where differential stress is larger, fault inclination should flatten to 60° in accordance with the Mohr-Coulomb criterion.

The potential ambiguity inherent in interpreting seismic reflections from normal faults is a theme addressed by **Yielding, Badley & Freeman**. They present well constrained depth migrations of fault plane reflections from large normal faults in the northern North Sea. Fault plane reflections are common in deltas and continental margins where clay and shale is smeared along the surfaces of low-angle, gravity-driven extensional faults. However, most faults are not expressed as a discrete seismic reflection but rather as a diffuse discontinuity along which adjacent sedimentary layers are truncated. Consequently, the identification of fault plane reflections from tectonically driven normal faults allows direct assessment of their geometry to depths of several kilometres.

In the examples of Yielding *et al.*, fault plane reflections are generated by the juxtaposition of footwall basement with Triassic half-graben fill, at which levels the faults are inclined at 30–35°, sometimes as gentle as 25°. At shallower levels, however, where these faults cut through the Jurassic sedimentary column overlying basement, they attain steeper inclinations of 40–50°. This marked change in dip is interpreted in terms of the well known two-stage rifting history of the North Sea. One consequence of stretching and thinning a volume of crust containing 'rigid' faults is that the fault planes will rotate to progressively gentler inclinations with increasing thinning. This is often envisioned in terms of an arrangement of initially vertical dominoes rotating within a horizontal shear couple. In the examples of Yielding *et al.*, earlier Triassic extension resulted in initially steep normal faults, progressively flattening with advancing extension, according to the 'domino' effect. Subsequent Late Jurassic rifting rejuvenated the Triassic faults whose tips propagated through the Jurassic syn-rift basin fills at steeper, Andersonian angles of 50–55°. Two lessons arise from this work:

(1) Relatively low-angle fault 'splays' that appear to cut through an unfaulted hanging wall succession can look like reflected refractions or even diffractions. However, in these North Sea examples, migration at higher velocities demonstrates that they are primary reflections from the deeper segments of faults.

(2) Without recognition of a multi-stage rifting history, faults that flatten with depth could be mistakenly interpreted as listric.

Given the rarity of reflections from tectonically driven faults, the most readily available guides to fault plane geometry are still (i) hanging wall and footwall cutoffs, and (ii) the geometry of fault-related hanging wall deformation. Conservation of hanging wall cutoff angle means that translation of a hanging wall succession over a curviplanar fault surface will generate predictable deformation in the overlying units. The precise relation between the geometry of this deformation and the shape of the bounding fault is examined by **White & Yielding**.

Starting with the 'Chevron' construction (Verrall 1981), White & Yielding modify it to take account of the geologically more realistic effects of inclined (rather than vertical) hanging wall simple shear, and differential compaction. The Chevron construction provided seismic interpreters with a straightforward method for constraining fault geometry at depth by assuming that a hanging wall sequence will collapse along vertical flow lines (with respect to a hanging wall reference point) onto a curviplanar normal fault plane. In White & Yielding's method, fault geometry is reconstructed from information on the geometry and heave of at least two marker horizons, and the geometry of the portion of the fault between the two markers.

Unlike previous solutions to this problem, White & Yielding's scheme requires no *a priori* knowledge of the angle of inclined simple shear, nor the amount of differential compaction. Instead, they employ a new algorithm to compute fault shape from hanging wall geometry using the two (or more) hanging wall markers. The optimum fault shape is computed from minimizing the difference between the solutions from each of the markers. In cases where differential compaction is negligible, the difference between solutions computed from different markers will also be negligible. Testing the algorithm on data from 3-D seismic surveys and analogue models of faulting suggests that hanging walls deform by antithetically inclined simple shear (i.e. shear couple inclined toward the fault), and that differential compaction is usually significant.

Walsh & Watterson provide practical tools that enable maximum objectivity to be applied to the manner in which seismic interpreters choose to link faults. This work has significant implications for the way in which fault maps are drawn from a series of vertical seismic sections – still the most common procedure by which seismic surveys are interpreted – and it has important bearings on the hydraulic connectivity of faulted reservoir successions.

Using data derived from interpretation of a 3-D seismic survey, Walsh & Watterson project displacement values at points on a fault surface onto a plane parallel to the strike of the fault. They show that displacement varies smoothly from a maximum in the centre of a fault, decreasing toward its tip. This simple observation provides seismic interpreters with a powerful tool for quality controlling their fault maps – incorrect correlations of fault cutoffs between adjacent seismic profiles will often show up as implausibly large changes in fault displacement on strike projections.

This kinematic coherence applies also to fault arrays, over several orders of magnitude of scale, and Walsh & Watterson introduce terms 'hard' versus 'soft' linkage to describe fault systems in which linkage between faults within an array is achieved respectively by physical connection of the fault planes, and ductile strain of the intervening rock. They show that each individual fault or splay contributes to the overall coherence of an array. If kinematic coherence is maintained during the evolution of an array, it necessitates a high degree of synchronous movement from the earliest stages of development of a fault array.

A quasi-philosophical implication of the kinematic coherence of fault systems over orders of magnitude of scale is the scale-dependence of maps. For example, an array of faults on one scale will often be represented by a single fault on a smaller scale map. Similarly, 'ductile strain' in this context describes deformation occurring on scales less than the smallest fault displacement shown on a map. Ductile strain during normal faulting is particularly important to seismic interpreters because it accommodates significant proportions of the total strain yet its presence is often difficult to resolve on seismic data. **Walsh et al.** present examples of ductile strain effects from coalfield data, seismic surveys and outcrop which account for as much as 50% of the total strain. They show that ductile effects are most pronounced at the intersections of conjugate faults (i.e. penecontemporaneous, opposing faults arranged symmetrically with respect to one of the principal stress planes), and in fault overlap zones.

Appreciation of overlap zones, today termed relay ramps, has been enormously important to the petroleum industry. This is largely because of the control they exert over the locations of topographic lows along fault-bounded basin margins, thereby acting as entry points for footwall-sourced sediment into the depositional hanging wall. In addition, overlap zones may act as migration pathways and 'weak' points during petroleum production that allow communication between otherwise isolated fault-bounded compartments in an oilfield. Drawing on examples from 3-D seismic datasets, **Childs, Watterson & Walsh** define overlap zones as sub-parallel fault traces arranged en echelon. The width of the overlap and the distance separating the overlapping faults is small compared to the total length of the fault traces.

Relay ramps are shown to be transient features that form during the growth of an initially unconnected array of faults, but they may be preserved in the rock record in situations where evolution of the fault array is arrested. For example, some relay ramps probably originate as helical shear planes (Riedel shears) splaying from an underlying basement lineament oriented slightly obliquely to the prevailing principal stress axes. Consequently, the size of the overlap will vary according to the level at which the relay ramp is mapped and Childs *et al.* present an illustrative example of this from the Gulf of Mexico. As well as their sedimentological significance, relay ramps have important bearings on trap integrity because of their tendency to compartmentalize reservoirs. Clearly, the size of the overlap zone needs to be less than the height of the petroleum column if trap integrity is to be maintained.

Active normal faulting and seismicity

For many years, exploration in rifted continental margins and aulacogenic basins like the North Sea continued without petroleum geoscientists having a good modern analogy for the basins in which they were working. It was clear that many of these ancient basins evolved in response to considerable amounts of lithospheric and crustal stretching and that, in the brittle upper crust, stretching was accommodated by normal faulting. However, until seismological and geological information from actively extending regions like the Aegean and the western USA came to light, there was little understanding of what petroliferous basins such as the East Shetland Platform looked like, and the link between their faulting, subsidence and sedimentary infilling.

The reviews of active normal faulting by **Sibson** and **Jackson** highlight two recurrent themes:

(1) Irrespective of altitude (and therefore gravitational potential), and the magnitude and nature of the forces driving actively extending regions, seismically active normal faulting looks much the same wherever it is encountered on the continents.

(2) The mechanical behaviour of continental fault zones at different crustal levels is governed largely by the temperature-dependent response of quartzo-feldspathic lithologies to deviatoric stress (the difference between measured stress and average, or hydrostatic, stress).

Drawing on examples from the Aegean, Turkey, Tibet, East Africa and the western USA, Jackson demonstrates the remarkable uniformity in fault spacing (*c.* 20 km–100 km), trace length (*c.* 15 km–20 km), inclination (40–60° to the horizontal) and geometry (planar). From examination of 20 years of fault plane solutions and focal depth records, Jackson is unable to cite a single example of a large continental earthquake (magnitude >5.2) that nucleated on a fault inclined at 20° or less and he devotes much space to discussion of the puzzling origins of low-angle extensional faulting.

The significance of Jackson's seemingly arbitrary definition of a large earthquake is that such earthquakes originate on faults of length 10–15 km, the thickness of the seismogenic layer of continental crust. This is the uppermost layer of relatively cool (<450°C), brittle crust that can store and release elastic strain energy (i.e. seismicity). Sibson uses graphs of shear resistance versus depth to characterize the seismogenic layer, the base of which is signified by an abrupt drop in shear resistance. These graphs provide semi-quantitative support for many common observations from modern and ancient extensional settings. For example:

- Rock deformation textures reflect the transition from episodic, pressure-sensitive cataclasis and frictional sliding within the seismogenic layer, to a quasi-plastic regime of largely aseismic, continuous shearing localized within mylonite belts.
- The depth of the transition from frictional (coseismic; slip rates of 10–100 cm/sec) to quasi-plastic (aseismic; slip rates of 0.1–10 cm/a) deformation is highly sensitive to geothermal gradient and strain rate – high geothermal gradient and slow strain rate promotes shallowing of the base of the seismogenic layer and vice versa.
- Large earthquake ruptures tend to nucleate near the base of the seismogenic layer in the region inferred to have the highest shear resistance and concentration of distortional strain energy. Beneath this level, seismic energy diminishes abruptly as quasi-plastic, aseismic creep becomes the dominant mode of strain accommodation.

As to the implications of this work for petroleum geoscience, it is the integration of fundamental data on fault mechanisms addressed by Sibson and Jackson, with information on the geomorphology and drainage of actively extending regions that have provided the most important insights. Thus, maps of fault traces, topography and bathymetry presented by **Roberts & Jackson** from a rapidly extending (10–20 mm/a) region of neotectonic half-graben in central Greece demonstrate the first order control exerted by major faults on topography and bathymetry. These maps offer a striking analogy for North Sea palaeogeography during the Kimmeridgian, showing submerged hanging wall basins juxtaposed across major fault scarps with emergent, actively eroding footwalls.

Analogue modelling

Given the preceding discussions on the importance to seismic interpreters of geological analogy and precedent, the familiar results of analogue deformation models exemplified by **McClay** and **Vendeville et al.** demonstrate the impressive success achieved by modern modelling methods in replicating a diverse range of geological phenomena. Thus, these papers include graphic examples of laboratory simulations of ramp-flat listric faults and associated hanging wall deformational geometry, en echelon fault segmentation above obliquely reactivated 'basement' lineaments, relay ramps, compressional reactivation of formerly extensional faults (basin inversion), systematic relations between brittle (seismogenic) layer thickness and fault spacing (c.f. Sibson, Jackson), fault trace length increasing with maximum displacement, gravity-driven faulting and spreading, the domino effect, and even isostatic uplift of rift shoulders.

Clearly there are limits to how far these analogues may be taken. Laboratory strain rates of *c.* 36^{-3}/sec, or 10% strain/60 mins (calculated from McClay's strain rate of 4.16 × 10^{-3}/sec in a 120 cm-long deformation rig), contrast with geological strain rates of 10^{-13}–10^{-14}/sec (10% strain/30–300 Ka). Vendeville *et al.* acknowledge this problem implicitly in their models of whole-lithosphere deformation which respectively employ silicone putty and honey to replicate the near-Newtonian behaviour of mantle lithosphere and aesthenosphere at geological strain rates. Also, in spite of the introduction of different grades of sand and mica to the models of McClay, it is impossible to simulate fully the orders of magnitude range in scale on which the crust exhibits anisotropic behaviour. In addition, as McClay points out, analogue models do not adequately simulate flexural, isostatic and thermal effects, compaction, synrift sediment starvation, fault displacement gradients, fluid flow and overpressure. Nonetheless, these models represent a major advance in our understanding of the geometric and kinematic evolution of fault and fold structures, and they can provide valuable templates for seismic interpretation in regions of poor data quality.

Most the models reported in McClay employ a rigid footwall in experiments simulating listric faulting and/or fault reactivation. Presumably this reflects the inability of the dry sand and mica layers to undergo displacement along low-angle surfaces unless the fault planes are near-frictionless. For basins which have experienced more or less uninterrupted uniaxial stretching and concurrent faulting, a rigid footwall is probably a reasonable approximation on the 'seismic' scales addressed by McClay's models (widths of *c.* 100 km). Thus, examples of fault geometry from the Gulf Coast, SE Asia and East Africa probably exhibit the greatest correspondence with McClay's sandbox models.

The assumption of a rigid footwall is, however, a poor approximation to reality in inverted basins. This is because normal faults are unfavourably steeply inclined during horizontal compression which favours gently inclined faults, at Andersonian angles of *c.* 30° to the horizontal. A common solution to this problem during basin inversion is for compressional faults to propagate into the footwalls of normal faults creating the well known 'shortcut fault' with its typical arrowhead geometry.

Eisenstadt & Withjack present the results of analogue modelling of basin inversion using wet clay to replicate more satisfactorily situations in which hanging wall and footwall are involved in the deformation. Their specific objective is to examine the relationship between the magnitude of shortening strain in inverted basins and the spectrum of structural geometries that evolve with progressively more shortening. The rationale for their work comes from the extreme difficulty in placing even the most qualitative constraints on shortening strain in inverted basins. For example, seismic interpreters may refer to their basins having undergone mild inversion when, as Eisenstadt & Withjack are able to demonstrate, seemingly modest amounts of buckling or reverse fault displacement may reflect substantial shortening.

Reasons for the deceptiveness of compressional structures in inverted basins are principally two-fold:

(1) Assuming that all shortening is accommodated by reverse-sense fault displacement, formerly extensional faults will need to have all their normal-sense displacement cancelled before they begin to accumulate reverse displacement.

(2) In reality, a plethora of small, sub-seismic scale structures will accommodate large amounts of shortening strain alongside major fault displacement. Furthermore, inverted basins will undergo 'horizontal compaction' in which significant shortening strain is accommodated by volume (porosity) reduction.

Eisenstadt & Withjack conclude that most methods for quantifying shortening strain in inverted basins yield gross underestimates. Expressing inversion magnitude as a percentage of the original extensional strain, they demonstrate that seismic-scale manifestation of compressional deformation is mild even when inversion has attained 100%. Thus, classic inversion structures exemplified by seismic examples from the southern North Sea, the Irish Sea basins and SE Asia may record extreme inversion of 200% or more.

Intra-reservoir faulting and fluid flow

For many years, fault-controlled juxtaposition seals have been among the most frequently targetted trapping configurations. Until recently, however, there has been a dearth of information on the key factors that control the permeability, capillary displacement pressure and porosity of *intra*-reservoir faults across which reservoir/reservoir juxtaposition is maintained. In many siliciclastic reservoirs, intra-reservoir faults constitute the major heterogeneity and will often segment a field into fault-bounded compartments, each with differing hydrocarbon-water contacts. The recent spate of work on intra-reservoir faulting has therefore provided geoscientists and engineers with essential new data with which to optimize their production strategies.

Based on extensive datasets, **Fisher & Knipe** and **Gibson** address the information shortfall with general descriptions of the microstructural characteristics of siliciclastic fault rocks and the principal factors controlling their petrophysical properties. Fault rock permeability and capillary displacement pressure can vary by over six and three orders of magnitude, respectively, according to the effective stress, temperature and sediment composition at the time of deformation. Consequently, faults are best considered as membranes, or flow retarders, with variable transmissibility properties (Ottesen Ellevset *et al.* 1998) and it is therefore over-simplistic to classify a range of faults, from minor production baffles to those which have supported large hydrocarbon columns over geological time, as either sealing or non-sealing.

Permeability is the fundamental factor controlling the rate of fluid flow in response to an applied differential hydraulic head and it is the most important parameter in dictating fluid flow patterns within a reservoir during its production. Capillary displacement pressure controls the ability of a lithological boundary to act as a barrier to fluid flow. Fisher & Knipe highlight five fault seal processes that serve to decrease porosity and permeability, and increase capillary displacement pressure:

- Deformation-induced mixing
- Pressure solution and quartz cementation
- Cataclasis
- Cementation
- Clay smear

The compositional purity of sandstones is a fundamental control on the extent to which these processes operate. Thus, clay-free sandstones tend to seal by cataclasis (physical diminution of framework grains by pervasive brittle fracturing) and post-deformation quartz cementation whereas impure (>15% phyllosilicates) sandstones tend to seal by deformation-induced mixing (particulate flow without grain-fracturing), cataclasis and post-deformation pressure solution. For a given sediment composition, the relative importance of these sealing processes depends upon the interaction between the stress, strain and temperature history of the reservoir. In the future, therefore, fully integrated studies of individual fields have the potential to deliver powerful fault seal predictive capability.

The idea of faults acting as membranes is reinforced in the case study of the Sleipner Vest Field by **Ottesen Ellevset *et al.***, in which marked pressure-compartmentalization occurs in a siliciclastic reservoir (Callovian Hugin Formation) whose diagenetic history is uniform across the extent of the field. 3-D seismic data reveal, however, that the field is dissected by faults across which offsets in hydrocarbon-water contacts of tens of metres occur between some dozen fault-bounded compartments. As Fisher & Knipe demonstrated from core data, phyllosilicate content of undeformed reservoir facies exerts a primary control over fault sealing capacity through its influence on the distribution of phyllosilicate smears. But it is the predictive capabilities of fault seal studies that give this branch of structural geology its uniquely applied edge. For example, the assumption of Ottesen Ellevset *et al.* of a correlation between the volume of shale in the undeformed reservoir and the volume of phyllosilicates in the resultant fault rock allows them to produce maps of Shale Gouge Ratio (SGR) – essentially the clay volume fraction in an interval equal to the fault throw – as a proxy for fault sealing capacity.

Clearly, different geological scenarios (such as fault reactivation), issues of data resolution and factors such as the type of phyllosilicate and the differential incorporation of phyllosilicates into fault rocks from footwalls and hanging walls mean that the robustness of SGR as a proxy for fault sealing capacity is uncertain. Nevertheless, the approach of Ottesen Ellevset *et al.* provides a consistent explanation for the observed differences in hydrocarbon-water contacts across Sleipner Vest and highlights the potential contribution of this methodology to the quantitative evaluation of risk associated with fault seal.

Many geologists will have examined zones of anastamosing networks of prominently weathered, very small-displacement faults in outcrops of high-porosity, often poorly indurated sandstone (e.g. the famous examples in the Entrada and Navajo sandstones in the USA described by Aydin & Johnson 1978). Frequently termed granulation seams, they generally exhibit individual displacements of millimetres or less and occur exclusively in quartz-rich sandstone. Granulation seams exemplify the process of strain-hardening in which a material becomes progressively stronger with increasing strain. Given that porosity exerts a primary control over rock strength, cataclasis-induced porosity reduction (up to 80% of initial porosity) leads inevitably to strengthening, and many granulation seams represent incipient faults that 'locked up' before a throughgoing slip surface developed.

The global importance of quartz-rich siliciclastic reservoir facies and the relative ease with which cataclasis can be modelled has meant that granulation seams have been the subject of numerous studies. Whilst Gibson presents extensive data to support orders of magnitude differences in the petrophysical properties of granulation seams according to their deformation mechanism (namely cataclasis, pressure solution and combined cataclasis/diffusive mass transfer) and undeformed formation characteristics, **Manzocchi, Ringrose & Underhill** use 2-D flow simulations to investigate the influence of geometrical architecture of granulation seams on effective permeability. Thus, their models assume self-similar granulation seams (i.e. no fault-scaling relationships) of identical thickness and permeability, with the connectivity of the seams exerting the strongest influence on fluid flow.

Modelling single-phase systems in this way, Manzocchi *et al.* show that effective permeability is linearly related to reservoir compartmentalization, itself a function of the spacing and clustering of granulation seams and the geometry of their intersections, or nodes (e.g. I-, Y- or X-type nodes). They also investigate the flow properties of granulation seams in two-phase systems in which oil recovery is largely controlled by the ratio of viscous to capillary forces. Thus, when the viscous/capillary ratio is high, the system tends toward viscous-dominated flow and recovery is enhanced accordingly. Because the viscous/capillary ratio is proportional to permeability, capillary forces will become more significant at smaller length scales (i.e. smaller seam-bounded compartments), hence optimum oil recovery is promoted by larger length scales. Comparing these results with field data from intensely faulted reservoirs, it is clear that typical compartment dimensions (e.g. 30 m-wide compartments bounded by 5 mm granulation seams) are such that capillary trapping effects will be minor. However, it is at pressures just in excess of the percolation thresholds of faults and granulation seams (i.e. fluid pressures just sufficient to overcome the capillary displacement pressure of the largest pores) where oil recovery may drop significantly. This occurs because water flood will be concentrated into narrow channels exploiting tortuous flow paths through the reservoir, leaving behind undrained accumulations of so called attic oil.

Whilst most of the faulting discussed in this volume owes its existence to tectonic and gravitational forces, faults are known also to form in response to other processes such as differential compaction and expansion/contraction during mineral phase changes. A fairly recent addition to this list are the remarkable polygonal networks of small-scale normal faults first described from Tertiary shale-dominated slope and basin floor depositional systems in the Central North Sea (Cartwright 1994). **Lonergan *et al.*** ascribe their origin to water loss and compaction during sedimentation and early burial, and they present spectacular maps from 3-D seismic interpretation of polygonal fault networks that resemble the surface of the brain. Polygonal faulting appears to be ubiquitous across the productive areas of the Central North Sea and has particular implications for the exploitation of basin floor fans like the Forties-Montrose system and the Nelson Field. This is because:

(1) Given that reservoir sandstone bodies are commonly 25–100 m thick, the existence of a dense 3-D fault network with extensional throws of 10m-100 m may account for the compartmentalization of reservoirs.

(2) Poor fault interpretation can lead to erroneous reservoir correlations.

(3) Polygonal faults will have acted as vertical migration pathways for Upper Jurassic-sourced hydrocarbons into the isolated Tertiary reservoirs.

Acknowledgements

We thank Tim Needham and Alan Roberts for their reviews of this article and numerous colleagues for suggestions as to which papers we should include.

References

ANDERSON, E.M. 1951. *The Dynamics of Faulting.* Oliver & Boyd, Edinburgh.

AYDIN, A. & JOHNSON, A.M. 1978. Development of faults as deformation bands in porous sandstones. *Pure and Applied Geophysics*, **116,** 931–942.

CARTWRIGHT, J.A. 1994. Episodic basin-wide fluid expulsion from geopressured shale sequences in the North Sea Basin. *Geology,* **22,** 447–450.

HAYWARD, A.B. & GRAHAM, R.H. 1989. Some geometrical characteristics of inversion. *In:* COOPER, M.A. & WILLIAMS, G.D. (eds) *Inversion Tectonics.* Geological Society, London, Special Publications, **44**, 17–39 [also published in Holdsworth & Turner 2002].

HOLDSWORTH, R.E. & TURNER, J.P. 2002. *Extensional Tectonics: Regional-scale Processes.* Geological Society, London, Key Issues in Earth Sciences, **2(1)**.

HOLDSWORTH, R.E., BUTLER, C.A. & ROBERTS, A.M. 1997. The recognition of reactivation during continental deformation. *Journal of the Geological Society, London,* **154,** 73–78.

LEEDER, M.R. & GAWTHORPE, R.L. 1987. Sedimentary models for extensional tilt-block/half graben basins. *In:* COWARD, M.P., DEWEY, J.F. & HANCOCK. P.L. (eds) *Continental Extensional Tectonics,* Geological Society, London, Special Publications, **28,** 139–152 [also published in Holdsworth & Turner 2002].

LEEDER, M.R., SEGER, M.J. & STARK, C.P. 1991. Sedimentation and tectonic geomorphology adjacent to major active and inactive normal faults, southern Greece. *Journal of the Geological Society, London,* **148,** 331–344 [also published in Holdsworth & Turner 2002].

MCCLAY, K. & PRICE, N.J. 1981. *Thrust and Nappe Tectonics.* Geological Society, London, Special Publications, **9.**

MORLEY, C. K. 1995. Developments in the structural geology of rifts over the last decade and their impact on hydrocarbon exploration. *In:* LAMBIASE, J.J. (ed) *Hydrocarbon Habitat in Rift Basins.* Geological Society, London, Special Publications, **80,** 1–32 [also published in Holdsworth & Turner 2002].

SIBSON, R.H. 1995. Selective fault reactivation during basin inversion: potential for fluid redistribution through fault-valve action. *In:* BUCHANAN, J.G. & BUCHANAN, P.G. (eds) *Basin Inversion.* Geological Society, London, Special Publications, **88,** 3–21 [also published in Holdsworth & Turner 2002].

VERRALL, P. 1981. *Structural interpretation with application to North Sea problems.* Course Notes **3,** Joint Association for Petroleum Exploration Courses (UK).

WILLIAMS, G.D., POWELL, C.M. & COOPER, M.A. 1989. Geometry and kinematics of inversion tectonics. *In:* COOPER, M.A. & WILLIAMS, G.D. (eds) *Inversion Tectonics.* Geological Society, London, Special Publications, **44**, 3–15 [also published in Holdsworth & Turner 2002].

Geometry and spatial attributes of normal faults

Structural evolution of extensional basin margins

A. D. Gibbs

SUMMARY: A new model for the structural development of extensional basins is outlined. The model suggests a close similarity in geometry between faults in high extension basins such as the North Sea and those in contractional zones. The geometric analogues are discussed and where necessary new terms are introduced to describe the structural style in the context of basin formation. Parallels are noted between contractional and extensional systems in deep crustal control on ramp and flat deformation zones which are now being interpreted from deep reflection work, e.g. MOIST and COCORP. The model is based on interpretation of industry seismic records particularly in the North Sea, and on field analogues of extensional faulting in the Basin and Range province of Nevada.

Extensional basins are characterized by faulted margins with tilted horst and terrace systems developed on both listric and planar normal faults. Structural elements of the margins are offset by lateral transfer ramps and folds which allow complex integrated subsidence to occur along the basin margin. Following the fault-dominated extensional phase of the basin development, thermal subsidence can result in large scale rotation and reactivation of some of the earlier geometries. The analysis of extensional features developed on the UK Continental Shelf has led to an empirical structural model, discussed in the paper, which integrates the varied elements observed in areas such as the Viking Graben, Buchan Graben and Western Shelf. Values for stretching in the brittle zone above a shallow detachment as derived by application of this model are comparable to those predicted by subsidence models. The new model has been extensively tested using techniques of section balancing similar to those used for thrust faulting. This approach highlights the analogy between the current models for contractional and extensional faulting. Regionally consistent values for the depth-to-detachment on the normal faults derived from cross-section balancing are similar to those observed for contractional thrusting. The model has particular significance in basin analysis and has predictive possibilities both for North-Sea-type aulacogens and passive margins.

Overall basin geometry

Cross-sections of areas such as the North Sea (Fig. 1) commonly show complex terrace and horst margins stepping towards a central axial graben. Whereas seismic data and well control can give reliable information on the shape of individual seismic horizons the 3-D form and deeper structural framework are necessarily conjectural. Structural techniques such as balancing the cross-sections can give additional geometric information (Gibbs 1983).

The margins of extensional basins are characterized by two groups of faults. Firstly, 'down to the basin' normal faults with associated antithetic faults define the margins and run parallel to the basin trend (Fig. 1). A second, little recognized, group trend at high angles to, and offset the longitudinal elements of the basin (Fig. 2). Both of these groups of faults are necessary to allow discontinuous slip and rotation of the main basin elements. The model considers firstly the 'down to the basin' extension faults, then the effect of the cross-faults.

The development of an extensional basin has been described theoretically (McKenzie 1978; Le Pichon & Sibuet 1981) by considering the subsidence history and post-rifting infill of the basin. Such models have not described in detail the structural evolution of the rifting stage of basin formation. Indeed extensions by a factor of 1.4 to 2.4 (the beta value), which are a consequence of McKenzie-type basin models, are incompatible with many published cross-sections of extensional basins such as the North Sea (cf. Blair 1975; Hay 1978; Ziegler 1982). β values for a basin of, say, 1.4 imply that locally very large extensions are necessary on faults which must be represented in interpretations of North Sea type grabens.

Cross-sections of an extensional basin emphasize basement continuity across relatively steeply-dipping faults. Simple geometrical constraints (e.g. Wernike & Burchfiel 1982) show that to achieve the required extensions, models with shallow-dipping fault planes similar to those proposed for the Basin and Range province are necessary. A geometric model for the evolution of a rift basin is developed below which provides a rationale for extension faulting analogous to that proposed for compressional low angle faulting (Elliott 1977; Boyer & Elliott 1982).

Listric faults and detachment geometry

Extension of a homogeneous crustal section (Fig. 3) can take place on an initial listric fault which subsequently forms the upper bounding fault, a 'roof

From Holdsworth, R.E. & Turner, J.P. (compilers) *Extensional Tectonics: Faulting and Related Processes*. The Geological Society, Key Issues in Earth Sciences, **2 (2)**, 13–24.

Frist published in "*Journal of the Geological Society* **141**, 609–620".

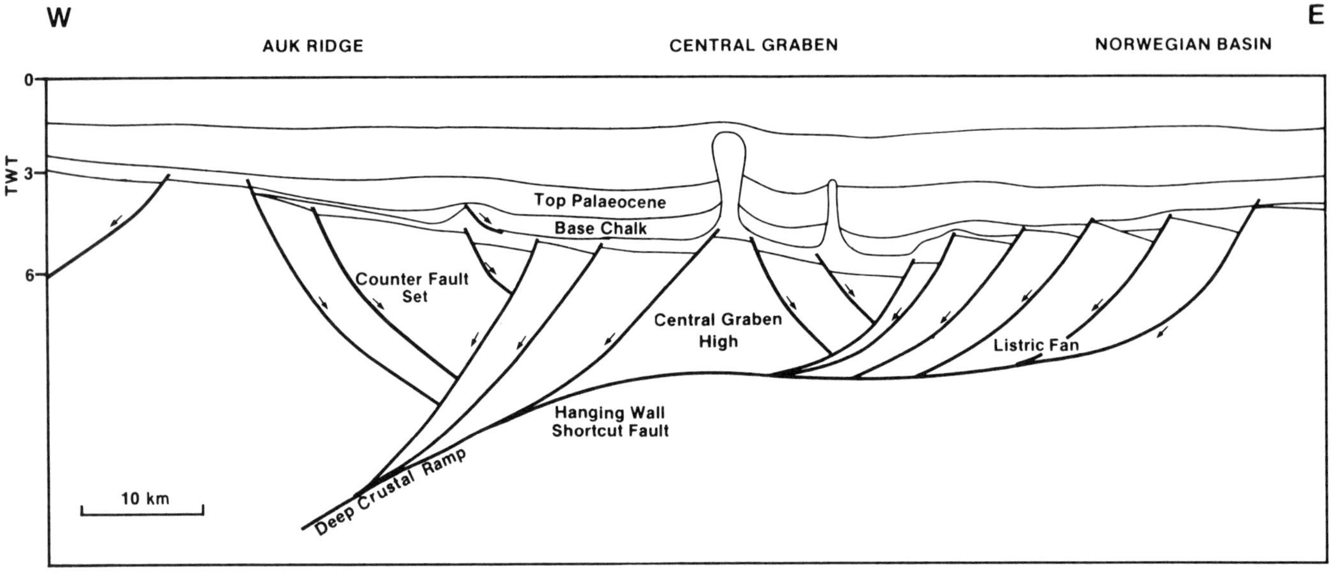

FIG. 1. Composite cross-section across the Central Graben, southern North Sea based on a regional seismic survey. The vertical scale is in two-way time and the interpretation is based on seismic and well control down to the 'basement'. Interpretation of the faulting is based on theoretical considerations discussed in text.

fault' *sensu lato*, of the extensional duplex (cf. Dahlstrom 1969). The extension on the initial and subsequent faults can be derived from considerations of heave and rotation across the fault plane (Wernike & Burchfiel 1982; Gibbs 1983). Wernike developed a model with flat-lying planar faults which introduces considerable volume problems, not only along the basal detachment, but adjacent to the undeformed blocks (Fig. 4). For these reasons, a listric model with faults curving towards a mid-crustal detachment within a zone of brittle–ductile transition is used. Space problems which arise from a listric model can be resolved by the generation of geometrically necessary hanging-wall folds and faults.

The development of a listric fault curving down to a mid-crustal detachment is seen on many seismic sections, but only in the upper part of the crust where the fault is interpreted as a single zone. Analogy with compressional tectonics suggests that the fault trajectory may in detail be in a series of flats and ramps or risers (Fig. 5). If this is the case, complex hanging-wall folding is a geometrical consequence of the fault geometry. The apparent lack of such geometry both in detail (Beach 1983) and on interpretations of regional deep seismic lines such as the MOIST profile (Smythe *et al.* 1982) indicates that such a ramp and flat geometry for extensional faults is uncommon or is an interpretational artefact. Deeper in the crustal section, the narrow brittle zone probably broadens into a complex brittle–ductile transition (Wernike & Burchfiel 1982; Sibson 1977) which gradually becomes asymptotic to a sub-horizontal ductile shear zone. Where this transition occurs, a complex 'chaos zone' consisting of separated riders (see below) in an anastomosing shear zone (Wernike & Burchfiel 1982) is to be expected. Few field examples of such detachment zones for extensional faults have yet been identified. Indeed, it is likely that the 'simple shear' deformation resulting from both contractional and extensional faults at depth results in zones which are indistinguishable outwith their regional context. The geometric effect of the deeper cataclastic and ductile part of the listric system on hanging-wall geometry is adequately described by a single sole fault. Cyclic re-utilisation of these detachment zones is in fact probable during the evolution of an extensional margin, followed by subsequent compression. The MOIST profile suggests that such major listric surfaces can be reworked repeatedly in different deformations.

The conclusion that the early history of areas such as the NW European shelf is crucial in developing anisotropies which are re-used in the formation of the shelf is inescapable. In effect many of the geometric elements are keyed in by the basement structures.

Antithetic faults and roll-over

Curvature and depth to detachment on the listric fault can be derived from calculating bed length and excess area balance in cross-section as well as graphically from the roll-over (Gibbs 1983). The geometrical restrictions imposed by rotation across fault planes constrains the possible cross-sectional interpretations. Coupled with well control, unique balanced section interpretations are possible. As with compressional ramp folds (Elliott 1977; Dahlstrom 1969), folding is a consequence of listric faulting, and 'reverse drag'

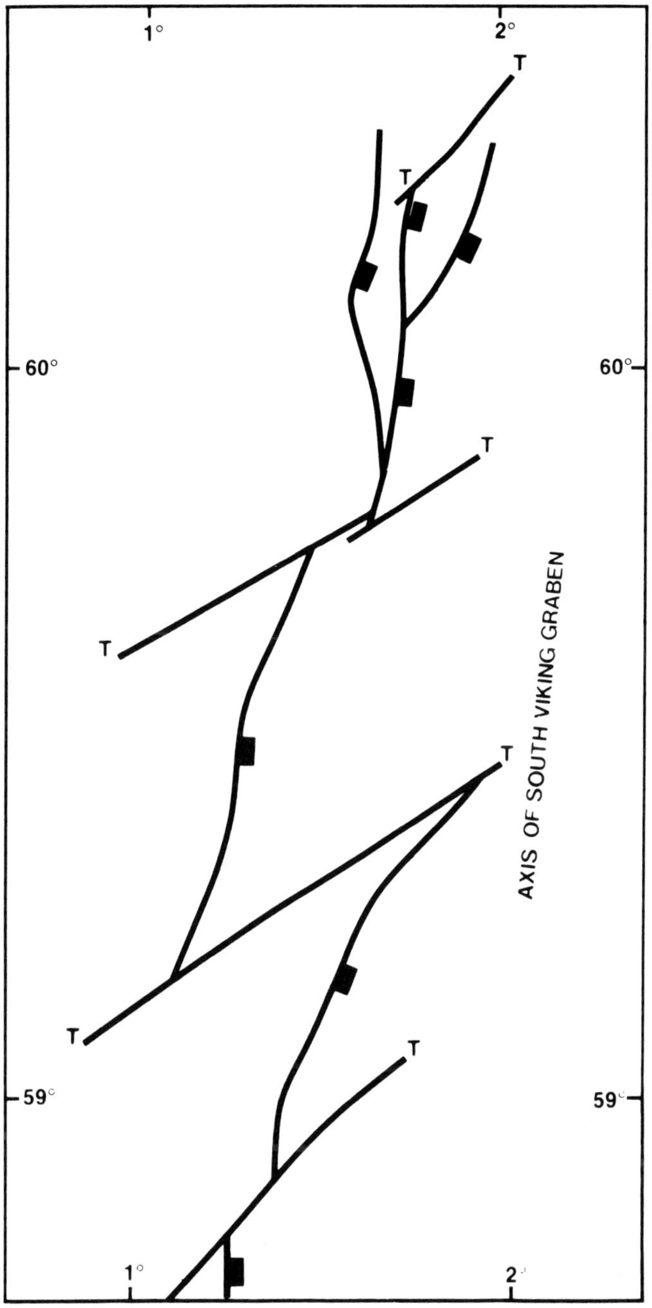

FIG. 2. Transfer faults (T) stepping the graben margin (S Viking Graben). The graben boundary faults are the floor or sole faults of the extensional graben.

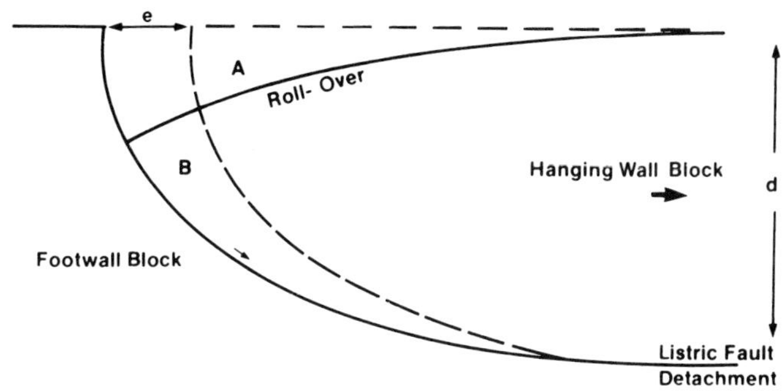

FIG. 3. Listric fault model showing roll-over as a consequence of extension e above a detachment at depth d. Areas *A* and *B* are equal.

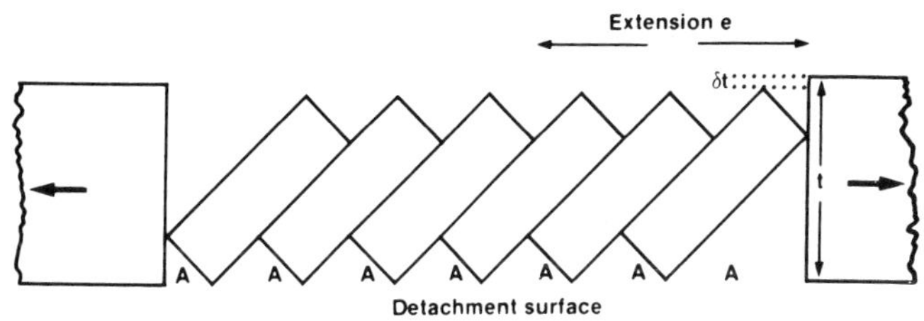

FIG. 4. Block fault model. Note that very large extensions are required before appreciable thinning (dt) can take place and that large space problems (ΣA) are created.

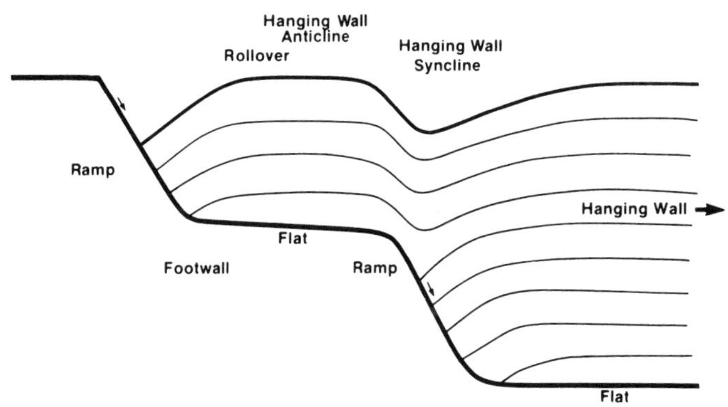

FIG. 5. Necessary folds in the hanging wall produced above footwall ramps and flats.

(Hamblin 1965) or 'roll-over' occurs. Vertical thinning of the roll-over on the hanging wall must occur. This is illustrated in Fig. 6*a* with a single step model which is easier to visualize than a continuous listric curve. The principle remains unchanged: bed length in the roll-over must extend and the thickness decrease if cross-sectional area is to remain constant. The only exception to this is if interlayer slip is efficient and the area in the roll-over can be accommodated elsewhere in the hanging wall (Fig. 6*b*). This second mechanism potentially allows out-of-sequence segmentation of the hanging wall and triangular half-graben zones to form. Stretching of the section in the roll-over as a consequence of the listric shape of the footwall fault renders drag at any appreciable scale unlikely as it would necessitate even further thinning of the hanging wall.

The need to thin the section in the roll-over and to generate the roll-over geometry leads to the formation of 'antithetic' or 'counter' faults (Fig. 7) with a variety of dip geometries. The term 'antithetic' is not strictly

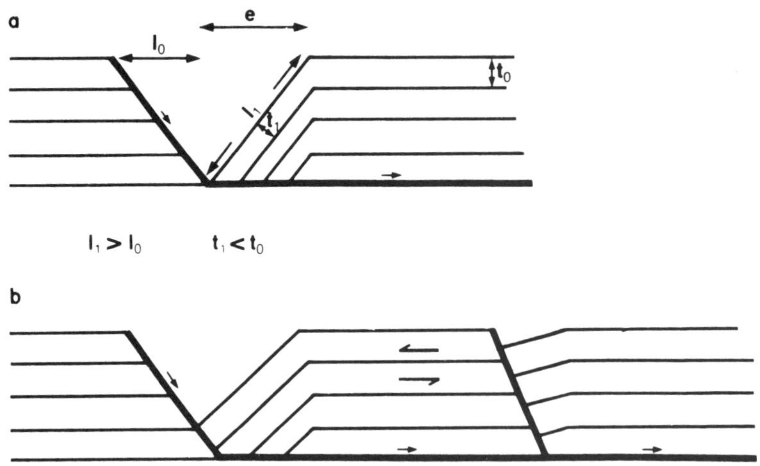

FIG. 6. (*a*) Cartoon showing thinning of layers in a roll-over. (*b*) Bedding slip maintains bed thickness in a roll-over, but results in out of sequence half graben forming on the hanging wall.

correct in this sense (Denis & Kelley 1980) but follows current usage. The antithetic faults will be listric and will tend to detach in any low strength or over-pressured zone in the hanging wall. Migration of the antithetic faults towards the hanging wall will occur as displacement continues on the main controlling synthetic fault and as the wavelength of the roll-over increases with further extension.

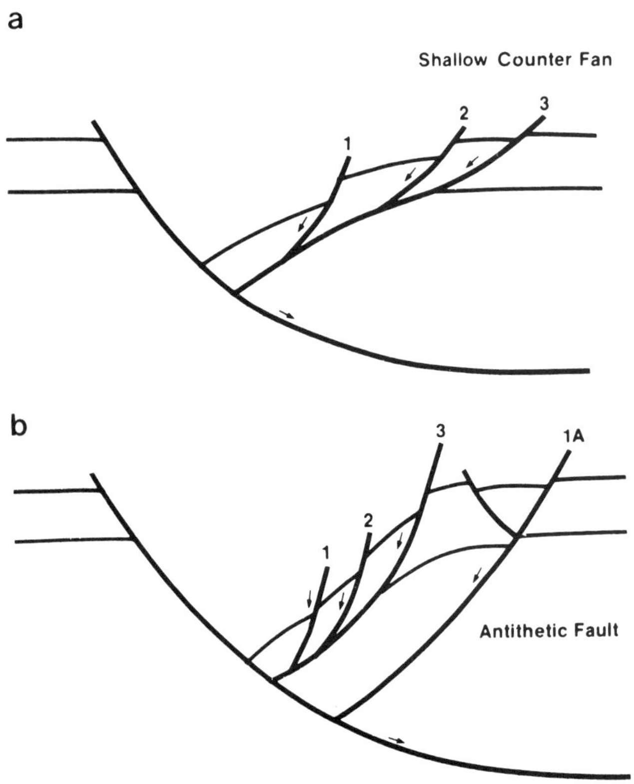

FIG. 7. (*a*) Roll-over thinned and extended on low angle faults detaching on bedding surfaces. (*b*) Roll-over thinned by both shallow and deep counter faults.

A second steeper set of antithetic faults may also form essentially as 'conjugate' faults to the main synthetic shear. These will characteristically have steeper dips and detach on the footwall fault plane. Such faults are more likely to develop in indurated rock failing in a brittle mode. They are most commonly observed in the basement and in pre-rifting sediments. These are the main set of antithetic faults apparent on geo-seismic sections. Where such antithetic faults are listric they will tend to reduce, but not to cancel, the roll-over dip towards the footwall fault by counter rotation. In some circumstances extreme curvature of the antithetic fault can result in local compressional stresses and the generation of drag, *sensu stricto*, or minor reverse faulting (Gibbs 1983). The geometrical requirement to thin and extend the roll-over makes such compressional structures unlikely in simple cases and they probably only occur where flats are present in the footwall fault and where rotational strains are locally dominant. Such compressional structures are certainly seen in experimental systems and in small-scale growth faults and landslides (Crans *et al.* 1980; Kirk 1982). At larger scales in areas such as the Viking Graben it is probable that their presence can be used to infer footwall ramps at depth although this remains to be demonstrated. Apparent reverse faults also occur as a consequence of later rotation of steeply-dipping, early formed synthetic and antithetic faults (Hamblin 1965; Gibbs 1983).

Sequence of faulting

Formation of a second, and then sequential, floor faults cutting back towards the undeformed footwall as extension proceeds is a result of unloading of the footwall. This back-sapping of the listric fan towards the footwall is probably largely gravity driven as it is difficult to envisage tensile stresses being transmitted across the sole fault to the footwall. Formation of a

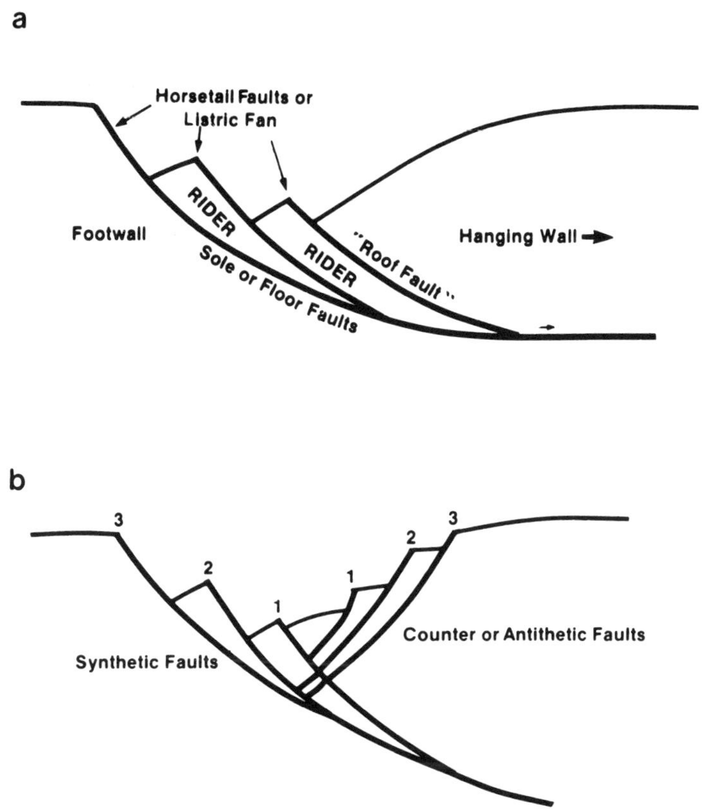

FIG. 8. (*a*) Horsetail or listric fan formed by the sequential development of riders on a sole fault migrating towards the footwall. The roof faults on the riders are passively rotated on the lower, active fault. (*b*) Listric fan and counter or antithetic fan.

second fault on the footwall gives rise to listric fan- and wedge-shaped segments in cross-section between the faults. These segments are 'emergent imbricate wedges' in thrust terminology but the colloquial term 'rider' is used here as being less cumbersome, and descriptive of the geometry of the wedges riding in sequence on the sole fault. The later, footwall sole faults carry the earlier structures upon them and displacement on the early-formed faults ceases. 'Piggy-back' faulting is well documented for contractional faults (e.g. Dahlstrom 1969) and this case leads to the formation of riders on the footwall at the exposed end of a listric fan with considerable separation of the footwall sole fault and the 'roof fault' (Fig. 8). The riders must deform to accommodate changes in sole fault geometry in a manner similar to that in piggy-back thrusting. The sole fault is at any time the active fault, and sequential development of faults occurs both with the synthetic and antithetic sets. In many cases this sequence of faulting can be clearly demonstrated both on seismic sections and from well data.

Where a flat and ramp exist in the footwall a duplex geometry can result (Fig. 9) with a floor and roof fault. Continuation of this process can give rise to chaos zones such as those proposed by Wernike & Burchfiel (1982). Only the sole fault is active, and the roof fault of the extensional duplex consists of sequential faults. Geometrically, both fanning listric riders at the upper surface, and a deeper duplex zone (Fig. 9) can exist and both may be associated with antithetic or counter faults and hanging wall folds. Where rapid sedimentation above the listric fan takes place the fan may be reactivated as an upper duplex (see Heybroek 1975; Gibbs 1983) with a new decoupling zone above what was once the pre-extension surface.

Out-of-sequence faults are possible with changes in footwall mechanics and pore pressures. This may be commonly the case in anistropic basement when overpressuring occurs due to syn-fault deposition above the hanging and footwalls. A further geometrical complication can arise with sequential faulting in the hanging wall only (Fig. 10). This situation is probably controlled again by pre-existing anistropy, for example when the counter faults can take advantage of an earlier fracture system, and is analogous to 'underthrusting' *sensu stricto* in compressional tectonics. This structural style can possibly be seen on some of the MOIST interpretations, with back slip on the Caledonian thrusts (Smythe *et al.* 1982) to form Devonian

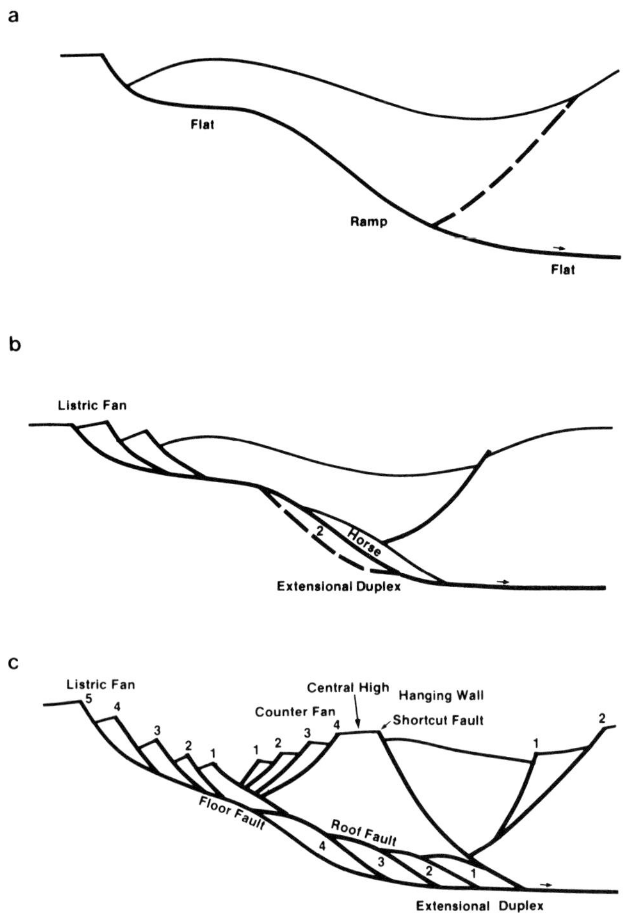

FIG. 9. (*a*) Listric fault with flat and ramp on footwall with necessary hanging wall anticline and syncline. Antithetic fault developing on hanging wall in response to ramp. (*b*) Listric fault with listric fan developed on trailing edge of hanging wall. Migration of the footwall fault on the ramp has formed a duplex zone with an extensional horse. (*c*) Fully developed system with a listric fan and counter fan on an extensional duplex formed by migrating footwall faults. A central high is developed above the ramp by formation of a short cut fault on the hanging wall.

basins. As a general mechanism it appears to be rare, and further examples should be sought.

An important consideration from the model is that whilst faults can be generated at relatively high angles close to the surface, the fault plane dip rapidly decreases with depth and the piggy-back mechanism rotates the synthetic faults to progressively shallower dips. In extreme cases (Fig. 11) the curved fault plane may become 'unfolded' as the listric, fault-bounded riders pass onto the flat detachment, and with high extensions out-of-sequence reactivation can result in the riders becoming detached on the floor fault. The unfolding process must inevitably result in some minor slip, reactivation of earlier fault planes, and reversals

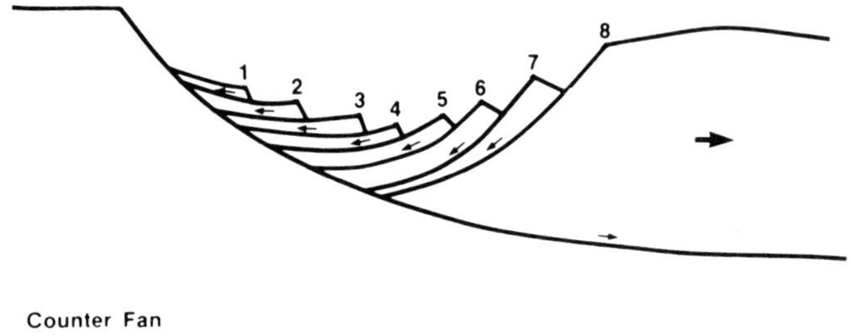

FIG. 10. Reverse listric fan with riders developed on the hanging wall only.

of the displacement (inversion) on second order structures. Detached riders generated in this manner may be modified by syn-extension erosion (De Charpal *et al.* 1978) and assume a flat topped geometry (Fig. 11*b*) with some confusion as to the true dip of the basement.

Graben geometry

A common feature of graben is that both 'half graben' and symmetric graben are present. Central graben highs are present in the southern North Sea (Ziegler 1982).

Symmetric graben have an inbuilt space problem in that classical graben models result in a 'key stone' misfit at the centre as the margins generate outwards onto the undeformed basement. One possible solution to this problem is that the central part of the graben develops by progressive slippage on both members of a conjugate fault set. Horsfield (1980) proposed a model, based on observations of model systems and faulting over salt domes, for crossing conjugate faults

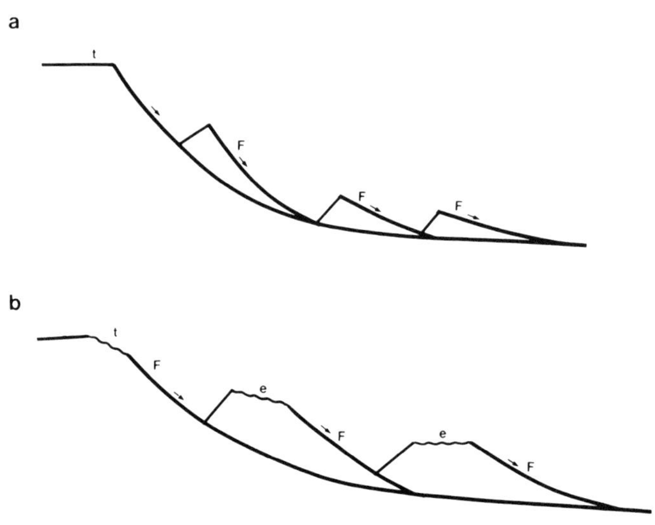

FIG. 11. (*a*) Riders detached at high extensions. (*b*) Erosionally modified riders and footwall terrace (t). Note that the shallow-dipping surfaces on the hanging wall are erosion surfaces (e) and fault planes (f).

which in part solves this problem. To account for observed central graben highs his model must involve both extensive microfaulting and ductile creep to form what would be a completely transposed fabric. There is no evidence that this occurs and an alternative model is necessary.

At a crustal scale, Figs 8 and 9 suggest that an apparently symmetric graben can evolve in a number of ways. Firstly, equal extension on the synthetic and counter faults (Fig. 8) can generate a graben without a central rise. One margin of the graben detaches on the sole and roof fault system of the main synthetic system. A significant difference in scale and timing of the formation of the margins should be observed, as one margin can develop only as a consequence of the presence of the other.

In Fig. 9 a roughly symmetric graben evolves from a hanging wall syncline over a mid-crustal ramp. Such a model 'keys' a central graben into the system from the initiation of the sole fault. Faulting on the monoclinal margins of the syncline then develops the second graben. The synthetic fault on the margin of the secondary graben, by linking down to the sole fault, shortens the deformation path and may be energetically favoured. More complex possibilities arise for parallel development of the paired graben with out-of-sequence faulting by this mechanism. The central high becomes a rider and may be regarded as a hanging wall anticline on a crustal scale.

Development of a second detachment

A number of workers in the Basin and Range Province have shown that secondary, deep, detached listric faults can become dominant after a certain stage and redeform the earlier faults (Fig. 12) (Davis & Hardy 1981; Proffett 1977). The generation of a second deep detachment may be indeed a second discrete deformation event but it is possible that it may be the result of continued progressive extension. This problem can be addressed in several ways. If the sole fault has a major crustal ramp as suggested in Fig. 9 the formation of a lower sole fault and a large scale horse can result in the deep sole fault cutting up through the upper detachment on the higher flat. This would result in inactivation of the upper set of horsetails and riders to give the geometries indicated in Fig 13*a* & *b*. Although this is geometrically probable it does not indicate why this process might happen. Thinning of the crust above a brittle–ductile transition, especially if relatively rapid, will cause migration of the brittle–ductile transition downwards in the crustal section where the transition is pressure- and temperature-controlled for constant deformation rate (Sibson 1977; Elliott 1976). In such a case, switching of the sole fault to occupy a deeper zone, possibly on an earlier subhorizontal shear zone, could be mechanically favoured.

Consideration of a fault system such as that in Fig. 4 suggests another mechanism by which switching to a deeper, new detachment can occur at high extensional strains. De Sitter (1958) showed that where the strain is by slip, with the intervening slabs lacking internal deformation, that slip increases with constant stress up to some value, and then decreases as the system 'locks up'. Ramsay (1974) developed this model to account for the geometry of chevron folds and to show the relationship between shortening, slip and relative 'ease of deformation'. This mechanism is directly applicable to extensional faulting models which can be regarded as rotational slip systems and a comparable set of curves can be derived. Following Ramsay's (1974) reasoning, with a linear stress–strain relationship, stress required to continue thinning must increase after a certain value and the system will lock-up. Using Wernike & Burchfiel's (1982) modification for listric, rather than straight planar, faults gives a similar result. The lock-up value for the system is therefore related to frictional properties of the fault surface and the geometry of the riders. Whilst this theory is not fully developed, Basin and Range work (Davis & Hardy 1981; Proffett 1977) provides some empirical notion that the lock-up and refaulting can occur after extension by a factor of about 1.5 (McKenzie 1978). Clearly large variations in this value might be expected especially if the pre-extension crust has inherited anisotropy from an earlier compressional or extensional episode. This may explain why we do not apparently see such multiple detachments on the COCORP, MOIST or DSDP seismic profiles (Cook 1981; Smythe

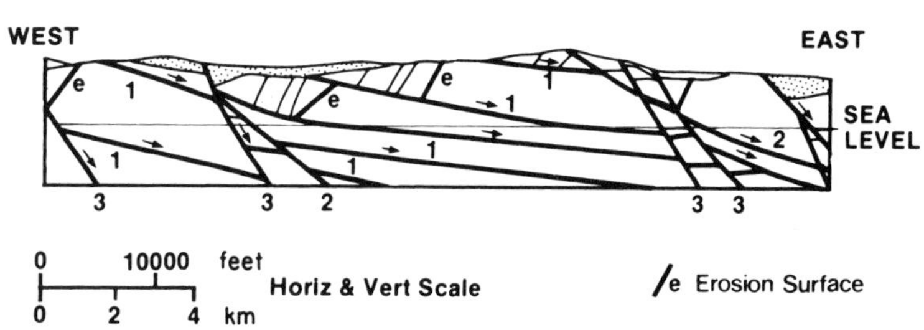

FIG. 12. Multiple faulting (Proffett 1977). Local extension factor = 2, with three phases of fault generation 1–3. Note that the pre-faulting erosion surface *e* has very high dips. Compare with Fig. 10.

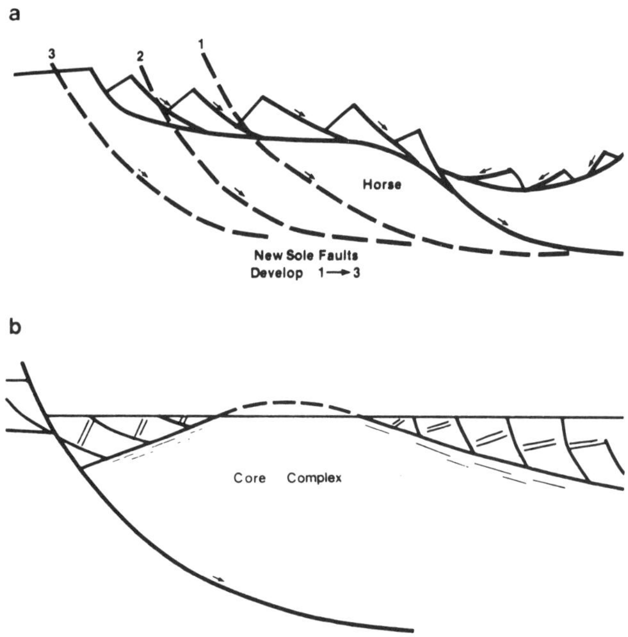

FIG. 13. (*a*) Model for formation of second detachment by evolution of a duplex and the rotation of an inactive overlying listric fan. (*b*) Roll-over on a deeper fault causing back rotation of a high level extensional system. This model illustrates the concept of a 'core complex' with a deformed outer sheath.

et al. 1982; De Charpel *et al.* 1978). However Sclater & Christie (1980) have suggested that only the later, steeper faults have been interpreted on most industry seismic data.

Transfer faults

Graben usually show fault trends at a high angle to the margins (Fig. 2) (see Johnson & Dingwall 1981) and these have been interpreted in many cases as later faults. Harding & Lowell (1979) suggested, however, that such cross faults may be an integral part of the extensional system. Oertel (1965), Reches (1978) and Aydin & Reches (1982) amongst others have suggested that multiple slip systems in faulting are necessary. By analogy, both with compressional thrusting (Dahlstrom 1969) and with oceanic plate deformations at spreading sites (McKenzie & Morgan 1969), some form of cross-fault system is to be expected in extensional faulting. These cross-faults are indeed characteristic of all extensional terranes and allow 'leakage' between extension faults with differing slip rates. The analogy with lateral ramps in thrust tectonics is exact: complex rotational, synthetic dip and strike slip components allow the extension to transfer style and activity along the graben (see Boyer & Elliott 1982). It is important that such faults are not confused with strike-slip or rotational faults in the Andersonian sense (Anderson 1951) and to realize that they are part of the extension system. For this reason the new term 'transfer fault' is proposed. The geometry of transfer faults may be controlled by earlier anisotropy, and in the Viking Graben this set of faults has been correlated with Caledonian trends (Johnson & Dingwall 1981). Both lateral and oblique transfers are possible; again the analogy with lateral and oblique thrust ramps is exact (Butler 1982). Similarly, both compressive (transpressional) and extensional (transtensional) structures may develop depending on the relative orientation of the transfer and normal extensional components of faulting. Where the transfer is oblique to the local or regional extension direction, dip-slip components may be important and roll-over will result parallel to the transfer trend. Highly curved listric faults will commonly give reverse displacements along part of the length of the transfer fault (Figs 14 & 15) and a complex linked system of listric extension faults and transfer systems can provide local compressional features and Reidel-type fold and minor fault geometries. This effect is also noted in shallow detachment growth faults such as those described from the Niger delta (Crans *et al.* 1980).

The presence of a strike-slip component on transfer faults is important, as such faults will have displacements much larger than the dip-slip component apparent on a single geo-seismic line. Clay smearing and the build-up of fault rock and possible sealing structures are enhanced in these faults. It is also

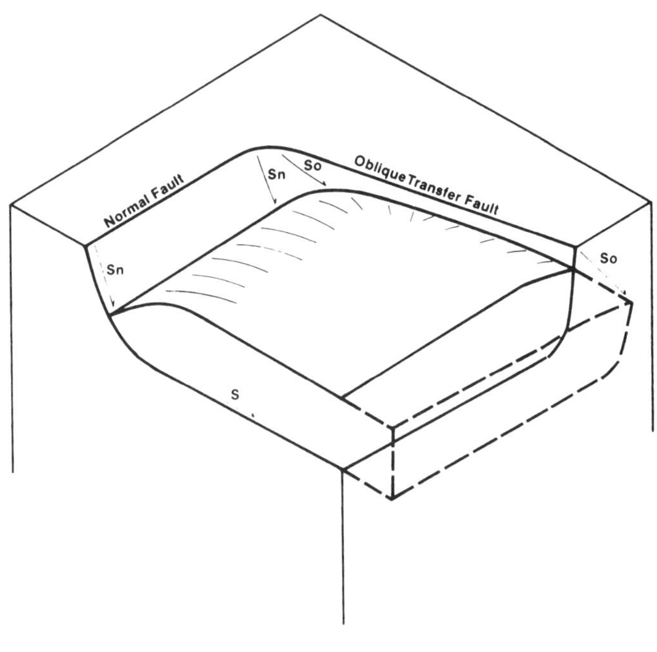

FIG. 14. Isometric sketch of a normal fault with normal slip on fault plane (Sn) and an oblique transfer fault with oblique slip (So).

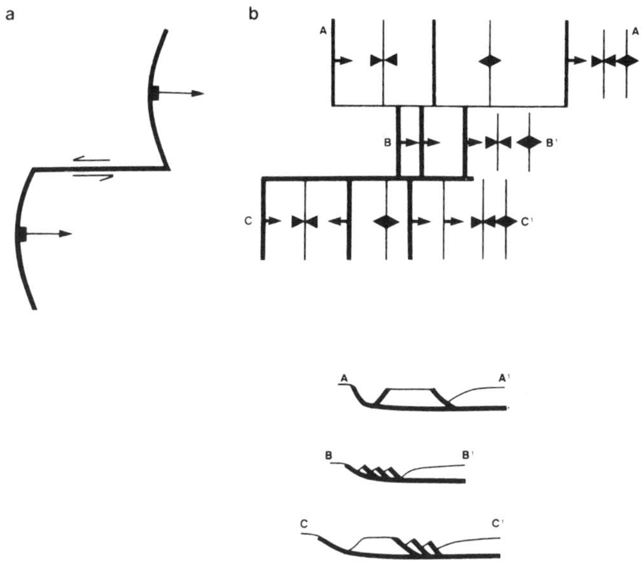

FIG. 15. (*a*) Plan of a simple transfer fault connecting two normal listric faults. (*b*) Plan of two transfer faults separating three zones of different deformation style, sections A–A′, B–B′ and C–C′.

important to note that the apparent sense of displacement on such oblique slip faults can change at different levels in the section, and along the fault trace. The effect of this is to make such structures difficult to interpret on the seismic sections, and their capacity to act as reservoir seals may be unrecognized.

Deformation style and number of components may be switched across transfer fault zones (Fig. 15). This may give rise to problems in fault trace mapping on seismic surveys where an apparently jumbled mixture of components cannot be resolved without mapping the transfer components. The use of stratigraphic separation diagrams (Elliott & Johnson 1980) and footwall maps may be useful tools in identifying jumps in displacement associated with such faults. These techniques are commonly used to investigate lateral ramps in thrust tectonics where similar problems occur.

Erosional and sedimentary factors

Basin models propose that the extension faulting is an early, relatively rapid component in the evolution of the basin. The geometry of the basin is controlled by the geometry of the extension faults and transfer components. In the early stages of formation, modification by erosion of parts of the uptilted riders, and sedimentation in back-rotated fault blocks, can cause considerable problems in identifying and mapping the early geometry. The effect of syn-rifting erosion shown in Fig. 12 can be applied to a composite section derived from published work in the Bay of Biscay and Goban Spur (De Charpal *et al.* 1978; Steckler & Watts 1982) to demonstrate this effect (Fig. 16). Basin subsidence must also be taken into acount (see: Sclater & Christie 1980; Beach 1984). One effect of such subsidence is to steepen-up faults which were more gently dipping immediately following the extension phase.

Decompaction using the techniques of Magara (1978) and Sclater & Christie (1980) may be required before the fault geometry can be established. A quick-look method involving 'flattening' the first continuous post-faulting event may be sufficient to establish the model. When dealing with seismic sections and geoseismic sections in time, the problems are more complex. Corrections should be made for both distortion of the geometry, as outlined by Gibbs (1983), as well as for compaction and subsidence.

Upper detachment—the stacked duplex

A second result of the subsidence of the margin and contemporaneous build-up of sediments on the margin

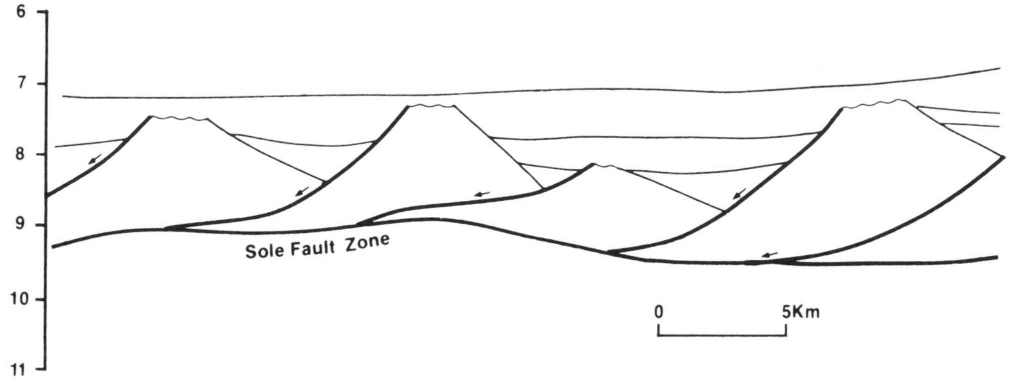

FIG. 16. Profile of continental margin based on De Charpal *et al.* (1978). Note the erosionally modified riders on the deformed sole fault zone.

edge is to generate a new and upper set of listric faults. Todd & Mitchum (1977) showed such growth faulting off the coast of Africa above a sole fault (Fig. 17*a*) and similar geometries can be identified off the NW European Shelf (see De Charpal *et al.* 1978). This is the stage at which the listric fan can evolve into a 'partial' or 'apparent' upper duplex by reactivation. A new listric fan can then develop above the upper detachment and the extension is transferred to it by differential displacement on the faults in the lower fan.

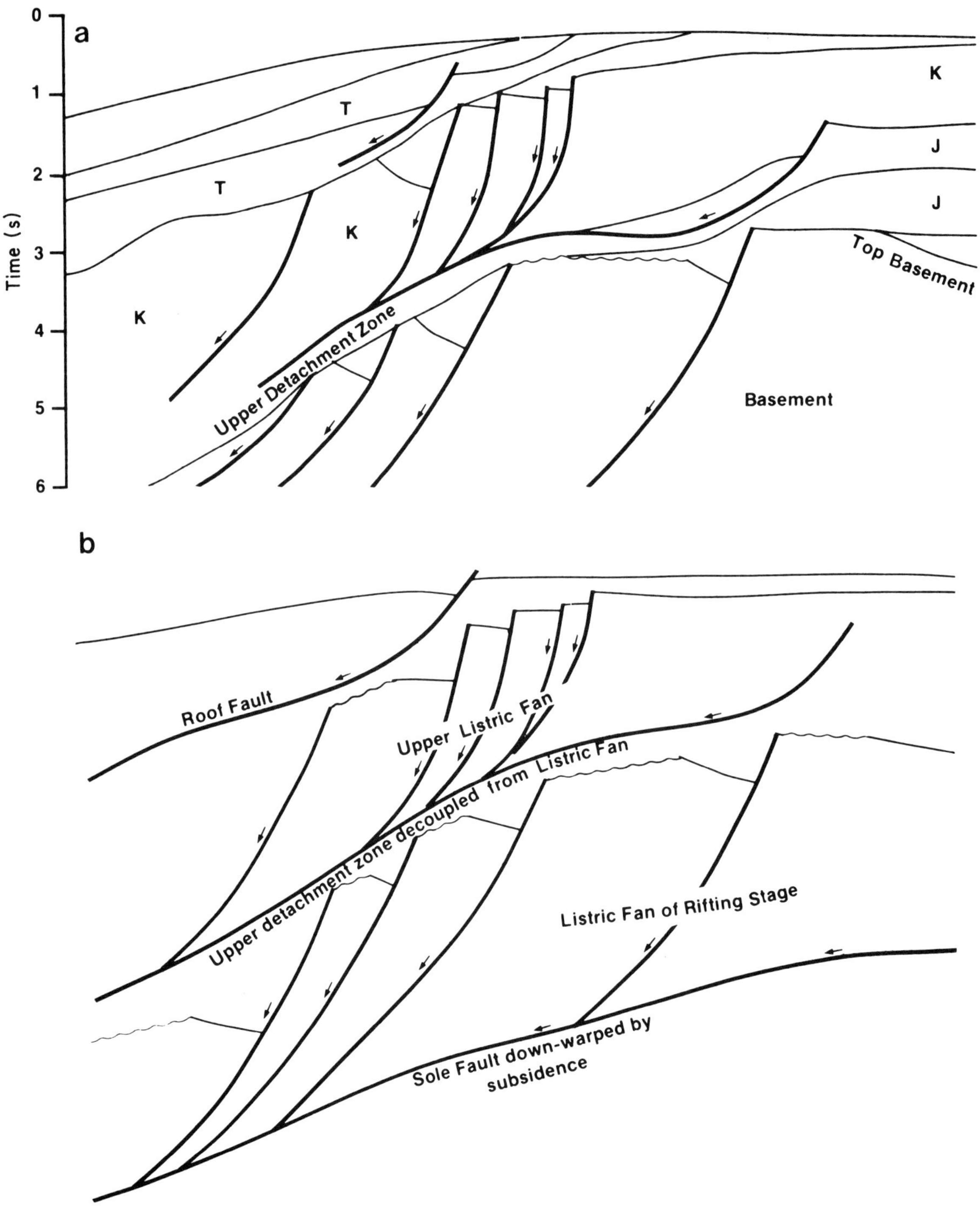

FIG. 17. (*a*) Schematic geoseismic section after Todd & Mitchum (1977). T, Tertiary; K, Cretaceous; J, Jurassic. (*b*) Geoseismic section interpreted following the model discussed in text. Continued extension and subsidence results in formation of stacked listric fans with erosionally modified riders decoupled on low strength sedimentary units to form a series of 'partial duplexes' in the sedimentary carapace.

Fig. 18*a* shows such a situation based on an interpretation by Todd & Mitchum (1977) along with a cartoon (Fig. 17*b*) illustrating the model. Where upper zone faults are present which are associated with continued basin extension, or extension across the margin due to subsidence, rather than sedimentary loading (growth faults *sensu stricto*), the upper zone faults must be integrated into the general model. The term 'partial duplex' is used to distinguish the over-running of the upper listric fan from a deep crustal duplex.

Conclusions

Whereas the geometry of extension both in continental aulacogens such as the North Sea and on continental passive margins is not fully understood the application of simple geometrical models and analogues is of considerable importance. The sequential superposition of a number of simple fault elements, with some syn- and post-deformational modification by erosion, sedimentation, compaction and subsidence, can give rise to very complex geometries. Such deformation, is, however, understandable by the application of a set of geometrical rules analogous to those successfully used for compressional tectonics. The thin-skinned extensional model of deformation on low-angle listric faults with ramp and transfer geometries enables the complex seismic and well data to be integrated successfully with models for subsidence and thermal maturation of the basin.

ACKNOWLEDGEMENTS. I wish to thank Britoil plc for permission to publish this work and to thank my colleagues for their helpful comments and discussion.

References

ANDERSON, E. M. 1951. *The Dynamics of Faulting*, Oliver & Boyd, Edinburgh.

AYDIN, A. & RECHES, Z. 1982. Number and orientation of fault sets in the field and in experiments. *Geology* **10,** 107–12.

BEACH, A. 1984. The structural evolution of the Witch Ground Graben. *J. geol. Soc. London.* **141,** 621–628.

BLAIR, D. G. 1975. Structural styles in North Sea oil and gas fields *In*: WOODLAND, A. W. (ed.) *Petroleum Geology and the Continental Shelf of North-West Europe.* Inst. Petroleum, London, 327–35.

BOYER, S. E. & ELLIOTT, D. 1982. Thrust systems. *Bull. Am. Assoc. Petrol. Geol.* **66,** 1196–230.

BUTLER, R. W. H. 1982. A structural analysis of the Moine Thrust Zone between Loch Eriboll and Foinaven, NW Scotland. *J. struct. Geol.* **4,** 19–29.

CRANS, W., MANDL, G. & HAREMBOURNE, J. 1980. On the theory of growth faulting; a geomechanical delta model based on gravity sliding. *J. Petrol. Geol.* **2,** 265–307.

DAHLSTROM, C. D. A. 1969. Balanced cross sections. *Can. J. Earth Sci.* **6,** 743–57.

DAVIS, D. H. & HARDY, J. J. 1981. The Eagle Pass Detachment; southeastern Arizona; product of Mid-Miocene listric normal faulting in the southern Basin and Range. *Geology.* **93,** 749–62.

DE CHARPAL, O., MONTADERT, L., GUENNOC, P. & ROBERTS, E. G. 1978. Rifting, crustal attenuation and subsidence in the Bay of Biscay. *Nature, London,* **275,** 706–10.

DENIS, J. G. & KELLEY, U. C. 1980. Antithetic and homothetic faults. *Geol. Rdsch.* **69,** 186–93.

ELLIOTT, D. 1976. The energy balance and deformation mechanisms of thrust sheets. *Philos. Trans. R. Soc. London, Ser. A. 283,* 289–312.

—— 1977. Some aspects of the geometry and mechanics of thrust belts. Part 1 *8th Annual Seminar Can. Soc. Petrol. Geol.* Univ. Calgary, **1,** 2.

—— & JOHNSON, M. R. W. 1980. Structural evolution in the northern part of the Moine Thrust Zone. *Trans. R. Soc. Edinburgh,* **71,** 69–6.

GIBBS, A. D. 1983. Balanced cross-section construction from seismic sections in areas of extensional tectonics. *J. struct. Geol.* **5,** 153–60.

HAMBLIN, W. K. 1965. Origin of 'reverse drag' on the down-thrown side of normal faults. *Bull. geol. Soc. Am.* **76,** 1145–64.

HARDING, T. P. & LOWELL, J. D. 1979. Structural styles, their plate tectonic habitats, and hydrocarbon traps in petroleum provinces. *Bull. Am. Assoc. Petrol. Geol.* **63,** 1016–59.

HAY, J. T. C. 1978. Structural development in the north North Sea. *J. Petrol. Geol.* **1,** 65–77.

HEYBROEK, P. 1975. On the structure of the Dutch part of the central North Sea Graben. *In*: WOODLAND, A. W. (ed.) *Petroleum and the Continental Shelf of North-West Europe,* Inst. Petroleum Geology, London, 338–52.

HORSFIELD, W. T. 1980. Contemporaneous movement along crossing conjugate normal faults. *J. struct. Geol.* **2,** 305–10.

JOHNSON, R. J. & DINGWALL, R. G. 1981. The Caledonides: their influence on the stratigraphy of the North West European continental shelf. *In*: ILLING, L. V. & HOBSON, G. D. (eds) *Petroleum geology of the continental shelf of North-West Europe.* Heyden, London, 85–97.

KIRK, M. A. 1982. Gravity slide in the Lower Coal Measures at Swallow Craig, near Kirkconnel, Dumfries and Galloway. *Scott. J. Geol.* **18,** 317–22.

LE PICHON, X. & SIBUET, J. C. 1981. Passive margins: a model of formation. *J. geophys. Res.* **86,** 3708–20.

MAGARA, K. 1978. Compaction and fluid migration. *Dev. Petrol. Sci.* **9,** Elsevier, New York.

MCKENZIE, D. P. 1978. Some remarks on the development of sedimentary basins, *Earth Planet. Sci. Lett.* **40,** 25–32.

—— & MORGAN, W. J. 1969. Evolution of triple junctions. *Nature, London,* **224,** 125–33.

OERTEL, G. 1965. The mechanism of faulting in clay experiments. *Tectonophysics,* **2,** 343–93.

PROFFETT, J. M. 1977. Cenozoic geology of the Yerington District, Nevada and its implications for the nature and origin of Basin and Range faulting. *Bull. geol. Soc. Am.*

88, 247–66.

Reches, Z. 1978. Analysis of faulting in three-dimensional strain field. *Tectonophysics,* **47,** 109–29.

Ramsay, J. G. 1974. Development of chevron folds. *Bull. geol. Soc. Am.* **85,** 1741–54.

Sclater, J. G. & Christie, P. A. F. 1980. Continental stretching and explanation of the post mid-Cretaceous subsidence of the central North Sea Basin. *J. geophys. Res.* **85,** 3711–39.

Sibson, R. H. 1977. Fault rocks and fault mechanisms. *J. geol. Soc. London,* **133,** 191–213.

DeSitter, L. U. 1958. Boudins and parasitic folds in relation to cleavage and folding. *Geol. Mijnbouw.* **20,** 272–86.

Smythe, D. K., Dobinson, A., McQuillon, R., Brewer, A., Matthews, D. H., Blundell, D. J., & Kelk, B. 1982. Deep structure of the Scottish Caledonides revealed by the MOIST reflection profile. *Nature, London,* **299,** 338–40.

Steckler, M. S. & Watts, A. B. 1978. Subsidence of the Atlantic margin of New York. *Earth Planet. Sci. Lett.* **41,** 1–13.

Todd, R. G. & Mitchum, R. M. 1977. Seismic stratigraphy and global changes of sea level, Part 8; identification of Upper Triassic, Jurassic and Lower Cretaceous seismic sequences in Gulf of Mexico and offshore West Africa; *In*: Payton, C. E. (ed.) *Seismic Stratigraphy—Applications to Hydrocarbon Exploration. Mem. Am. Assoc. Petrol. Geol.* **26,** 145–63.

Wernike, B. & Burchfiel, B. C. 1982. Modes of extensional tectonics. *J. struct. Geol.* **4,** 104–15.

Ziegler, P. A. 1982. *Geological Atlas of Western and Central Europe.* Shell Internationale Petroleum Mij. BV, The Hague.

Received 29 January 1983; revised typescript accepted 10 November 1983.

A. D. Gibbs, Britoil plc, 150 St Vincent Street, Glasgow G2 5LJ.

Dips of normal faults in British Coal Measures and other sedimentary sequences

J. J. WALSH & J. WATTERSON
Department of Earth Sciences, University of Liverpool, P.O. Box 147, Liverpool L69 3BX, UK

Abstract: Fault dip measurements ($n = 3291$) from 122 faults in British Coalfields are presented. Two classes of normal fault are recognized: (a) gently dipping (<55°) synsedimentary faults of gravity driven or, less commonly, tectonic origin and (b) steeply dipping (>55°) post-coalification tectonic faults formed at depths of about 2.5 km which comprise 90% of the faults analysed. Dip data from the steeply dipping faults are variable (standard deviation = 9°), both on single faults and between faults, and 70° is a better approximation of fault dip than the commonly accepted value of 60°. Data on tectonic normal faults, with comparable depths of formation (<3–4 km), from other sedimentary sequences show that dips of greater than 60° are quite usual. Systematic changes of dip on single fault surfaces have predictable effects on the curvature of fault traces.

Using the modified Griffith fracture criterion the dips of faults at varying depths (up to 4 km) are calculated for both unsaturated (dry) and saturated rocks. The dips of normal faults are expected to vary from 90° at the surface to 60° at 3–4 km and this variation is consistent with the limited field data available. The coalfield data are consistent with fault formation at standard hydrostatic pore fluid pressures at depths of about 2.5 km. Below depths of a few kilometres normal faults are expected to flatten out gradually to a dip of 45°, but such flattening is not necessarily an indication of an underlying detachment.

The dip of normal faults is an important element in fault models applied to the analysis and geological reconstruction of extensional terrains, but relatively few quantitative data are available. For many years the dip of 60° suggested by Anderson (1951) was accepted as the usual value but, more recently, the recognition of listric faults as significant structural elements has tended to overshadow the concept of high-angle planar faults. Normal faults are nowadays mostly viewed by geologists on seismic sections on which the true dip is not seen unless the section has been accurately depth-converted; this can be a difficult procedure on the shallow sections acquired by the hydrocarbon industry and even more difficult on deep seismic sections. The outstanding questions to which an answer is required can be summarized as follows: (i) is there a representative dip for normal faults at or near the surface, (ii) does this angle vary with the material properties of the faulted rocks, (iii) is there a systematic change of fault dip with depth, which is independent of material properties, (iv) are there systematic differences in dip according to fault size, (v) are there systematic differences in dip between tectonic faults and faults which are gravity driven?

This paper does not aim to provide a definitive answer to all of these questions but presents data which are intended to provide a quantitative basis for discussion of some of them. The new data presented and analysed are mainly from faults which are recorded on mine plans of the East Pennine Coalfield and which intersect two or more worked coal seams.

Coalfield faults

Fault dip data have been collected from mine plans (1:2500) of the East Pennine Coalfield in Derbyshire, North Yorkshire and Nottinghamshire. In this coalfield, there are over 25 productive coal seams, several of which have been extensively worked. The main productive sequence is up to 600 m thick and the stratigraphic intervals between worked seams range from 5 to 250 m (mean stratigraphic thickness = 60 m). Up to about 10 seams have been worked in individual collieries and seam dips are usually less than 10°. For faults which are recorded on two or more worked seams (the mean value for analysed faults is 3.5 seams), fault dips between adjacent seams were calculated at 25 m intervals along the fault. A diagram of dips on a vertical projection of a fault surface, a fault dip diagram, was constructed for each fault studied (Fig. 1). Fault dip diagrams of complete fault surfaces are rare (Rippon 1985*a*, 1985*b*), and are generally restricted to faults with displacements of less than 5 m (i.e. larger faults usually extend through the entire Coal Measure sequence). Nevertheless, as some faults terminate upwards within the workings and other faults are seen to terminate downwards, representative dips of all positions on faults with maximum displacements of up to about 10 m have been sampled. Maximum displacements on the studied faults range from 1 to 10 m, but are generally less than 5 m. 122 faults were analysed from Derbyshire ($n = 77$), Yorkshire ($n = 34$) and Nottinghamshire ($n = 11$), and over 3000 dip readings were calculated. Supplementary data from three other studies, in Derbyshire, Nottinghamshire and South Wales (unpublished NCB reports of Elliott 1973, Vowles 1978 and Raine 1978), are also referred to.

The methods employed provide an accurate estimate of fault dip only if the fault is a single discrete surface. Some errors in calculated dips may arise from imprecise surveying and it is thought likely that for seam intervals of less than 10 m errors could be as large as 10°. A further shortcoming of the methods used is that for large seam intervals the pairing of fault/seam intersections from one seam to another is more difficult for gently dipping faults. However, for seam intervals of less than 10 m, faults with dips of between 10°

From Holdsworth, R.E. & Turner, J.P. (compilers) *Extensional Tectonics: Faulting and Related Processes.* The Geological Society, Key Issues in Earth Sciences, **2 (2)**, 25–39.
1476-315X/02/$15.00
First published in: "*Journal of the Geological Society* **145**, 859–874"

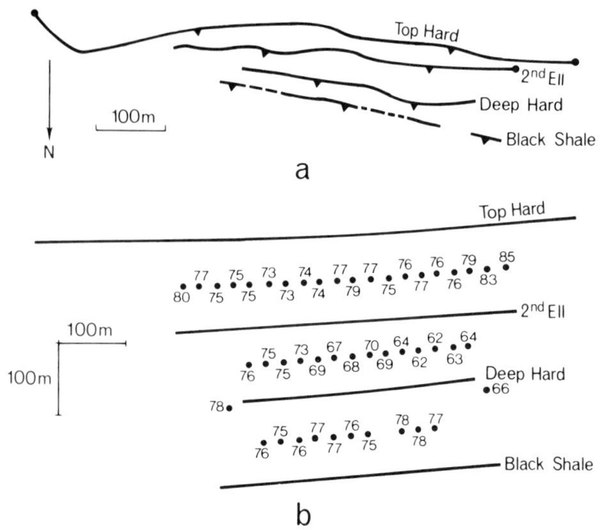

Fig. 1. (**a**) Plan view of four fault/seam intersections of a normal fault, as shown on mine plans from the East Pennine Coalfield; dip and throw directions arrowed. Solid circles are fault tips encountered in workings. Fault trace in unworked ground is shown by broken line. (**b**) Fault dip diagram showing measured dips on fault in (a) projected onto a vertical plane parallel to the strike of the fault surface.

and 90° would, if present, be routinely identified. Where displacement across a fault is effected within a narrow zone of fractures (<5 m) a reasonable estimate of the fault dip may be made, particularly if the inter-seam interval is large. Fault dip estimates for broader zones of fracture have been avoided. Even where a fault is represented by discrete fault/seam intersections on different seams it may nevertheless be discontinuous and comprise a number of discrete surfaces which are offset in vertical section. Such offsets cannot be identified on mine plans and may be responsible for small errors (<5°) in the data.

Derbyshire

The geological structure of the Westphalian Coal Measures of North-East Derbyshire is relatively simple (Smith *et al.* 1967; Rippon 1985*a*, 1985*b*). Coal Measure sequences are exposed in the west and dip gently (<5°) to the east to lie beneath Permian and younger rocks; gentle folds (limb dip <3°) occur occasionally. Most faults are normal and fault throws are generally less than 100 m (Rippon 1985*a*). Faults rarely extend upwards into the Permian, but where they do, a much decreased throw (i.e. less than 25%) is recorded in the younger rocks (Smith *et al.* 1967).

77 faults have been studied from the North-East Derbyshire area (Fig. 2). The following fault types are distinguished:

(i) Strike-slip faults ($n = 4$), which have vertical mean dips, the direction of dip changing along the fault strike. In some cases, the direction of vertical displacement changes along the length of individual fault traces.

(ii) Antithetic faults ($n = 3$), which are clearly antithetic to larger faults.

(iii) Soft-sediment faults ($n = 4$), which may be of compactional or slump (gravity slide) origin. Faults of this type, of which only four examples have been found extending from one seam to another, are not always identified as such on mine plans, and some may have been included in the next category of fault.

(iv) Normal faults ($n = 66$), which have consistent dip directions and a normal dip-slip component of displacement. Although slip vectors for individual faults are not recorded on mine plans, field evidence from open-cast mines indicates that faults with similar geometries are pure dip-slip.

Cumulative frequency curves (CF curves) of all dip measurements for each of these four fault types, and of the calculated mean dips of individual normal faults, are shown in Fig. 3a.

Possible effects of fault rotation concomitant with extension (domino-style extension, Wernicke & Burchfiel 1982) or post-fault tilting on the dips of normal faults have been examined by subdividing the data set into groups of similar orientation, and by comparing the dips of each group. Three groups are compared: northeast-striking faults with dips to the south (group 1, 11 faults), northeast-striking faults with dips to the north (group 2, 10 faults), and north–south-striking faults which dip to the west (group 3, 7 faults). Mean values of the mean dips of individual faults in each of these groups are 71.52°, 68.37° and 73.19°, respectively. The mean values of all dip readings measured in each of these groups are: group 1, 69.2° ($n = 361$, standard deviation = 8.7°); group 2, 69.4° ($n = 221$, standard deviation = 8.8°); group 3, 71.6° ($n = 178$, standard deviation = 11.9°). Groups 1 and 2, dipping south and north, respectively, have similar mean fault dips and indistinguishable CF curves of dip (Fig. 3b), and are therefore unlikely to have been much affected by rotation during extension or by subsequent tilting, a conclusion substantiated by low easterly regional dips of the Coal Measures (<5°) and of the overlying Permian (1°). However, northerly striking Group 3 faults have slightly higher mean fault dips and a slightly broader distribution of dip readings, features which could be due to minor tilting and gentle folding about N–S axes but which could also be due to the small sample size. The similarity between the CF curves of individual dip readings for both Group 1 and Group 2 faults and the CF curve for the mean dips of all normal faults suggests that the latter is representative of the distribution of primary fault dips of tectonic normal faults in this area.

Yorkshire

The dominant structure of this area is a regional NW–SE-trending gentle fold. Bed dips on the flanks of this structure are generally 3–4° and the overlying Permian has a regional dip of 1.35° towards the east. All of the faults studied are normal faults; a large proportion are thought to be of tectonic origin and some may be soft-sediment faults but are not distinguished as such on mine plans. Cumulative frequencies of the mean dips of all faults ($n = 34$) and of all of the dip values measured on these faults ($n = 1428$) are shown in Fig. 3c. The mean fault dip is 67.7° (standard deviation = 9.7°) and the median fault dip is also 67.7°.

Nottinghamshire

The Nottinghamshire Coal Measures are concealed beneath Permian cover but otherwise have a broadly similar geological structure to the North Derbyshire area. Only 11 faults were studied, 10 of which were normal and one of which was a soft-sediment fault. The CF curve of all dip measurements ($n = 148$) on normal tectonic faults is not significantly different from those of the Derbyshire and

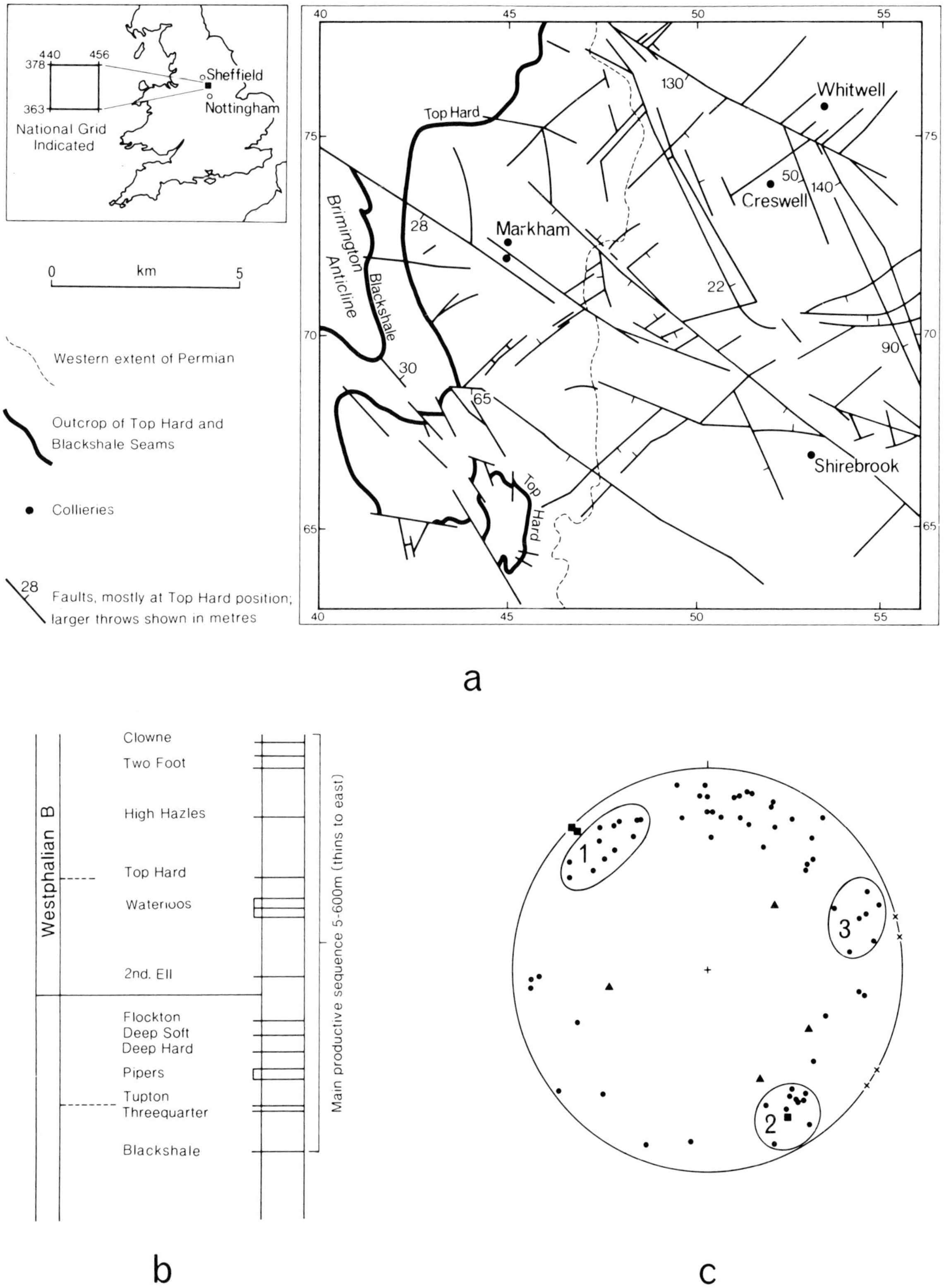

Fig. 2. (**a**) Generalized structure of the Coal Measures of North-East Derbyshire (modified after Rippon 1985*a*).
(**b**) Generalized stratigraphy of the North-East Derbyshire area showing the main worked coal seams.
(**c**) Stereographic projection of poles to faults studied in the Derbyshire area. (●) normal faults ($n = 66$), (■) antithetic faults ($n = 3$), (▲) soft-sediment faults ($n = 4$), (×) strike-slip faults ($n = 4$). Dip plotted is the mean dip of each fault. The three fields shown represent Groups 1, 2 and 3 referred to in the text.

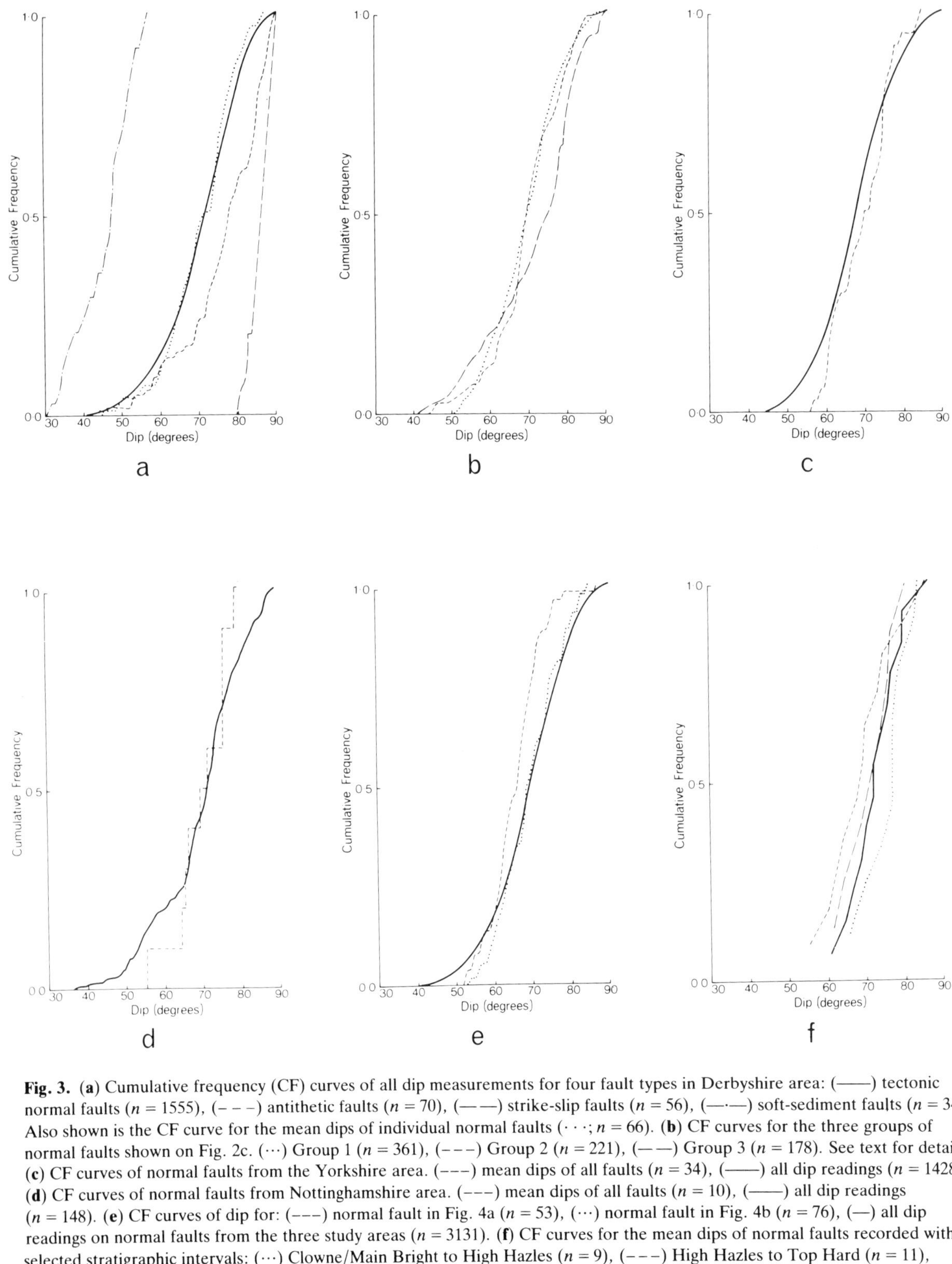

Fig. 3. (**a**) Cumulative frequency (CF) curves of all dip measurements for four fault types in Derbyshire area: (——) tectonic normal faults ($n = 1555$), (- - -) antithetic faults ($n = 70$), (— —) strike-slip faults ($n = 56$), (—·—) soft-sediment faults ($n = 34$). Also shown is the CF curve for the mean dips of individual normal faults (· · ·; $n = 66$). (**b**) CF curves for the three groups of normal faults shown on Fig. 2c. (···) Group 1 ($n = 361$), (---) Group 2 ($n = 221$), (— —) Group 3 ($n = 178$). See text for details. (**c**) CF curves of normal faults from the Yorkshire area. (---) mean dips of all faults ($n = 34$), (——) all dip readings ($n = 1428$). (**d**) CF curves of normal faults from Nottinghamshire area. (---) mean dips of all faults ($n = 10$), (——) all dip readings ($n = 148$). (**e**) CF curves of dip for: (---) normal fault in Fig. 4a ($n = 53$), (···) normal fault in Fig. 4b ($n = 76$), (—) all dip readings on normal faults from the three study areas ($n = 3131$). (**f**) CF curves for the mean dips of normal faults recorded within selected stratigraphic intervals: (···) Clowne/Main Bright to High Hazles ($n = 9$), (---) High Hazles to Top Hard ($n = 11$), (— —) Top Hard/1st Waterloo to Deep Soft/Deep Hard ($n = 8$), (——) Deep Soft/Deep Hard to Pipers/Tupton/Threequarters ($n = 13$).

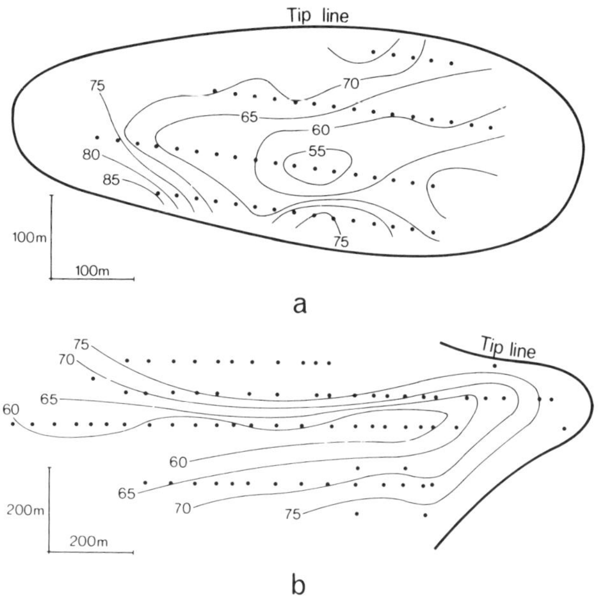

Fig. 4. Contoured dip diagrams for two faults from Markham Colliery, North-East Derbyshire. Heavy solid line, tip line (zero displacement contour) of fault; solid circles, sampling points for dip calculation. Contour interval 5°. Fault (b) is from Rippon (1985*a*, fig. 6).

Yorkshire areas (Fig. 3d). The mean fault dip is 69.3° (standard deviation = 10.7°) and the median fault dip is 70.5°.

Single faults

On fault dip diagrams small faults are sometimes seen to terminate upwards and downwards within the workings. In these cases it is possible to test for systematic variations in dip on a single fault surface by examination of the fault dip contour diagram. There is often an appreciable variation of dip on a given fault surface and standard deviations of 10° are not uncommon (Fig. 3e). Indeed, as previously noted by Rippon (1985*a*, 1985*b*), some faults show systematic radial or vertical changes in dip which are independent of lithology (Fig. 4): in such cases the dips always decrease toward the centre of the fault ellipse.

Dip changes with lithology

Sedimentary cycles, or cyclothems, within the Coal Measures are usually about 10 m thick (Rippon 1985*a*), and sandstones thicker than 10 m are rare. As the stratigraphic interval between the worked seams sampled averages about 60 m, which is larger than this cyclicity, variations in fault dip as shown on dip contour diagrams are not a reflection of lithological changes (Rippon 1985*a*). Analysis of dip changes with lithology would have to be conducted on the scale either of individual cycles or of stratigraphic series.

Dip changes with depth

To test for possible dip changes with depth, CF curves have been drawn for the mean dips of faults recorded over four selected depth intervals within the worked sequence: (1) Clowne/Main Bright to High Hazles, (2) High Hazles to Top Hard, (3) Top Hard/1st Waterloo to Deep Soft/Deep Hard and (4) Deep Soft/Deep Hard to Pipers/Tupton/Threequarter. Comparison of these curves indicates that there is no systematic flattening of fault surfaces with depth (Fig. 3f). Although the uppermost stratigraphic interval is apparently characterized by slightly steeper fault dips, more data are required to establish whether or not this is statistically significant.

Soft-sediment faults

The coalfield data demonstrate the existence of a group of faults which consistently dip at angles less than 55° and with a median dip value of 47°. These faults have a limited vertical persistence of less than 50 m (Fig. 5) and in some cases are identified as soft-sediment faults (i.e. compactional or gravity-slide origin). They are believed to have formed in unlithified sediments, with displacement transferred first to a sub-horizontal detachment and subsequently to the free surface at the toe of the fault. The examples we have identified are probably exceptional, in that most of the soft-sediment faults recorded are restricted to one seam or cyclothem (P. Guion pers. comm.; Crans *et al.* 1980; Shepherd *et al.* 1981).

Synsedimentary tectonic faults

There are some larger scale synsedimentary faults, with maximum displacements in the order of hundreds of metres, which dip between 40° and 50° and on which we have little information. These faults are laterally and vertically persistent and are characterized by marked changes in stratigraphy from footwall to hanging-wall (e.g. Cinderhill and Kirkby North Faults; Barnett *et al.* 1987; Frost & Smart 1979). These are synsedimentary faults, i.e. they intersected the contemporary surface, and their upper parts at least would have developed in unlithified sediment where conditions are expected to approximate to those of the plastic regime, with consequent gentle fault dips (see below). The primary dip of such faults is likely to have been further reduced by compaction.

Depth of faulting

Field observation of faults in opencast coal workings show that faulting took place in already lithified sediments, probably when the Coal Measures had mechanical properties similar to those which they have now. The faulted coal seams behaved as relatively competent units and hard sandstones have angular breccias and closely spaced fractures: development of cleat and other joints in the coals clearly preceded the faulting, as in the Western Coalfield of New South Wales (Shepherd *et al.* 1981), because they are reoriented in ductile drag zones adjacent to faults. Shale units were very incompetent at the time of faulting and remain so: some units are still sufficiently plastic to mould in the hand. As the faulting is extensional, it is likely to have occurred before the uplift and erosion which preceded deposition of the overlying Permian rocks. These considerations suggest that the faulting took place at or about the time of maximum depth of burial during the late Carboniferous. That the depth of burial at this time was greater than at any later time is shown by the distribution of gas contents of the coals. Contours of equal gas content

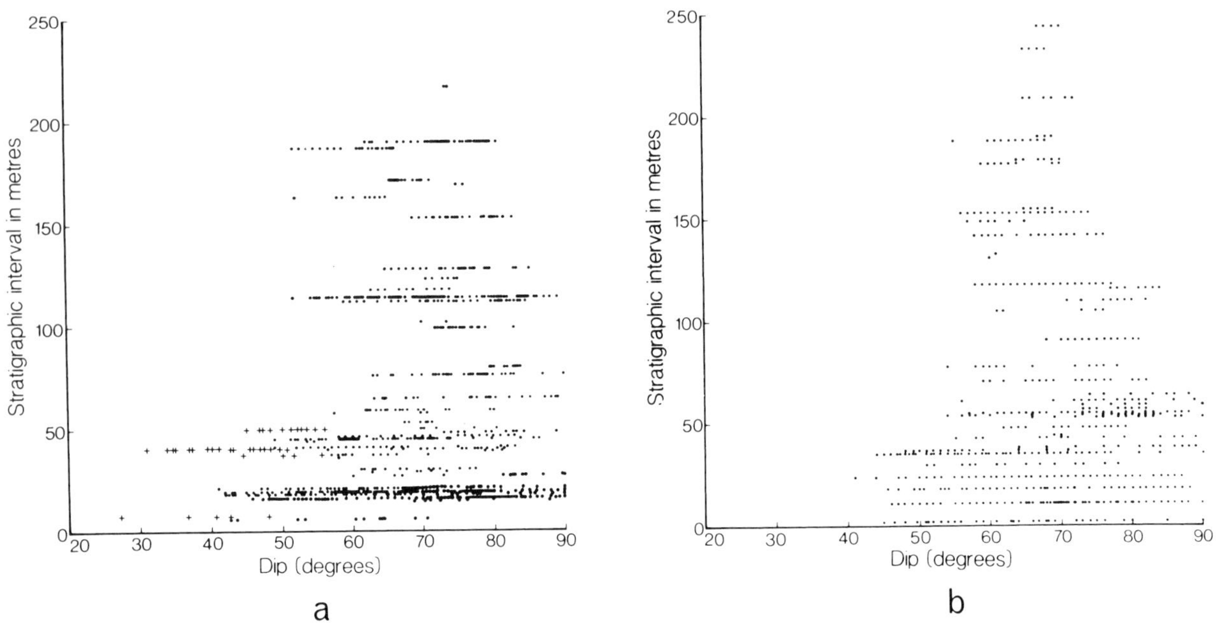

Fig. 5. Plots of fault dip against the stratigraphic interval (metres) between the two seams used for a dip measurement. (**a**) Derbyshire, (●) tectonic normal faults ($n = 1555$), (+) soft-sediment faults ($n = 34$), (**b**) Yorkshire, normal faults ($n = 1428$). Most of the Yorkshire faults are believed to be tectonic but those faults which dip less than 50° and are restricted to stratigraphic intervals of less than 50 m may be soft-sediment faults.

appear to have been intersected by the sub-Permian erosion surface and concentrations of gas do not occur below the sub-Permian unconformity: such concentrations would be expected if maximum depth of burial, and rate of outgassing, has post-dated deposition of the Permian rocks (Creedy 1988).

An indication of the maximum depth of burial of the coals can be obtained from their calorific values, which are depth related: within the Coal Measures calorific value, on a dry ash-free basis, typically has gradients of 3.6–4 $\mathrm{kJ\,kg^{-1}\,m^{-1}}$ (Creedy 1988). The fault data are from seams over a vertical range of about 500 m and approximate calorific values for the top and bottom of this range are 32 and 34 $\mathrm{MJ\,kg^{-1}}$, i.e. a gradient of 4 $\mathrm{kJ\,kg^{-1}\,m^{-1}}$, as shown by British Coal core data. Assuming a value of 33 $\mathrm{MJ\,kg^{-1}}$ for the middle of the range and a linear gradient of calorific values of 4 $\mathrm{kJ\,kg^{-1}\,m^{-1}}$, from 23 $\mathrm{MJ\,kg^{-1}}$ for peat at zero depth, the maximum depth of burial of the middle seams was 2500 m. The assumption of linear gradient from the surface probably represents an oversimplification.

The moisture contents of coals in the East Pennine coalfield vary systematically with calorific values and coals of 33 $\mathrm{MJ\,kg^{-1}}$ have moisture contents in the range 9–16% (Creedy 1988). According to Suggate (1974, fig. 8) such moisture contents could correspond to temperatures of 70–100° which, with a geothermal gradient of 25° $\mathrm{km^{-1}}$ would correspond to maximum depths of burial of 2–3 km.

Representative dips of coalfield normal faults

Figure 6a shows cumulative curves for all dip measurements on tectonic normal faults in each of the three areas. There is no significant difference between either the median values or the distributions in the three areas. A cumulative curve for data from all three areas has a median dip value of 69.3° ($n = 3131$ for 110 faults; mean = 68.9°, stand. dev. = 10.0°; Fig. 6b), and this is the value which we take as representative for tectonic normal faults in the areas examined.

Similar values of fault dips in British Coalfields have been reported by other workers. A preliminary survey of faulting in the concealed coalfield of Nottinghamshire and North Derbyshire showed that normal faults, with maximum displacements in excess of 6 m, have fault dips ranging from 64° to 72° ($n = 36$), with a mean value of 67–68° (Elliott 1973). Dips of normal faults in Warsop Main Colliery, North Derbyshire, range from 80° to 50°, with a mean value of 67° (Raine 1978). From a study of normal faulting in the eastern-central part of the South Wales Coalfield, Vowles (1978) concluded that the dips of normal faults, with maximum displacements of 18–75 m, are generally between 65° and 72°.

Regional context

Within the East Pennine coalfield the Westphalian is about 1 km thick but incomplete, often lacking the entire Upper Coal Measures. The top of the Dinantian is almost 1.5 km below the lowest worked seam in the study area (Smith *et al.* 1967; Frost & Smart 1979) and although borehole data are available for only the uppermost 700 m, the thickness of the Dinantian is probably no more than 1200 m. There is no evidence to suggest that any of the major Carboniferous faults of the East Pennine Coalfield have detachment surfaces, either above or below the base of the Carboniferous, which is at a depth of 3 km.

Regional studies of the Carboniferous of the British Isles provide support for a model in which the Rheno–Hercynian zone represents a back-arc extensional basin, of Upper Devonian to Upper Carboniferous age, lying to the north of

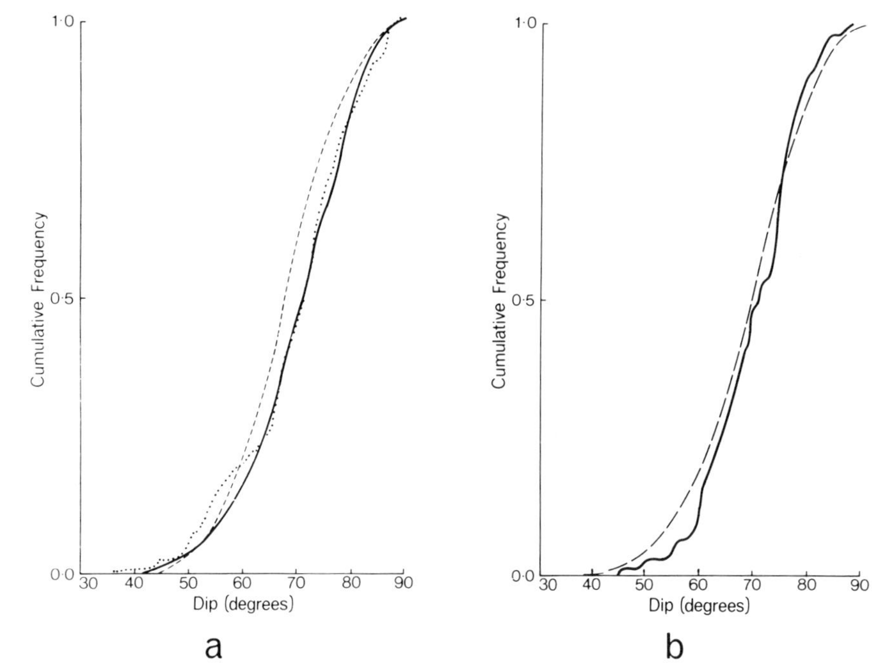

Fig. 6. (**a**) Cumulative frequency curves for all dip measurements on tectonic normal faults in each of the three areas: (—) Derbyshire (n = 1555), (---) Yorkshire (n = 1428), (···) Nottinghamshire (n = 148). (**b**) Cumulative frequency curves of tectonic normal fault dip data from all areas: (—) all dip measurements (n = 3131), (---) mean fault dips of all faults (n = 110).

the metamorphic Hercynide orogen which is partly exposed in SW England (Leeder 1982; Leeder & McMahon 1988). A late Carboniferous regional inversion episode was characterized by progressive northward migration of thrust faults during the Namurian, Westphalian and Stephanian (Leeder & McMahon 1988). Within the East Pennine Coalfield, however, thrust faults are very rare and inversion-related structures take the form of gentle folds. Although our data do not unequivocally point to a post-faulting age for these folds, we suggest that, in view of the regional considerations, the normal faults studied, which show no apparent spatial association with folding, are pre-inversion structures. Although some of the larger faults in these coalfields were reactivated in post-Carboniferous times, with up to 25%, but generally less than 10%, of the maximum displacement being recorded in Permian rocks, fault reactivation is unlikely to have significantly affected the dip data. This situation is different from that in the Southern North Sea, where thicker Westphalian sequences are recorded and substantial Mesozoic reactivation of Carboniferous faults is recognized (Glennie 1984).

Representative dips of normal faults: brittle regime

The data show that the dips of coalfield faults are variable, both on a single fault and between faults, and that 70° is a better approximation of fault dip than the commonly accepted value of 60° (Price 1966). Data on faults in other sedimentary sequences show that primary fault dips greater than 60° are quite usual and that there is no reason to suppose that the coalfield faults are in any way exceptional (Table 1).

Available dip data for other coalfields in Alberta, Limburg, New South Wales and Peninsular India show similar ranges of fault dip (i.e. 50–80°) but generally have average dips between 60° and 65° (Table 1). A mode of 70–75° is given by Shepherd *et al.* (1981) for 336 faults, formed at less than 1.5 km depth, in the Wallerawang No. 2 Colliery, New South Wales. Although limited variations in fault dip are the rule, some intrabasinal faults in the Indian coalfields are strongly listric (Chowdhury *et al.* 1975; Laskar 1979): these faults are restricted to Coal Measure sequences and are probably growth faults of gravitational origin, in the sense of Crans *et al.* (1980).

A mean fault dip of 69° is given by Gudmundssen (1984) for 68 faults in NW Iceland. Fault dip data for narrow (<100 km) continental rifts, such as the Ethiopia, Kenya, Rhine and Suez rifts, generally range from 50° to 80° and average 60–70° (Table 1). These may be minimum values, as extension within such rifts is usually attributed to domino-style faulting with progressive rotation of original fault dips to gentler dips with increasing regional extension (Morton & Black 1975). There is some indirect evidence for limited flattening and, less commonly, steepening of fault dips with depth (Garfunkel & Bartov 1977; Illies 1977 Angelier & Colletta 1983). Primary fault dips of less than 50° are rare and are characteristic of growth faults or gravity slides (Illies 1977; Mohr 1983). Data assembled by Golombek *et al.* (1983) for the Espanola Basin segment of the Rio Grande Rift give a mean dip of 69.7° for 85 faults.

Quantitative data for normal fault dips in oil provinces are scarce because most of the published seismic sections are not depth-converted. In the northern North Sea fault dips generally range from 50° to 80° with modal dips of close to 60° (Halstead 1975; Hamar 1975), but primary dips, prior to extension-related rotation and/or the imposition of a regional dip, may have been between 65° and 75° (Halstead 1975). Mesozoic faults in East Greenland generally dip between 60° and 80°, but occasional shallower dipping gravity slides on Triassic gypsum layers are present (Surlyk 1977).

Field and well data from the Gulf Coast indicate that faults of regional extent have dips within the range 30–65° but are usually between 40° and 60° (Quarles 1953; Murray 1961; Perkins 1961). These faults are synsedimentary and their shallow dips are a reflection both of flattening due to

Table 1. *Fault dip data from various sources. In some cases the usual range and the entire range (in brackets) of fault dip are presented. Where available, the mean dip of data sets is presented. Three categories of data type are noted: f, field or mine, s, seismic, e, experimental*

Location	Source	Range		Mean	Data
Kenya Rift	Baker & Wohlenburg (1971)			63	f
Ethiopian Rift	Morton & Black (1975)	60–70			f
Suez Rift	Chenet & Letousky (1983)	60–80			fs
Suez Rift	Angelier (1985)	60–75	(40–80)		f
Suez Rift	Garfunkel & Bartov (1977)	50–70			f
Rhine Graben	Illies (1977)	60–65	(50–80)		fs
Rio Grande Rift	Golombek *et al.* (1983)	60–70	(50–85)	70	f
Indian Coalfields	Chowdhury *et al.* (1975)	50–80			f
Limburg Coalfield	Sax (1946)			63	f
New South Wales Coalfield	Shepherd *et al.* (1981)	70–90	(50–90)		f
Alberta Coalfields	Norris (1958)			58	f
Northern North Sea	Hamar (1975)	50–80			s
Northern North Sea	Halstead (1975)	65–75			s
East Greenland	Surlyk (1977)	60–80			f
Basin and Range Province	Best & Hamblin (1978)	55–70			f
Basin and Range Province	Pardee (1950)	70–72			f
Basin and Range Province	Mackin (1960)	60–70			f
Basin and Range Province	Dunne (1977)	>60			f
Basin and Range Province	Stewart (1971)	55–70			f
Basin and Range Province	Sharp (1939)	60–70			f
Basin and Range Province	Christensen (1966)	60–75	(45–90)		f
Basin and Range Province	Wernicke *et al.* (1985)	60–90			fs
General	Angelier & Colletta (1983)	60–65			fs
Welsh Caledonides	Phillips (1972)	55–65			f
Sand Box Experiments	Hubbert (1951)			61	e

sedimentary compaction and of the poorly-lithified (in some cases plastic) nature of the faulted sediments (see below; Roux 1979). These factors have produced faults which may shallow or, less commonly, steepen with depth. Seismic reflection evidence, however, suggests that many of the regional Gulf Coast faults have listric geometries with detachment surfaces, at depths of 6–8 km and commonly associated with shale horizons, which are consistent with a gravity-driven origin for these faults. Field and well data show that smaller faults within the topmost 3–4 km generally have steeper dips and do not shallow to a detachment. Interpretation of these small faults as shallowing with depth is largely based on seismic reflection profiles which have not been depth-converted (faults which are planar will appear listric on these profiles, see Withjack & Pollock 1984) and on the belief, which we regard as mistaken (see below; Barnett *et al.* 1987), that hanging-wall rollover requires a listric fault geometry. It is possible that the smaller faults are planar (Ocamb 1961) and lose displacement rapidly within shale horizons (Roux 1979), without a requirement for detachment surfaces. Such faults would be characterized by uplift within the footwall (Barnett *et al.* 1987). A subordinate category of faults, the high angle faults of Ocamb (1961), has dips of between 60° and 75° and occurs in association with shallow piercement salt domes and, less commonly, deep-seated domes.

The Basin and Range Province differs from narrow continental rifts in that it is wider and is characterized by larger amounts of extension. Faults within this province fall into two main categories: (i) low angle detachment faults (dip <25°) and (ii) high angle faults (60–90°; see Table 1). The principal models proposed for the province differ only in the presence or otherwise of low angle detachment faults. Low angle faults occur along pre-existing mechanical interfaces, such as earlier thrusts or zones of ductile shear, and occasionally have a gravity slide origin (Brun & Choukroune 1983). According to Wernicke's model (1981), these faults are crust-penetrating structures above which planar high angle faults (with primary dips of 60–90°) rotate domino-style with increasing extension. Other models involve high angle faulting at the surface with fault dips shallowing at greater depths either to be replaced by anastomosing shear zones (Hamilton 1982; Kligfield *et al.* 1984) or to detach on a subhorizontal decoupling zone (Smith 1978; Miller *et al.* 1983). Our principal concern, however, is with the primary fault dips within the uppermost 3–4 km of the crust. The dips of high angle faults in the Basin and Range are consistent with British coalfield data (Table 1). The primary low angle faults are anomalous only if they are not reactivated structures or gravity slides (see discussion in Brun & Choukroune 1983).

Mechanical controls on fault dip

The modified Griffith fracture criterion

The reviewed data on faults in other sedimentary sequences show that the faults from which our data are derived are not exceptional and, in sedimentary rocks at least, 70° dips are more common than those of 60°.

When normal faults form within 3–4 km of the surface, values of maximum principal stress (σ_1) are low and, at a given depth, remain constant during faulting: given these

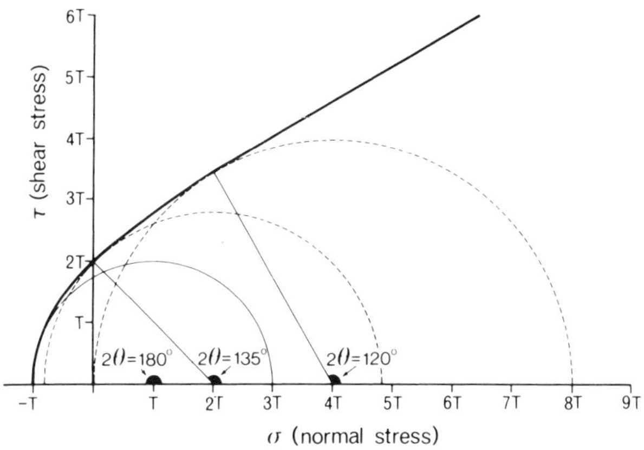

Fig. 7. The composite failure envelope for the modified Griffith fracture criterion (Secor 1965) with partial Mohr circles for stress differences of $4T$, $5.66T$ and $8T$, corresponding to fault dips (θ) of 90°, 67.5° and 60°.

two conditions, it can be shown that fault dips of 60° are unlikely to predominate. As the three groups of faults of different orientation (Fig. 2c) have similar mean dips, it is concluded that σ_1 did not depart significantly from the vertical during the period in which the faults formed.

The significance of low values of σ_1 can be seen by reference to the modified Griffith fracture criterion as presented by Phillips (1972), following Secor (1965), and illustrated in Fig. 7. For mechanically isotropic materials, the failure envelope in the tensile region is parabolic and in the compressive region is a straight line, the two sections being joined where their slopes are equal. The inclination of the straight line is 30°, which is the most common experimentally determined 'angle of internal friction' or, more correctly, the angle which represents the 'rate of increase of shear strength with normal stress' (Byerlee 1969). T is the intercept of the failure envelope with the normal stress axis; although T is referred to as the tensile strength it does not necessarily have any direct relationship to laboratory measurements of tensile strength and, like the angle of internal friction, is best regarded as an empirical number. Similarly, the intersection of the failure envelope with the shear stress axis at $2T$, often referred to as 'cohesive strength', is best referred to as 'the shear stress on a plane across which the normal stress is zero' (Byerlee 1969).

Phillips (1972) showed that the minimum stress difference ($\Delta\sigma = \sigma_1 - \sigma_3$, in a plane stress system) at which 60° dipping normal faults will form is $8T$. At values of $\Delta\sigma$ of $4T$ or less only vertical tension fractures will form. At values of $\Delta\sigma$ between $4T$ and $8T$ shear fractures dipping between 90° and 60° will form (see Fig. 7). Faults with dips of 90° initiate as tensile fractures and any shear displacement is a secondary response to the vertical component of displacement on the fault at a lower level, where the dip is less than 90°. Predicted fault dips in dry or unsaturated rock are considered first followed by evaluation of the dips expected in saturated rock.

Faulting in dry or undersaturated rock

The lithostatic pressure gradient (L_g) is ($h \cdot \rho \cdot g \cdot 10^{-6}$) bars km^{-1} and σ_1 may be taken as equal to the lithostatic pressure; L_g is 230 bars km^{-1} for an assumed rock density of 2.35. The minimum depth at which 60° faults can form is $8T/Lg$ (1.74 km if $T = 50$ bars) and the minimum depth for shear fracture is $3T/L_g$ (0.88 km if $T = 50$ bars). The value of σ_3 must be less than zero, i.e. tensile, for any fracture to form at a depth less than $8T/L_g$ and the fracture will then dip more steeply than 60°.

A fault dip of 69° $\Delta\sigma = 5.38T$, which at the depth at which the coalfield faults are thought to have formed, 2.5 km, would require $T = 128$ bars and a tensile σ_3 (-112 bars). Tensile horizontal stresses at this depth are improbable and σ_3 is unlikely even to approach zero (Hubbert 1951); when faulting occurs σ_3 is likely to be about $0.3 \cdot \sigma_1$ (du Rouchet 1981), although the σ_3/σ_1 ratio cannot be constant. Measurements of the minimum normal stress in the Gulf Coast, a region of active normal faulting, show σ_3 values about one-third or half of σ_1, i.e. 190–280 bars at a depth of 2.5 km (Gretener 1978).

Given these requirements, the coalfield faults are unlikely to have formed in dry rock.

Faulting in saturated rock

Rocks undergoing faulting in an active extensional depositional environment are likely to be saturated: in this circumstance the effective stresses have to be considered. The effective principal stress $\sigma' = \sigma - p$ where $p =$ pore water (fluid) pressure. As σ_1' decreases, failure will occur at progressively lower values of $\Delta\sigma$ and intersection with the parabolic part of the failure envelope may occur (Fig. 8).

Under standard hydrostatic conditions the pore fluid pressure, p, at a given depth, is equal to the pressure which would be exerted by a column of water extending to the surface:

$$p = 0.42 \cdot L_p$$

where L_p is the lithostatic pressure.

Although values of pore fluid pressure may differ from standard hydrostatic pressures, such that $\lambda = p/L_p$ where λ may vary from 0 to 1, for depths of 2.5 km it is reasonable to assume standard hydrostatic pressures, $\lambda = 0.42$ (see below).

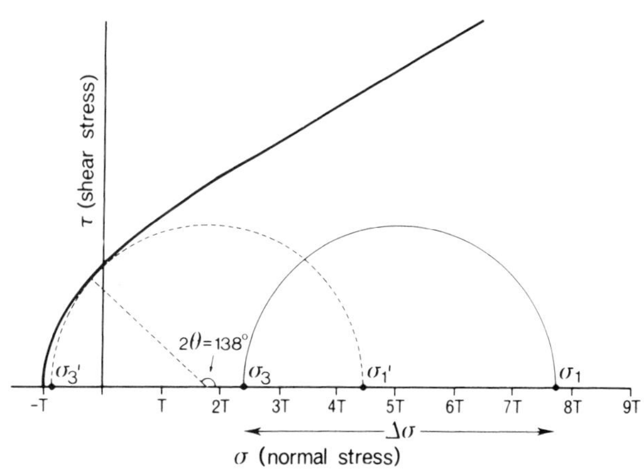

Fig. 8. Mohr construction with effective stresses (broken circle) for faults dipping 69° and stress difference = $5.38T$. Solid circle has maximum principal stress corresponding to lithostatic pressure at 2.5 km depth (575 bars = $7.75T$) for $T = 74$ bars. $p = \sigma_1 - \sigma_1' = 241$ bars, $\lambda = 0.42$.

Table 2. *Depths of formation and values of* T *corresponding to fault dips of 69°, 90° and 60° (λ = 0.42)*

Depths of 69° faults (km) $\frac{5.38T}{L_g(1-(\lambda))}$	Maximum depth of 90° fractures (km) $\frac{3T}{L_g(1-(\lambda))}$	Minimum depth of 60° faults (km) $\frac{8T}{L_g(1-(\lambda))}$	T (bars)
2.00	1.33	3.30	59.3
2.50	1.66	4.40	74.0
3.00	2.08	5.56	92.7

Using the modified Griffith fracture criterion (Phillips 1972) and assuming $\lambda = 0.42$, fault dips will be 69° at 2.5 km depth with $\Delta\sigma = 5.38T$. These values correspond to a stress ratio $\sigma_3/\sigma_1 = 0.2$, which is within the normal range (du Rouchet 1981), and $T = 74$ bars. An estimated value for T of 24.1 bars, based on extrapolation from compressive testing using the modified Griffith relationship, is given by Watts (1983) for Cretaceous and Tertiary chalks from the North Sea.

Maximum depths of vertical fractures and minimum depths of 60° faults are given in Table 2 for different depths of 69° faults, and the characteristic parabolic profile is shown in Fig. 9a. The profile is based on the assumption that fracture orientation is determined solely by the stress distribution in the bulk material and is not influenced by the direction of propagation: this assumption may not be valid.

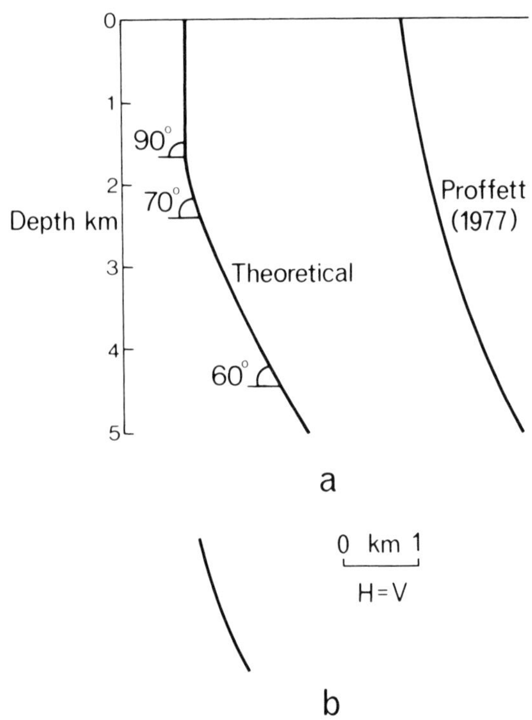

Fig. 9. Fault dip profiles. (**a**) Theoretical profile showing the changes in dip with depth at standard hydrostatic pore fluid pressures ($T = 74$ bars), and a representative profile for faults in Nevada, Basin and Range Province (from Proffett 1977). (**b**) Partial fault profile for Hurricane Fault, Utah, Basin and Range Province (Hamblin 1965). The top of the profile is at the present surface but the depth at the time of faulting is not known.

The modified Griffith criterion predicts that rates of change of fault dip will vary with depth. For faults which dip 69° at 2.5 km the rate of change of dip with depth is 1.09°/100 m, varying from 2.7°/100 m between 90° and 70°, to 0.5°/100 m between 70° and 60°. Rates of change of dip at the upper levels of normal faults in the Basin and Range Province are given by Proffett (1977) as varying from 0.3 to 0.7°/100 m in Western Nevada, 1.3° to 2.0°/100 m on the Hurricane fault, Utah (Hamblin 1965) where it changes from 90° to 60°, and 0.7° to 1.3°/100 m on faults near Las Vegas (Hamblin 1965). The downward decreases in dip which characterize the higher levels of normal faults, as observed (Fig. 9; Proffet 1977; Hamblin 1965) and as predicted by the modified Griffith failure envelope, cannot be taken as an indication of continued shallowing of faults at deeper levels and they are not diagnostic of flattening to a detachment at depth (Proffett 1977). Detachments may be invoked because of a model-dependent belief that they are geometrically necessary but 'few field examples of such detachment zones for extensional faults have yet been identified' (Gibbs 1984). Nevertheless, shallowing from 90° to 60° in the upper parts of normal faults will complicate the deformation geometry in the rocks surrounding the fault: enhancement of hanging-wall reverse drag and absence or reduction of footwall reverse drag, as observed by Hamblin (1965), are possible consequences.

Even allowing for some uncertainty in the depth estimate for the formation of the coalfield faults it is evident that the interval between 2 and 3 km depth will contain faults dipping more steeply than 60° and is also likely to show significant variation in fault dip. In the interval over which a fault changes dip from 90° to 60° the mean dip of the fault is 67.5°. This is a consequence of the parabolic form of the dip profile (Fig. 9a). Thus a random sample of fault dips within a depth interval in which this change of dip occurs, will have a mean value close to 67.5° if σ_1 is vertical.

The interpretation given and the dip profile (Fig. 9a) are based on the assumption of standard hydrostatic pore fluid pressures. However, an alternative explanation for the frequency of fault dips between 65° and 70°, and for joint sets with conjugate angles of 45° (Hancock 1986; Hancock *et al.* 1987), is that 135° is a preferred value for 2α: this value corresponds to zero normal stress across the fracture surface (Fig. 7). So long as the normal stress on a pre-existing or newly formed fracture is zero, the permeability of the rock is enhanced and over-pressures will be restricted to values which maintain, ideally, a fault dip of 67.5°.

Changes in dip during fault growth

As a fault propagates downwards and/or upwards dip changes will occur according to the mechanisms described earlier, if they are independent of the propagation process. Vertical changes in dip may also result from gross lithological changes but, for the reasons given earlier, such changes are not seen in the coalfield data.

As fault dips are sensitive to changes in T, λ and depth, a change in one of these parameters during lateral propagation of a fault may result in lateral changes of dip along a fault surface. A change in either T or λ during growth of a fault, or a change in depth due either to erosion or to burial, would produce a radial dip change on the fault surface, which can be taken to be elliptical (Barnett *et al.* 1987, fig. 8). The radial dip changes described in an earlier

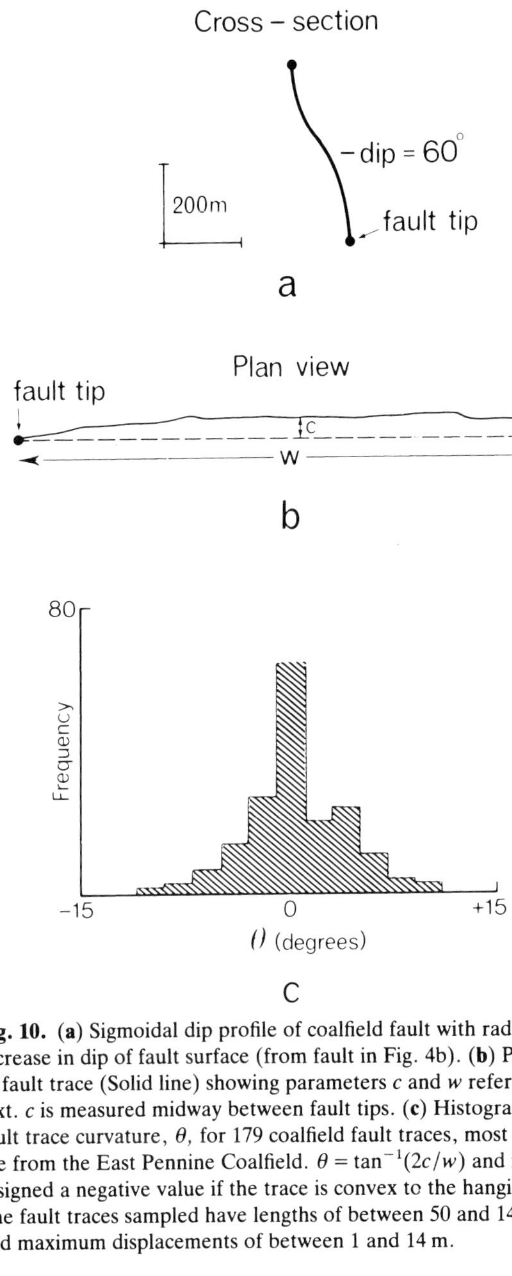

Fig. 10. (**a**) Sigmoidal dip profile of coalfield fault with radial increase in dip of fault surface (from fault in Fig. 4b). (**b**) Plan view of fault trace (Solid line) showing parameters c and w referred to in text. c is measured midway between fault tips. (**c**) Histogram of fault trace curvature, θ, for 179 coalfield fault traces, most of which are from the East Pennine Coalfield. $\theta = \tan^{-1}(2c/w)$ and is assigned a negative value if the trace is convex to the hanging-wall. The fault traces sampled have lengths of between 50 and 1450 m and maximum displacements of between 1 and 14 m.

section invariably indicate a steepening of the dip during fault growth. In vertical section such faults have a characteristic sigmoidal dip profile (Fig. 10a).

Progressive radial increase in dip could be due to decrease in overburden by erosion during fault growth: this would require erosion of approximately 1 km for a dip change of 10° and is thought to be improbable during the growth of the small coalfield faults sampled. Significant changes in T during growth of a single small fault are unlikely, especially because lithification was already well advanced when the coalfield faults were initiated. The preferred explanation is for a progressive change in λ during fault growth: at 2.5 km depth, a change of λ from 0.42 (standard hydrostatic) to 0.58 would require an increase in pore water pressure of 90 bars and produce a change in fault dip of 10°. Such increases in λ at these depths may result either from porosity reduction or from hydrocarbon generation (Spencer 1987). The depth of formation of the coalfield faults is about that at which pore fluid pressures would be expected to start to exceed the standard hydrostatic pressure. Overpressures in the Gulf Coast, a region of active normal faulting, develop at depths of 2.4 km (Bruce 1973; Magara 1975).

Radial dip changes have predictable geometrical effects on the form of the fault surface, because such changes allow only one horizontal chord on the fault surface to remain straight. Assuming this to be the chord passing through the centre of the fault, i.e. the horizontal diameter, horizontal chords on the upper half of the fault surface will be concave towards the hanging-wall and those on the lower half convex towards the hanging-wall. The material expression of such curved chords will be curved fault traces on horizontal or near horizontal layers. Curved fault traces are common in the coalfield data we have examined and this feature has been remarked upon by Rippon (1985*a*, 1985*b*), and is particularly noticeable on small faults of length less than 1 km.

A sample of 179 complete fault traces, i.e. those where both fault tips are mapped, has been tested for curvature. The ratio $2c/w$ (see Fig. 10b) was measured and for fault traces convex to the hanging-wall the ratio was assigned a negative value. The angle θ (where $\theta = \tan^{-1}(2c/w)$), provides a measure of the curvature of fault traces. A histogram of θ values (Fig. 10c) shows approximately equal numbers of traces to be concave and convex, respectively, to the hanging-wall, with the majority showing little curvature. Moore (1960) examined surface traces of 55 faults in the Basin and Range which are 'generally from 10 to 30 miles long with a radius of curvature of 20 to 40 miles', of which 41 are concave to the hanging-wall; these faults also show dips decreasing downwards in their upper parts but it is not known whether the faults also show lateral dip changes. It is clear, however, that fault traces which are concave to the hanging-wall are not necessarily an indication of listric faults.

Superficially a change in fault strike towards the tips of a fault trace, as illustrated in Fig. 10b, is similar to the curving of a crack which often precedes the crack tip (Lawn & Wilshaw 1975; Beach 1977) and is a dynamic feature. Faults, unlike cracks, are not products of a single dynamic event (Watterson 1986), and the interpretation of curving fault traces as a necessary geometric consequence of progressive change in fault dip, is consistent with the slow growth of faults through many cycles of loading and unloading: it also allows the shapes of fault traces to be interpreted systematically in terms of variable dip. An improved understanding of fault trace curvature would be particularly useful because traces concave to the hanging-wall are almost essential for the formation of simple hanging-wall closures (Barnett *et al.* 1987) and the volume of footwall closure may be increased when the fault trace is convex to the hanging-wall.

Dip changes can lead to fault surfaces which are curved either in vertical section only or in both vertical and horizontal section. The additional compatibility strains required to accommodate slip on fault surfaces with complex morphology could be a reason for such fault surfaces to be abandoned and replaced by new planar faults elsewhere, or for new planar fractures to form inside the fault tip-line loop (see Davison 1987) and isolate a horse block.

Detachments at shallow crustal levels

Our main concern has been to analyse the data relevant to tectonic normal faults in the brittle regime, and this has shown that such faults are expected to be vertical near to the surface and to shallow out to 60° at depth, i.e. they show a limited listric form but with no requirement for a detachment. If, on the other hand, slip on a detachment surface or within a detachment layer occurs this will result in kinematically controlled faulting at the trailing edge of the detachment and these faults must necessarily transfer slip from the detachment surface upwards to the surface. The driving force of such systems is gravitational and the sliding surface is likely to be a layer in which the value of λ approaches 1 as in the case of the large-scale growth faults of the Gulf Coast and other deltaic oil provinces (Crans *et al.* 1980). In the East Pennine coalfield such faults occur on a small scale, affecting only the uppermost 50 m, similar to that of growth faults in the Namurian of Western Ireland (Crans *et al.* 1980).

Changes in dip below 3–4 km

On faults which are not associated with shallow detachment, downward shallowing to 45° is anticipated at depths where the displacement is accommodated plastically. There is persuasive field evidence (Platt & Vissers 1980; Ramsay 1980; Park 1981) that, within the plastic regime, conjugate angles containing the σ_1 direction can be obtuse, in contrast to the acute angles which characterize the brittle regime. This observation suggests that the downward transition from the brittle to the plastic regime is characterized by shallowing of dips from 60°.

Source mechanism studies of earthquakes along major normal faults, with cumulative displacements of greater than 1 km, indicate fault dips of between 30° and 65°, to depths of up to 12 km (Smith & Sbar 1974; Jackson 1983, 1987; Fletcher *et al.* 1984; Barrientos *et al.* 1985; Okaya & Thompson 1985). Examples of earthquakes from four areas are briefly described below:

(a) *Basin and Range, USA*. Some of the most intensively studied dip-slip earthquakes are from the north-eastern Basin and Range Province (Scott *et al.* 1985). For the 1983 Borah Peak earthquake on the Lost River fault (a normal fault with a small to moderate strike-slip component) estimates of fault dip range between 45° and 53° (Barrientos *et al.* 1985; Boatwright 1985), and the seismological and geodetic data are consistent with a planar fault which extends to a depth of 15 km; listric-like curvature of the fault is unacceptable (Barrientos *et al.* 1985; Stein & Barrientos 1985). Fault plane solutions for the dip-slip 1954 Fairview Peak and the 1959 Hebgen Lake earthquakes are 62° and 64°, respectively (Smith & Sbar 1974; Okaya & Thompson 1985); available data indicate that the Fairview Peak earthquake had a focal depth of 15 km (Okaya & Thompson 1985).

(b) *Appennines, Italy*. Focal mechanisms of earthquakes along two normal faults from intra-Appennine sedimentary basins in Southern Italy are indicative of fault dips of 65° and both occurred at around 10 km depth (Westaway 1987).

(c) *Sierra Nevada range, USA* Hypocentres for a dip-slip earthquake within the Sierra Nevada of California are consistent with a steeply dipping fault plane (50°–60°) to depths of about 10 km; this fault appears to steepen to more than 70° above depths of about 3 km (Lahr *et al.* 1976, fig. 5) and may also steepen below depths of 10 km (Fletcher *et al.* 1984, fig. 3).

(d) *Aegean Sea*. Seismogenic normal faults from the Aegean have dips which range from 40° to 60° at depths shallower than about 8 km (Jackson *et al.* 1982; Soufleris *et al.* 1982). For a given fault, the dip shows little variation with depth although there is some steepening very close to the surface, with in some cases vertical fault dips at the surface (Jackson 1983). Recently, however, a model invoking motion along shallow seismic faults below the brittle–plastic transition has been proposed to explain the surface deformation and distribution of aftershocks associated with three earthquakes which occurred near Corinth, in 1981 (King *et al.* 1985).

The evidence from earthquakes indicates that major seismogenic faults are steeply dipping (generally between 45° and 60°) throughout the entire 'brittle' (seismogenic) crust and show only limited dip decrease with depth (90–45°) with the highest rates of dip change at depths of less than 3 km (see Jackson 1987). The main evidence pointing to the possibility of normal faults flattening with depth, below the brittle–plastic transition, is the occurrence of aftershocks and antithetic faults in the hanging-walls of normal faults: these features are inconsistent with the extension model in which steep faults remain planar but are rotated domino-style, with no requirement for deformation within the domino-shaped blocks which behave as though 'floating' in a liquid (Eyidogan & Jackson 1985). A soft-domino model (Barnett *et al.* 1987), which is consistent with earthquake data, assumes that faulting is accommodated by ductile strain within the volume surrounding faults (i.e. rigid fault bounded blocks do not exist except in special circumstances related to a free surface) without any requirement for a major horizontal detachment.

Experimental work shows that the conjugate angle for ductile shears is 110° for axisymmetric compression ($\sigma_1 > \sigma_2 = \sigma_3$) in rock (Friedman & Logan 1973; Anderson 1974) and in other materials (Bauwens 1967; Hagan 1980; Reid 1973). This angle is close to the conjugate angle (109.4°) between zero infinitesimal strain rate directions in the axisymmetric compressive strain ellipsoid. For plastic shear zones to maintain continuity with the surrounding matrix, they must propagate parallel to an instantaneous zero strain rate direction of the bulk strain (Bowden & Jukes 1972). A conjugate angle of 110° would correspond to dips of 35° for ductile shears with normal fault displacement. However, crustal extension will usually be a plane strain rather than an axisymmetric strain; for plane strain the conjugate angle between zero strain rate directions is 90° (Bowden & Jukes 1972; Van Iterson 1947), corresponding to dips of 45° for ductile normal faults. The downward transition from the brittle to the plastic regime is not abrupt and gradual shallowing of faults from 60° to 45° is to be expected and is not an indication of further shallowing to a crustal detachment. Most sedimentary sequences will contain lithological units which behave more plastically than underlying basement rocks and rheology is depth-controlled only in a very general way. Where faults continue downwards from a sedimentary sequence into metamorphic basement the angle of dip may be unpredictable, because of the mechanical anisotropy in the basement (Brun & Choukroune 1983), with new displacements following zones of older faulting or ductile shear. Most of the coalfield faults examined in this study probably die out within the Carboniferous succession, but those which intersect the base of the succession could root in older faults within the Lower Palaeozoic basement.

Conclusions

The mean dip of the coalfield faults examined is 69° but dips are variable both on single faults and between faults. There is no significant difference in the dips of faults from the different coalfield areas, or between faults of different orientation. Dips have not been significantly affected by later tilt. The data are derived from workings extending over a vertical interval of about 500 m and the faults formed at a depth of about 2.5 km. No systematic change of dip with depth is evident. Data on fault dips from other areas show that dips greater than 60° are quite usual. Lower dips are indicative either of faulting in poorly lithified sediments with subsequent compaction or of reactivation of an original low angle thrust.

The modified Griffith fracture criterion is used to show that the observed dips are consistent with pore fluid pressures close to standard hydrostatic values. The dips of normal faults in a sedimentary succession are expected to vary from 90° at the surface to 60° at 3–4 km but this shallowing is not an indication of an underlying detachment. The uppermost 3 km is the economic zone within which 90% of known recoverable oil reserves (Gretener 1978), and all of the recoverable coal, occurs and within this zone tectonic fault dips will generally be steeper than 60°, but due allowance must be made for lithological variations and thickness of cover removed during denudation.

Some faults show radial dip changes and such changes necessarily lead to curved fault traces. At deeper levels than those observed in the coalfields normal faults are expected to flatten out gradually to a dip of 45°, in line with the seismological evidence.

J. Mortimer, I. Williams, P. Gillespie and C. Childs assisted with data collection and processing. We thank J. Gibson for computing and other advice, D. Creedy for discussion and advice on estimation of depth of burial, and other British Coal staff for help and advice. We are grateful to P. Hancock for a constructive review which led to several modifications to the manuscript. The work was funded by British Coal/European Coal and Steel Commission contract No. 7220 AF825. Publication is with the permission of British Coal. The views expressed are those of the authors and are not necessarily those of British Coal.

References

Anderson, E. M. 1951. *The Dynamics of Faulting.* Oliver & Boyd, Edinburgh.

Anderson, T. B. 1974. The relationship between kink-bands and shear fractures in the experimental deformation of slate. *Journal of the Geological Society, London,* **130,** 367–82.

Angelier, J. 1985. Extension and rifting: the Zeit region, Gulf of Suez. *Journal of Structural Geology,* **7,** 605–12.

—— & Colletta, B. 1983. Tension fractures and extensional tectonics. *Nature,* **301,** 49–51.

Baker, B. H. & Wohlenberg, J. 1971. Structure and evolution of the Kenya rift valley. *Nature,* **229,** 538–42.

Barnett, J. A. M., Mortimer, J., Rippon, J. H., Walsh, J. J. & Watterson, J. 1987. Displacement geometry in the volume containing a single normal fault. *Bulletin of the American Association of Petroleum Geologists,* **71,** 925–37.

Barrientos, S. E., Ward, S. N., Gonzalez-Ruiz, J. R. & Stein, R. 1985. Inversion for moment as a function of depth from geodetic observations and long period body waves of the 1983 Borah Peak, Idaho earthquake *In*: Stein, R. & Buckham, R. C. (eds) *Proceedings of Workshop XXVIII on the Borah Peak, Idaho, Earthquake.* United States Geological Survey Open File Report 85-290, 495–518.

Bauwens, J. C. 1967. Deformations plastique des fauts polymeres vitreux soumis a un systeme de contraintes quelcouque. *Journal of Polymer Science,* **5,** 1145–56.

Beach, A. 1977. Vein arrays, hydraulic fractures and pressure solution structures in a deformed flysch sequence, SW England. *Tectonophysics,* **40,** 201–225.

Best, M. G. & Hamblin, W. K. 1978. Origin of the northern Basin and Range Province: implications from the geology of its eastern boundary. *In*: Smith, R. B. & Eaton, G. P. (eds) *Cenozoic tectonics and regional geophysics of the western Cordillera.* Geological Society of America Memoirs, **152,** 313–40.

Boatwright, J. 1985. Characteristics of the aftershock sequence of the Borah Peak, Idaho, earthquake determined from digital recordings of the events. *Bulletin of the Seismological Society of America,* **75,** 1265–84.

Bowden, P. B. & Jukes, J. A. 1972. The plastic flow of isotropic polymers. *Journal of Materials Science,* **7,** 52–63.

Bruce, C. H. 1973. Pressured shale and related sediment deformation: Mechanism for development of regional contemporaneous faults. *Bulletin of the American Association of Petroleum Geologists,* **57,** 878–86.

Brun, J. P. & Choukroune, P. 1983. Normal faulting, block tilting, and decollement in a stretched crust. *Tectonics,* **2,** 345–56.

Byerlee, J. D. 1969. Letter. *Journal of Geophysical Research,* **74,** 5349–50.

Chenet, P.-Y. & Letousky, J. 1983. Tectonics of the area between Abu Durba and Gebel Mezzazat (Sinai, Egypt) in the context of the evolution of the Suez Rift. *Bulletin of the Centre of Exploration–Production Research Elf-Aquitaine,* **7,** 201–15.

Christensen, M. N. 1966. Late Cenozoic crustal movements in the Sierra Nevada of California. *Bulletin of the Geological Society of America,* **77,** 163–82.

Chowdhury, M. K., Laskar, B. & Mitra, N. D. 1975. Tectonic control of lower Gondwana sedimentation in peninsular India. *In*: Campbell, K. S. W. (ed.) *Gondwana Geology.* National University Press, Canberra, 675–80.

Crans, W., Mandl, G. & Haremboure, J. 1980. On the theory of growth faulting: a geomechanical delta model based on gravity sliding. *Journal of Petroleum Geology,* **2,** 265–307.

Creedy, D. P. 1988. Geological controls on the formation and distribution of gas in the British Coal Measure strata. *International Journal of Coal Geology* (in press).

Davison, I. 1987. Normal fault geometry related to sediment compaction and burial. *Journal of Structural Geology,* **9,** 393–401.

Dunne, G. C. 1977. Geology and structural evolution of Old Dad Mountain, Mojave Desert, California. *Bulletin of the Geological Society of America,* **88,** 737–48.

Elliott, R. E. 1973. *Preliminary survey of faults.* Internal report, East Midlands Regional Geological Service, National Coal Board.

Eyidogan, H. & Jackson, J. 1985. A seismological study of normal faulting in the Demirci, Alasehir and Gediz earthquakes of 1969–70 in western Turkey: implications for the nature and geometry of deformation in the continental crust. *Geophysical Journal of the Royal Astronomical Society,* **81,** 569–607.

Fletcher, J., Boatwright, J., Haar, L., Hanks, T. & McGarr, A. 1984. Source parameters for aftershocks of the Oroville, California, earthquake. *Bulletin of the Seismological Society of America,* **74,** 1101–123.

Friedman, M. & Logan, J. M. 1973. Luders bands in experimentally deformed sandstone and limestone. *Bulletin of the Geological Society of America,* **84,** 1465–76.

Frost, D. V. & Smart, J. G. O. 1979. Geology of the country north of Derby. *Memoirs of the Geological Survey of Great Britain,* **125.**

Garfunkel, Z. & Bartov, Y. 1977. The tectonics of the Suez Rift. *Bulletin of the Geological Survey of Israel,* **71,** 1–44.

Gibbs, A. D. 1984. Structural evolution of extensional basin margins. *Journal of the Geological Society, London,* **141,** 609–20.

Glennie, K. W. (ed.) 1984. *Introduction to the Petroleum Geology of the North Sea.* Blackwell Scientific Publications, Oxford.

Golombek, M. P., McGill, G. E. & Brown, L. 1983. Tectonic and geologic evolution of the Espanola Basin, Rio Grande Rift: structure, rate of extension, and relation to the state of stress in the Western United States. *Tectonophysics,* **94,** 483–507.

Gretener, P. E. 1978. Pore pressure: fundamentals, general ramifications and implications for structural geology. *American Association of Petroleum Geologists, Continuing Education Course Note Series,* **4.**

GUDMUNDSSON, A. 1984. *A study of dykes, fissures and faults in selected areas of Iceland*. PhD thesis, University of London.

HAGAN, J. T. 1980. Shear deformation under pyramidal indentations in soda-lime glass. *Journal of Materials Science,* **15,** 1417–24.

HALSTEAD, P. H. 1975. Northern North Sea Faulting. *In*: FINSTEAD, K. G. & SELLEY, R. C. (eds) *Proceedings of the Jurassic Northern North Sea Symposium 1975.* Norwegian Petroleum Society, Stavanger, 10/1–10/38.

HAMAR, G. P. 1975. A Jurassic complex in the North Sea. *In*: FINSTEAD, K. G. & SELLEY, R. C. (eds) *Proceedings of the Jurassic Northern North Sea Symposium*. Norwegian Petroleum Society, Stavanger, 17/1–17/18.

HAMBLIN, W. K. 1965. Origin of 'reverse drag' on the downthrown side of normal faults. *Bulletin of the Geological Society of America,* **76,** 1145–64.

HAMILTON, W. B. 1982, Structural evolution of the Big Maria Mountains northeastern Riverside county, southeastern California. *In*: FROST, E. G. & MARTIN, D. L. (eds) *Mesozoic-Cenozoic Tectonic Evolution of the Colorado River Region, California, Arizona and Nevada.* Cordilleran Publishers, San Diego, Calif., 1–28.

HANCOCK, P. L. 1986. Joint Spectra. *In*: NICHOL, I. & NESBITT, R. W. (eds) *Geology in the Real World—the Kingsley Dunham Volume.* Institution of Mining and Metallurgy, London, 155–64.

——, AL-KALIDI, A., BARKA, A. A. & BEVAN, T. G. 1987. Aspects of analysing brittle structures. *Annales Tectonicae,* **1,** 5–19.

HUBBERT, M. K. 1951. Mechanical basis for certain familiar geologic structures. *Bulletin of the Geological Society of America,* **62,** 355–72.

ILLIES, J. H. 1977. Ancient and recent rifting in the Rhine Graben. *Geologie en Minjbouw,* **56,** 329–50.

JACKSON, J. A. 1983. The use of earthquake source studies in continental tectonic geology. *In*: *Earthquakes*: *observation, theory and interpretation.* LXXXV Corso, Societe Italia di Fisica, Bologna, Italy.

—— 1987. Active normal faulting and crustal extension. *In*: COWARD, M. P., DEWEY, J. F. & HANCOCK, P. L. (eds) *Continental Extensional Tectonics.* Special Publication of the Geological Society, London, **28,** 3–18.

——, GAGNEPAIN, J., HOUSEMAN, G., KING, G. C. P., Papadimitriou, P., Soufleris, C. & Virieux, J. 1982. Seismicity, normal faulting, and the geomorphological development of the Gulf of Corinth (Greece): the Corinth earthquakes of February and March 1981. *Earth and Planetary Science Letters,* **57,** 377–97.

KING, G. C. P., OUYANG, Z. X., PAPADIMITRIOU, P., DESCHAMP, A., GAGNEPAIN, J., HOUSEMAN, G., JACKSON, J. A., SOUFLIERS, C. & VIRIEUX, J. 1985. The evolution of the Gulf of Corinth (Greece): an aftershock study of the 1981 earthquakes. *Geophysical Journal of the Royal Astronomical Society,* **80,** 677–93.

KLIGFIELD, R., CRESPI, J., NARUK, S. & DAVIS, G. H. 1984. Displacement and strain patterns of extensional orogens. *Tectonics,* **3,** 557–609.

LAHR, K. M., LAHR, J. C., LINDH, A. G., BUFE, C. G. & LESTER, F. W. 1976. The August 1975 Oroville earthquakes. *Bulletin of the Seismological Society of America,* **66,** 1085–99.

LASKAR, B. 1979. Evolution of Gondwana coal basins. *In*: LASKAR, B. & RAO, R. C. S. (eds) *Fourth International Gondwana Symposium,* Vol. 1. Hindustan Publications, Delhi, 224–37.

LAWN, B. R. & WILSHAW, T. R. 1975. *Fracture of Brittle Solids.* Cambridge University Press, London.

LEEDER, M. R. 1982. Upper Palaeozic basins of the British Isles—Caledonide inheritence versus Hercynian plate margin processes. *Journal of the Geological Society, London,* **139,** 479–91.

—— & MCMAHON, A. H. 1988. Upper Carboniferous (Silesian) basin subsidence in northern Britain. *In*: BESLY, B. M. & KELLING, G. (eds) *Sedimentation in a Synorogenic Basin Complex—the Upper Carboniferous of NW Europe.* Blackie, London, 43–52.

MACKIN, J. H. 1960. Structural significance of Tertiary volcanic rocks in southwestern Utah. *American Journal of Science,* **258,** 81–131.

MAGARA, K. 1975. Importance of aquathermal pressuring effect in Gulf Coast. *Bulletin of the American Association of Petroleum Geologists,* **59,** 2037–65.

MILLER, E. L., GANS, P. B. & GARING, J. 1983. The Snake Range decollement: an exhumed mid-Tertiary ductile-brittle transition. *Tectonics,* **2,** 239–63.

MOHR, P. 1983. The Morton–Black hypothesis for the thinning of continental crust; revisited in Western Afar. *Tectonophysics,* **94,** 509–28.

MOORE, J. G. 1960. Curvature of normal faults in the Basin and Range province of the Western United States. *United States Geological Survey Professional Paper,* **400-B,** 409–11.

MORTON, W. H. & BLACK, R. 1975. Crustal attenuation in Afar. *In*: PILGER, A. & ROSLER, A. (eds) *The Afar Region of Ethiopia and Related Rift Problems.* Scheizerbert, Stuttgart, 55–65.

MURRAY, G. E. 1961. *Geology of the Atlantic and Gulf Coastal Province of North America.* Harper & Brothers, New York, 167–201.

NORRIS, D. K. 1958. Structural conditions in Canadian coal mines. *Bulletin of the Geological Survey of Canada,* **44,** 1–54.

OCAMB, R. D. 1961. Growth faults of South Louisiana. *Transactions of the Gulf Coast Association of Geological Societies,* **11,** 139–73.

OKAYA, D. A. & THOMPSON, G. A. 1985. Geometry of Cenozoic extensional faulting: Dixie Valley, Nevada. *Tectonics,* **4,** 107–25.

PARDEE, J. T. 1950. Late Cenozoic block faulting in western Montana. *Bulletin of the Geological Society of America,* **61,** 359–406.

PARK, R. G. 1981. Shear-zone deformation and bulk strain in granite–greenstone terrain of the Western Superior Province, Canada. *Precambrian Research,* **14,** 31–47.

PERKINS, H. 1961. Fault closure-type fields, Southeast Louisiana. *Transactions of the Gulf Coast Association of Geological Societies,* **11,** 177–96.

PHILLIPS, W. J. 1972. Hydraulic fracturing and mineralization. *Journal of the Geological Society, London,* **128,** 337–59.

PLATT, J. P. & VISSERS, R. L. M. 1980. Extensional structures in anisotropic rocks. *Journal Structural Geology,* **2,** 397–410.

PRICE, N. J. 1966. *Fault and Joint Development in Brittle and Semi-Brittle Rock.* Pergamon Press, Oxford.

PROFFETT, J. M. 1977. Cenozoic geology of the Yerington district, Nevada and implications for the nature and origin of Basin and Range faulting. *Bulletin of the Geological Society of America,* **88,** 247–66.

QUARLES, M. Q. 1953. Salt-ridge hypothesis on origin of Texas Gulf Coast type of faulting. *Bulletin of the American Association of Petroleum Geologists,* **37,** 489–508.

RAINE, J. D. 1978. *Fault pattern analysis at Warsop Main Colliery, North Derbyshire Area.* Internal report, National Coal Board.

RAMSAY, J. G. 1980. Shear zone geometry: a review. *Journal of Structural Geology,* **2,** 83–99.

REID, C. N. 1973. *Deformation Geometry for Materials Scientists.* Pergamon, London.

RIPPON, J. H. 1985*a*. Contoured patterns of the throw and hade of normal faults in the Coal Measures (Westphalian) of North-east Derbyshire. *Proceedings of the Yorkshire Geological Society,* **45,** 147–61.

—— 1985*b*. New methods of forecasting throws and hade of faults in some North Derbyshire collieries. *Transactions of the Institute of Mining Engineers,* **145,** 198–204.

DU ROUCHET, J. 1981. Stress fields, a key to oil migration. *Bulletin of the American Association of Petroleum Geologists,* **65,** 74–85.

ROUX, W. F. JR 1979. The development of growth fault structures. *American Association of Petroleum Geologists Structural Geology School Course Notes,* **33.**

SAX, H. G. J. 1946. De Tectoniek van het Carboon in het Zuid-Limburgsche Mijngebied. *Mededeelingen van de Geologische Stichting, Serie C-1-1,* **3,** 1–77.

SCOTT, W. E., PIERCE, K. L. & HAIT, M. H. 1985. Quaternary tectonic setting of the 1983 Borah Peak earthquake, central Idaho. *Bulletin of the Seismological Society of America,* **75,** 1053–66.

SECOR, D. T. 1965. Role of fluid pressures in jointing. *American Journal of Science,* **263,** 633–46.

SHARP, R. P. 1939. Basin-Range structure of the Ruby–East Humboldt range, northeastern Nevada. *Bulletin of the Geological Society of America,* **50,** 881–920.

SHEPHERD, J., CREASEY, J. W. & HUNTINGTON, J. F. 1981. Geometric and dynamic analysis of some faults in the Western Coalfield of New South Wales. *Proceedings of the Australasian Institute of Mining and Metallurgy,* **279,** 19–32.

SMITH, E. G., RHYS, G. H. & EDEN, R. A. 1967. Geology of the country around Chesterfield, Matlock and Mansfield. *Memoirs of the Geological Survey of Great Britain,* **112.**

SMITH, R. B. 1978. Seismicity, crustal structure and intraplate tectonics of the western Cordillera. *In*: SMITH, R. B. & EATON, G. P. (eds) *Cenozoic tectonics and regional geophysics of the western Cordillera.* Geological Society of America Memoir **152,** 111–44.

—— & SBAR, M. L. 1974. Contemporary tectonics and seismicity of the western United States with emphasis on the intermountain seismic belt. *Bulletin of the Geological Society of America,* **85,** 1205–18.

SOUFLERIS, C., JACKSON, J., KING, G., SPENCER, C. & SCHOLZ, C. 1982. The 1978 earthquake sequence near Thessaloniki (northern Greece). *Geophysical Journal of the Royal Astronomical Society,* **68,** 429–58.

SPENCER, C. W. 1987. Hydrocarbon generation as a mechanism for overpressuring in Rocky Mountain Region. *Bulletin of the American Association of Petroleum Geologists,* **71,** 368–88.

STEIN, R. S. & BARRIENTOS, S. E. 1985. The 1985 Borah Peak, Idaho, earthquake: Geodetic evidence for deep rupture on a planar fault. *In*: STEIN, R. S. & BUCKHAM, R. C. (eds) *Proceedings of Workshop XXVIII on the Borah Peak, Idaho, Earthquake.* United States Geological Survey Open File Report 85-290, 459–484.

STEWART, J. H. 1971. Basin and Range structure: a system of horsts and grabens produced by deep-seated extension. *Bulletin of the Geological Society of America,* **82,** 1019–44.

SUGGATE, R. P. 1974. Coal ranks in relation to depth and temperature in Australian and New Zealand oil and gas wells. *New Zealand Journal of Geology and Geophysics*, **17**, 149–67.
SURLYK, F. 1977. Mesozoic faulting in East Greenland. *Geologie en Mijnbouw*, **56**, 311–27.
VAN ITERSON, F. K. 1947. *Plasticity in Engineering*. Blackie, London.
VOWLES, D. B. 1978. *Trends and patterns of faulting in the eastern central part of the South Wales Coalfield.* Internal report, National Coal Board.
WATTERSON, J. 1986. Fault dimensions, displacements and growth. *Pure and Applied Geophysics*, **124**, 365–73.
WATTS, N. L. 1983. Microfractures in chalks of Albuskjell Field, Norwegian Sector, North Sea; possible origin and distribution. *Bulletin of the American Association of Petroleum Geologists*, **67**, 201–34.
WERNICKE, B. 1981. Low angle normal faults in the Basin and Range Province: nappe tectonics in an extending orogen. *Nature*, **291**, 645–8.
—— & BURCHFIEL, B. L. 1982. Modes of extensional tectonics. *Journal of Structural Geology*, **4**, 105–15.
——, WALKER, J. D. & BEAUFAIT, M. S. 1985. Structural discordance between Neogene detachments and frontal Sevier thrusts, central Mormon Mountains, southern Nevada. *Tectonics*, **4**, 213–46.
WESTAWAY, R. 1987. The Campania, southern Italy, earthquakes of 1962 August 21. *Geophysical Journal of the Royal Astronomical Society*, **88**, 1–24.
WITHJACK, M. O. & POLLOCK, D. J. D. P. 1984. Synthetic seismic-reflection profiles of rift-related structures. *Bulletin of the American Association of Petroleum Geologists*, **68**, 1160–78.

Received 24 September 1987; Revised typescript accepted 19 January 1988.

Seismic reflections from normal faults in the northern North Sea

G. YIELDING, M. E. BADLEY & B. FREEMAN

Badley Ashton & Associates Limited, Winceby House, Winceby, Horncastle, Lincolnshire LN9 6PB, UK

Abstract: Fault-plane reflections from gravity-driven, listric faults in thick sedimentary sequences are well known. However, reflections from normal faults associated with crustal extension are much rarer and less well studied. We discuss a number of examples of such reflections, present on commercial seismic reflection data from the northern North Sea. Some of these examples have previously been described as reflected refractions, but we show that this requires unrealistic velocities within the sedimentary column. Instead, we consider it more likely that these events are true fault-plane reflections. Depth migration and image-ray migration are used to determine the probable geometry of the faults in both cross section and map view. Estimates of the true geological dip remain sensitive to the velocity structure used, particularly that applied to the Triassic half-graben fill. Despite this uncertainty, the depth-migrations show that these faults are composed of two near-linear segments in cross section. The upper parts (in the Jurassic and Upper Triassic section) dip at 40–50°, but the deeper parts (Triassic hangingwall against basement in the footwall) typically dip at 30–35°. Occasionally, reflections indicate that faults continue into the basement with a 30–35° dip. We interpret this fault geometry as originating from two phases of fault movement. Triassic extension caused rotation of sets of initially steep planar faults. Following Late-Triassic-to-Middle-Jurassic thermal subsidence, renewed extension in the Late Jurassic caused the faults to cut up through the overburden at a steeper angle. Without recognition of a two-phase extension history, such faults might be mistakenly interpreted as forming with a listric geometry.

In recent years, many papers have been published on the prediction of normal-fault geometry from the shape of deformed beds in the hangingwall (e.g. Gibbs 1983; White *et al.* 1986; Williams & Vann 1987; Wheeler 1987). Each of these studies assumes a rigid footwall, with hangingwall deformation being a direct consequence of curvature of the fault plane. Such a model of fault-related deformation appears to be appropriate for gravity-driven listric faults in thick sedimentary successions such as the Gulf Coast and Niger Delta. These listric faults are commonly observed directly on seismic reflection data (e.g. Wernicke & Burchfield 1982, fig. 15; Jackson & Galloway 1984; White & Yielding, this volume). However, studies of fault-related deformation in basement and compacted sediment indicate that hangingwall rollover can be produced by *planar* faults. The shape of the rollover is controlled by factors such as the extent of the fault plane, the slip distribution on the fault plane, and the mechanical properties of the upper lithosphere (e.g. Stein & Barrientos 1985; King *et al.* 1988; Kusznir *et al.* 1988; Gibson *et al.* 1989). Application of conventional 'section-balancing' to such hangingwall rollovers would erroneously predict a curved fault plane. With such a plethora of models available, for both curved and planar faults, there is generally not a unique solution to the problem of deducing fault geometry from a limited number of deformed beds around a fault.

The purpose of this paper is to present some *observations* of basement-involved faulting, using commercial seismic reflection data. We therefore avoid using a structural model to tell us about fault geometry, but rather use the observations to indicate which of the many models may be the most appropriate for the faults we describe. The uncertainty in the fault geometry is then dependent on the ability of the seismic reflection method to produce a depth image, and this uncertainty can be constrained within limits.

The structures we discuss are fault-plane reflections in the northern North Sea, a Mesozoic extensional basin now covered by Cretaceous and Tertiary post-rift sediments. We have chosen examples that show clear, dipping seismic events associated with major normal faults with several kilometres of throw. These faults are basement-involved, though do not always produce a reflection from within the basement. Our strategy has been to obtain depth sections (using the Sattlegger Interpretive Seismic Processing package) for selected lines across the structures. We have chosen lines approxi-

From Holdsworth, R.E. & Turner, J.P. (compilers) *Extensional Tectonics: Faulting and Related Processes*. The Geological Society, Key Issues in Earth Sciences, **2 (2)**, 41–51.
1476-315X/02/$15.00
First published in: "Roberts, A.M., Yielding, G. & Freeman, B. (eds) *The Geometry of Normal Faults*, The Geological Society, London, Special Publications **56**, 61–78".

mately normal to the fault traces, to avoid out-of-plane reflections that cannot be migrated.

For each line, we have performed depth migration on the stack section and image-ray migration on the time-migrated section. 'Depth migration' moves dipping subsurface reflectors into their true subsurface positions. Conventional industry migrated sections, however, are 'time-migrated', a process which is computationally cheaper than depth migration, but neglects refraction caused by dipping structure within the overburden. However, if the overburden is simple and flat-lying, time-migration should produce acceptable results (Yilmaz 1987). The resulting section is displayed in the time domain. 'Image-ray migration' attempts to move time-migrated events to their correct depth position. The 'image ray' is the notional ray that emerges vertically at the surface, and it can be used to account for the refraction that is neglected by time migration (Hubral 1977; Larner *et al.* 1981). Conventional 'vertical depth conversion' is conversion of time to depth along vertical rays, and is only strictly correct in the complete absence of dipping structure in the overburden. By performing both depth migration and image-ray migration on each line we can check on the extent to which standard time migration has adequately imaged the structure.

The first examples we discuss are both from the Horda Platform, in the Norwegian sector of the northern North Sea. We then examine a fault in the East Shetland Basin (UKCS). Finally, we summarize our observations and use them to show that although the fault geometries may superficially be interpreted as 'listric', they are in fact approximately planar in the basement.

Horda platform faults

Figure 1(a) shows an unmigrated (stack) section across the Øygarden Fault, offshore Norway. The east end of the line is close to the Norwegian coast, and Caledonian basement is at less than 1 s two-way-time (TWT) (horizon picks are shown in Fig. 1(b)). Basement is then downfaulted to >3 s TWT across the Øygarden Fault, forming the first of a series of tilted basement blocks which underlie the Horda Platform (Eynon 1981; Badley *et al.* 1984, 1988). This faulted basement surface has only been reached by one well 31/6−1, which found Early Triassic sediments resting on gneiss (Lervik *et al.* 1989). Much of the half-graben fill comprises poorly reflective Triassic clastics, passing upwards into more reflective Jurassic, Cretaceous and Tertiary sequences. Most of the fault movement affecting the basement probably occurred in the latest Permian or early Triassic, followed by thermal subsidence which was interrupted by minor fault reactivation in the Late Jurassic and Early Cretaceous (Badley *et al.* 1988).

Table 1. *Velocities used for Horda Platform profiles*

Interval	Velocity (m s^{-1})
1. Water column	1500
2. Plio-Quaternary	1650
3. Post-Palaeocene	1850
4. Palaeocene	1900
5. Cretaceous	2540
6. Draupne Fm.	2340
7. Viking Gp clastics	3380
8a. Triassic−Middle Jurassic	3000/3800/4500
8b. Triassic−Middle Jurassic	$2070 + 0.48z$

Of interest to us here is the prominent dipping event which appears to emanate from the fault zone. Day & Edwards (1983) interpreted this event as a reflected refraction, i.e. a refraction along one of the layers in the hangingwall which has been reflected at the fault surface (see Fig. 2). The two-way traveltime associated with such an event will increase linearly with distance from the fault, at a rate dependent on the velocities of the refractor and the over-burden. For a flat refractor, the slope of the event on an unmigrated time section is simply $2/V_R$ where V_R is the velocity of sound in the refractor. Day & Edwards (1983) showed that if the potential refractor dips away from the fault (as in Fig. 1), a higher refractor velocity is required to produce an event of the same slope. The event in Fig. 1 has a slope of *c.* 0.36 s km^{-1}. This would imply a refractor velocity of 5.5 km s^{-1} for a flat refractor, or an even higher velocity for a refractor dipping away from the fault, as in this case. Such a velocity is completely unrealistic for the succession present here (Jurassic sandstones and shales with velocities of 2.5−3.5 km s^{-1}). We therefore reject the interpretation of the dipping event as a reflected refraction. Instead, we consider it to be simply a reflection from the fault plane where it separates basement (to the east) from Triassic half-graben fill (to the west). As a reflection, the event can be migrated to give us the true dip of the fault surface.

Figure 1(c) shows a depth migration of the picks shown in Fig. 1(b), using the interval velocities listed in Table 1. These velocities are well-constrained by local wells, with the excep-

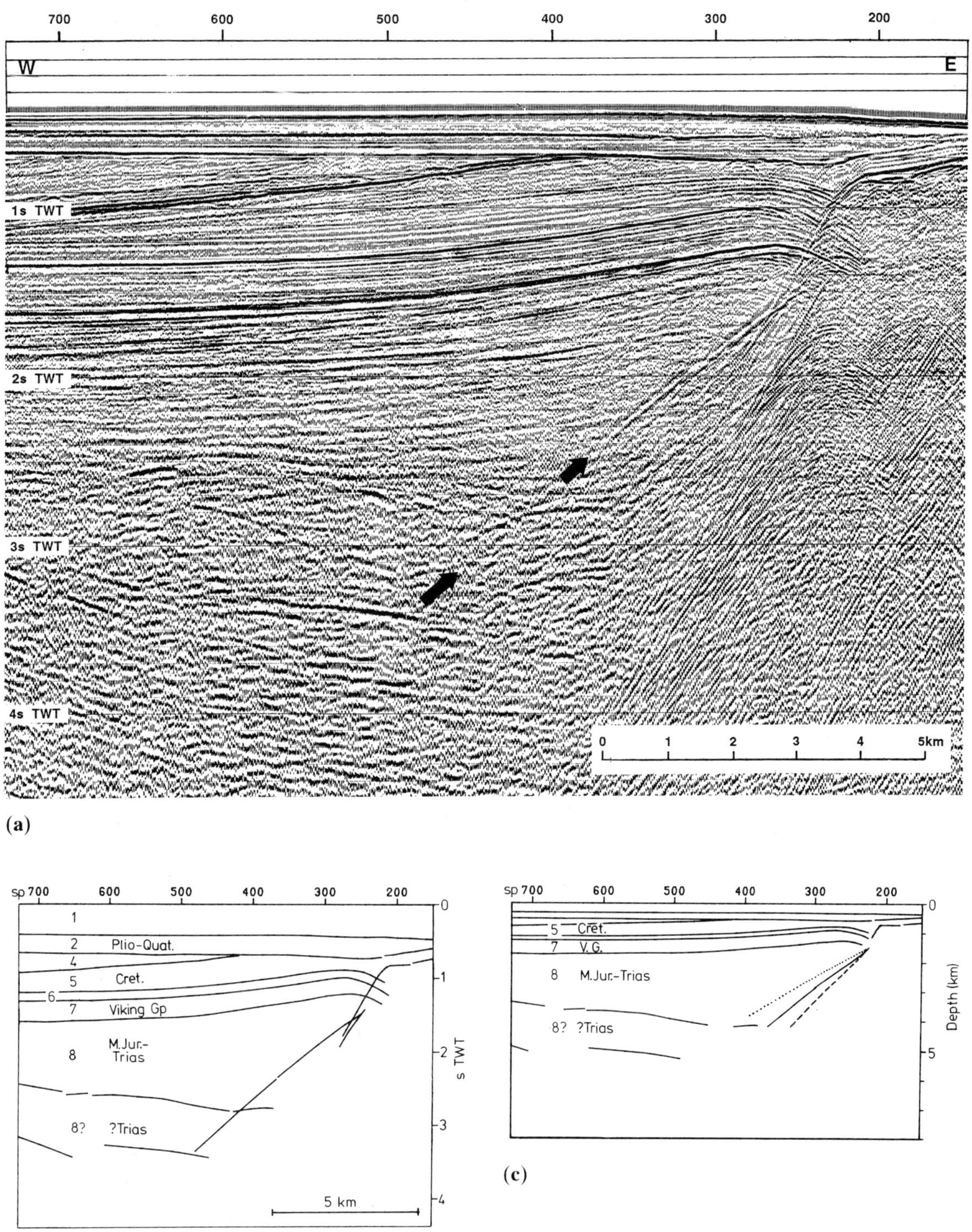

Fig. 1. (**a**) Unmigrated (stack) seismic section across the Øygarden Fault at the margin of the Horda Platform (Norwegian North Sea). Fault-plane reflection is arrowed (**b**) Horizon picks for the section in (a). Intervals are numbered as in Table 1. (**c**) Depth migration of the horizon picks in (b), using the velocities listed in Table 1. Dotted line (for fault-plane reflection) shows migration using lower velocity for half-graben fill, dashed line shows migration using a higher velocity.

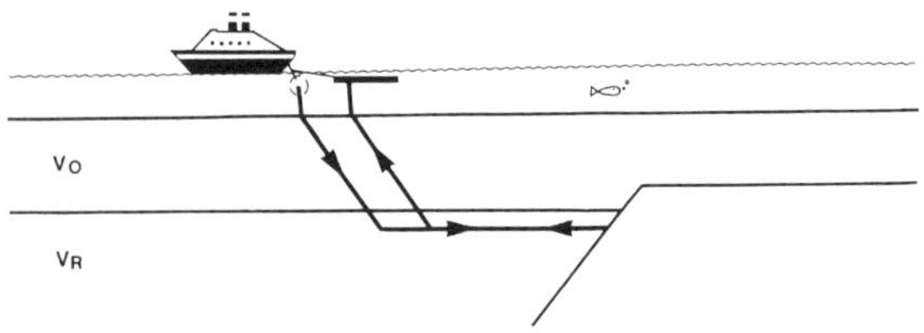

Fig. 2. The generation of 'reflected refractions', by reflection (at a fault) of a ray refracted along a bed in the hangingwall.

tion of the lowermost interval (Triassic-Middle Jurassic). As stated earlier, only one well fully penetrates this interval, and gives a velocity of 3800 m s^{-1}. Because of the uncertainty in the velocity of this interval, which occupies about half of the sedimentary column, we have also performed the depth conversion for both higher (4500 m s^{-1}) and lower (3000 m s^{-1}) velocities. It is extremely unlikely that the real average velocity of the half-graben fill is outside these limits, and therefore the resulting depth structures will represent bounds on the real geometry of the fault plane. These bounds are also plotted on the depth section of Fig. 1(c). The main features of the fault plane in Fig. 1(c) are that it is approximately planar, and it has a true dip of *c.* 40° (>30° and <45°). There is some suggestion that the uppermost part may be rather steeper (*c.* 50°). The principal reflection segment arises from the contrast between basement in the footwall and Triassic sedimentary rock in the hangingwall. No reflection is seen from the fault within basement, presumably because of the lack of acoustic impedance contrast. However, this fault has a throw of over 4 km at top basement level, and is planar over this range. Thus before the fault moved, it was planar in the uppermost 4 km of basement (and more if footwall erosion is taken into account). There is no indication that the fault dip decreases at greater depths.

The next major fault to the west of the Øygarden Fault also gives rise to a fault-plane reflection, and this is shown in Fig. 3(a). The slope of the event on the stack section is only 0.17 s km^{-1}; if this were a reflected refraction, this slope would imply a refractor velocity of 11.5 km s^{-1}, which is clearly absurd. We are therefore confident that this event is also a true fault-plane reflection rather than a reflected refraction. Figs 3(b) & 3(c) show horizon picks and the depth-migrated section, respectively. Again, the principal uncertainty in the depth migration is the velocity of the half-graben fill. We have again used high and low velocities to constrain the range of possible fault dips. However, as the Triassic fill is even thicker in this half-graben, we also repeated the depth migration using a velocity gradient (see Table 1), to simulate the effect of compaction. This velocity function is based on a regional compilation of interval velocities (Marsden *et al.* 1990), and gives a good fit to local well data. The resulting depth profile is also shown on Fig. 3(c).

The fault in Fig. 3 has significantly lower dip than the Øygarden Fault (Fig. 1). The constant interval velocities give dips of 15–25°, whereas the velocity gradient gives a dip of 25–30°. The gradient effectively 'unbends' the slightly curved reflection on the time section. We consider the velocity gradient to be geologically more realistic and therefore favour the dip estimate of 25–30°. As with the Øygarden Fault, the lower part of the observed fault profile is approximately planar, between the footwall and hangingwall basement cut-offs. At least 5 km of displacement is accommodated at top basement level. The upper part of the fault, in the Jurassic and Cretaceous section, is well defined by the cut-off positions (though not by a fault-plane reflection), and has a dip of *c.* 50°. Displacement at this level is <1 km. Following Badley *et al.* (1988), we would interpret the basement offset as resulting mainly from early Triassic fault movement, with renewed but lesser movement in the latest Jurassic to early Cretaceous. The 'later', upper, part of the fault cuts up through the Jurassic–Cretaceous sequence at a higher angle than the 'older' pre-existing fault where it penetrates basement. We return to this point in the later discussion.

We have also performed image-ray migration and vertical depth conversion on the time-migrated versions of the lines shown in Figs 1 & 3. The results were almost indistinguishable from the full depth migrations of the stacks (shown in the figures). The reason for this is the extremely simple overburden structure. In such cases time migration works well, and only vertical velocity variations need to be considered in the time-to-depth conversion. However, we consider it prudent always to investigate the effect of the velocity structure rather than to assume its simplicity at the outset.

East Shetland Basin

Our principal example from the East Shetland Basin is the main boundary fault to the Hutton Field (see Fig. 4a). Many regional lines across this structure show a reflection dipping eastwards from the observed fault offsets at Jurassic levels. Often the reflection can be seen continuing at the bottom of the section (6–7 s TWT).

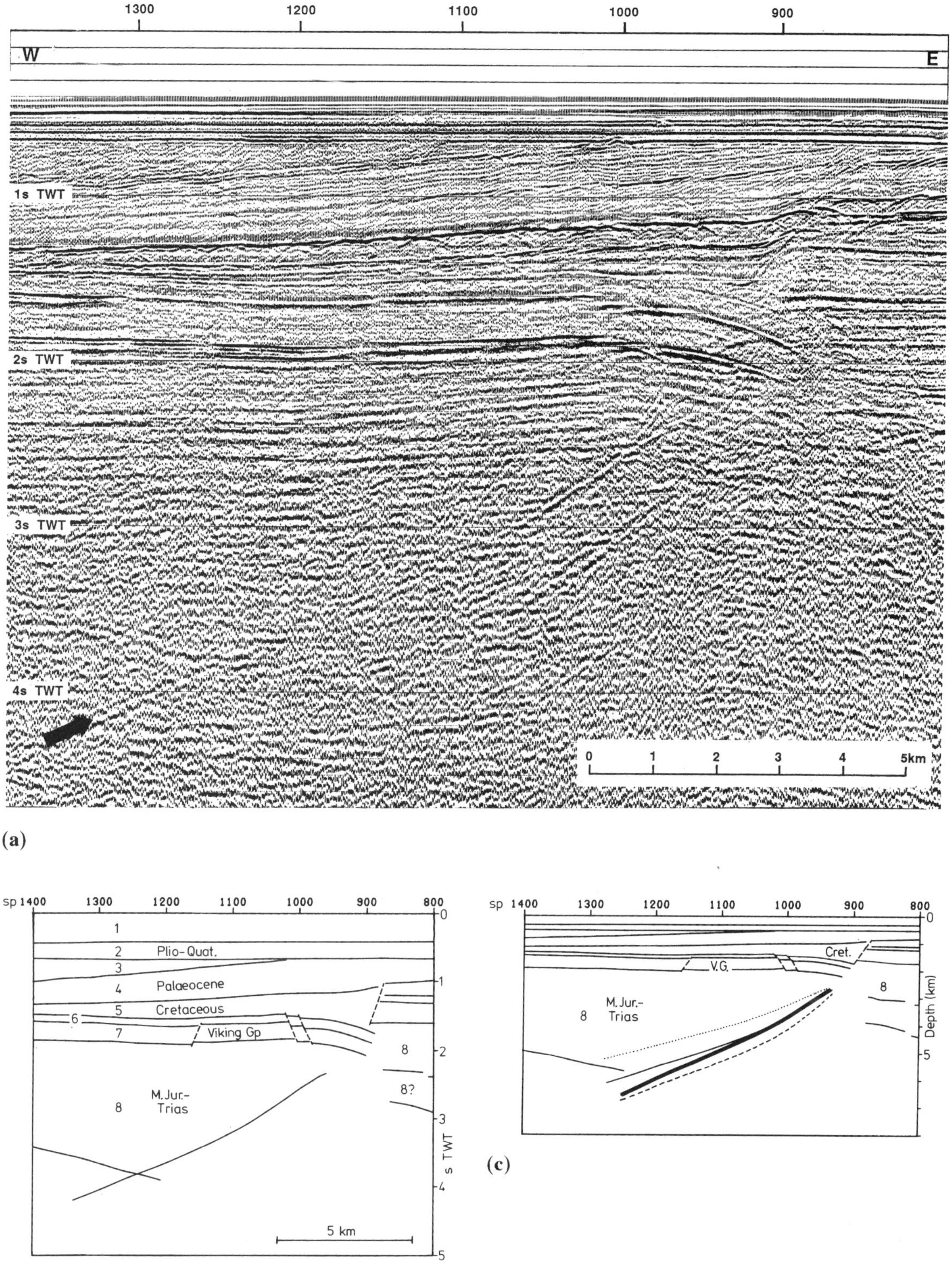

Fig. 3. (**a**) Unmigrated (stack) seismic section across a fault on the Horda Platform. This section is a westwards continuation of that shown in Fig. 1. Fault-plane reflection is arrowed. (**b**) Horizon picks for the section in (a). Intervals are numbered as in Table 1. (**c**) Depth migration of the horizon picks in (b), using the velocities listed in Table 1. Dotted line (for fault-plane reflection) shows migration using lower velocity for half-graben fill, dashed line shows migration using higher velocity, and bold line shows migration using a velocity gradient.

Table 2. *Velocities used for Hutton Fault*

Interval	Velocity (m s^{-1})
1. Water column	1500
2. Post-Palaeocene	$1855 + 0.215\,z$
3. Palaeocene	2460
4. Cretaceous	$1280 + 0.624\,z$
5. U. Jurassic	$2067 + 0.27\,z$
6a. Triassic–Middle Jurassic	3800/4500
6b. Triassic–Middle Jurassic	$2072 + 0.48\,z$
7. Basement	5500

Horizon picks from a representative (unmigrated) line are shown in Fig. 5a. On the Hutton–Ninian fault-block, top basement is known from released well data to be *c.* 0.8 s TWT below the crest of the field. On the Brent fault-block to the east, basement has not been reached. However, well over 2 km of Triassic–Middle Jurassic sequence has been penetrated, and from this we can make a conservative estimate of the top basement position on the Brent block. (The intra-Triassic Lomvi Formation gives rise to the deepest reflection that can be positively identified, *c.* 1 s below the top of the Middle Jurassic (Brent Gp).) Despite this uncertainty in the depth to basement, the deeper parts of the fault-plane reflection (> 5 s TWT) must be from within basement, as it is extremely unlikely that the Triassic succession here would extend that deep.

Velocities used for depth migration (derived from local wells) are shown in Table 2. Again, we use a variety of velocity functions for the Triassic–Middle Jurassic interval, as it is the most poorly-constrained. A velocity of 5.5 km s^{-1} has been assumed for the basement of the Brent fault-block.

The depth-migrated picks are shown in Fig. 5b. We would again favour the gradient as the most realistic velocity model. However, regardless of the exact velocities used, the imaged fault plane is approximately planar rather than listric. Its average dip is about 30–35°. Its

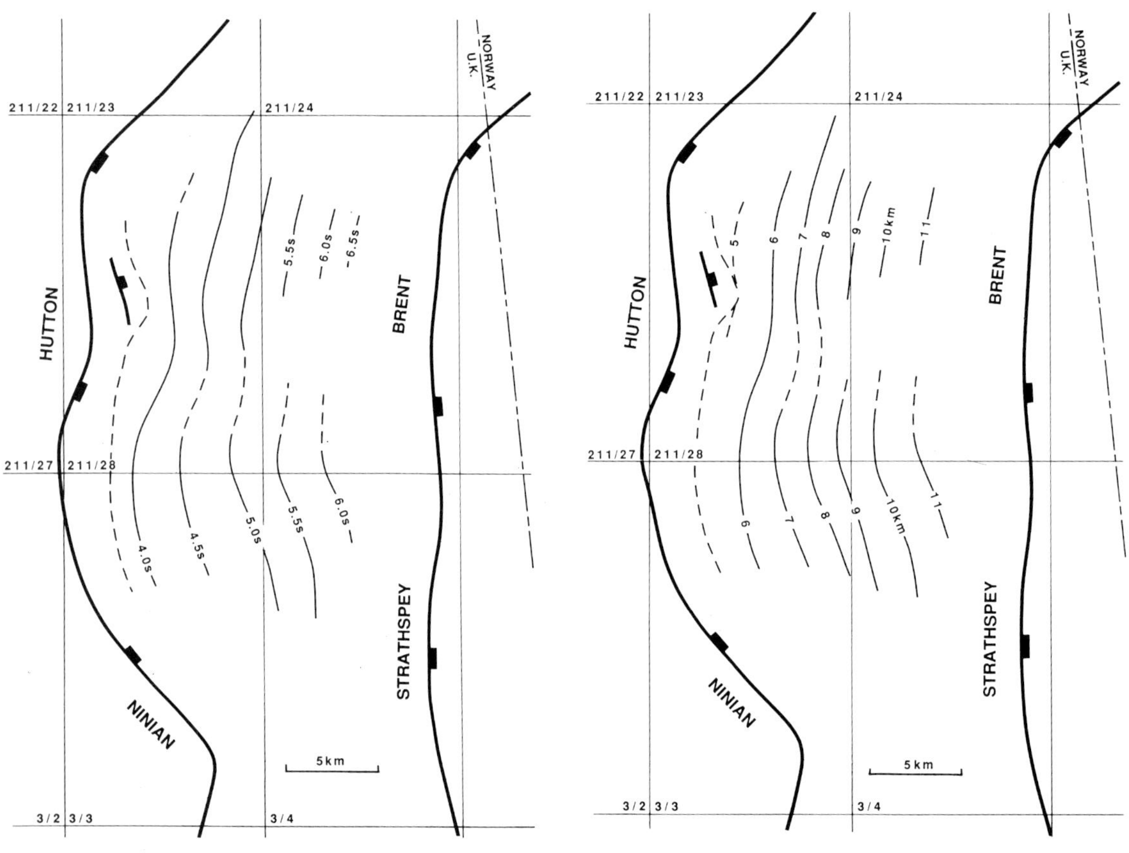

(a) (b)

Fig. 4. (**a**) Two-way-time map of the Hutton Fault reflection, compiled from time-migrated regional seismic lines. (**b**) Depth map of Hutton Fault, based on depth migrations of the lines used to construct (a).

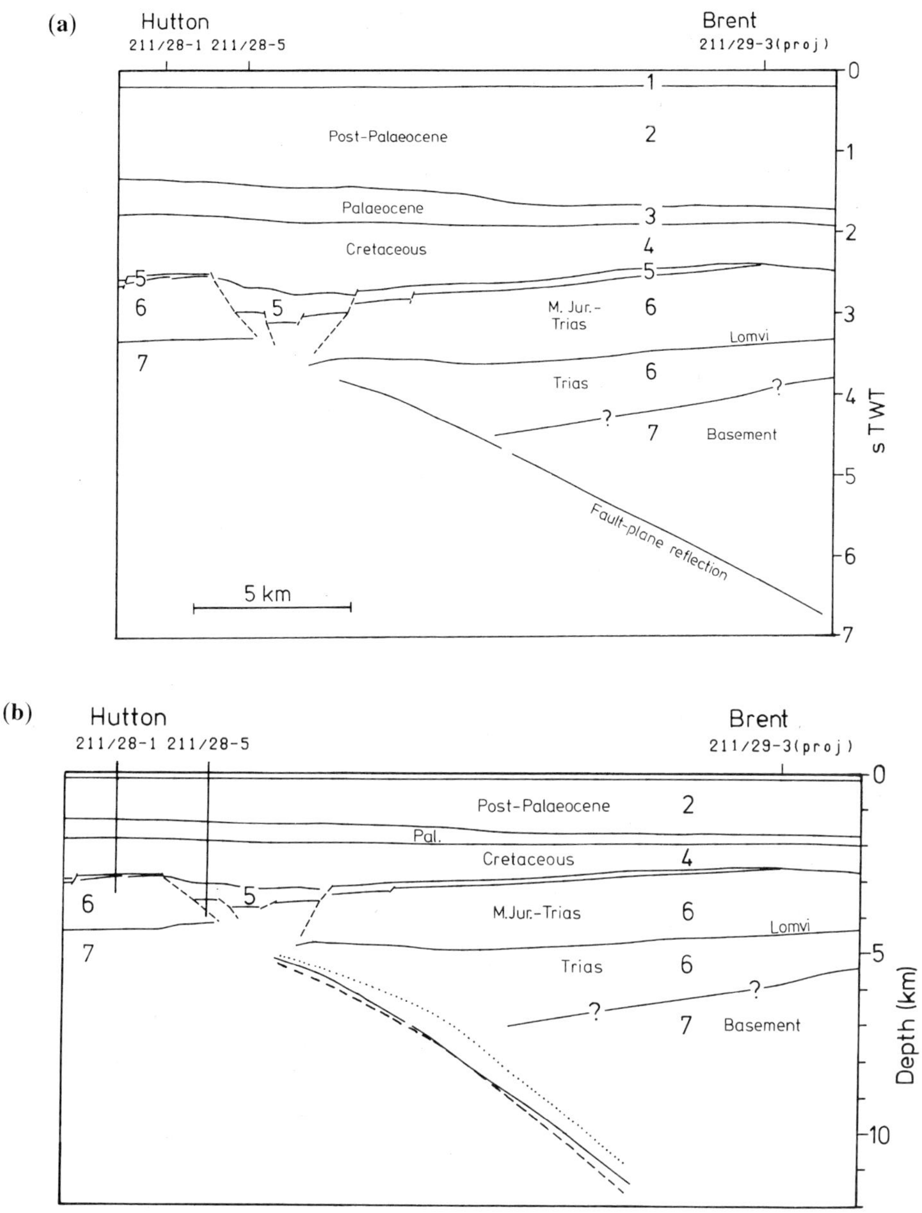

Fig. 5. (**a**) Horizon picks on a representative stack section across the Hutton Fault. Intervals are numbered as in Table 2. (**b**) Depth migration of the horizon picks in (a), using the velocities listed in Table 2. Dotted line (for fault-plane reflection) shows migration using 3800 m s^{-1} for velocity of half-graben fill, continuous line shows migration using a velocity gradient, and dashed line shows migration using a higher velocity.

down-dip extent from the basement footwall cutoff is *c*. 13 km. The displacement at top basement level is *c*. 6 km, compared to 1–2 km at Middle Jurassic levels. This again suggests significant Triassic movement followed by a smaller amount of later (mostly Late Jurassic) offset.

The upper part of the fault, cutting the Jurassic sedimentary section, does not give rise to a reflection, but the cut-off positions on our depth-migrated section imply a present dip of *c*. 40°. This is confirmed by data from released wells on the Hutton Field. In particular, well 211/28–5, targeted at a down-thrown terrace, penetrated the main fault-plane almost 1 km below the crest of the field, from which a fault-plane dip of 41° can be calculated. This upper part of the fault has a footwall composed of Triassic–Middle Jurassic sediments, and is now buried beneath *c*. 3 km of post-rift sediments. Decompacting the footwall sediments changes the fault dip to 50–55°. This is the dip that the fault would have had when active, i.e. prior to burial by the Cretaceous–Tertiary sediment pile. It is interesting to note that this is some 20° steeper than the estimated dip in the basement. We return to this point in the discussion.

As with the Horda Platform faults, we have produced depth sections by image-ray migration and vertical depth conversion. Again, the results are practically indistinguishable from the full depth migration, because the velocity structure of the overburden is so simple.

By depth migrating a number of lines, we are able to produce a depth map of the imaged fault plane, shown in Figure 4b. The regular spacing of the contours attests to the approximately planar nature of the fault, over its observed depth range of 5–11 km. However, it is worth noting that in map view the trace of the fault at Base Cretaceous level (shown in Fig. 4b) is broadly arcuate. Fault-plane reflections are absent to the north and south of the contoured area of Fig. 4, where the fault trace swings northeast and southeast, respectively. There is no obvious reason for this change of reflectivity along strike.

In our experience of seismic data from the East Shetland Basin, the Hutton Fault is unusual in generating a clear fault-plane reflection. It is plausible that most other faults are significantly steeper, and therefore less likely to be imaged. However, Nelson & Lamy (1987, their figs 4 & 5) show a fault-plane reflection from the Snorre Fault, very similar to those we observe from the Hutton Fault. Their depth conversion of the Snorre Fault shows a present dip in basement of 25–30°, broadly similar to that we observe for the Hutton and Horda Platform faults. Again, early Triassic movement appears to account for a significant proportion of the total fault offset.

Discussion

The principal observation we make from our depth-migrated sections is that the major normal faults are planar where they cut down into basement. In the case of the Horda Platform faults the fault-plane reflection is derived from the basement footwall juxtaposed against Triassic half-graben fill. In the case of the Hutton Fault the reflecting portion continues deep into probable basement with the same attitude. The present-day dips of these basement faults are typically 30–35°, though in some cases possibly as shallow as 25° or as steep as 45°.

The upper parts of the faults, where they cut through the sedimentary column overlying basement, tend to be steeper than the basement portions, typically 40–50°. These upper parts do not give rise to reflections, probably because there is insufficient acoustic impedance contrast between footwall and hangingwall. In addition, they are too steep for conventional finite difference migration algorithms. However, their present dips are constrained by horizon cut-offs and well information. Original fault dips through Jurassic sediment were probably 50–55° before compaction beneath the Cretaceous–Tertiary sediment pile.

Thus the upper parts of the faults show a downward decrease in dip as they pass from sedimentary column to basement. However, it is important to stress that the dip does not continue to decrease downwards; the basement portions are planar. The faults are not listric in the sense of passing downwards into a horizontal detachment surface, at least over the depth range visible on industry seismic data.

Within each fault-block, top basement and Mesozoic sequences are tilted towards the underlying fault. Thus the overall structure comprises tilted fault-blocks bounded by planar faults. The most appropriate description of the gross structure is some form of 'domino' model, either of 'rigid' dominoes (Barr 1987; Jackson *et al.* 1988; Yielding, 1990) or 'soft' dominoes (Kusznir & Egan 1989; Gibson *et al.* 1989; Walsh & Watterson, this volume).

A common feature of all the faults we have examined is that offsets at top basement are much larger than at top Middle Jurassic (*c.* 5 km and *c.* 1 km, respectively). We interpret this to indicate major fault movement in the early Triassic, followed by reactivation in the late Jurassic (see also Badley *et al.* 1988; Marsden *et al.* 1990). Our interpretation is summarized in cartoon form in Fig. 6. Fault-block rotation in the early Triassic produced a tilted-block relief at top basement level. Major planar faults originally dipping at perhaps 45–50° were rotated by up to 10°, along with the intervening blocks. In terms of a rigid-domino model, this would correspond to an extension of $\beta = 1.2$. Late Triassic and Lower–Middle Jurassic sediments represent a thermal subsidence phase after this initial extension. Documented thickening of the Brent and Dunlin Groups over the sites of major faults (Brown 1986; Badley *et al.* 1988) may be partly related to compaction of the underlying Triassic half-graben fill (see Fig. 6a). However, renewed fault movement in the Late Jurassic caused a further 5–10° of fault-block rotation. Reactivation of the basement faults led to them cutting up through the overlying sediment, at an angle of perhaps 50–55°.

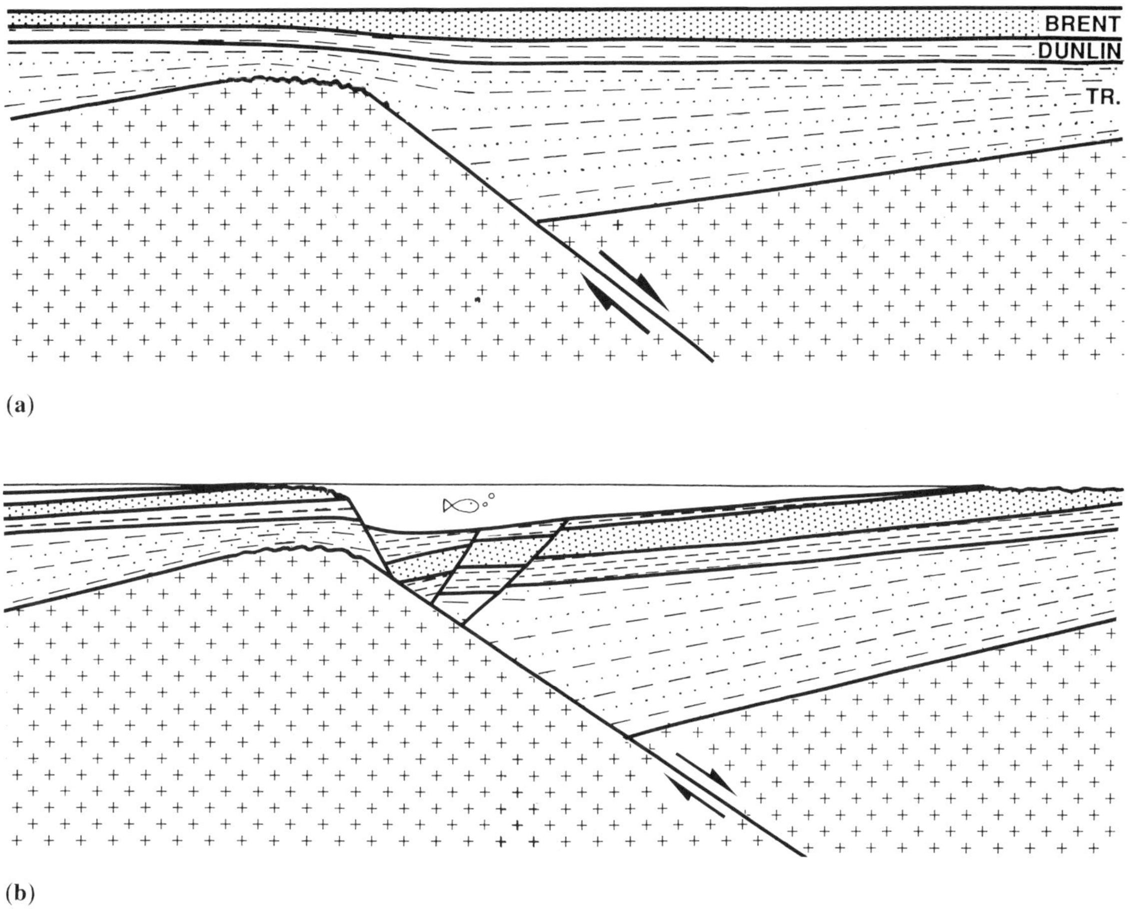

Fig. 6. Cartoon interpretation of the observations detailed in this paper. (**a**) shows the fault-block geometry at the end of the Middle Jurassic. Following early Triassic extension on planar basement faults, late Triassic–Middle Jurassic thermal subsidence has covered the basement topography. (**b**) Further fault movement and fault-block rotation in the Late Jurassic causes new fault segments to cut up through the sedimentary cover, at a steeper angle than the now-rotated basement faults.

Treating the basement as rigid dominoes, β for the Late Jurassic extension on the basement faults was again about 1.2, giving a total Mesozoic extension factor of about 1.45. This is in good agreement with crustal thickness observations for the terrace areas flanking the graben centre (Klemperer 1988; Marsden *et al.* 1990).

Upward steepening of active faults has been predicted in theoretical models by Barr (1987) and in analogue models by Vendeville & Cobbold (1988). Both these studies adopted the premise that although domino-style faults will rotate to shallower angles, any new fault development in overlying sediment will occur at a fixed angle, say 60°. In this way, the interaction of fault rotation and sedimentation can produce concave-upwards faults. This is the case in the examples we have shown, though two discrete periods of fault movement are separated by a period of sedimentation. The abrupt bend in the fault above the footwall basement cut-off appears to have produced an accentuated roll-over in the hangingwall (cf Fig. 6b with Figs 1, 3 & 5).

Conclusions

Dipping events on seismic sections across basement faults are (in the cases we have studied) real fault-plane reflections, rather than refractions. The probable geometry of the fault plane can be obtained by depth migration of the fault-plane reflection. Velocity structure has a significant effect on the estimated fault geometry.

The examples we have studied (Horda Platform faults, and the Hutton Fault) comprise two near-linear segments in cross section. The upper parts of the faults (cutting through the sedimentary column), now dip at 40–50°; prior to compaction during Cretaceous and Tertiary burial this dip would have been 50–55°. The lower parts of the faults (involving pre-Triassic basement) now dip at 30–35°, though they probably formed in the early Triassic with an initial dip of 40–50°. The observed geometry is consistent with Triassic rotation of near-planar faults, followed by reactivation of these faults in the Late Jurassic, cutting up through overlying sediment.

We are grateful to Alan Roberts and Oz Yilmaz for helpful discussions during the course of this work. The permission of Statoil to publish the seismic data is gratefully acknowledged.

References

BADLEY, M. E., EGEBERG, T. & NIPEN, O. 1984. Development of rift basins illustrated by the structural evolution of the Oseberg feature, Block 30/6, offshore Norway. *Journal of the Geological Society, London*, **141**, 639–649.

——, PRICE, J. D., RAMBECH DAHL, C. & AGDESTEIN, T. 1988. The structural evolution of the northern Viking Graben and its bearing upon extensional modes of basin formation. *Journal of the Geological Society, London*, **145**, 455–472.

BARR, D. 1987. Structural/stratigraphic models for extensional basins of half-graben type. *Journal of Structural Geology*, **9**, 491–500.

BROWN, S. 1986. Jurassic. *In*: GLENNIE, K. W. (ed.) *Introduction to the Petroleum Geology of the North Sea*, Blackwell, Oxford, 133–159.

DAY, G. A., & EDWARDS, J. W. F. 1983. Reflected refracted events on seismic sections. *First Break*, 14–17.

EYNON, G. 1981. Basin development and sedimentation in the Middle Jurassic of the northern North Sea. *In*: ILLING, L. V. & HOBSON, G. D. (eds). *Petroleum Geology of the Continental Shelf of North West Europe*, Heyden, London, 196–204.

GIBBS, A. D. 1983. Balanced cross-section construction from seismic lines in areas of extensional tectonics. *Journal of Structural Geology*, **5**, 153–160.

GIBSON, J. R., WALSH, J. J. & WATTERSON, J. 1989. Modelling of bed contours and cross-sections adjacent to planar normal faults. *Journal of Structural Geology*, **11**, 317–328.

HUBRAL, P. 1977. Time migration — some ray theoretical aspects. *Geophysical Prospecting*, **25**, 738–745.

JACKSON, M. P. A., & GALLOWAY, W. E. 1984. *Structural and Depositional Styles of Gulf Coast Tertiary Continental Margins: Application to Hydrocarbon Exploration*. AAPG Continuing Education Course Note Series **25**, AAPG, Tulsa, 1–226.

JACKSON, J. A., WHITE, N. J., GARFUNKEL, Z. & ANDERSON, H. 1988. Relations between normal-fault geometry, tilting and vertical motions in extensional terrains: an example from the southern Gulf of Suez. *Journal of Structural Geology*, **10**, 155–170.

KING, G. C. P., STEIN, R. S. & RUNDLE, J. B. 1988. The Growth of Geological Structures by Repeated Earthquakes, 1, Conceptual Framework. *Journal of Geophysical Research*, **93**, 13307–19.

KLEMPERER, S. 1988. Crustal Thinning and Nature of Extension in the Northern North Sea From Deep Seismic Reflection Profiling. *Tectonics*, **7**, 803–822.

KUSZNIR, N. J. & EGAN, S. 1989. Simple-shear and pure-shear models of extensional sedimentary basin formation: application to the Jeanne d'Arc Basin, Grand Banks of Newfoundland. *In*: TANKARD, A. J. & BALKWILL, H. R. (eds) *Extensional Tectonics and Stratigraphy of the North Atlantic Margins*. American Association of Petroleum Geologists Memoir **46**, 305–322.

LARNER, K. L., HATTON, L., GIBSON, B. S. & HSU, I. C. 1981. Depth migration of imaged time sections. *Geophysics*, **46**, 734–750.

LERVIK, K. S., SPENCER, A. M. & WARRINGTON, G. 1989. Outline of Triassic Stratigraphy and Structure in the central and northern North Sea. *In*: COLLINSON, J. D. (ed.) *Correlation in Hydrocarbon Exploration*, Graham & Trotman London, 173–189.

MARSDEN, G., YIELDING, G., ROBERTS, A. M. & KUSZNIR, N. J. 1990. Application of a flexural cantilever simple-shear/pure-shear model of continental lithosphere extension to the formation of the northern North Sea Basin. *In*: BLUNDELL, D. J. & GIBBS, A. D. (eds), *Tectonic evolution of the North Sea Rifts*, Oxford University Press.

NELSON, P. H. H. & LAMY, J. M. 1987. The Møre/West Shetland area: a review. *In*: BROOKS, J. & GLENNIE, K. W. (eds). *Petroleum Geology of North West Europe*. Graham & Trotman, London, 775–784.

STEIN, R. S. & BARRIENTOS, S. E. 1985. Planar High-Angle Faulting in the Basin and Range: Geodetic Analysis of the 1983 Borah Peak, Idaho, Earthquake. *Journal of Geophysical Research*, **90**, 11355–11366.

VENDEVILLE, B. & COBBOLD, P. R. 1988. How normal faulting and sedimentation interact to produce listric fault profiles and stratigraphic wedges. *Journal of Structural Geology*, **10**, 649–659.

WALSH, J. J. & WATTERSON, J. (This volume). Geometric and kinematic coherence and scale effects in normal fault systems.

WERNICKE, B. & BURCHFIELD, B. C. 1982. Modes of extensional tectonics. *Journal of Structural Geology*, **4**, 105–115.

WHEELER, J. 1987. Variable-heave models of deformation above listric normal faults: the importance of area conservation. *Journal of Structural Geology*, **9**, 1047–1050.

WHITE, N. & YIELDING, G. (This volume). Calculating normal fault geometries at depth: theory and examples.

——, JACKSON, J. A. & MCKENZIE, D. P. 1986. The relationship between the geometry of normal faults and that of the sedimentary layers in their hangingwalls. *Journal of Structural Geology*, **8**, 897–909.

WILLIAMS, G. & VANN, I. 1987. The geometry of listric normal faults and deformation in their hangingwalls. *Journal of Structural Geology*, **9**, 789–795.

YIELDING, G. 1990. Footwall uplift associated with Late Jurassic normal faulting in the northern North Sea. *Journal of the Geological Society, London*, **147**, 219–222.

YILMAZ, Ö. 1987. *Seismic data processing. Investigations in Geophysics 2*, Society of Exploration Geophysicists, Tulsa.

Calculating normal fault geometries at depth: theory and examples

NICKY WHITE[1] & GRAHAM YIELDING[2]

[1]*Bullard Laboratories, Madingley Rise, Madingley Road, University of Cambridge, Cambridge CB3 0EZ, UK*

[2]*Badley Ashton & Associates Ltd, Winceby House, Winceby, Horncastle, Lincs LN9 6PB, UK*

Abstract: In recent years, a large number of methods relating the shape of a normal fault at depth to the shape of a sedimentary horizon in its hangingwall have been described. Such methods are best used in areas where it can be assumed that the footwall remains relatively undeformed during extension. This assumption may not always be valid, especially in the case of large-scale, basement-extending normal faults. Conservation of area (or solid area when compaction occurs) must be the fundamental constraint. The advantages and disadvantages of the various methods are briefly outlined. A general model which assumes that hangingwall deforms by arbitrarily inclined simple shear and differential compaction is discussed. Fault geometry, the inclination of simple shear, and compaction parameters may all be determined from $\mathcal{N}$ beds using a simple inversion scheme based on this general method. The algorithm has previously been tested on synthetic data using a range of fault geometries. Such testing indicates that all unknown parameters including fault geometry can, in general, be uniquely determined provided that the geometries of two or more beds within the hangingwall are known. It is important that hangingwall and footwall stratigraphies are accurately known, though uncertainty in either could be included as a bounded variable within a formal inversion scheme. In this paper, the method is applied to laboratory-modelled normal faulting and to depth-converted seismic reflection data. Results suggest that hangingwalls deform by bulk, antithetically inclined, simple shear. Differential compaction of hangingwall sediments is often important and should be taken into account.

Over the last twenty years there has been considerable interest in determining the shape of normal faults at depth. For large-scale active faults, the best method is to use focal mechanism solutions combined with surface break information (e.g. Jackson 1987). In sedimentary basins, seismic reflection data are used while field mapping can be useful in well exposed regions (e.g. the Basin and Range Province; Wernicke & Burchfiel 1982). However, there are circumstances when none of these methods can be applied and geometric techniques must be employed to calculate the likely fault geometry.

Since Verrall (1981) first introduced the so-called 'Chevron construction' to the geological community, a large number of section-balancing schemes for normal faulting have been proposed. Most of these use the geometry of one horizon within the hangingwall to determine the fault geometry at depth. Such methods fall into three broad categories: (a) shear models, (b) flexural-slip models and (c) constant-slip models. All methods assume that deformation occurs within the plane of the section. They also assume that the footwall remains rigid and so are applicable principally to detached fault systems within the sedimentary fill of a basin. Footwalls of major basement-penetrating normal faults can rotate and/or deform internally (see, e.g. Jackson & White 1989). In such cases, geometrical methods should be applied with great care.

Shear models assume that the hangingwall deforms by bulk simple shear which can be either vertical (Verrall 1981) or arbitrarily inclined (White *et al.* 1986). Flexural-slip models assume that well developed plane-parallel stratification within the hangingwall allows deformation to occur by flexural slip (Davison 1986). Constant-slip models assume that the amount of slip at any point along the fault plane during extension remains constant (Williams & Vann 1987). More recently, Waltham (1989) has argued in favour of a much more general approach where the particle displacement field within the hangingwall is arbitrarily defined.

The most useful models are those which can solve both the forward and inverse problems. The forward problem is where bed geometries are calculated from a given fault geometry and is mainly used to get an idea of the sort of bed

From Holdsworth, R.E. & Turner, J.P. (compiler) *Extensional Tectonics: Faulting and Fault Related Processes*. The Geological Society,
Key Issues in Earth Sciences, **2 (2)**, 53–62.
1476-315X/02/$15.00
First published in: "Roberts, A.M., Yielding, G. & Freeman, B. (eds) *The Geometry of Normal Faults*, The Geological Society, London, Special Publications **56**, 251–260".

geometries to be expected for a whole range of fault geometries. The more practical inverse problem is where the fault geometry is calculated from one or more bed geometries. The latter problem is of some interest to the oil industry.

The purpose of this short paper is firstly to discuss briefly these different types of model. A general modelling scheme is then proposed, based on the shear model. This scheme is applied to laboratory-modelled normal faulting and to depth-converted seismic reflection data.

Which model?

The problem with having a plethora of models for hangingwall deformation is that in order to calculate a unique fault geometry, it is first necessary to decide which model is most appropriate. Clearly, this *ad hoc* situation is rather unsatisfactory: it would be much simpler and far more useful if just one model could be used in a whole range of situations. The purpose of this short paper is to make the case for such a model. A general section-balancing scheme for normal faulting should have a relatively small number of parameters which describe the deformation in a simple but realistic way. Its suitability must be assessed by rigorous testing on models and data where the fault geometry is accurately known. Before discussing one possible method, each of the three sets of models will examined further.

In the shear model, inhomogeneous simple shear leads to the development of shear (similar) folds in the hangingwall block. Shear folds can occur in the presence or absence of well developed stratification. Shear planes may be infinitely close and are not necessarily visible in naturally deformed rock (Ramsay 1967). The development of a shear fold in the hangingwall of a normal fault can be considered in terms of a constant horizontal component of extension and a varying vertical component of collapse (Verrall 1981).

The shear model was first proposed by Verrall (1981), who used a simple graphical technique (the 'Chevron construction') to relate fault and bed geometries (Fig. 1a). Since then, Gibbs (1983, 1984) has applied the method to large-scale normal faulting in the North Sea. The usual objection to the model in this form is that vertical simple shear is unrealistic since minor faults within hangingwalls are usually inclined. Subsequently, White *et al.* (1986) derived a general solution which does not fix the inclination of the planes along which simple shear occurs and which can be modified to include differential compaction and footwall rotation. Vertical simple shear is clearly just one special case within this more general scheme.

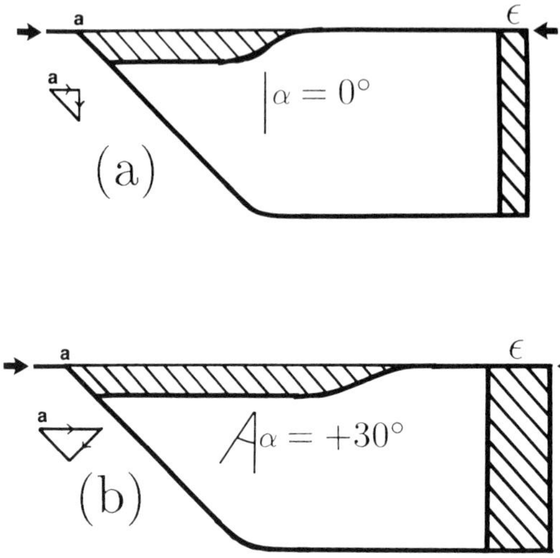

Fig. 1. Difference between vertical shear and inclined shear. Heave (apparent horizontal displacement) is identical in both cases. Depth of bed prior to extension (i.e. regional) indicated by arrows. (**a**) Vertical shear ($\alpha = 0°$), (**b**) Inclined shear ($\alpha = +30°$). Note that the area of depression in the hangingwall and the excess area due to extension, ϵ, (both shaded) are equal in each case. The vector diagrams show horizontal and vertical components of displacement for point a in each case.

The most important feature of shear models is that cross-sectional area is preserved. When compaction is taken into account, *solid* area is preserved. Bed-length does not remain constant and the orthogonal thickness of beds decreases with increasing dip of the strata. A simple relationship between extension and heave (apparent horizontal displacement) can be derived, where, depending on the inclination of the shear planes, the amount of extension is greater or less than the heave (Fig. 1b, White *et al.* 1986). Heave is equal to extension only when simple shear is vertical (Fig. 1a).

The flexural-slip model assumes that when the hangingwall has a well developed plane-parallel stratification, extension is accommodated by flexural-slip folding. As folding proceeds, the individual layers are flexed such that the uppermost layers slip over the lower layers towards the fold axis. Bed-length and the orthogonal thickness of layers remain constant. Suppe (1983) and Davison (1986) have proposed graphical methods for solving both the forward and inverse problems. Unfortunately, certain geometries are only possible provided arbitrary amounts of horizontal shear are applied to the hangingwall (Suppe 1983).

The flexural-slip model has several significant disadvantages (White 1987). Firstly, antithetic faulting is often observed within the hangingwall block. Since this disrupts layering it will inhibit inter-bed slip. Secondly, thickness variations such as those which occur as a result of growth-type faulting may cause problems since plane-parallel stratification is required for efficient inter-bed slip (Ramsay 1967). Thirdly, it is not clear how the problem of area changes in certain circumstances can be avoided (Wheeler 1987). Flexural-slip models will not be considered any further although it is important to note that there may be circumstances when small amounts of inter-bed slip occur as beds deform by, say, bulk simple shear.

Both the flexural-slip model and the shear model do not, in general, conserve slip along the fault. In the shear model, slip decreases if simple shear is vertical and the fault shallows out with depth. Williams & Vann (1987) argue that this is unreasonable since it implies that the fault initiated at the surface and propagated downwards. Alternative models which conserve slip are proposed instead. However, as Williams & Vann (1987) state, a consequence of conserving slip is that, in their model at least, area is no longer preserved (see also Wheeler 1987). In addition, their propagation argument does not apply to the generalized shear model of White *et al.* (1986) where, depending on the inclination of the shear, slip can increase or decrease down the fault.

Waltham's (1989) finite difference scheme, where the particle displacement field within the hangingwall is arbitrarily defined, has several disadvantages. Only the less important forward problem can be solved and differential compaction of hangingwall sediments has not been allowed for.

In summary, we would emphasize that conservation of mass (i.e. area or solid area in the case of compaction) must be a fundamental constraint for any model. The proposed constant-slip model fails in this regard and should therefore be rejected. While inter-bed slip may well be an important mechanism in some situations, flexural-slip models are generally difficult to justify. In contrast, the shear model conserves mass, predicts faulting within the hangingwall, and can be easily modified to allow for compaction.

A general method

Apart from conserving mass, any general method should allow the hangingwall to deform in an intuitively reasonable way that agrees with observation on the appropriate scale. Some allowance for differential compaction of hangingwall material as it is progressively buried (e.g. growth faulting) is also desirable. To be of general use, such a model must allow us to solve the inverse problem (i.e. calculate the fault geometry from the bed geometry) and not just the forward model. Here we attempt to show that the arbitrarily inclined simple shear model (White *et al.* 1986), which allows for compaction, adequately accounts for hangingwall deformation. As before, the method assumes that the footwall is rigid and undeformable and that deformation occurs by plane strain.

White *et al.* (1986) derived a generalized shear model where the inclination of planes along which simple shear occurs is not fixed (Fig. 1). A more usable version of their scheme (i.e. one which allows for large displacements) was proposed by White (1987) and is used here. In both cases, it was suggested that differential compaction should also be taken into account since sediments within the hangingwall may contain considerable amounts of water. As deformation proceeds, this water is expelled and the sediments compact, often accounting for up to 30% of the strain (Ramsay 1967).

Most workers assume that the strain caused by compaction is uniaxial, the axis of shortening being vertical (e.g. Sclater & Christie 1980; Wood 1981). This assumption is reasonable provided that lateral changes in facies and thickness of sedimentary layers can be neglected. However, it is unlikely to be an accurate description of the strain field in a region which is undergoing burial and tectonic deformation at the same time. The equations governing the behaviour of such a system are discussed by McKenzie (1984).

No attempt has been made here to solve this more general problem. Instead, an empirical expression, which relates porosity, $\phi(z)$, to depth, z (e.g. Sclater & Christie 1980), is used:

$$\phi(z) = \phi_0 \exp(-z/\lambda) \qquad (1)$$

where ϕ_0 is the initial porosity and λ is the porosity decay length.

The equation relating the deformation of the hangingwall to the geometry of the fault (White 1987) is:

$$\begin{aligned} F'(x'+h') = F'(x') - R'(x') + B'(x'+h') \\ + \phi'_0\lambda' \{ \exp(-F'(x')/\lambda') - \exp(-R'(x')/\lambda') \\ - \exp(-F'(x'+h')/\lambda') \\ + \exp(-B'(x'+h')/\lambda')\} \qquad (2) \end{aligned}$$

where $F'(x')$ and $B'(x')$ are y' co-ordinates of the fault and bed at the point x' respectively, all in a reference frame rotated through α (the inclination of shear); h' is the rotated heave, ϕ'_0, the rotated initial porosity, and λ' the rotated compaction decay length (White 1987, Fig. 2). $R'(x')$ is the rotated shape of the bed before extension (i.e. the 'regional'). Equation (2) may be solved by iteration, either for F when B is given, or for B when F is given. When $\phi_0 = 0\%$, equation (2) reduces to a much simpler form:

$$F'(x' + h') = F'(x') - R'(x') + B'(x' + h'). \quad (3)$$

This equation provides the initial solution. Five iterations usually yield a solution accurate to ~ 1 cm.

Equation (2) assumes that the hangingwall deforms by simple shear alone. If the strain field is more complicated, such equations do not hold. Thus if the shear is inclined and compaction is vertical, the fault geometry cannot be calculated since removing the effect of vertical compaction would require *a priori* knowledge of the fault geometry. This difficulty is avoided by assuming that compaction occurs by uniaxial shortening parallel to the shear direction within the hangingwall. Provided α is less than 45°, this is unlikely to be any worse than assuming that compaction occurs by uniaxial shortening in a purely vertical direction. This assumption is more fully justified in a later paper.

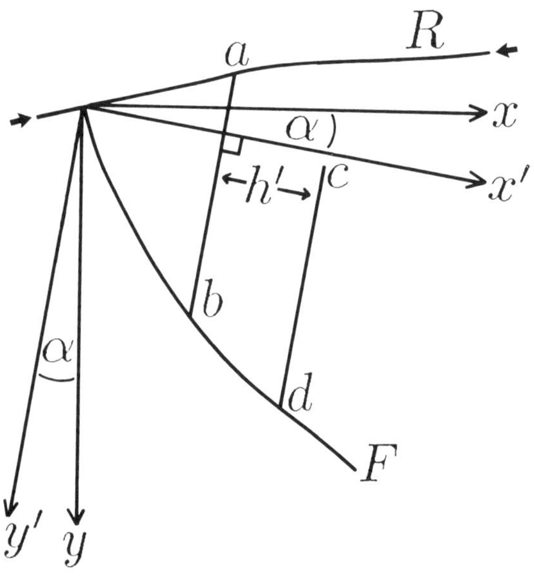

Fig. 2. Co-ordinate system and geometrical relationships used to derive Eqn. 2. Generalised shape of bed prior to extension (regional) indicated by arrows. In the rotated frame, $a = R'(x')$, $b = F'(x')$, $c = B'(x' + h')$, and $d = F'(x' + h')$. Solid lengths $|ab|$ and $|cd|$ are equal. For clarity, the deformed position of a single point (c) on the bed is shown.

Equation (2) has four unknowns: the fault geometry (F), the inclination of simple shear (α), the initial porosity (ϕ_0), and the porosity decay length (λ). In certain circumstances, depending on the quality of the data, one or more of these parameters may be fixed. In the most general case, however, only broad constraints can be applied:

(a) α is unlikely to be less than 0° and is probably not greater than 60° (Anderson 1951);
(b) ϕ_0 cannot be less than 0% and is given an arbitrary upper limit of 90%;
(c) for the main categories of sediment, λ varies between 2 and 10 km (Sclater & Christie 1980).

α can be assumed to be identical for different beds within the hangingwall block. Since hangingwall sediments are generally not uniform but consist of interbedded layers of different sediment types, ϕ_0 and generally represent the bulk compaction parameters of the hangingwall as a whole.

Unfortunately, these generalizations concerning shear inclination and compaction mean that the geometry of a single bed in the hangingwall does not uniquely determine the geometry of the fault unless both the inclination of shear and the amount of compaction are known (Figs 3a & 3b). Sometimes, shear inclination may be estimated (White *et al.* 1986) but in general all three parameters are poorly determined. Not knowing α is worse than not knowing ϕ_0 while changing λ makes very little difference at all.

To calculate the geometry of the fault and the values of α, ϕ_0 and λ, the geometries of at least two beds must be known. In addition, the initial portion of the fault (down to the point where each bed meets the fault) and the appropriate heaves (h) and regionals (R) are required.

As is schematically illustrated in Fig. 4a, each bed can be used to calculate a fault geometry for any given values of α, ϕ_0 and λ. $\mathscr{A}$, which is the areal difference between the two faults, is then calculated by numerical integration. This areal difference could be weighted along the calculated fault traces in many different ways. For given bed geometries, $\mathscr{A}$ is a function of *three* parameters: α, ϕ_0, and λ (Fig. 4b). To calculate the correct fault geometry, $\mathscr{A}$ must be minimized with respect to all three. Many algorithms are now available which can perform this minimization process automatically (Scales 1985). It can be demonstrated that a unique minimum exists for the

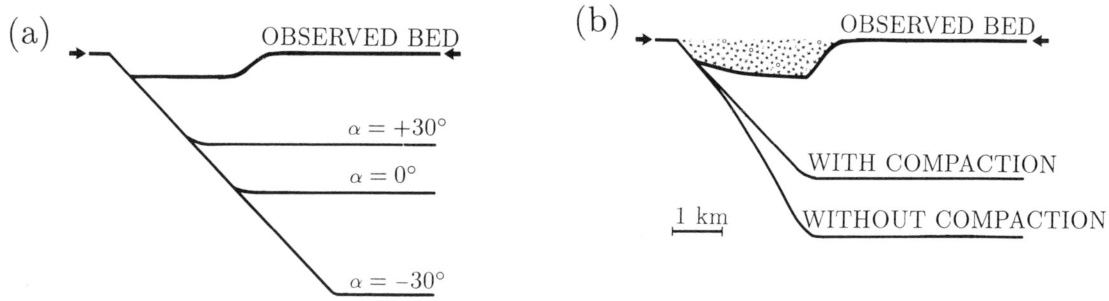

Fig. 3. (**a**) Three different faults are predicted for the same observed bed geometry by changing α, the inclination of simple shear. The depth to décollement increases as α increases. (**b**) Two different faults are predicted for the same observed compacted bed ($\alpha = 0°$, $\phi_0 = 60\%$, and $\lambda = 2$ km). The correct fault is found if only the correct values of ϕ_0 and λ used. Note the consequences of ignoring compaction. Arrows indicate the regionals in each case.

whole range of fault geometries from planar through to listric (White 1987).

Although only two beds are actually needed to find the correct fault geometry, the solution will be much better constrained if a larger number is used. In general, $\mathcal{N}$ beds can be used to calculate $\mathcal{N}$ faults for any given values of α, ϕ_0, and λ. The total area, $\Sigma_{i=1}^{\mathcal{N}-1} \Sigma_{j=i+1}^{\mathcal{N}} \mathcal{A}_{ij}$, which is the sum of the areas between each and every pair of calculated faults, must then be minimized. $\mathcal{N}(\mathcal{N} - 1)/2$ areas are calculated altogether.

Applications

The algorithm can be tested in a variety of different ways. The most obvious way is to apply it to synthetic data (i.e. run the forward model to calculate a set of beds for a particular fault geometry and then inverse model the same beds). It is useful for testing the self-consistency of the scheme in the early stages. More importantly, this testing has demonstrated that a unique solution can be obtained for a range of fault geometries from planar through to listric (White 1987). Thus, in general, there exists a single minimum value of $\mathcal{A}(\alpha, \phi_0, \lambda)$ within α–ϕ_0–λ space (Fig. 4b).

Laboratory models

The second method of testing is to apply the algorithm to laboratory-produced normal faulting where the geometry of the fault plane is accurately known and where the hangingwall deformation can be observed in detail. Here just one example of wet clay modelling taken from Cloos (1968) is discussed although the conclusions drawn are based on applying the method to the sand-box models of McClay &

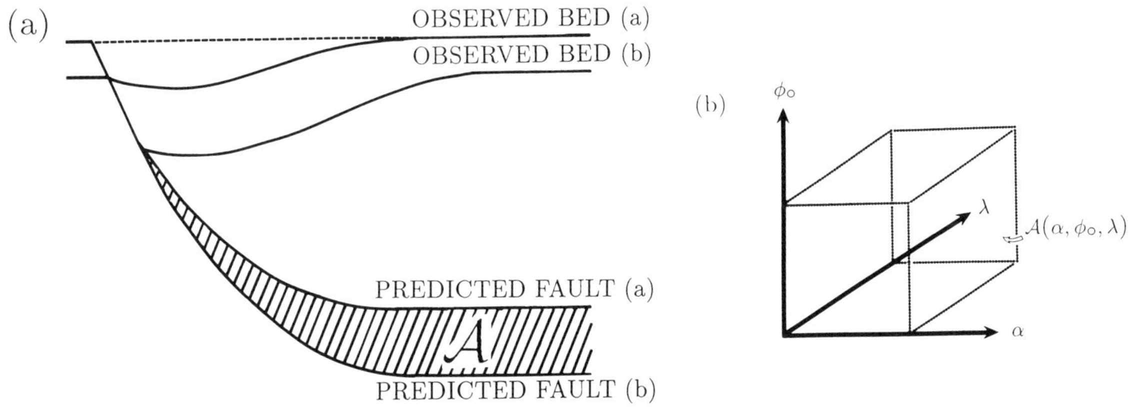

Fig. 4. The inverse problem: (a) in general, for the given values of α, ϕ_0, and λ, faults (a) and (b) can be calculated using beds (a) and (b). The dashed line indicates the regional for bed (a); the regional for bed (b) is similar. To find the correct fault geometry and the correct values of α, ϕ_0, and λ, the area $\mathcal{A}$ between two calculated faults must be minimized by varying α, ϕ_0, and λ. (b) Area $\mathcal{A}$ is a function of α, ϕ_0, and λ. A minimum value of $\mathcal{A}$ can be determined by systematically searching α-ϕ_0-λ space.

Ellis (1987) and Ellis & McClay (1988) as well.

Figure 5a shows that although some antithetic faulting is visible in parts of the hangingwall, the detailed deformation is superficially more complicated. Clearly, planar inclined simple shear does not, at first glance, seem to be particularly appropriate.

Four beds were used to calculate the fault geometry. Each bed was horizontal prior to extension, at a depth given by the layering within the footwall. If vertical shear is assumed (as in the 'Chevron construction' of Verrall (1981)), the four calculated faults fail to coincide with either each other or with the actual fault plane (Fig. 5b). The best solution (i.e. when $\mathcal{A}$ is a minimum) is obtained when $\alpha = 41°$ (Fig. 5c). In this case, the fit between calculated and observed geometries is surprisingly good. Allowing for compaction, whether the shear be inclined or not, gives poor results (Figs 5d & 5e) as would have been expected since no compaction has occurred.

Figures 5b–5e suggest that although hangingwall deformation appears, in detail, to be complex, *it can be represented by bulk inclined simple shear*. In fact, the complex small-scale rotations and synthetic faults in the lower part of the hangingwall are quite consistent with bulk simple shear: these details of the deformation are a consequence of the vorticity generated by bulk simple shear. The angular velocity of a small circular disk embedded within the hangingwall is easily calculated (Fig. 6). If $\mathbf{v}$ is the velocity vector at a point in the hangingwall then, since the y component of the velocity, v_y, is zero, we have:

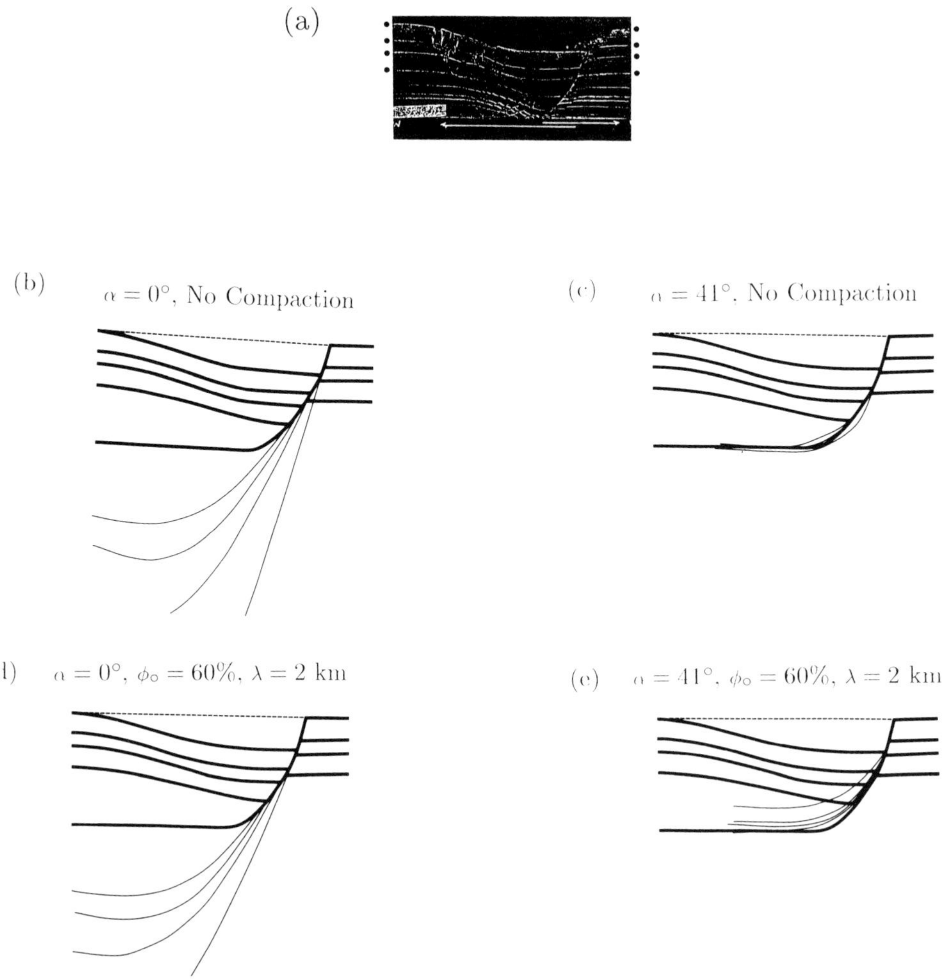

Fig. 5. (a) Wet clay model of normal fault (Cloos 1968). Note complex, small-scale deformation in the lower half of the hangingwall. Solid circles indicate regionals (pre-extension levels) of four beds used in the inverse modelling. **(b)–(e)** Inverse models of (a) using geometry of four hangingwall horizons to determine the fault geometry. Thick lines, observed beds and fault; thin lines, calculated faults; dashed line, regional for shallowest bed (other regionals are similar). Note best fit is example (c).

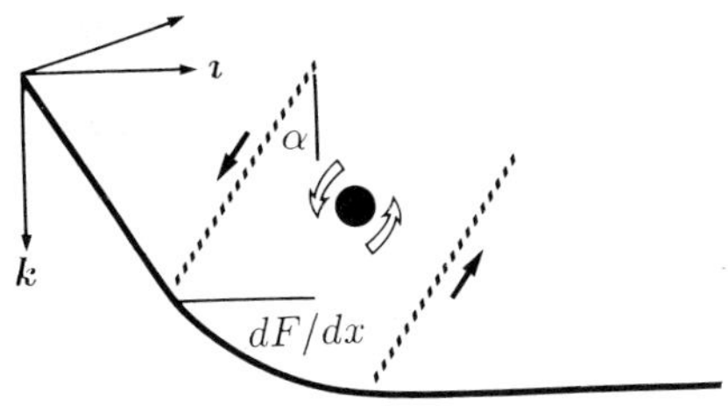

Fig. 6. Coordinate system and geometrical relationships used to determine vorticity, and hence rotation, of small circular disk inscribed in hangingwall.

$$\nabla \times \boldsymbol{v} = -\frac{\partial v_z}{\partial x}\boldsymbol{j}. \quad (4)$$

Following White *et al.* (1986), in a frame rotated through α we get:

$$\omega = -\frac{1}{2}U_0'\frac{\mathrm{d}^2 F'}{\mathrm{d}x'^2} \quad (5)$$

where ω is the angular velocity, U_0' is the horizontal velocity, and F' is the z' coordinate of a point on the fault, all measured in the rotated frame. Thus the rotation, θ, of a small disc embedded at position $x_0' + \epsilon'$ in the hangingwall is given by:

$$\theta = \frac{1}{2}\left\{\left[\frac{dF'}{dx'}\right]_{x'_0} - \left[\frac{dF'}{dx'}\right]_{x'_0+\epsilon'}\right\} \quad (6)$$

where ϵ' is the amount of extension in the rotated frame. The sense of rotation is anticlockwise if the dip of the fault decreases with depth in the positive x direction and vice versa.

In Fig. 5a, small fault-bounded blocks within the lower part of the hangingwall appear to have rotated coherently as extension proceeded rather than deforming internally. The deepest of these blocks has rotated by *c.* 26°. The expected rotation, calculated using Equation (6), is the same. Some of the sand-box models of McClay & Ellis (1987) and Ellis & McClay (1988) provide excellent examples of such rotations.

Seismic reflection data

The scheme described here can also be applied to depth-converted seismic reflection data. Testing can be carried out on sections where the fault plane reflection has been clearly imaged although the method will obviously be of most value when there is no knowledge of the fault plane. Here, two examples, both from areas of surficial, gravity-driven, faulting rather than areas of basement extension, are presented. Methods used to calculate normal fault geometries can only be applied with great care to basement-extending normal faults since the assumption of a rigid footwall may not be valid.

The first example (Fig. 7a) is from the Gulf of Mexico and has previously been interpreted by Wernicke & Burchfield (1982) and White *et al.* (1986). However, in both cases only the time section was used. To depth-convert, we assume that the published section is of standard aspect ratio (no horizontal scale is given). Image-ray migration was carried out using a velocity variation given by:

$$V = 1800 + 0.22z \quad (7)$$

where V is the velocity in m s^{-1} at a depth of z m. This velocity gradient is typical of that observed in sand-shale sequences, reflecting the increase in velocity with compaction. As it turns out the fault-calculation method works equally well on the time section so inaccuracies in the horizontal scale and depth conversion are not important in this particular example. In general, however, it is better to work with accurately depth-converted sections.

Figure 7a shows the uninterpreted time section. It is clear that a considerable amount of 'growth' has occurred within the hangingwall. Note the presence of a small-amplitude, large-wavelength hangingwall syncline, indicative of differential compaction (White *et al.* 1986). The fault plane reflection is especially clear.

The Wernicke & Burchfield (1982) interpretation is shown in Fig. 7b and shows a series of planar rotational normal faults in the hangingwall. As discussed in the previous section, the detailed deformation does not concern us here. What is important is whether or not the deformation can be adequately represented by *bulk* inclined simple shear.

Five depth-converted beds have been used to determine the fault geometry at depth (Fig. 7c–f). Since this is a growth fault structure, the regional or pre-extensional depth of each bed is the same, dipping 3° to the right. If vertical shear and no compaction are assumed, all five calculated faults overestimate the depth of the observed fault and fail to coincide with each other (i.e. the area difference, $\mathcal{A}$, is large) (Fig. 7c). Combinations such as inclined shear and no compaction (Fig. 7d) or vertical shear and compaction (Fig. 7e) also fail to predict the correct fault geometry. A systematic search through $\alpha-\phi_0-\lambda$ space yields the best solution: $\alpha = 34°$, $\phi_0 = 50\%$, and $\lambda = 2$ km (Fig. 7f). Not only do the calculated faults coincide quite well with each other but they also agree well with

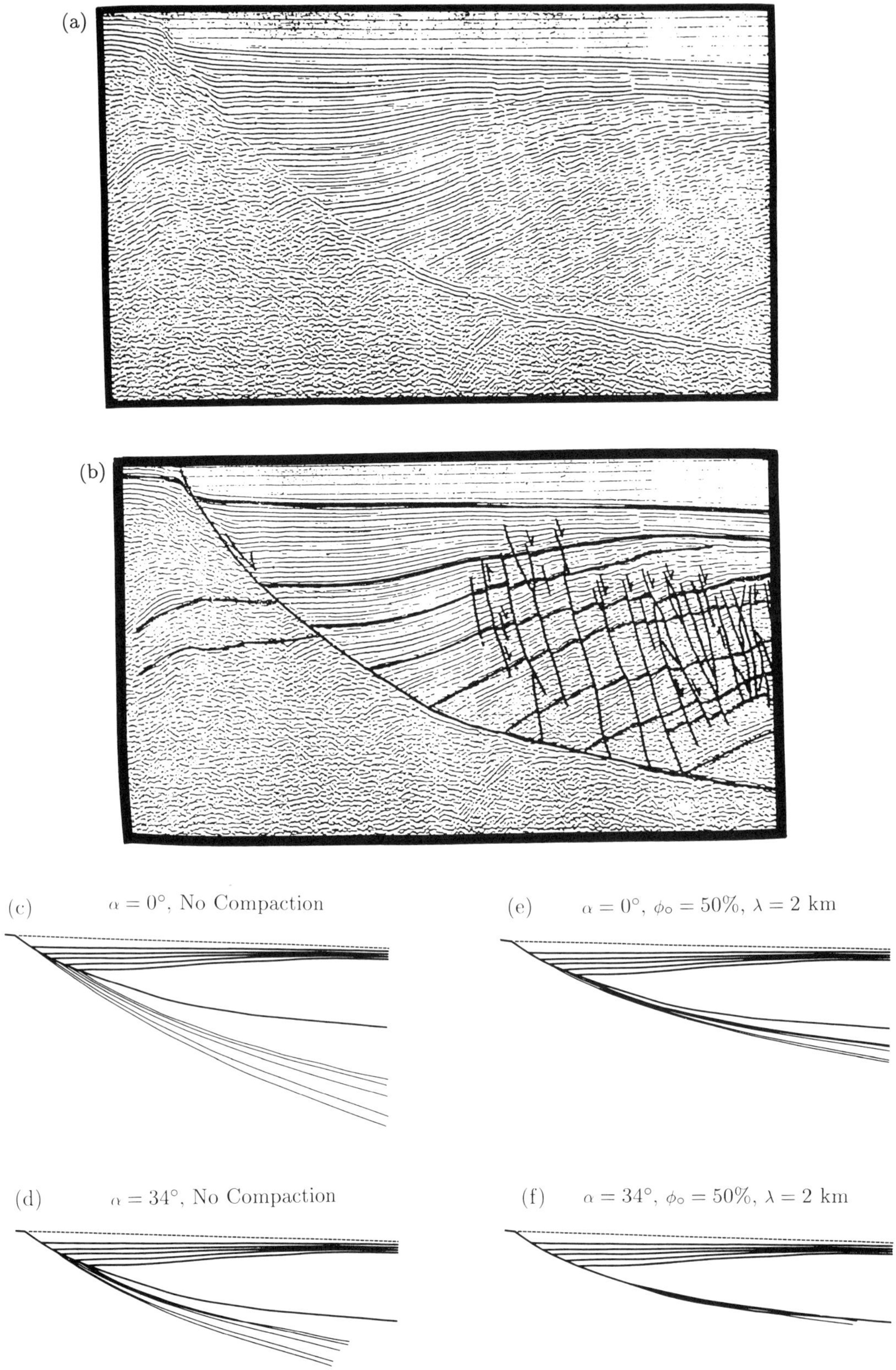

Fig. 7. (**a**) Uninterpreted seismic reflection section from the Gulf of Mexico (from Wernicke & Burchfiel 1982). (**b**) Interpretation by Wernicke & Burchfiel. (**c**)–(**f**) Inverse models of depth-converted version of (a) using the geometry of five hangingwall horizons. The regional of each bed is the same (dashed line) and dips 3° to right. Note best fit is example (f).

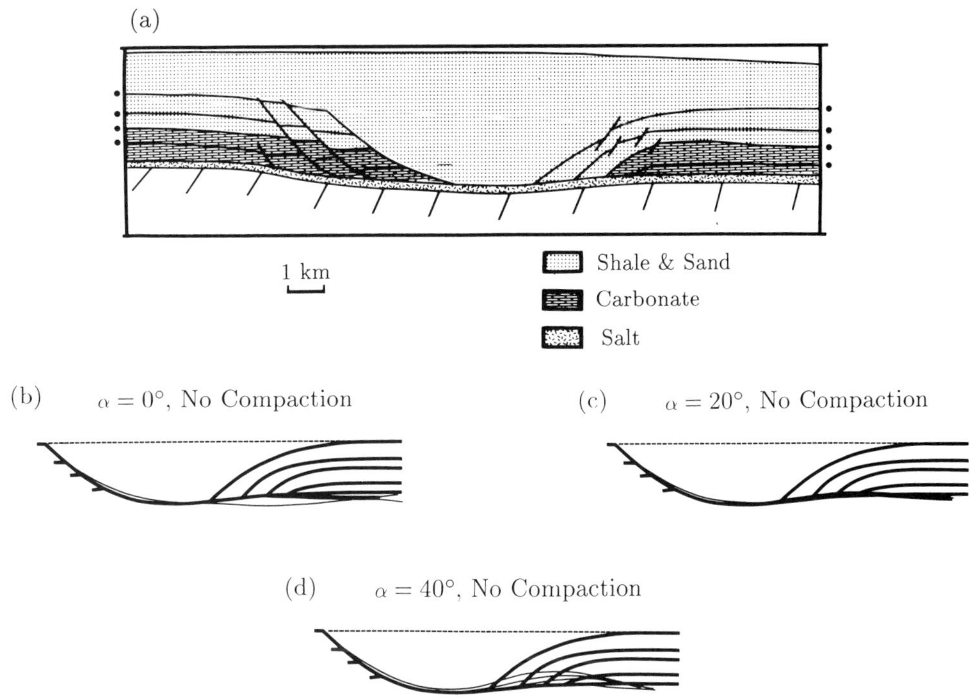

Fig. 8. (**a**) Interpreted, depth-converted seismic reflection section from a passive continental margin setting similar to Fig. 7. Solid circles indicate regionals of four beds used in inverse modelling. (**b**)–(**d**) inverse models of (a) using the geometry of four hangingwall horizons. Dashed line indicates regional of shallowest bed, other regionals being similar. Note best fit is example (c).

the observed fault geometry. If the next five beds down the fault are used, an identical solution is obtained giving confidence in the method. It is worth noting that the solution to this example presented by White *et al.* (1986) is only approximately correct since their scheme is only accurate provided the heave is small.

The second example (Fig. 8) comes from a similar passive continental margin setting to the first. The profile is depth-converted already. The fault is listric, soling out on a thin layer of salt. The example differs from both Figs 5 & 7 in that the hangingwall sediments (shales and carbonates) have been displaced a considerable distance down the fault (Fig. 8a). Therefore only a small portion of the observed fault requires to be predicted using the algorithm. All four observed beds from the hangingwall are used to calculate the actual fault geometry. The regional for each bed is assumed to be horizontal prior to extension, the depth being given by the footwall stratigraphy. Vertical shear gives a poor fit (Fig. 8b), the best fit being for $\alpha = 20°$. A larger α of, say, 40° gives a very poor fit. Both hangingwall and footwall strata are buried under about 3 km of sediments which means that *differential* compaction within the hangingwall is small and its inclusion does not alter the results.

Conclusions

The most important conclusion is that, despite the fact that hangingwall deformation is often quite complex in detail, it can be represented as a combination of inclined simple shear and, where appropriate, differential compaction. We emphasise that such simple shear may not always be very clearly manifested since it represents the *bulk* deformation: in detail the vorticity caused by bulk simple shear introduces local complexity in the form of small-scale rotation and synthetic faulting.

The inclination of simple shear and the presence or absence of differential compaction (and its degree) do not have to be known in order to calculate fault geometries at depth. Simply by assuming that several beds have been deformed in a similar fashion due to extension on the same fault, all unknown parameters including the fault geometry can, in theory, be determined. Throughout we have assumed that the hangingwall and footwall stratigraphies are accurately known. In circumstances where this assumption is invalid, stratigraphic cutoffs can be included as bounded parameters within a more formal inversion scheme.

This work is supported by a Natural Environment Research Council Grant to the British Institutions Reflection Profiling Syndicate (BIRPS). NW also gratefully acknowledges a British Council Studentship and generous support from Merlin Geophysical Ltd. Depth conversion of Fig. 7a was carried out on the Sattlegger I.S.P. system installed at Badley, Ashton & Associates. Jake Hossack and John Wheeler provided helpful reviews. Department of Earth Sciences Contribution Number 1479.

References

ANDERSON, E. M. 1951. *The Dynamics of Faulting*. Oliver & Boyd, Edinburgh.

CLOOS, E. 1968. Experimental analysis of Gulf Coast fracture patterns. *American Association of Petroleum Geologists Bulletin*, **52**, 420–444.

DAVISON, I. 1986. Listric normal fault profiles: calculation using bed-length balance and fault displacement. *Journal of Structural Geology*, **8**, 209–210.

ELLIS, P. G. & MCCLAY, K. R. 1988. Listric extensional fault systems — results of analogue model experiments. *Basin Research*, **1**, 55–70.

GIBBS, A. D. 1983. Balanced cross-section construction from seismic sections in areas of extensional tectonics. *Journal of Structural Geology*, **5**, 153–160.

—— 1984. Structural evolution of extensional basin margins. *Journal of the Geological Society, London*, **141**, 609–620.

JACKSON, J. A. 1987. Active normal faulting and crustal extension. *In*: COWARD, M. P., DEWEY, J. F. & HANCOCK, P. L. (eds) *Continental Extensional Tectonics*. Geological Society, London, Special Publication, **28**, 3–17.

—— & WHITE, N. J. 1989. Normal faulting in the upper continental crust: observations from regions of active extension. *Journal of the Geological Society, London*, **11**, 15–36.

MCCLAY, K. R. & ELLIS, P. G. 1987. Analogue models of extensional fault geometries. *In*: COWARD, M. P., DEWEY, J. F. & HANCOCK, P. L. (eds) *Continental Extensional Tectonics*. Geological Society, London, Special Publication, **28**, 109–125.

MCKENZIE, D. 1984. The generation and compaction of partially molten rock. *Journal of Petrology*, **25**, 713–765.

RAMSAY, J. G. 1967. *Folding and Fracturing of Rocks*. McGraw-Hill, San Francisco.

SCALES, L. E. 1985. *Introduction to Non-linear Optimisation*. McMillan London.

SCLATER, J. G. & CHRISTIE, P. A. F. 1980. Continental stretching: An explanation of the post-mid-Cretaceous subsidence of the Central North Sea Basin. *Journal of Geophysical Research*, **85**, 3711–3739.

SUPPE, J. 1983. The geometry and kinematics of fault-bend folding. *American Journal of Science*, **283**, 684–721.

VERRALL, P. 1981. *Structural interpretation with applications to North Sea problems*. Course Notes No. **3**, JAPEC (UK).

WALTHAM, D. 1989. Finite difference modelling of hanging wall deformation. *Journal of Structural Geology*, **11**, 433–437.

WERNICKE, B. & BURCHFIEL, B. C. 1983. Modes of extensional tectonics. *Journal of Structural Geology*, **4**, 105–115.

WHEELER, J. 1987. The geometry of listric normal faults and deformation within their hanging walls: a discussion. *Journal of Structural Geology*, **9**, 811–815.

WHITE, N. 1987. Constraints on the measurement of extension in the brittle upper crust. *Norsk Geologisk Tidsskrift*, **67**, 269–279.

——, JACKSON, J. A. & MCKENZIE, D. P. 1986. The relationship between the geometry of normal faults and that of the sedimentary layers in their hanging walls. *Journal of Structural Geology*, **8**, 879–909.

WILLIAMS, G. & VANN, I. 1987. The geometry of listric normal faults and deformation in their hanging walls. *Journal of Structural Geology*, **9**, 789–795.

WOOD, R. J. 1981. The subsidence history of CONOCO well 15/30–1, central, North Sea. *Earth & Planetary Science Letters*, **54**, 306–312.

Geometric and kinematic coherence and scale effects in normal fault systems

JOHN J. WALSH & JUAN WATTERSON

Department of Earth Sciences, University of Liverpool, P.O. Box 147, Liverpool L69 3BX, UK

Abstract: Fault displacements derived from a seismic reflection survey of an offshore oilfield are projected onto a vertical plane parallel to the fault strike and the displacement values contoured. In addition to those for single faults, displacement diagrams are also constructed for fault arrays by aggregating the displacement values on selected faults. Displacement contours form regular and systematic patterns, even when there is no continuity between the fault surfaces in the array. A system in which linkage between the elements of a fault array is achieved by ductile strain of the intervening rock is referred to as soft-linked and is regarded as the general case. Geometrical coherence, with regular and systematic displacement patterns, exists at all stages in the growth of a fault array, implying a kinematic coherence requiring a high degree of synchronous movement, as opposed to sequential development, on individual elements in the array.

Conventional maps and diagrams allow representation of only two orders of magnitude of fault displacement rather than the five or more which may occur. A soft-domino model is presented in which the role of ductile strains is acknowledged both in accommodating varying displacements on fault surfaces and in extension on structures too small to be represented individually, and in which the rotation of rigid fault blocks is of reduced importance when compared with rigid-domino models.

To the extent that any seismic interpretation is based on incomplete data it incorporates, either explicity or implicity, some elements of a preferred conceptual model. The manner in which faults are linked with neighbouring faults, if linked at all, is one such element and determines, for example, the way in which a fault map is drawn from a series of interpreted seismic sections. Although the concept of geometric coherence in arrays of faults is not a novel one, the consequences of such coherence are not always acknowledged in interpretation of fault geometry and timing. The degree and types of fault linkage represent a basic difference between fault models and have an important bearing on the hydraulic connectivity of a faulted reservoir formation.

Examination of the geometric coherence in fault systems requires consideration of scale effects and of the range of fault sizes which should be taken account of in a conceptual fault model. A model should give due recognition to the existence and effect of faults on a wide range of scales even though only a restricted range of fault sizes can be represented on maps, cross sections and block diagrams. Soft-domino models (Barnett *et al.* 1987; Gibson *et al.* 1989) are characterized by non-rigid behaviour of the rock volume between faults and allow for incorporation of the effects of ductile strain. A ductile strain is a systematic change in shape due to structures too small to be individually represented on the scale of a particular map, cross section or diagram; it is therefore characterized by an absence of geometrical discontinuities. The scale range of structures contributing to ductile strain therefore varies with the scale of the map or diagram and extends upwards from that of structures due to grain scale processes to that of faults with maximum displacements of 100 m or greater in the case of regional cross sections and maps.

Analysis of the geometry of fault arrays

Strike-projections

In most of the figures in this article the geometries of individual faults and fault arrays are represented by strike-projections (Fig. 1). No single graphical projection allows the full three-dimensional geometry of a fault to be represented but a strike-projection illustrates some aspects of the geometry not apparent on conventional map and cross section projections. Strike-projections are constructed by projecting displacement values at points on a fault surface onto a surface parallel to the fault strike (Rippon 1985a,b, Barnett *et al.* 1987, Childs *et al.* 1990);

From Holdsworth, R.E. & Turner, J.P. (compilers) *Extensional Tectonics: Faulting and Related Processes*. The Geological Society, Key Issues in Earth Sciences, **2 (2)**, 63–73.
1476-315X/02/$15.00
First published in: "Roberts, A.M., Yielding, G. & Freeman, B. (eds) *The Geometry of Normal Faults*, The Geological Society, London, Special Publications **56**, 193–203".

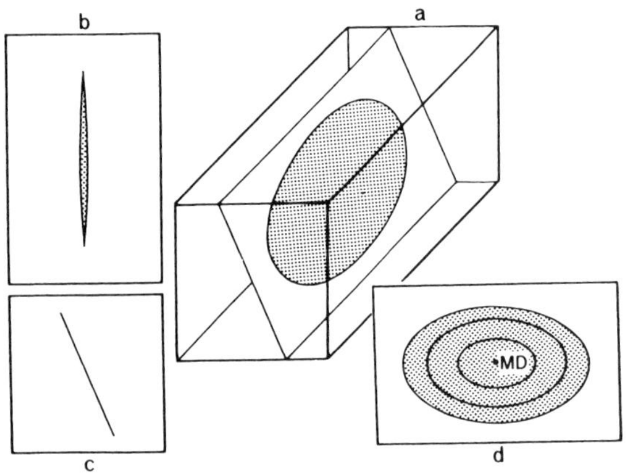

Fig. 1. (**a**) Rock volume with elliptical fault surface. (**b**) Map projection of fault. (**c**) Cross section projection. (**d**) Strike-projection on vertical plane parallel to the fault strike with contours of equal displacement centred on maximum displacement (MD).

plane-vertical projection surfaces are used for all the projections illustrated here. The displacement values are contoured to allow a visual appreciation of the pattern of displacement variation over the fault surface. Some degree of regularity of form and spacing of the contours would be expected *a priori* in a natural system and this regularity has been demonstrated on small-scale faults (Roux 1979; Rippon 1985a,b; Barnett *et al.* 1987). Abrupt changes in displacement, shown as very close spacing of displacement contours, are not normally to be expected because of the compatibility problems to which they give rise, but they can of course occur at boundaries of salt or other incompetent materials and at unconformities and free surfaces. On an ideal single blind fault the displacement is a maximum at the centre of the elliptical fault surface and reduces systematically to the zero displacement contour, or tip-line, of the fault (Roux 1979; Barnett *et al.* 1987; Walsh & Watterson 1987). Examination of the displacement variation on fault surfaces is useful when considering the relationship between two or more adjacent faults (Pfiffner 1985) and for quality control of seismic interpretations (Barnett *et al.* 1987, Childs *et al.* 1988). The strike-projections illustrated have been constructed using Fault Analysis Projection System (FAPS) software developed by the Liverpool Fault Analysis Group.

Data

The fault projections illustrated are derived from 151 seismic reflection lines at 100 m spacing from a single oilfield. The lines are each 6.125 km long and the total area covered is *c.* 92 km^2. The seismic lines are sub-perpendicular to fault strike and fault displacements, expressed as the two-way time (TWT) difference between the hanging-wall and footwall cut-offs, are derived from six mapped reflectors. For reasons of confidentiality, maps and complete cross sections of the field are not available for publication. The interpretation used is that made by company staff with some reinterpretation based on the results of FAPS-generated projections. Seismic velocity variations are sufficiently small as not to be significant for the purposes of this article. Fault displacement values are throw values unless stated otherwise; strike-projection displacement diagrams can be plotted for throw, heave or total dip-slip displacement but those illustrated are plotted using throw values.

Fault linkage

Two end member types of linkage are distinguished. *Hard-linked* faults are those in which the fault surfaces are linked on the scale of the map or cross section in use. Interpretations assuming hard linkage have become usual in thrust complexes (Boyer & Elliott 1982) and similar interpretations have subsequently been advocated for normal fault arrays (Gibbs 1984). *Soft-linked* faults are those between which a mechanical and geometric continuity is achieved by ductile strain of the rock volume between them, rather than by continuity of their fault surfaces. Solf-linked faults will appear to be isolated from one another, on the scale of the map or cross section in use.

Where the geometries of adjacent faults are interdependent, as shown for example by complementary variations in displacement, the faults are considered to be geometrically linked. As seen on a map or cross-section faults may be strike-linked i.e. one is essentially a lateral continuation of the other, dip-linked i.e. one fault surface is essentially an upwards or downwards continuation of the other, or occluding i.e. the fault surfaces overlap on a strike-projection. In most cases strike-linked and dip-linked faults also have some degree of overlap producing relay zones (Ramsay & Huber 1987, Larsen 1988, Walsh & Watterson 1990). The heaves of occluding faults are additive when the local or regional extension is measured. This article considers mainly the linkage of occluding and strike-linked faults.

Geometric coherence

Single fault strike-projections

Figure 2 shows a strike-projection of what is represented on a 1:12 500 scale map as a single isolated fault. The ordinate of the projection is represented in two-way time and displacement contours in (msec × 10); assuming a seismic velocity of 3 km s^{-1}, the maximum displacement is *c*. 60 m and the vertical exaggeration *c*. 1.5. The zero displacement contour represents the limit of detectable displacement on the seismic lines which is estimated to be no more than 10 m. The contours are derived from 62 displacement measurements on four mapped reflectors and show the displacement variation to be systematic with the fault dying out upwards, downwards and laterally. The fault, when mapped on this scale, closely approaches the idealized single fault (Fig. 1). The contour spacing below the centre of the fault is less than that above the centre; this may be due to a real difference in rates of change of displacement above and below the fault centre but could also be due to increase in seismic velocity with depth. A velocity change will make the TWT ordinate scale non-linear with respect to depth and will also represent similar displacements by shorter travel times below the fault centre.

The maximum dimension of the fault as mapped is 1800 m, corresponding to a 16 cm trace on a 1:12 500 map. However, to represent the fault by a single 80 cm trace on a 1:2500 map would almost certainly be misleading. Had this fault been mapped by methods which allowed realistic representation on a scale of 1:2500 (i.e. the scale of standard mine plans for UK coalfields which are constructed from direct observations of faults), it would very likely be shown to consist of a discontinuous array of fault traces and splays. On a scale of 1:1, as in outcrop, a single fault trace 1800 m long would not be expected. That a multiplicity of fault traces observed on one scale is routinely represented as a single fault on another scale reflects a belief that a multiplicity of small faults comprises a geometrically coherent system in which each individual element is necessary. The systematic displacement variation seen when this array is mapped as a single fault shows that a high degree of order must exist in the array of smaller elements. In this case, representation as a single fault is inevitable because the seismic method does not permit resolution of the, probably numerous, individual elements within the array.

The routine assumption of geometric coherence referred to above can be tested by quantitative construction of a small-scale projection from data on a larger-scale projection, as is done below.

Aggregate strike-projections

Aggregate strike-projections are constructed, on each seismic line, by summing the displacements on selected faults along each horizon and plotting as for a single fault. The method is illustrated on a diagrammatic seismic section in

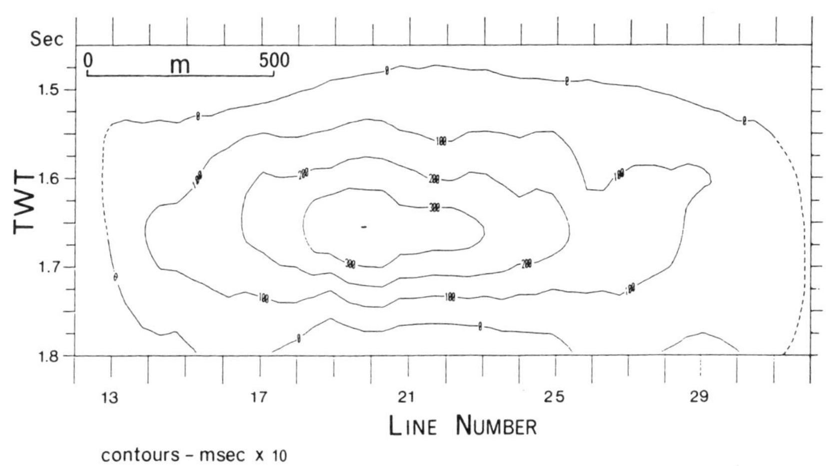

Fig. 2. Strike-projection showing contours of displacement for a single isolated fault. Displacement values (n=62) are derived from measurements on four mapped reflectors. Assuming a velocity of 3 km s^{-1}, the maximum displacement is *c*. 60 m and the vertical exaggeration is 1.53. The absence of computer-derived lateral tip-lines is an artifact of the contouring routine used, but as the fault is not seen on lines 12 and 32, manually-drawn lateral tip-lines are shown (broken lines).

Fig. 3. The aggregate displacement on horizon X for the closely spaced array of faults, A, B & C, is a + b + c. Other combinations of faults can be selected at will and an aggregate displacement strike-projection plotted for each combination. An aggregate displacement plot for an array of closely spaced faults, spanning *c.* 200 m across the strike of the faults, is shown in Fig. 4. The displacement contours show a regularity similar to that on the single fault (Fig. 2) but with the important difference that the contour spacing is *c.* 15 m for the single fault and is *c.* 75 m for the aggregate plot. The aggregate plot is constructed from 1280 individual displacement values on 153 faults and fault splays, which all lie within 200 m of the single major fault. Only the upper part of the fault array is imaged on the seismic sections. The systematic pattern and spacing of the contours are similar to those of a 'single' fault and, as the spaces between the faults are small relative to the lengths of their traces, demonstrate that representation of the array, shown in Fig. 4, by a single 6 cm fault trace on a 1:250000 map would be realistic even though several smaller faults are represented on the original 1:12500 map. The importance of individual small faults and splays to the geometric coherence of the array is illustrated in Fig. 5 which shows a small area of the strike-projection of Fig. 4. Figure 5a shows the aggregate displacement plot with all elements included and the cross section (Fig. 5c) shows the elements present on one seismic line. Fig. 5b shows the same aggregate plot but with a single splay excluded (Fig. 5c). Removal of this splay results in a markedly increased irregularity of the displacement contours demonstrating that this splay is necessary for geometrical coherence of the system. The splay is close to the upper tip-line of the array and, as shown in Fig. 5c, is hard-linked to other elements of the system.

The significance of aggregating displacements on faults which are not hard-linked is shown in Fig. 6. Figure 6b shows an aggregate displacement plot of a major fault and associated hard-linked splays one of which is shown in the cross section (Fig. 6a). Between lines 3 & 17 the displacement contours of 100 ms and greater show a prominent irregularity. A small fault (Y) is isolated from the main fault both on the cross section illustrated and on all other seismic lines on which it has been identified. When displacements on this isolated fault are included

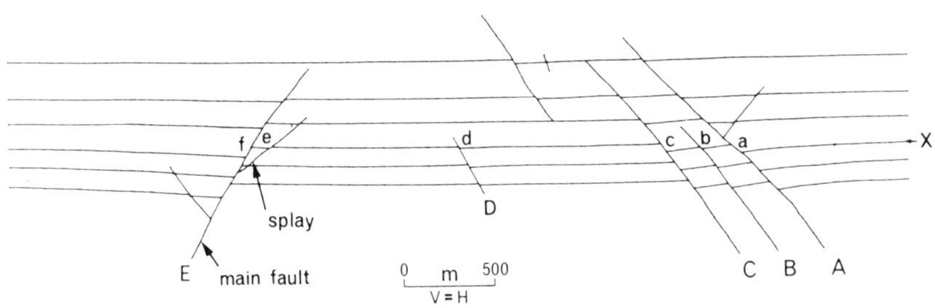

Fig. 3. Schematic cross section showing method of aggregating displacements on fault arrays (see text for details). The aggregate displacement on horizon X for all faults on the cross section is a + b + c + d + e + f. Two fault segments of fault E are identified i.e. the main fault and a fault splay.

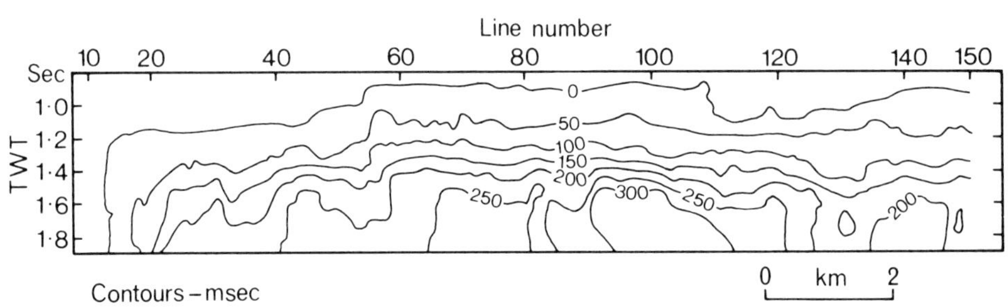

Fig. 4. Strike-projection showing contours of aggregate displacement for a fault system comprising a main fault and 153 dominantly hard-linked splays. Displacements are derived from six mapped reflectors. 736 aggregate displacement values are contoured and these were obtained from a total of 1280 individual displacements. Assuming a velocity of 3 km s^{-1} the maximum displacement is *c.* 450 m and the vertical exaggeration is 1.38.

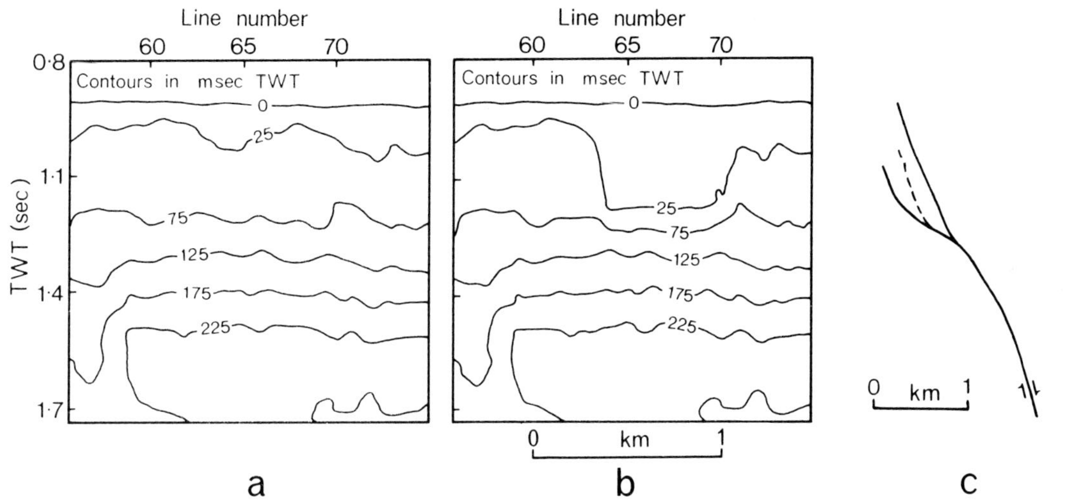

Fig. 5. (**a**) Enlargement of a small area of the strike-projection shown in Fig. 4 incorporating 120 aggregate displacement values, including displacements (n=219) from all splays, as in Fig. 4. For a velocity of 3 km s^{-1} the maximum displacement is *c*. 415 m and the vertical exaggeration is 1.38. (**b**) As (a) but with displacement data from one splay subtracted (n=211). (**c**) Cross section on line 65 with the splay subtracted in (b) shown as dashed line.

in the aggregate plot (Fig. 6c) the contour irregularity is greatly reduced, although not eliminated. Although isolated, the fault is clearly necessary to the geometrical coherence of the array of which it must be regarded as a part. The residual irregularity of the displacement contours is thought to be due to some of the displacement at this point being accommodated by plastic shear deformation in the relatively thin slice of rock between the isolated fault and the main fault.

Figure 7a shows the aggregate displacement plot for all the faults in the field which have the dominant dip direction. The maximum displacement is *c*. 600 m and 4082 individual displacement readings are incorporated in the plot. The faults included in the aggregate plot are distributed over 6 km in the direction normal to the fault strike and the main faults are clearly isolated from one another, at least within the imaged depth range and in respect of the minimum imaged fault size. The numbers of main faults and of fault segments (main faults plus splays) included in the aggregate plot at differ-

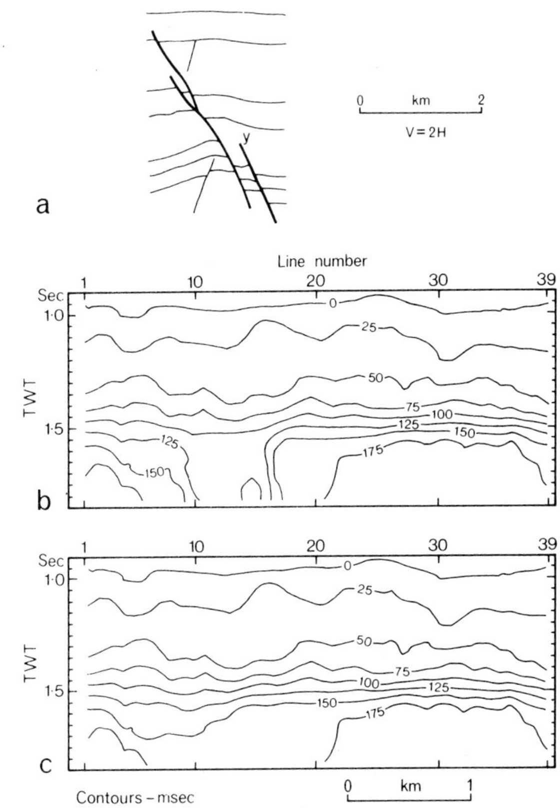

Fig. 6. (**a**) Cross section showing the faults (heavy lines) plotted in (**b**) and (**c**). (b) Strike-projection showing contours of aggregate displacement for a fault system comprising a main fault and 19 hard-linked splays. (c) Strike-projection of main fault and splays as in (b) but with addition of the isolated fault (y) in the hangingwall of the main fault system shown in (a). Displacements are derived from six mapped reflectors. 240 aggregate displacement values are contoured in each strike-projection and these values were obtained from a total of 292 (b) and 312 (c) individual displacements. Assuming a velocity of 3 km s^{-1} the maximum displacement is *c*. 270 m and the vertical exaggeration is 0.73.

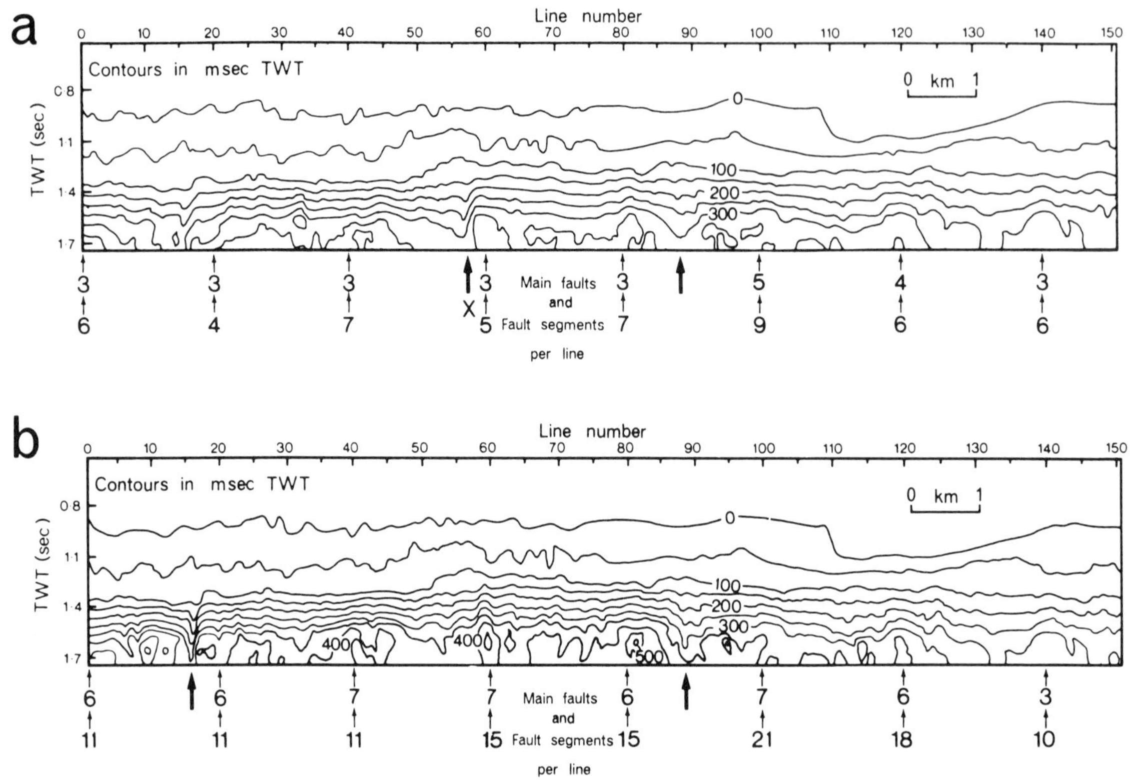

Fig. 7. (**a**) Aggregate strike-projection of all faults with the dominant dip-direction within the field: prominent contour irregularities are shown (heavy arrows). (**b**) Strike-projection of all faults in the field: residual sharp contour irregularities are shown (heavy arrows). Displacements are derived from six mapped reflectors. 872 aggregate displacement values are contoured in each strike-projection and these values were obtained from a total of 4082 (a) and 4949 (b) individual displacements. Assuming a velocity of 3 km s^{-1} the vertical exaggeration is 1.55 and the maximum displacements are 675 m (a) and 750 m (b). The numbers of main faults and the numbers of fault segments aggregated are shown for selected lines.

ent points on the strike-length of 15 km are shown on Fig. 7. The overall contour pattern is again similar to that of a single fault but with some prominent irregularities (arrowed) the system, in respect of the main faults at least, is demonstrably soft-linked. The prominent irregularity at X is greatly reduced in Fig. 7b which includes all the faults in the field, i.e. both the dominant set and a set with an opposed dip; there are no cross faults in this field. Aggregating throws on faults with opposing dips produces values of no obvious geological significance but a similar pattern would be expected for aggregate fault heaves, representing aggregate extension. Residual sharp contour irregularities (Fig. 7b, arrowed) occur at positions corresponding to lateral tips of main faults in the array. It is not yet known whether these irrëularities represent interpretational artefacts or places where an unusually high proportion of displacement is accommodated by ductile shear, i.e. a continuous displacement, with a corresponding reduction in the discontinuous displacement accommodated along fault surfaces. Figure 8a shows a small part of the aggregate displacement plot of Fig. 7 and the same area is shown in Fig. 8b but with the displacements on a single fault within the array subtracted. The reduction in the regularity of the displacement contours caused by the removal of this single fault demonstrates a reduced geometric coherence of the system and it is concluded that this single fault is an essential part of the system rather than an arbitrary or unrelated structure.

The principal conclusion drawn is that the pattern of aggregate displacement on an array of faults, the traces of which are up to 6 km apart and which have no direct physical continuity with the main rupture surfaces, is both systematic and similar to that which would be expected on a small part of the surface of a very large single fault surface. Each fault in the field is therefore an integral and essential part of a single, geometrically coherent fault system which, for some purposes and on some scales,

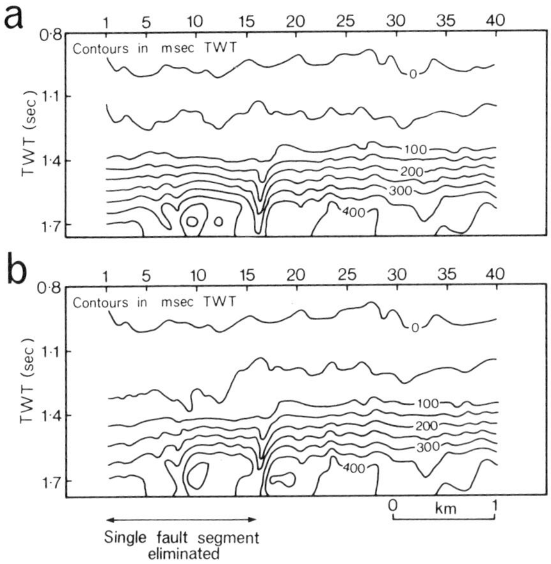

Fig. 8. (**a**) Enlargement of part of the strike-projection shown in Fig. 7b (**b**) As (a) but with a single fault subtracted. Displacements are derived from six mapped reflectors. 234 aggregate displacement values are contoured in each strike-projection and these values were obtained from a total of 1417 (a) and 1344 (b) displacements. Assuming a velocity of 3 km s^{-1} the maximum displacement is 630 m and the vertical exaggeration is 1.41.

can properly be regarded as a single fault. The system is largely soft-linked and such linkage demands a significant degree of ductile deformation in the rock volume containing the array of faults. Further evidence for significant ductile strain is provided by the displacement variation on fault surfaces which is evident on all scales (Watterson 1986, Barnett *et al.* 1987, Gibson *et al.* 1989). On no fault surface for which a displacement plot has been constructed, in any of the several data sets which we have examined, has displacement been shown to be constant. The variable displacements on all the individual fault segments in a fault array, on all the scales illustrated, are thus dependent on the precise positions, sizes and displacement distribution patterns of all other fault elements within the array. To this extent, the omission of a fault segment from a seismic interpretation potentially can be detected. Faults formed at different times and with unrelated displacements could also be distinguished by this method.

Kinematic coherence

In the cases illustrated, a geometric coherence can be demonstrated to exist in an array of faults on which movement has ceased. It can then be inferred that a comparable degree of coherence characterised the array throughout all stages of its development. If so, then the timing and rates of displacement at each point on the surfaces of all faults and splays in the array are related to the timing and rates of displacement at all other points in the array i.e. the array is kinematically coherent. Not all points on all fault surfaces will be active throughout the life of the array but the time at which each fault surface is initiated, the rate at which it grows and the time at which it ceases to grow are fixed in relation to the overall growth history of the array. A considerable degree of synchronous growth is necessary in such a system.

Interpretations of the sequential growth of individual fault segments and splays based on their geometrical relationships as seen in 2D sections are inconsistent with the evidence for a high degree of synchronous growth of the individual surfaces and the progressive increase in displacement on each surface. Sequential forward propagation of either normal fault or thrust arrays (Butler 1982; Gibbs 1984) is unlikely. For any individual fault elements, forward, rearward (Eisenstadt & De Paor 1987; Price 1988) and lateral propagations are of equal significance for any individual fault element even though the overall dimensions of a regional scale array may be fixed at the time of initiation. Four separate surfaces, linked on three branch-lines, are shown diagramatically in Fig. 9. The variable displacement on each is not shown. In the cross section represented by the plane ABCD, three of the individual elements are represented and, given this single cross section, it might be tempting to deduce a sequence of formation. However, when lateral propagation of each surface is taken into account the order in which each element first appears in any one cross section is not necessarily related to the order in which each new displacement surface is first generated. It is also possible that displacement on all four surfaces took place synchronously. Given that the lateral boundaries of only one of the surfaces are shown in the block diagram (Fig. 9b) it is not possible to say which of the other three surfaces has the greatest extent and displacement. Given the degree of interdependence of displacements on all individual elements in the array and the three-dimensional nature of the geometric coherence which has been demonstrated, it is unrealistic to make a simple sequential kinematic interpretation of any single cross section.

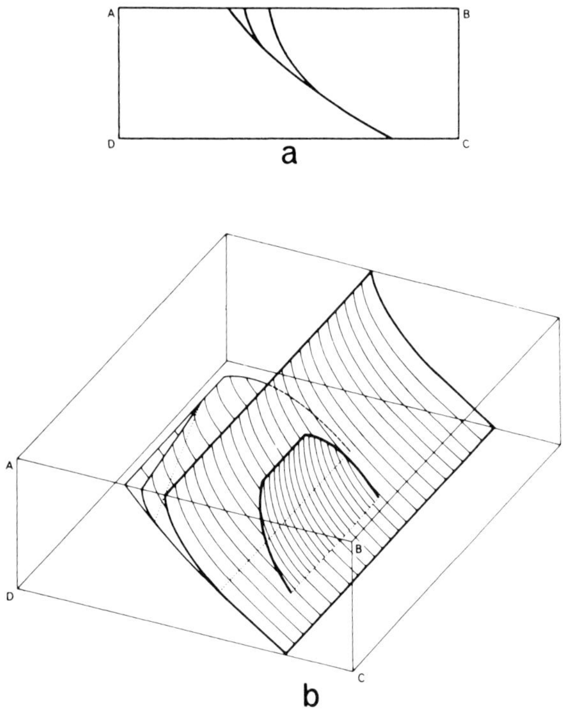

Fig. 9. (**a**) Cross section of hypothetical fault with two hard-linked splays. (**b**) Block diagram showing the geometry of part of the fault array of which (a) is a cross section on the front of the block (ABCD).

Effects of scaling

The range of scales on which fault arrays are geometrically coherent is not known. We believe that coherence is likely to persist on the basin scale but have not yet been able to assemble sufficient data to test this assertion which assumes that faults with displacements up to at least 5 km form part of coherent arrays. The smallest scale of fault which contributes to coherent arrays has, likewise, not been demonstrated but fault population studies (Childs *et al.* 1990) suggest that displacements down to 10 cm or less are systematic elements in a coherent, larger-scale geometry. We believe it likely that faults with maximum displacements ranging over at least five orders of magnitude may form elements of a single coherent array whereas even a good map can represent faults with displacements ranging over only *c.* two orders of magnitude. This raises the question of what is actually represented on a map.

On a map on scale 1:250 000 the minimum length of fault trace represented may be 5 cm, corresponding to 12.5 km; a fault of this length would have a maximum displacement of *c.* 1 km (Walsh & Watterson 1988a) and faults with smaller maximum displacements would not be represented. If the 1:250 000 map is derived from a larger-scale map on which smaller faults have been identified, then the faults represented on the smaller-scale map may be aggregated so that what is represented as a 1 km maximum displacement fault may be an array of smaller faults with aggregate maximum displacement of 1 km; most maps do incorporate some degree of aggregation as a convenient simplification which overcomes some of the scale limitations of graphical representaton. Whether a smaller fault is included in an aggregated array will depend to some extent on its position relative to a larger fault — if distant from a larger fault then a smaller fault is unlikely to be included in an aggregated array. These considerations are relevant to fault maps on all scales and even in fully aggregated maps there will be no record of faults which are below the limit of resolution of the method used to acquire the data from which the map is constructed. All maps and diagrams therefore under-represent the strain accommodated by faulting although for some purposes such under-representation may be of little significance. Faulting on a scale below that which is represented on a map or diagram constitutes a ductile strain.

Although fault sizes represented on maps of different scale have been referred to in terms of their trace lengths, the limits can also be referred to in terms of the minimum fault displacements which are represented. In general, the smallest fault displacement represented on a structural contour map is determined by the smallest displacement which can be detected as a contour offset, which in turn depends on the spacing of the structural contours. This limitation means that, as the displacement on a fault varies along the length of the fault trace, the fault trace on the map shows only that part of the apparent fault on which displacement is greater than the limit imposed by the contour spacing; the dimensions of a fault will therefore vary with the scale of map or section on which it is represented.

The fault model

A conceptual soft-domino fault model is shown in Fig. 10 which illustrates the principal conclusions derived from analysis of geometrical and kinematic coherence in several datasets and incorporates features previously described (Barnett *et al.* 1987; Gibson *et al.* 1989). The principal features shown in the block diagram are as follows. Each fault shows a systematic non-linear variation in displacement (Walsh & Watterson 1987). The sum of the displacements

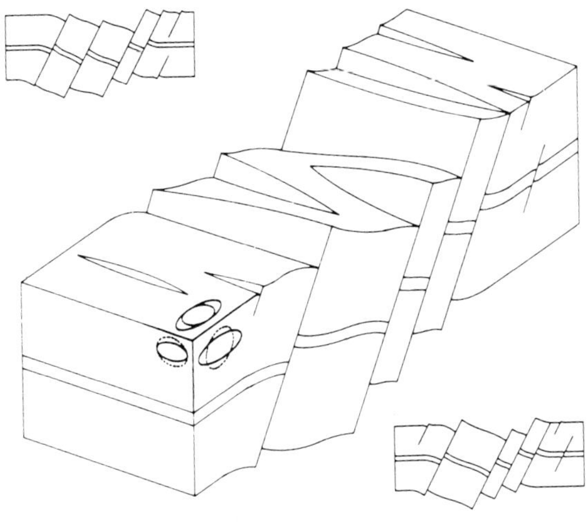

Fig. 10. Conceptual block diagram showing the principal features of the soft-domino fault model (see text for details). Insets show cross sections on sides of the block. Strain ellipses on the sides of the block show the ductile strain, for the block as a whole, accommodated by structures which are too small to be represented on the diagram (unstrained circles shown with broken lines): the amount of strain shown is arbitrary.

on all faults encountered on any line through the block and parallel to its length, is the same for all lines, i.e. the whole array shown can be considered as a small part of a single much larger fault. The block as a whole has not rotated relative to the horizontal and the marker horizon is at the same elevation at both ends of the block. The unchanged relative elevations of the marker horizon at both ends is achieved by reverse drag of horizons on both sides of each fault and, unlike the rigid-domino model, is not dependent on fault block rotation. The variable displacement on individual faults is accommodated by heterogeneous ductile strains. The ductile strain accommodated by structures too small to be individually represented on the diagram is shown by the strain ellipses. Plane strain is assumed in which there is no change of dimension in the direction parallel to the fault strike; the amounts of ductile strain shown by the strain ellipses is arbitrary. If the volume represented by the block diagram were to be illustrated on a much smaller scale than in Fig. 10, all of the fault-related deformation would be represented as a ductile extension by necking. For simplicity, faults dipping in only one direction are shown and the number of smaller faults shown relative to the number of larger faults is less than would be expected (Childs *et al.* 1990). The maximum displacements of faults illustrated range over *c.* two orders of magnitude. All material lines and planes within the block, other than lines parallel to the length of the block, have been rotated and faults have shallower dips than when initiated. Only the final geometry is shown but similarly constrained geometries existed at all stages in the evolution of the fault system and the faults are more nearly synchronous than sequential. Although blocks between individual pairs of faults have been rotated they have not been rotated as rigid blocks. No scale is shown because similar geometries will exist whatever the block length – except for features due to isostatic effects and thermal subsidence, which are not incorporated in the diagram. The displacement/dimension ratios of the largest faults shown are, however, typical for faults with maximum displacements of several kilometres rather than the much smaller ratios typical for faults with maximum displacements of several metres (Walsh & Watterson 1988a). Also not illustrated is the likely shallowing of fault dips to 45° (Walsh & Watterson 1988b), and the downward change from the discontinuous geometry of brittle faults to the continuous geometries of ductile shear zones (Bak *et al.* 1975; Grocott 1977; Sibson 1977; Ramsay & Huber 1987), both of which would be apparent on the larger scale. The soft-domino model is similar in many respects to that which has long been applied to ductile straining of single crystals (Taylor 1938) in which glide planes are bounded by dislocations along their tip-lines. In the crystal model, individual glide planes do not extend across the crystal, but, in common with soft-domino faults, rotate throughout the deformation. The rigid-domino model requires propagation of faults across the whole of the faulted layer before either regional strain or fault rotation can begin. When soft-domino faults have grown sufficiently to span the faulted layer, further deformation is characterized by elements of both the soft and rigid-domino models i.e. rigid block rotation and rotation by internal deformation of major fault blocks.

It is apparent from Fig. 10 that the question of whether the strain illustrated is a simple shear or a pure shear strain is scale-dependent. On the scale of an individual fault or shear zone the strain is a simple shear but on the scale of a block incorporating many faults the strain is a pure shear. The concept of simple shear in the upper crust contrasting with pure shear in the lower crust (Kusznir & Egan 1990) simply reflects an unstated assumption that upper crustal strains are accommodated on fewer and larger structures than equivalent strains in the lower crust. The evidence from mid-crustal rocks (Bak *et al.* 1975; Escher & Watterson 1974; Coward 1976) is that individual ductile shear zones accommodate displacements of similar scale to those of the largest upper crustal faults.

Conclusions

(1) Displacement contours on strike-projections of both single faults and fault arrays have simple and systematic patterns and spacing.
(2) An array of faults on one scale can be, and often is, represented by a single fault on a smaller-scale representation.
(3) Each individual fault or splay is necessary to the geometrical coherence of an array.
(4) A significant degree of soft linkage is usual in fault arrays but the type of linkage is strongly scale-dependent.
(5) Fault arrays are kinematically coherent during growth and this implies a high degree of synchronous fault movement within an array.
(6) Sequential fault propagation interpretations are of doubtful validity.
(7) Deformation by faulting may involve faults with sizes spanning five orders of magnitude. Because graphical representations can show only two orders of magnitude of fault size, normal fault models must incorporate the effects of ductile extension.
(8) Variable displacement on fault surfaces is accommodated by heterogeneous ductile strain.
(9) Soft-domino models are characterised by ductile rotation of faults and beds at all stages of deformation and by the reduced significance of rigid block rotation compared with rigid domino models.
(10) The distinction between simple and pure shear deformation when applied to regional stretching is scale-dependent.

We thank Conrad Childs, John Gibson, Chris Lavers and other members of the Liverpool Fault Analysis Group for their considerable efforts in developing FAPS software, data handling and preparation of diagrams and for much useful discussion. Thanks are due to Britoil plc for access to data. Funding by the European Coal and Steel Commission and British Coal (contract 7220 AF825), by Britoil (contract 6050) and the University of Liverpool Research Development Fund is gratefuly acknowledged. Publication is with the permission of Britoil and British Coal but the views expressed are those of the authors and do not necessarily correspond with those of either British Coal or Britoil.

References

Bak, J., Sørensen, K., Korstgård, J., Nash, D. & Watterson, J. 1975. Tectonic implications of Precambrian shear belts in western Greenland. *Nature*, **254**, 556–559.

Barnett, J. A. M., Mortimer, J., Rippon, J. H., Walsh, J. J. & Watterson, J. 1987. Displacement geometry in the volume containing a single normal fault. *The American Association of Petroleum Geologists Bulletin*, **71**, 925–937.

Boyer, S. E. & Elliott, D. 1982. Thrust systems. *The American Association of Petroleum Geologists Bulletin*, **66**, 1196–1230.

Butler, R. W. H., 1982. The terminology of structures in thrust belts. *Journal of Structural Geology*, **4**, 239–245.

Childs, C., Lavers, C., Walsh, J. J. & Watterson, J. 1988. A new system for quantitative fault analysis from seismic reflection data (Abs). *Interne Skrifter Institutt for Geologi*, Universitet i Oslo.

——, Walsh, J. J. & Watterson, J. 1990. A method for estimation of the density of fault displacements below the limit of seismic resolution in reservoir formations. *In: North Sea Oil and Gas Reservoirs II*. Norwegian Institute of Technology, Trondheim. Graham & Trotman, London, 309–318.

Coward, M. P. 1976. Strain within ductile shear zones. *Tectonophysics*, **34**, 181–197.

Eisenstadt, G. & DePaor, D. G. 1987. Alternative model of thrust-fault propagation. *Geology*, **15**, 630–633.

Escher, A. & Watterson, J. 1974. Stretching fabrics, folds and crustal shortening. *Tectonophysics*, **22**, 223–231.

Gibbs, A. D. 1984. Structural evolution of extensional basin margins. *Journal of the Geological Society, London*, **141**, 609–620.

Gibson, J. R., Walsh, J. J. & Watterson, J. 1989. Modelling of bed contours and cross-sections adjacent to planar normal faults. *Journal of Structural Geology*, **11**, 317–328.

Grocott, J. 1977. The relationship between Precambrian shear belts and modern fault systems. *Journal of the Geological Society, London*, **133**, 257–262.

Kusznir, N. J. & Egan, S. S. 1990. Simple shear and pure shear models of extensional sedimentary basin formation: application to the Jeanne D'Arc Basin, Grand Banks of Newfoundland. *The American Association of Petroleum Geologists Memoirs*.

Larsen, P-H. 1988. Relay structures in a Lower Permian basement-involved extension system, East Greenland. *Journal of Structural Geology*, **10**, 3–8.

Pfiffner, A. P. 1985. Displacements along thrust faults. *Eclogae Geologicae Helvetiae*, **78**, 313–333.

Price, R. A. 1988. The mechanical paradox of large overthrusts. *Geological Society of America*

Bulletin, **100**, 1898–1908.

RAMSAY, J. G. & HUBER, M. I. 1987. *The techniques of modern structural geology. Vol. 2, Folds and fractures*. Academic, London.

RIPPON, J. H. 1985a, Contoured patterns of the throw and hade of normal faults in the Coal Measures (Westphalian) of north-east Derbyshire. *Proceedings of the Yorkshire Geological Society*, **45**, 147–161.

—— 1985b, New methods of forecasting the throw and hade of faults in some North Derbyshire Collieries. *Transactions of the Institute of Mining Engineers*, 198–204.

ROUX, W. F. 1979. *The development of growth fault structures*. American Association of Petroleum Geologists, Structural Geology School Course Notes.

SIBSON, R. H. 1977. Fault rocks and fault mechanisms. *Journal of the Geological Society, London*, **133**, 191–213.

TAYLOR, G. I. 1938. Plastic strain in metals. *Journal of the Institute of Metals*, **62**, 307–324.

WALSH, J. J. & WATTERSON, J. 1987. Distributions of cumulative displacement and seismic slip on a single normal fault surface. *Journal of Structural Geology*, **9**, 1039–1046.

—— & —— 1988a. Analysis of the relationship between displacements and dimensions of faults. *Journal of Structural Geology* **10**, 239–247.

—— & —— 1988b. Dips of normal faults in British Coal Measures and other sedimentary sequences. *Journal of the Geological Society, London*, **145**, 859–873.

—— & —— 1989. Displacement gradients on fault surfaces. *Journal of Structural Geology*, **11**, 307–316.

—— & —— 1990. New methods of fault projection for coalmine planning. *Proceedings of the Yorkshire Geological Society*, **48**.

WATTERSON, J. 1986. Fault dimensions, displacements and growth. *Pure and Applied Geophysics*, **124**, 365–373.

Ductile strain effects in the analysis of seismic interpretations of normal fault systems

J. J. WALSH, J. WATTERSON, C. CHILDS & A. NICOL
Fault Analysis Group, Department of Earth Sciences,
University of Liverpool, Brownlow Street, Liverpool L69 3BX, UK

Abstract: The limited vertical resolution of reflection seismic data results in sub-resolution structures accommodating appreciable ductile strains which should be taken into account in structural restoration and balancing and in analysis of fault displacements. Non-ductile strains are expressed as discontinuities, such as fault offsets. A fault map of an intensively mined coal-seam is used to demonstrate the role of structures which would not be detected by a typical offshore seismic survey. Four examples are given of interpretations of seismic datasets in which significant ductile strain can be demonstrated; (i) an array of normal faults in which significant extension is accommodated by sub-seismic faults, (ii) a relay zone in which ductile shear strain accommodates significant displacement and change of bed-lengths, (iii) an intersection zone between conjugate normal faults which is effectively a zone of ductile pure shear strain with horizontal extension, and (iv) a hanging-wall fold which accommodates appreciable ductile displacement and ductile bed extension. In all these cases the ductile strains and displacements can be estimated from the seismic interpretation. Unless ductile strains are incorporated in restorations bed lengths will not balance. Balancing nevertheless remains the prime method for detection of grossly invalid sections.

Interpretations of seismic data now provide the basis for a high proportion of geometric analyses of normal fault systems, but the systematic resolution limitations of seismic data have to be considered when applying analytical methods originally developed for other types of data, although resolution effects occur with all types of data. Bed-length and/or area balance are convenient geometrical assumptions used in the construction, validation and restoration of cross-sections. Although the role of section balancing varies with the type of exercise to which it is applied, its usefulness as a tool is always dependent on the basic premise that 'valid sections' must balance. Valid results for section balancing or restoration and fault displacement analysis based on seismic interpretations may require quantification and explicit incorporation of ductile strains associated with faults and arising from limited resolution, in addition to the discontinuous displacements on seismically imaged faults. We define ductile strain as a change in shape produced by structures which are too small to be imaged individually by a particular technique and/or too small to be represented individually on a given map or cross-section. Non-ductile strains are expressed by discontinuities, such as fault offsets. The ductile strain threshold varies with the scale of the map or cross-section and is therefore a scale dependent, or resolution dependent, property. The deformation processes which give rise to ductile strains include sub-resolution faulting and plastic deformation processes, including compaction, that do not produce geometrical discontinuities other than on the microscopic scale. Homogeneous ductile shear strains can be simply related to ductile displacements but displacements may be more difficult to estimate when the ductile shear is heterogeneous and gives rise to folds. The ductile strains that are necessary to accommodate displacement variations or changes of fault surface geometry on faults and that are an integral part of the faulting process are not considered here (Verrall 1982; White *et al.* 1986; Gibson *et al.* 1989; Anders & Schlische 1994). Five examples of structures in which ductile strain is either significant or in which it would be if the structures were to be imaged seismically are reviewed; two of the examples are more fully described elsewhere (Nicol *et al.* 1995; Watterson *et al.* in press). These examples are used to briefly consider how ductile strains may effect the veracity of section restoration and validation.

Regional ductile strain

All maps and cross-sections derived from seismic interpretations have resolution, or truncation, values below which faults are not explicitly represented (Walsh & Watterson 1991). Truncation value

From Holdsworth, R.E. & Turner, J.P. (compilers) *Extensional Tectonics: Faulting and Related Processes*. The Geological Society, Key Issues in Earth Sciences, **2 (2)**, 75–88.

First published in: "Buchanan, P.G. & Nieuwland, D.A. (eds) Modern Developments in Structural Interpretation, Validation and Modelling, The Geological Society, London, Special Publications **90**, 27–40".

is determined by the quality of the seismic data and its interpretation; good quality North Sea seismic data, for example, rarely permit interpretation of all fault throws down to *c.* 20 m on pre-Cretaceous horizons.

Sub-resolution fault populations, and their fault-related ductile strains, can be predicted in many cases because the fault size distributions within many fault systems are systematic (Childs *et al.* 1990; Marrett & Allmendinger 1991; Jackson & Sanderson 1992; Walsh & Watterson 1992), allowing the population in one size range to be inferred from data for a different size range (Yielding *et al.* 1992). Extensional tectonic fault systems show the highest degree of ordering with fault size distributions conforming to a power-law such that,

$$\log N = a - d\log(S)$$

or

$$N = \log(a) \, . \, S^d,$$

where S is fault size, measured as either displacement or dimension, N is the cumulative number of faults with size greater than or equal to a given value, $-d$ is the power-law exponent and a is a constant depending on sample size. d is often referred to as the fractal dimension of the size distribution. Higher values of fractal dimension indicate higher proportions of small faults relative to larger faults (Walsh *et al.* 1991). Even in systems that do not have such a simple scaling law the fault size distributions usually indicate substantial numbers of faults below seismic resolution limits. Only datasets measured from outcrops of finely laminated sequences include the smallest faults in a system.

Coalfield dataset

The ductile strain contribution of sub-resolution faults is illustrated by reference to a fault map of a 12 km^2 area in the South Yorkshire coalfield (Fig. 1). The map was constructed from 1:2500 seam abandonment plans for the Lidgett seam in North Gawber and Woolley Collieries. The faults are of Carboniferous age and form a near orthogonal system of NE and NW striking normal fault sets. These data are a sub-set of a larger dataset constructed for investigation of the scaling properties of faults (Watterson *et al.* in press).

Because the population systematics of different fault sets should, where possible, be measured separately (Childs *et al.* 1990; Gauthier & Lake 1993), we consider only the main fault set in the area, i.e. the NE striking normal faults. The fault population is measured using a line sampling method in which the throws are recorded for all faults intersecting an array of parallel sample lines (Walsh *et al.* 1994). The throw population curve for 7 sample lines oriented normal to fault strike shows three main segments (Fig. 2a): a steep right-hand segment with length corresponding to the number of sample lines, an approximately straight line central segment which merges with a left-hand segment of low slope at a throw value of *c.* 30 cm. This value is therefore taken as the truncation value so the range of throws fully represented is 30 cm to 30 m. The central straight segment of the population curve is representative of the size distribution of the fault set and its slope, *c.* –0.55, corresponds to a fractal dimension of 0.55. This value lies at the lower end of the range of fractal dimensions, 0.5–1.0, established from other studies (Yielding *et al.* 1992), indicating a lower proportion of small to large faults than in many tectonic fault systems.

The proportion of extensional strain accommodated by small faults is nevertheless still significant (Fig. 2b). For example, fault throws between 30 cm and 3 m, which represent half the valid range of throw of the dataset, accommodate *c.* 14% of the cumulative throw on the NE fault set. Therefore, with a truncation value of 3 m the ductile extension accommodated by sub-resolution faults with throws down to 30 cm would be *c.* 14% of the extension accommodated by fault throws between 30 cm and 30 m. The proportion of extension represented by sub-seismic faults is dependent on the minimum and maximum fault sizes which are taken into account (Fig. 2c). The smallest fault size in the systematic population for this coalfield dataset will be in the range 1 mm–10 cm. The largest fault size will increase with size of the area considered up to the maximum sized fault in the fault system. This point is illustrated by throw population data from an 8 km long sample line which extends beyond but includes the study area and on which the largest throw intersected is 110 m (Fig. 2c) as opposed to the 30 m maximum in the study area. The fractal dimension of this population curve is also relatively low (*c.* 0.4). If the throw resolution limit along this sample line were 20 m, i.e. a value typical of high quality offshore seismic data, sub-resolution faults could accommodate a ductile extensional strain representing as much as 15% of the total fault-related extension.

The significance of resolution effects on strain estimates is easily appreciated by comparison of fault maps produced by removing fault traces with throws below a specified truncation value. Figure 1 shows truncated derivatives of the primary fault map for 5 m, 10 m and 20 m throw truncations. Degradation of the data resolution reduces the apparent compartmentalisation. At 20 m truncation, only one fault remains and the proportion of

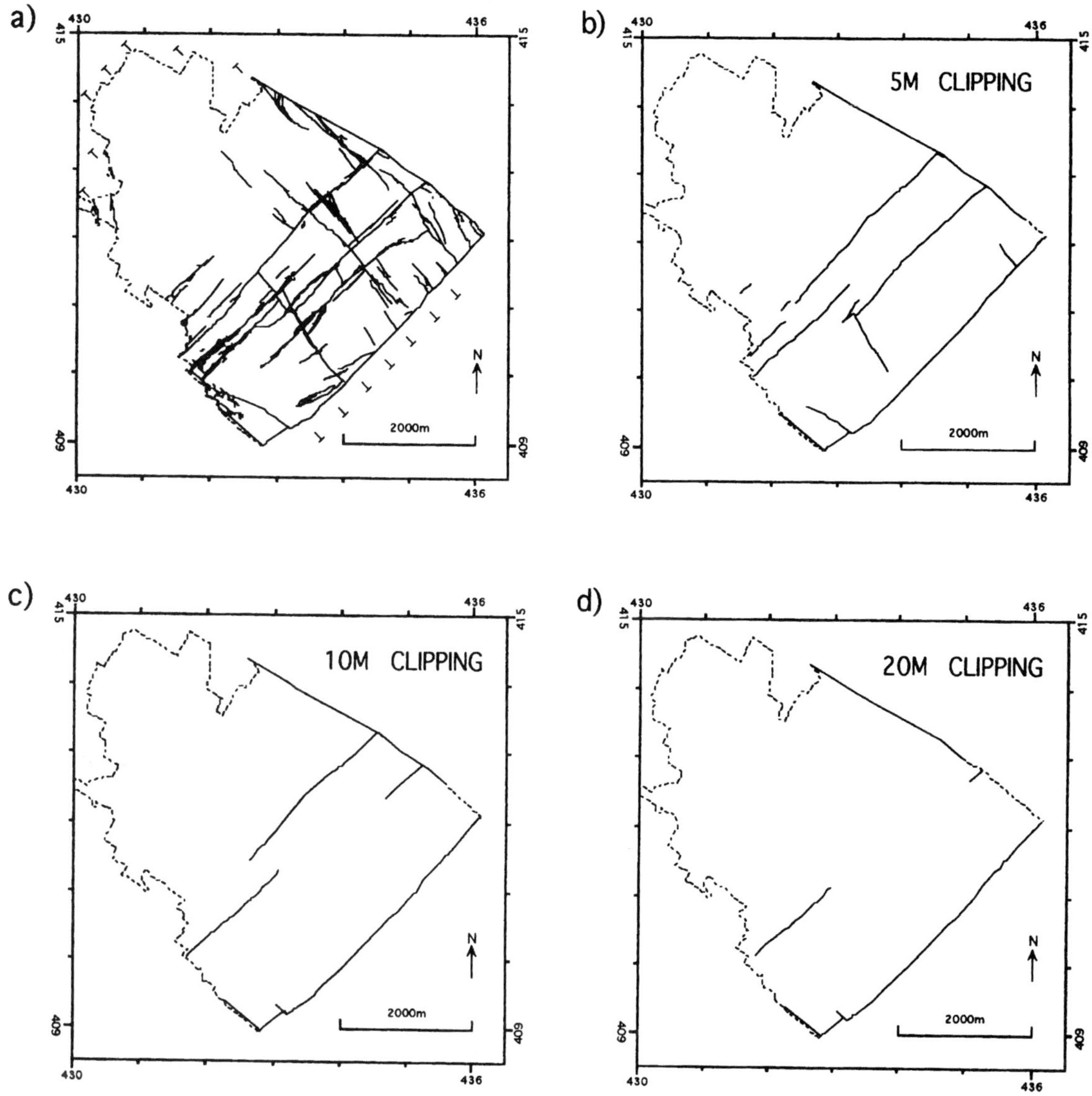

Fig. 1. Fault maps for the Lidgett seam in Woolley and North Gawber Collieries, South Yorkshire. All mapped faults are shown in (**a**), and maps showing only fault traces (or parts of fault traces) with throws greater than a given throw truncation value in (**b-d**): (**b**) 5 m, (**c**) 10 m, (**d**) 20 m. Faults are shown as solid lines and the boundary to the mapped area is defined either by the limits of the seam workings (broken line) or by faults with throws greater than 30 m (solid lines). The locations of NW-oriented sample lines (n = 7) for the multi-line throw population curve in Fig. 2a are shown in (a).

extension accommodated by sub-resolution faults (20 m to 30 cm fault throw) is *c*. 70%. Although such exercises reproduce, in a general way, fault maps appropriate to seismic data of different vertical resolutions, the effects of lateral seismic resolution are not incorporated. Limited lateral seismic resolution will effectively aggregate throws of closely spaced faults and therefore effectively image fault throws which would be sub-resolution if more widely spaced. The effect on Fig. 1b of incorporating lateral resolution would be slight and there would no effect on Figs 1c and d.

The conclusions to be drawn from the coal-mine data are that the vertical resolution limit of the data is a principal factor in determining ductile strain due to sub-resolution faults. With highly ordered size population systematics this ductile strain can be assessed and a reasonable estimate made even when the fault size distribution is non-fractal, as is sometimes the case for both gravity driven and tectonic fault systems (Gillespie *et al.* 1993). These estimates of ductile extension should be explicitly incorporated in restorations.

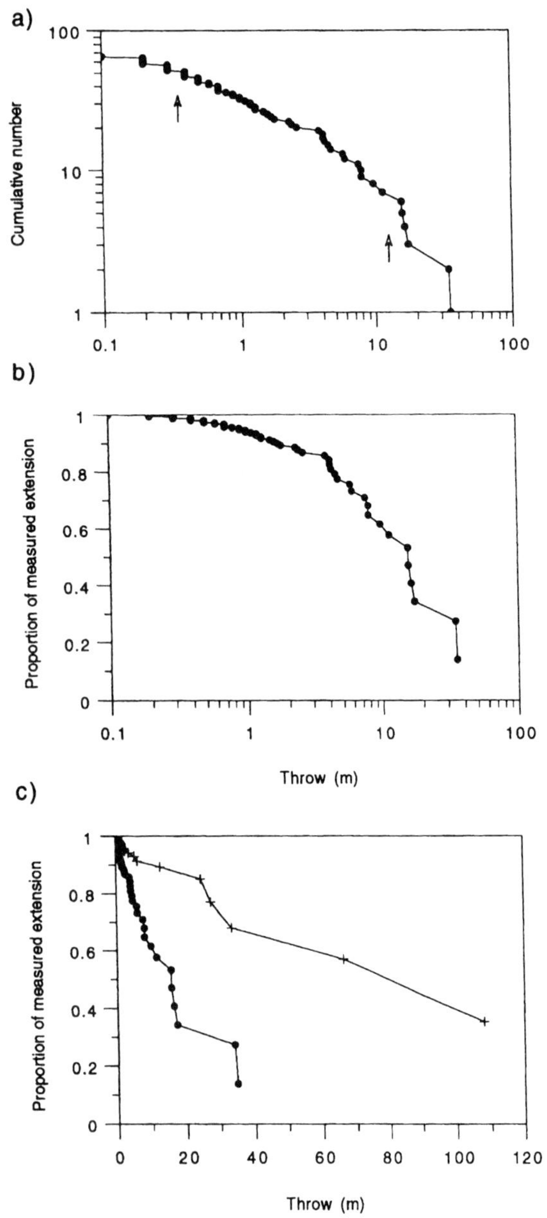

Fig. 2. Fault population curves from South Yorkshire coalfield. (**a**) Logarithmic plot of fault throw v. cumulative number for multi-line (n = 7) throw population of NE striking faults on the 12 km^2 fault map shown in Fig. 1. The limits of the central segment of the population curve are arrowed. (**b**) Throw v. proportion of measured extension for the fault data in (a). (**c**) Throw v. proportion of measured extension, for NE striking faults intersecting an 8 km sample line extending SE from the mapped area in Fig. 1. The curve (crosses) shows the contributions of faults of decreasing throw to the total extension accommodated by faults with throws between 30 cm and 110 m. The curve from (b) is shown for comparison.

Seismic dataset

Several studies have shown that fault populations of seismically imaged faults from offshore oil and gas fields are either fractal or otherwise highly systematic (Childs *et al.* 1990; Yielding *et al.* 1992; Gauthier & Lake 1993). These systematics provide the basis for determining the relative numbers of seismic and sub-seismic faults and for estimating the ductile strain contribution of the sub-seismic faults (Walsh & Watterson 1992). We use fault data from a 2D seismic dataset for a 900 km^2 area on the south-east edge of the Cartier Trough within the Timor Sea to illustrate this point (Figs 3 & 4). Seismic coverage consists of a rectangular array of seismic lines (32 NW–SE oriented dip-lines, with 0.5 km to 1 km spacing, and 11 strike-lines). The faults are assigned to 2 main phases of extension: Upper Jurassic–Base Cretaceous and Plio-Pleistocene (Woods 1992; Nicol *et al.* 1995). The most recent phase of extension is represented by a conjugate system of ENE–WSW striking normal faults (Fig. 4), some of which are reactivated Late Jurassic structures. Seismic data quality is very good within a strongly reflective Palaeocene–Recent sequence of shelf carbonates (Fig. 3), but deteriorates rapidly below the Top Cretaceous.

The contribution of faults of different sizes to fault-related Plio-Pleistocene extension is shown on a plot of fault throw versus cumulative β value (Fig. 5) using fault data from 3 pre-faulting horizons on a single sample line across the study area (Fig. 3). Seismically resolved faults show an approximately linear increase in β with decrease in the smallest included fault throw from 200 m to *c.* 20 m (Fig. 4). The decrease in curve slope at throws of *c.* 10–20 m marks the effective limit of resolution of the seismic data. Extrapolation of the straight-line segments (which span throws 20–200 m) of the curves to 1m throw shows that faults in the range 1–20 m, many of which are sub-seismic, may accommodate a ductile extension of 2–3%, representing *c.* 40% of the total extension (Fig. 5). The validity of linear extrapolation beyond the data remains untested as no quantitative well data are available. Inspection of the population curve (Fig. 5a) suggests however that even if a linear extrapolation is inappropriate, small faults (i.e. with displacements of 1–20 m) are likely to accommodate a significant component of the extension. Although this ductile extensional strain is a significant proportion of the whole, it will be represented by a thinning of the faulted sequence which is insignificant in comparison to likely compaction-induced thickness changes. The relative insignificance of fault-related thinning of the sequence will apply also to rifting events with larger beta values, as in the Late Jurassic

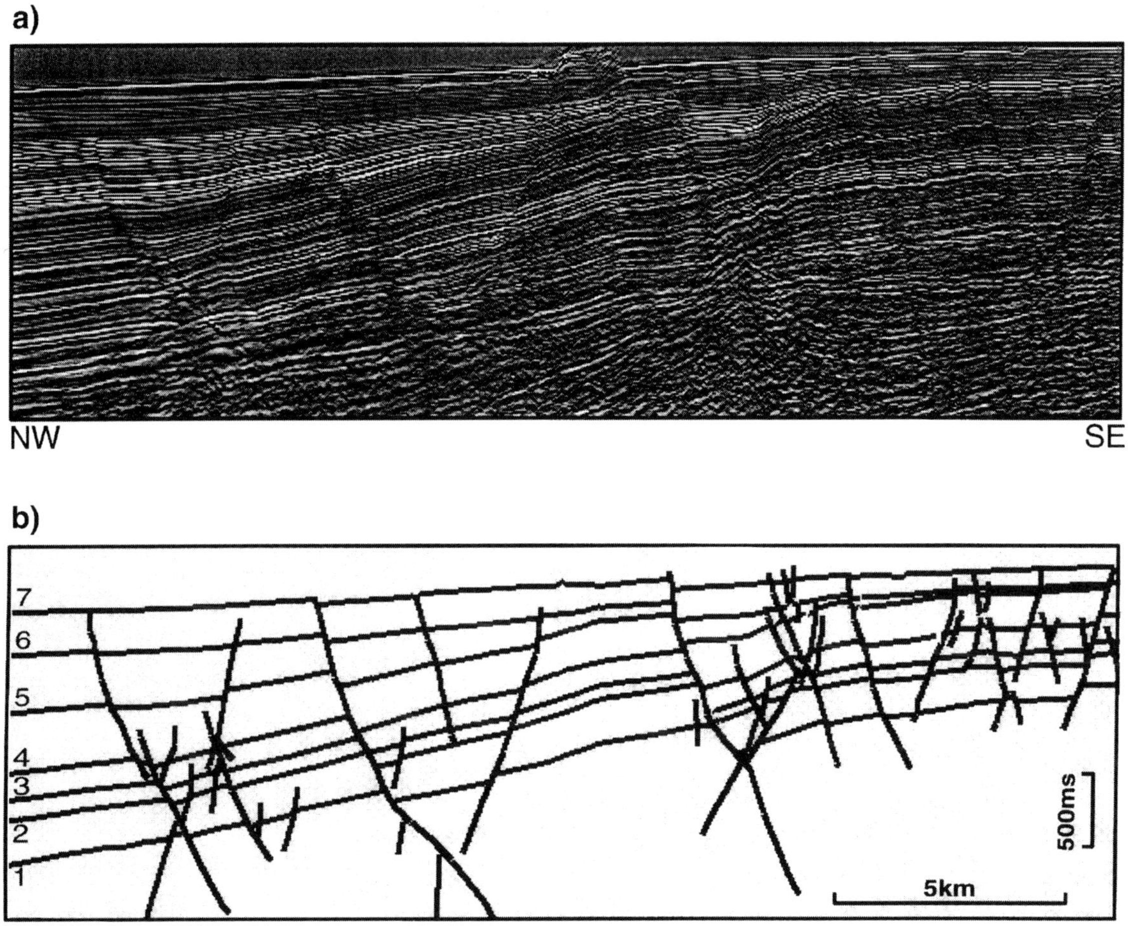

Fig. 3. Seismic (**a**) and interpretation (**b**) of 24 km section from the south-east edge of the Cartier Trough, Timor Sea. Interpreted horizons are: (1) Top Cretaceous, (2) Top Palaeocene, (3) Middle Eocene, (4) Base Miocene, (5) Top Miocene, (6) *c.* Middle Pliocene and (7) *c.* Lower Pleistocene. Vertical exaggeration is *c.* 2 (1 ms = *c.* 1.4 m). A carbonate mound is responsible for the seismic image distortion towards the centre of (a) (see Houland *et al.* 1994).

extension of the North Sea (*c.* β <1.2)(Roberts *et al.* 1993).

The resolution sensitivity of these fault data is illustrated graphically using fault maps with different throw truncations (Fig. 4). If this Timor Sea dataset had a throw resolution of 40 m, typical of poorer quality 2D and 3D North Sea seismic datasets, the numbers of imaged faults would decrease and those imaged would have shorter traces. For the seismic line shown in Fig. 3, a 40 m throw truncation would reduce the estimated β value from 1.035 to 1.027 unless account were taken of the ductile extensional strain. Seismically resolved faults in this dataset are not strongly clustered on scales relevant to lateral resolution effects (< 300 m), but some aggregation of small and large throws would be expected.

Restored bed lengths in datasets of this type will be inaccurate unless allowance is made for ductile extension. It might be argued that as the ductile extensions of all horizons is the same the true values are not required. If ductile strains are homogeneous then section restoration and validation may be valid. If however, the spatial distributions of sub-seismic faults, and therefore ductile strain, is heterogeneous, as is usually the case, then section restoration and validation will be subject to error. In addition, where some combination of pre-, syn- and post-faulting horizons is included in a restoration the ductile extension will be neither the same on every horizon nor be a constant proportion of the total extension on each horizon (Fig. 5b). The significance of the errors arising from these effects depends entirely on the use which is made of the results.

Localized ductile strain

Ductile strains accommodated by sub-seismic faults and/or plastic processes occur where there is

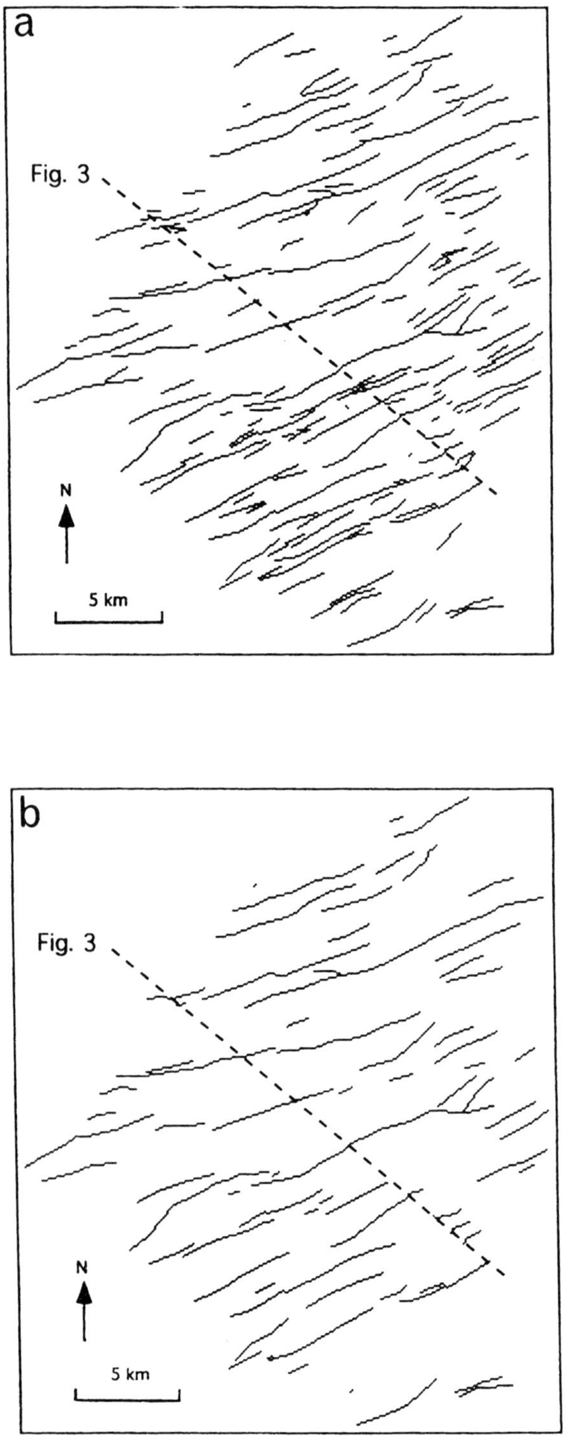

Fig. 4. Fault maps for a pre-faulting Base Miocene horizon (horizon 4, Fig. 3) in an area from the south-east edge of the Cartier Trough, Timor Sea. (**a**) All interpreted fault traces (i.e. all faults correlated on two or more seismic lines). (**b**) Fault traces with throws >40 m. Northwest oriented seismic lines are spaced at 0.5–1 km, and the location of the seismic line in Fig. 3 is shown (broken line).

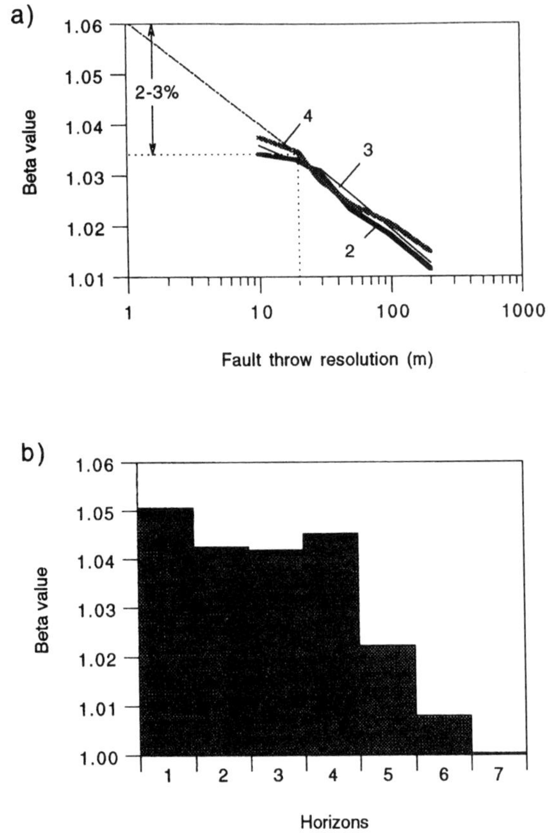

Fig. 5. (**a**) Fault throw resolution v. cumulative β value for three pre-faulting horizons (2, 3 and 4) on section shown in Fig. 3b. The seismic vertical resolution of *c.* 20 m is reflected in the shallowing of the left-hand segments of the curves. The cumulative β value at 1 m throw is indicated by the extrapolated straight segment of the data curve (broken line). The β value at 20 m throw is shown by the horizontal dotted line. (**b**) Histogram showing the extension accommodated by seismically imaged faults (stipple) and ductile strain at conjugate intersections (black) for seven horizons on the cross-section in Fig. 3b. Horizons 5–7 are syn-faulting and are less extended than the pre-faulting horizons.

interaction between adjacent coeval faults or where there is fault-related folding. Where the interacting faults have the same dip direction the commonest form of interaction is that associated with relay zones. Where the faults have opposed dips interaction is concentrated at conjugate intersections. Examples of ductile strains associated with these two circumstances are described below.

Relay zones

Relay zone geometries require high shear strains in the plane of the overlapping faults which are manifested as bed rotations and relay ramps (Fig. 6a). There may also be an element of shear strain in

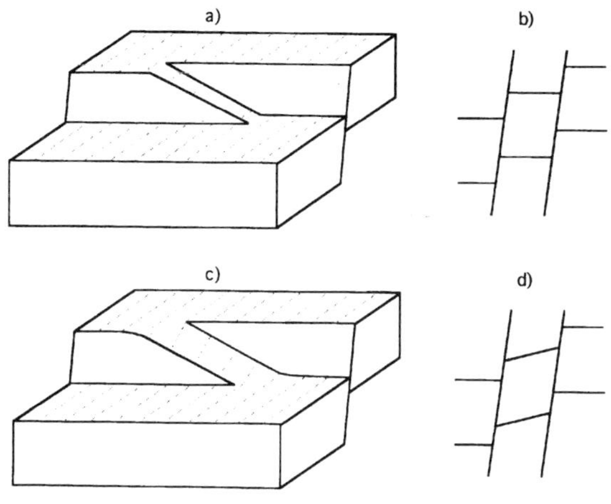

Fig. 6. Schematic block diagrams and cross-sections showing displacement and horizon geometries of relay zones without (**a** & **b**) and with (**c** & **d**) a component of shear displacement in the fault-normal plane within the relay.

the plane normal to the overlapping faults resulting in horizon dips towards the mutual hangingwall of the overlapping faults (Fig. 6c & d). Shear strain in the fault-normal plane is not a geometrically essential feature of relay zones but appears usually to be present (Peacock & Sanderson 1991, 1994; Huggins *et al.* in press). Fault-parallel and fault-normal shear strains together effect a non-plane strain deformation within a relay zone. However, it is shear strain in the fault-normal plane (Fig. 6c & d) which is of most significance in the context of section balancing and restoration, and in the analysis of fault displacements.

Figure 7a shows structural contours on a 1.5 km wide relay zone which provides lateral closure to a structural hydrocarbon trap lying to the west. The relay zone is bounded by two faults (A and B in Fig. 7a) with throws of up to *c.* 200 ms (≈ 300 m). Within the relay zone the rate of lateral change of throw on fault A is *c.* 5 times greater than elsewhere on this fault; no information is available on fault B beyond the ramp zone. Reorientation and tightening of the horizon depth contours within the relay give a contour azimuth 60° anti-clockwise of the regional ENE–WSW structural trend. On the cross-section in Fig. 7b the apparent dip of horizons within the relay zone is in the same direction as the fault dip. This direction of apparent dip is exceptional within this survey area where horizon apparent dips in the fault-normal plane are elsewhere opposed to the dip direction of the large faults.

The horizon geometry within the ramp zone is shown on the horizon separation diagram in Fig. 8a. The hangingwall cutoff of fault A and the

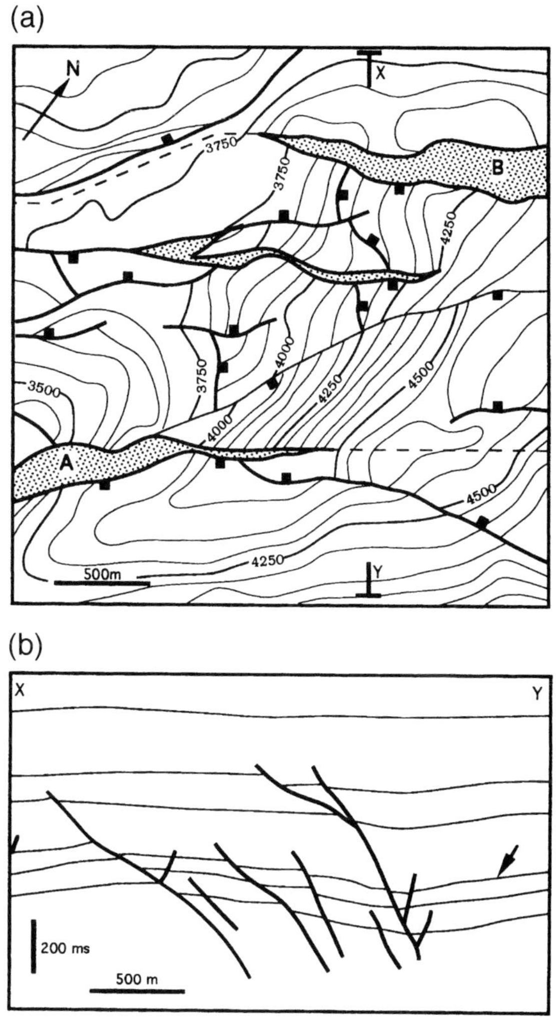

Fig. 7. (**a**) Structure contour map of a large relay zone from the North Sea: depth contours are in feet. Heave polygons of major faults are shown (stippled). The relay zone is bounded by faults A and B. Broken lines are the traces of cross-sections used to estimate differences in bed elevation across the relay zone (see Fig. 8b).
(**b**) Cross-section along X–Y (location shown in (a)). Mapped horizon in (a) is identified (arrow). The sequence above this horizon is syn- and post-faulting. Data are from seismic lines with 100m spacings oriented parallel to the vertical edge of (a).

footwall cutoff of B are both straight and approximately horizontal. Within the ramp zone the footwall cutoff of A and the hangingwall cutoff of B are parallel to one another and plunge *c.* 10° towards the east. Figure 8b shows the throw profiles for the two faults within the ramp zone. The profiles are projected horizontally onto a vertical plane parallel to the strike of the overlapping faults. As imaged, there is clearly very little actual overlap between these two faults and their aggregate throw profile has a central low (Fig. 8b).

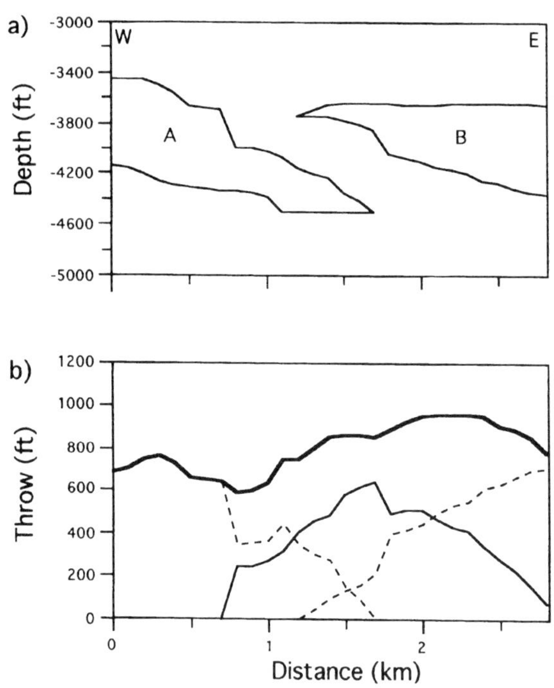

Fig. 8. (**a**) Horizon separation diagram for relay bounding faults A and B (see Fig. 7a), showing the traces of the mapped horizon cutoffs. Distance (km) is measured from the left-hand margin of the map in Fig. 7a and depths are in feet. (b) Throw profiles for (i) faults A and B (broken lines), (ii) elevation changes of horizon due to combined effect of throws on minor faults and bed rotations within the relay zone (light solid line; values represent a minimum estimate and are calculated from the depth difference between the two fault parallel sections shown in Fig. 7a), and (iii) the aggregate of curves (i) and (ii) (heavy solid line). Throws are in feet.

The horizon apparent dip towards the mutual hangingwall (Fig. 7b) is partly accommodated by small faults and partly by bed rotation. The apparent dip and its variation along strike can be estimated from differences in horizon elevations on lines bounding the relay zone and parallel to its margins. The map traces of the two sections are shown in Fig. 7a and the differences in horizon elevations on the two sections are shown in Fig. 8b. The depth difference between the two sections provides a minimum estimate of the combined effects of bed rotation and of throws on the minor faults, as in the profile in Fig. 8b. The imaged minor faults within the ramp are responsible for *c.* 40% of the difference in elevation across the relay zone. The bed rotation and minor faulting together largely compensate for the low aggregate throws on faults A and B, as shown by the more regular aggregate profile (Fig. 8b). Throw is therefore conserved across the ramp region but is partly accommodated by the ductile shear strain giving rise to the horizon rotation and increased bed length across the relay zone. Only after the significance and spatial distribution of this ductile strain is established can a 'valid' restoration of a given cross-section across this relay be performed using a method which explicitly incorporates the ductile strain (e.g. synthetic simple shear method; Rowan & Kligfield 1989). Displacement analysis of this structure must also take account of the displacements accommodated by the ductile shear strain.

Conjugate faults

Intersecting conjugate faults are common in areas where the fault density and the numbers of adjacent opposed-dipping faults are relatively high. Cross-sections of intersecting faults show conjugate, or hourglass, structures with opposed-dipping faults or fault arrays (Horsfield 1980; Woods 1992; Nicol *et al.* 1995). The notion of contemporaneous

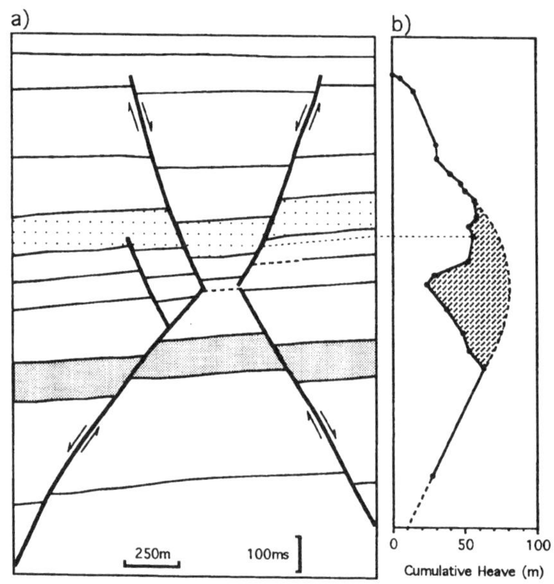

Fig. 9. (**a**) Line drawing of interpretation of a seismic section from the Timor Sea dataset, showing conjugate fault structure (see Nicol *et al.* 1995). The bounding faults cannot be traced across the intersection zone, because their displacements are believed to be accommodated by a more distributed network of sub-seismic faults. The horizons within the intersection zone are therefore continuous at the resolution of the seismic data (see text for details). The sequence above the light stippled unit is syn-faulting. Vertical and horizontal scales are approximately equal. (**b**) Aggregate heave profile for seismically imaged faults within the conjugate structure in (a). Heaves are aggregated along each horizon. Heavy broken line shows the extension (heave) profile expected for two faults with central displacement maxima, while the shaded area shows the inferred extension due to ductile deformation.

movements, over protracted geological time (millions of years), on mutually cross-cutting faults is difficult to envisage from a single cross-section of a conjugate structure (Fig. 9). However, where a conjugate structure is seen to be restricted to the lateral tip regions of the constituent faults with most of their surfaces not intersecting one another, contemporaneous movement is plausible (Fig. 10). Nicol *et al.* (1995) have shown that conjugate structures formed at overlapping tip regions of opposed-dipping faults of similar size, change polarity along the overlap region with an approximately symmetrical conjugate structure only towards the centre of the overlap (Fig. 10).

Significant interaction between faults and a structurally complex intersection zone are predictable consequences of synchronous movement on intersecting faults. Figure 9 shows a cross-section of a small conjugate structure within the highly reflective post-Base Cretaceous sequence of the Timor Sea (see Fig. 3). The bounding faults, which cannot be traced through the conjugate intersection zone, have displacements which decrease both upwards and downwards from the approximate centre of each fault trace (Fig. 9a). The aggregate heave profile (Fig. 9b) shows increasing heave upwards and downwards from the intersection zone but with a sharp decrease centred on thc intersection; the expected heave, or extension, profile is shown by the broken line in Fig. 9b. Nicol *et al.* (1995) have shown that the fault displacement variations are consistent with an increasing component of pure shear ductile strain towards the intersection zone. This ductile strain is expressed as a thinning and extension of the stratigraphic units between the conjugate faults by an amount approximately equivalent to the shortfall in aggregate extension on the observed heave profile (Fig. 9b). Although conjugate structures are common within the survey area the ductile strains associated with their intersections account for only a small proportion of the regional extension (Fig. 5b).

The decrease in imaged fault displacements towards the intersection zones of the conjugate structures is interpreted as being complemented by an increase in ductile extension arising from the changes in the fault slip geometries within the intersection zone (Nicol *et al.* 1995). Whereas the fault surfaces beyond the intersection zones

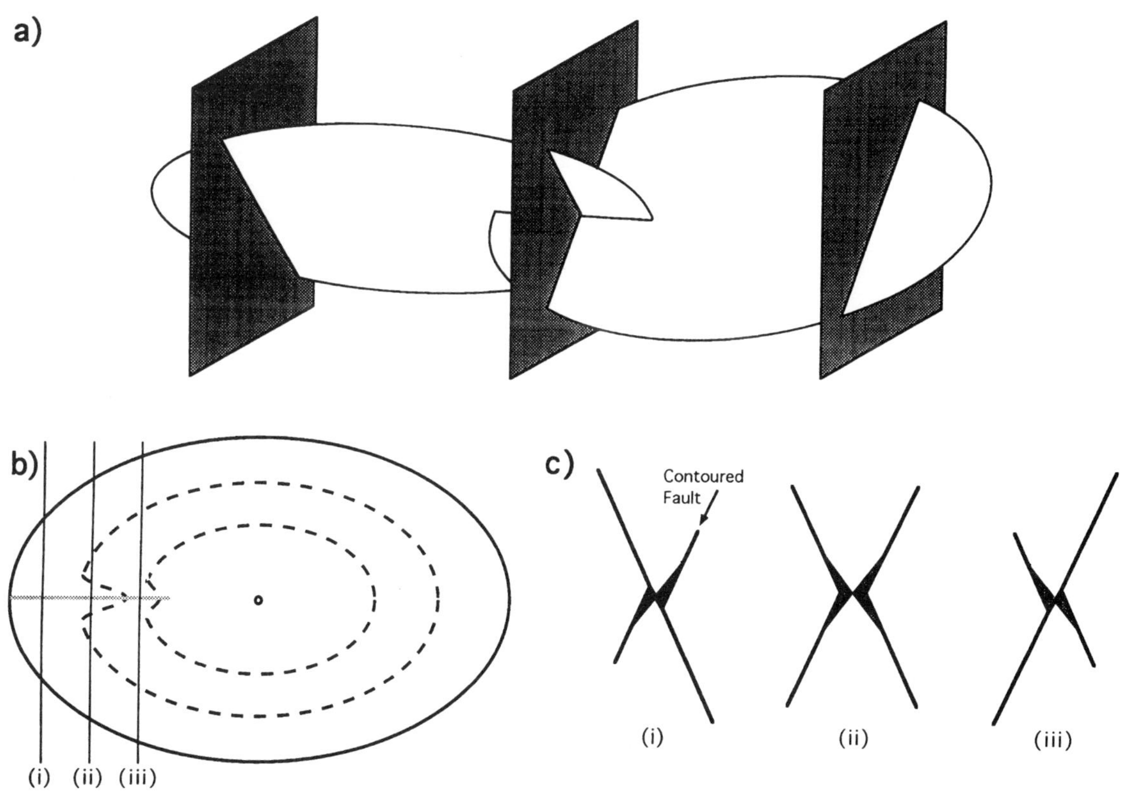

Fig. 10. (**a**) Schematic diagram showing a conjugate structure at the lateral overlap between ideal elliptical fault surfaces. (**b**) Throw contour pattern on right hand fault in (a), with regularity of throw contours (broken lines) modified by apparent decreases in throw towards the intersection line (stippled) of the two fault surfaces. (**c**) Cross-sections along lines (i), (ii) & (iii) in (b) showing symmetry. (a) and (b) modified from Nicol *et al.* (1995).

accumulated many increments of slip on essentially the same fault surfaces, which is the usual method of fault growth, the progressive offset of each fault by the other requires generation of numerous new slip surfaces to maintain continuity of each fault surface and related displacements. Thus within the intersection zones the cumulative fault displacements are believed to be high but accommodated by an array of sub-seismic structures which either splay from the bounding faults or form a more distributed network of minor faults. Deformation associated with this complex system of opposed-dipping sub-seismic faults is seen as a ductile strain. In these circumstances, bed length balancing will not be possible but, if no volume loss occurs, pre- and post-faulting areas will balance. Figure 11 shows a relatively simple outcrop example of the type of structure expected at the intersection of the seismically imaged conjugate faults.

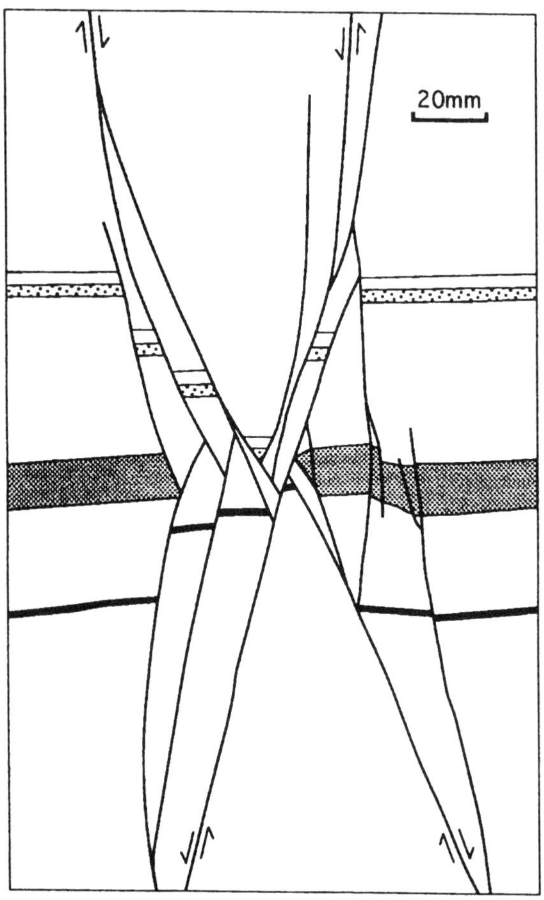

Fig. 11. Outcrop example of conjugate normal faults exposed in an outcrop of Pleistocene fluvio-glacial sands, Germany, showing the type of structure expected at the intersection of the seismically imaged conjugate faults. Drawing traced from a photograph provided by Dietmar Meier.

The concept of area balance associated with conjugate structures provides the basis for a simple geometric model (Fig. 12) which predicts heave variations on bounding faults similar to those described above and which could be used in restorations. The model assumes that the intersection point of conjugate faults bounding a graben lies at the top of a notional layer which is uniformly extended by ductile strain during extension. Area balance is maintained within the conjugate graben and initially horizontal horizons remain so during extension (Fig. 12a & b): the mutual hangingwall area extends by pure shear and no deformation occurs outwith the graben. Pre-faulting sequences are overlain by syn-faulting units which are modelled by assuming constant rates of both footwall (i.e. regional) sedimentation and fault slip: the effects of compaction are neglected. Calculated heave profiles are shown in Fig. 12c, for a model with parameters (extension, sequence thicknesses and fault dips) similar to those of the Timor Sea conjugates (Fig. 9; Nicol *et al.* 1995). Within the pre-faulting sequence, fault displacements decrease downwards from the surface to the intersection point (Fig. 12c), compensated for by an increase in ductile strain at the base of the mutual hangingwall wedge. With increasing extension, the extension accommodated by ductile strain increases. Heave profiles are similar to those of seismically imaged conjugates, mainly because the downward increase in ductile strain towards the intersection point is consistent with the notion that discrete bounding faults are replaced towards the intersection by an array of sub-seismic structures.

Bed lengths across conjugate structures do not balance, because the extension across each structure is accommodated by seismically imaged faults in some parts and by sub-seismic faults in other parts. However, if there is no volume change, area balance criteria can be used both for validating cross-sections and for estimating extensions (e.g. Rowan & Kligfield 1989).

Ductile folding

Distinction between different types of fault-related folds is critical to structural evaluation of fault systems. The fold types associated with normal faults include (see also Schlische in press) the following.

(i) Fault propagation folds which develop at fault tip-lines and are believed to be largely responsible for the presence of 'normal drag' adjacent to normal faults. 'Forced folds' within cover sequences above reactivated basement normal faults (Withjack *et al.* 1989) are an accentuated form of propagation fold. Normal drag folds

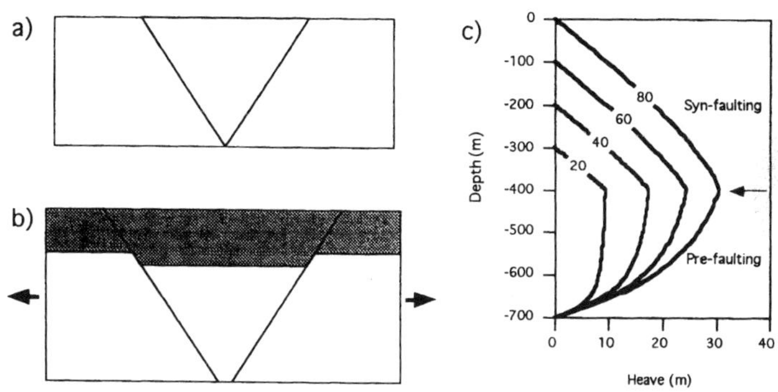

Fig. 12. Simple geometric model for conjugate faults. (**a**) Faults prior to extension. The intersection is at the top of a layer which is assumed to extend ductilely during extension. (**b**) Geometry after extension and sedimentation. The pre-faulting unit within the graben area balances: there is no deformation of the footwall blocks. The syn-faulting unit (stippled) infills the fault generated topography. (**c**) Curves of heave (m) vs. depth (m) for a model bounding fault in a model with parameters consistent with those of simple conjugates in the Timor Sea: fault dip = 75°, pre-faulting sequence thickness = 300 m, syn-faulting hangingwall sequence thickness = 400 m. Heave profiles for extensions of 20 m, 40 m, 60 m and 80 m are shown, based on a ratio of 1:10 between rate of footwall sedimentation and rate of extension.

accommodate a component of continuous displacement which is complementary and additional to the discontinuous component of displacement represented by the fault offset (Walsh & Watterson 1990). Faults which are surrounded by a cluster of numerous smaller sub-resolution synthetic faults may appear to be bordered by normal drag folds on seismic sections.

(ii) Reverse drag folds within the volume surrounding an individual fault accommodate reducing displacement with distance from a fault surface (Hamblin 1965; Barnett *et al.* 1987; Williams & Vann 1987). The wavelengths of 'reverse drag' folds are often an order of magnitude larger than 'normal drag' folds.

(iii) Compaction synclines in which the compaction of hangingwall sediments forms a syncline adjacent to the fault (White *et al.* 1986; Gibson *et al.* 1989).

Fault offsets and normal drag folds accommodate the discontinuous and the continuous, or ductile, components of fault-related displacement respectively (Walsh & Watterson 1990). The continuous component of displacement is no less significant than the discontinuous component for displacement analysis and for section balancing or restoration purposes. Indeed what is seen on one scale as folding may be faulting when seen on another scale. In this section we describe an example of changes in the proportions of discontinuous and continuous components of displacement which would not be apparent from examination of a single cross-section.

Figure 13 shows two cross-sections across a Middle–Upper Jurassic fault interpreted from a Northern North Sea 3D seismic dataset. Faulting initiated during the Brent with stratigraphic growth across the fault. Apart from some fault reactivation at the northern end of the mapped area, fault movement had largely ceased by the Base Cretaceous. Along the length of the fault, displacement is locally partitioned onto 3 sub-parallel, laterally discontinuous, synthetic hanging-wall faults (Fig. 14a). At the Base Brent these minor faults, which may branch from the main fault below the seismic data window, occur between the main fault and the axial trace of a hanging-wall syncline of variable amplitude (Fig. 13b & c). The main fault shows very irregular lateral variations in throw (Fig. 14b). Aggregation of throws on the main fault and the minor faults shows throw variations which are only slightly more regular (Fig. 14b): throws are aggregated along the regional extension direction, but aggregation along lines perpendicular to the local fault strike provides similar results. If the continuous throw component represented by the fold is included in the throw aggregation, the throw variations are regular (Fig. 14b): throws accommodated by folding are calculated by extrapolating the horizon trends beyond the fold hinge, up to the fault. Figure 13b shows that the proportion of displacement accommodated by folding can be as high as 50%.

This exercise suggests that the monocline is not primarily a result either of compaction or of fault surface irregularity at depth. The fold cannot be a fault-propagation structure, because the oldest syn-rift sequence does not increase in thickness towards the axis of the syncline. Instead, the systematic partitioning of displacements between faults and fold (i.e. discontinuous and continuous displacements), indicates that they are coeval and

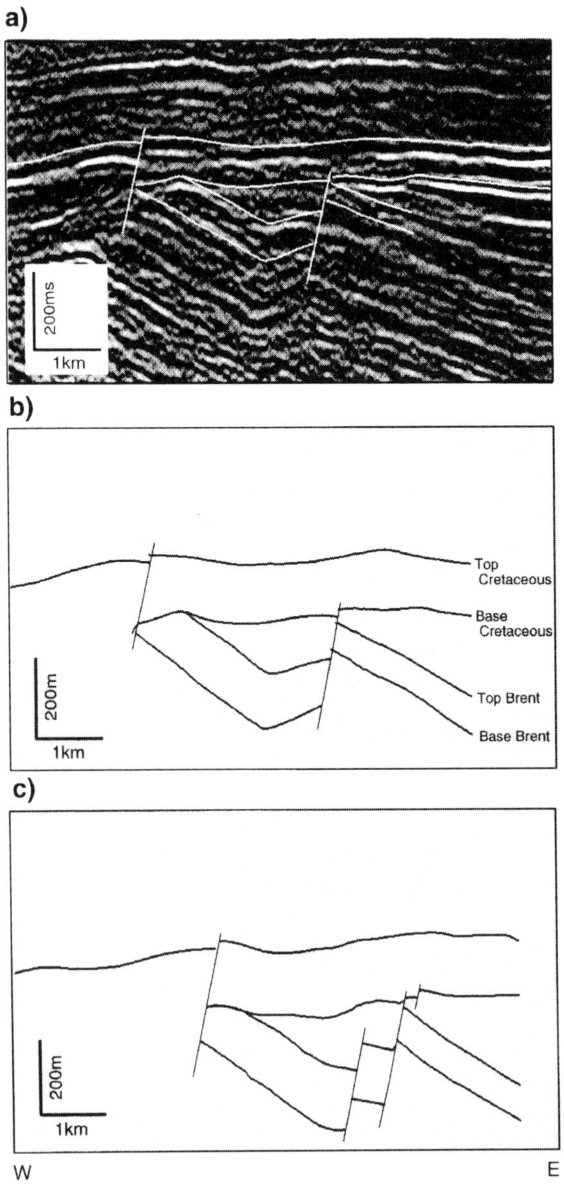

Fig. 13. (**a**) Seismic section across a fault from the Northern North Sea, showing four interpreted horizons (Base Brent, Top Brent, Base Cretaceous and Top Cretaceous). Brent sequence thickens in the hangingwall as a result of syn-faulting sedimentation. (**b**) Depth-converted interpretation of section in (a). Vertical exaggeration is *c*. 6. (**c**) Depth-converted interpretation of section *c*. 2.3 km along strike from (b). All sections are viewed from the south and Fig. 14a shows their locations.

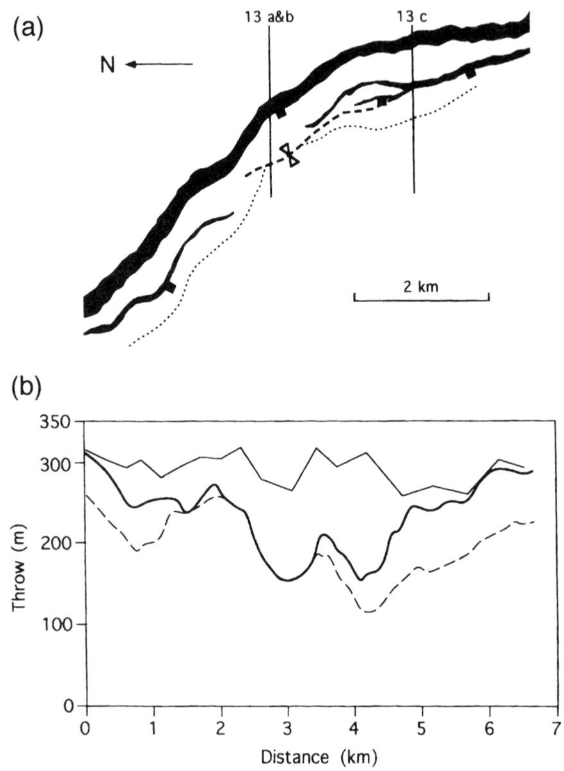

Fig. 14. (**a**) Fault map of the Base Brent horizon showing locations of sections in Fig. 13 which extend beyond the map area. The axial trace (heavy broken line) of a hangingwall syncline and the trace of the outer limit of the fault-related fold (light broken line) are also shown. (**b**) Throw profiles for the Base Brent horizon along the length of the fault array shown in (a): profile for main fault – broken line; aggregate profile for main fault and hangingwall minor faults – heavy solid line, aggregate throw profile for all faults and fold – light solid line. Errors on throw measurements are up to *c*. 20 m.

represent a kinematically coherent system. Whether or not the folding is accommodated by sub-seismic faults cannot be determined from the seismic data. Conclusions based on bed length restoration and or extension estimates based only on the discontinuous components of throw would depend on which cross-section(s) was analysed. Only by analysing the structure in 3D can the origin and displacement contribution of the fault-related fold be assessed.

Discussion

The restoration of depth-converted interpretations of seismic sections represents one of the principal methods used in the structural analysis of extensional basins. The technique provides a basis for both construction of geologically valid cross-sections and estimation of extensions. The two main approaches to section balancing, i.e. line length and area balancing, were originally applied to thrust belt tectonics (Bally *et al.* 1966; Dahlstrom 1969) and later modified for application to extensional tectonics (Verrall 1982; Gibbs 1983).

Explicit incorporation of ductile strain in balancing of contractional sections has sometimes been possible where outcrop data are available (Hossack 1979; Cooper & Trayner 1986). In extensional fault systems, the primary data generally comprise 2D or 3D seismic surveys, supplemented by well control, so the estimation of ductile strains requires different methods. In areas characterized by gravity-driven listric faulting (e.g. Gulf of Mexico and Niger Delta), various restoration techniques (incorporating fault shape and a variety of geometric constructions; see Williams & Vann 1987) can be used to provide estimates of the ductile strain accommodating hangingwall rollover geometries, but the choice of technique to apply can be difficult and often appears to be arbitrary. None of these techniques can be applied where faults are known not to be listric or where they accommodate regional extension or where 2nd order faults are not necessarily rollover accommodation structures. In this article we have described several examples of structures which come into this category, but for which ductile strain can be estimated by displacement and fault size population analysis techniques. The examples show that significant ductile strains occur on both regional and local scales and can be heterogeneous, varying both laterally and vertically. Line length balancing cannot be simply applied to sections within which there is heterogeneous stretching or shear. Where the ductile strain is homogeneous across a section, the section will line length balance without taking account of the ductile strain but extension estimates will be minimum values; the contribution of sub-resolution faulting should be estimated. Area balancing methods are more robust, particularly where a well constrained pre-deformation stratigraphic template with several mapped horizons is available. However, where the pre-deformation stratigraphy is poorly constrained, as is usually the case for syn-rift sequences, the results of area balancing are equivocal. Our examples suggest that validation or modification of seismic interpretations on the basis of conventional section restoration methods should be avoided. Even where several restoration techniques are utilized, selection of the most appropriate technique is subjective given the variety and complexity of possible fault-related ductile strains. In such circumstances, displacement analysis represents a supplementary, or alternative, method for testing the validity of 3D fault-related geometries.

Conclusions

(1) Ductile strains accommodated by sub-resolution faults can represent a significant proportion (*c.* 40–50%) of the total strain.

(2) The significance of fault-related ductile strain increases in the vicinity of interacting faults (e.g. intersecting conjugate faults and within relay zones).

(3) Displacements accommodated by ductile strain (i.e. continuous, as opposed to discontinuous displacements) can vary along the length of a fault system.

(4) The amount of fault-related ductile strain can be assessed from fault displacement and fault size population analysis. The coalfield map and other data suggest that prediction of sub-seismic fault numbers and throws is possible.

(5) While bed length and area balancing are useful methods for identification of grossly invalid sections, they should be supplemented by displacement and ductile strain analysis when used for more refined testing and reconstruction of sections.

The authors thank Ampolex Ltd, BP Exploration Operating Company and Norsk Hydro for providing the seismic data on which this work is based. They are grateful to Chris Bonson, Peter Bretan, Marie Eeles, Dan Ellis and Isabel Jones for preparation of diagrams and other members of the Fault Analysis Group for much useful discussion. They also wish to thank Dietmar Meier for providing the photograph from which Fig. 11 was traced. Reviews by Jack Filbrandt and Mark Rowan improved the paper. This research was part funded by Amoco (UK) Production Company (contract RAD 59 (90)), OSO/NERC Hydrocarbon Reservoirs LINK Programme (project 827/7053), NERC Petroleum Earth Sciences Programme (grant D1/G1/189/03) and E.C. JOULE II Hydrocarbons Programme (contracts JOU2-CT92-0182 and JOU2-CT92-0099).

References

ANDERS, M. H. & SCHLISCHE, R. W. 1994. Overlapping faults, intrabasin highs and the growth of normal faults. *Journal of Geology*, **102**, 165–180.

BALLY, A. W., GORDY, P L. & STEWART, G. A. 1966. Structure, seismic data, and orogenic evolution of southern Canadian Rocky Mountains. *Bulletin of Canadian Petroleum Geology,* **14,** 337–381.

BARNETT, J. A. M., MORTIMER, J., RIPPON, J., WALSH, J. J. & WATTERSON, J. 1987. Displacement geometry in the volume containing a single normal fault. *American Association of Petroleum Geologists Bulletin*, **71,** 925–937.

CHILDS, C., WALSH, J. J. & WATTERSON, J. 1990. A method for estimation of the density of fault displacements below the limits of seismic resolution in reservoir formations. *In*: BULLER, A. T., BERG, E., HJELMELAND, O., KLEPPE, J., TORSÆTER, O. and AASEN, J. O. (eds) *North Sea Oil and Gas*

Reservoirs II. Graham & Trotman, London, 309–318.
COOPER, M. A. & TRAYNER, P. M. 1986. Thrust-surface geometry: implications for thrust belt evolution and section balancing techniques. *Journal of Structural Geology,* **8,** 305–312.
DAHLSTROM, C. D. A. 1969. Balanced cross sections. *Canadian Journal of Earth Sciences,* **6,** 743–757.
GAUTHIER, B. D. M. & LAKE, S. D. 1993. Probabilistic modelling of faults below the limit of seismic resolution in the Pelican Field, North Sea, offshore U.K. *American Association of Petroleum Geologists Bulletin,* **77,** 761–771.
GIBBS, A. D. 1983. Balanced cross-section construction from seismic sections in areas of extensional tectonics. *Journal of Structural Geology,* **5,** 153–160.
GIBSON, J. R., WALSH, J. J. & WATTERSON, J. 1989. Modelling of bed contours and cross-sections adjacent to planar normal faults. *Journal of Structural Geology,* **11,** 317–328.
GILLESPIE, P. A., HOWARD, C. B., WALSH, J. J. & WATTERSON, J. 1993. Measurement and characterisation of spatial distributions of fractures. *Tectonophysics,* **226,** 113–141.
HAMBLIN, W. K. 1965. Origin of 'reverse drag' on the downthrown side of normal faults. *Bulletin of the Geological Society of America,* **76,** 1145–1164.
HORSFIELD, W. T. 1980. Contemporaneous movement along crossing conjugate normal faults. *Journal of Structural Geology,* **2,** 305–310.
HOSSACK, J. R. 1979. The use of balanced cross-sections in the calculation of orogenic contraction: A review. *Journal of the Geological Society, London,* **136,** 705–711.
HOULAND, M., CROKER, P. F. & MARTIN, M. 1994. Fault-associated seabed mounds (carbonate knolls?) off western Ireland and north-west Australia. *Marine and Petroleum Geology,* **11,** 232–246.
HUGGINS, P., WATTERSON, J., WALSH, J. J. & CHILDS, C. in press. Relay zone geometry and displacement transfer between normal faults recorded in coal-mine plans. *Journal of Structural Geology.*
JACKSON, P. & SANDERSON, D. J. 1992. Scaling of fault displacements from the Badajoz-Córdobe shear zone, SW Spain. *Tectonophysics,* **210,** 179–190.
MARRETT, R. & ALLMENDINGER, R. W. 1991. Estimates of strain due to brittle faulting: sampling of fault populations. *Journal of Structural Geology,* **13,** 735–738.
NICOL, A., WALSH, J. J., WATTERSON, J. & BRETAN, P. 1995. Three dimensional geometry and growth of conjugate normal faults. *Journal of Structural Geology,* **17,** 847–862.
PEACOCK, D. C. P. & SANDERSON, D. J. 1991. Displacements, segment linkage and relay ramps in normal fault zones. *Journal of Structural Geology,* **13,** 721–733.
—— & —— 1994. Geometry and development of relay ramps in normal fault systems. *American Association of Petroleum Geologists Bulletin,* **78,** 147–165.
ROBERTS, A.M., YIELDING, G., KUSZNIR, N.J., WALKER, I. & DORN-LOPEZ, D. 1993. Mesozoic extension in the North Sea: constraints from flexural backstripping, forward modelling and fault populations. *In*: PARKER, J. R. (ed.) *Petroleum Geology of Northwest Europe: Proceedings of the 4th Conference.* Geological Society, London, **2,** 1123–1136.
ROWAN, M. & KLIGFIELD, R. 1989. Cross section restoration and balancing as aid to seismic interpretation in extensional terranes. *American Association of Petroleum Geologists Bulletin,* **73,** 955–966.
SCHLISCHE, R.W. in press. Geometry and origin of fault-related folds in extensional settings. *American Association of Petroleum Geologists Bulletin.*
VERRALL, P. 1982. *Structural Interpretation with Applications to North Sea Problems.* Course Notes No. **3**. JAPEC, London.
WALSH, J. J. & WATTERSON, J. 1990. New methods of fault projection for coalmine planning. *Proceedings of the Yorkshire Geological Society,* **48,** 209–219.
—— & —— 1991. Geometric and kinematic coherence and scale effects in normal fault systems. *In*: A. M. ROBERTS, YIELDING, G. & FREEMAN, B. (eds) *The Geometry of Normal Faults.* Geological Society, London, Special Publication, **56,** 193–203.
—— & —— 1992. Populations of faults and fault displacements and their effects on estimates of fault-related regional extension. *Journal of Structural Geology,* **14,** 701–712.
——, —— & YIELDING, G. 1991. The importance of small-scale faulting in regional extension. *Nature,* **351,** 391–393.
——, —— & —— 1994. Determination and interpretation of fault size populations: procedures and problems. *In*: AASEN, J. O., BERG, E., BULLER, A. T., HJELMELAND, O., HOLT, R. M., KLEPPE, J. & TORSÆTER, O. (eds) *North Sea Oil and Gas Reservoirs III.* Kluwer Academic Publishers, London, 141–155.
WATTERSON, J., WALSH, J., GILLESPIE, P. A. & EASTON, S. in press. Scaling systematics of fault sizes on a large scale range fault map. *Journal of Structural Geology.*
WHITE, N. J., JACKSON, J. A. & MCKENZIE, D.P . 1986. The relationship between the geometry of normal faults and that of the sedimentary layers in their hangingwall. *Journal of Structural Geology,* **8,** 897–909.
WILLIAMS, G. & VANN, I. 1987. The geometry of listric normal faults and deformation in their hangingwalls. *Journal of Structural Geology,* **9,** 789–795.
WITHJACK, M. O., MEISLING, K. E. & RUSSELL, L. R. 1989. Forced folding and basement-detached normal faulting in the Haltenbanken area, offshore Norway. *In*: TANKARD, A. J. & BALKWILL, H. R. (eds) *Extensional Tectonics and Stratigraphy of the North Atlantic Margins.* AAPG Memoir, **46,** 567–575.
WOODS, E. P. 1992. Vulcan sub-basin fault styles – implications for hydrocarbon migration and entrapment. *APEA Journal,* **32,** 138–158.
YIELDING, G., WALSH, J. J. & WATTERSON, J. 1992. The prediction of small-scale faulting in reservoirs. *First Break,* **10,** 449–460.

Fault overlap zones within developing normal fault systems

C. CHILDS, J. WATTERSON & J. J. WALSH

Fault Analysis Group, Department of Earth Sciences, University of Liverpool, PO Box 147, Liverpool L69 3BX, UK

Abstract: Overlap zones between normal faults have been studied using a variety of 2D and 3D seismic reflection datasets. The overlaps are of two types, (i) relay zones in which displacement is transferred between the overlapping faults and (ii) non-relay overlaps in which displacement is not transferred.

Overlap zones are continually formed and destroyed during the growth of a fault system. Overlap zones are formed either by interference between initially isolated faults or as a result of bifurcation of a single fault. The mode of overlap formation is reflected in the 3D geometry of the overlapping faults which may be either unconnected or linked at a branch-line or branch-point.

Seismic reflection data from regions of growth faulting, and also sandbox analogue data, allow analysis of fault development through time. Reconstructions of the displacement distribution on some faults with sharp bends and associated hanging-wall splays, show that the bends originated as overlap zones which were later breached to form through-going faults.

Depending on the displacements of relay-bounding faults, the effect of relay zones on hydrocarbon reservoirs may be to (a) provide structural closure, (b) form gaps in otherwise sealing faults or (c) increase reservoir connectivity.

Keywords: normal faults, segmentation, kinematics, displacements, oil and gas fields.

Fault overlap zones, of which relay zones are one variety, are mostly temporary features of a fault system and, unless growth ceases, will inevitably be breached as displacement accumulates on the overlapping faults. Existing overlap zones are preserved when growth of the constituent faults ceases. The preserved geometry of an overlap or a breached overlap zone shows only one of the many stages in the evolution of that zone. The geometries of previous growth stages can be determined only indirectly, as can the geometries which might have developed had fault growth continued. The indirect methods include observations of overlap zones at different stages of their development (Peacock & Sanderson 1991) and displacement backstripping on individual overlap zones (Childs *et al.* 1993). Overlap zones occur and are preserved on a wide range of scales (Griffiths 1980; Larsen 1988; Peacock & Sanderson 1991, 1994; Stewart & Hancock 1991) but small overlaps cannot grow into large overlaps as the perpendicular distance between a pair of overlapping faults is fixed. It follows that not only are overlaps destroyed by breaching but that new overlaps must be generated during growth of a fault or fault system. It also follows from the widely accepted model in which faults lengthen during accumulation of displacement (Elliot 1976; Ranalli 1977; Watterson 1986; Marrett & Allmendinger 1991; Cowie & Scholtz 1992; Gillespie *et al.* 1992) that new overlap zones between fault pairs are continually generated during growth of a fault system. Ideally, therefore, a model for overlap zones should encompass their generation, propagation and kinematic development and should also incorporate an understanding of their scaling properties. As the 3D geometry of an overlap zone is complex, it cannot be determined by examination of a single map or cross-section. Progress towards development of a fault overlap model therefore requires analysis of good 3D data. These considerations provided the basis for this study of overlapping normal faults.

Seismic reflection survey data are used to describe three separate aspects of relay zones. The first of these is the 3D geometry of overlapping faults which are interpreted in terms of the likely causes of overlap formation. Secondly, the geometries of horizon deformations associated with overlap zones are described. Thirdly, some examples of branching faults are demonstrated to have formed by relay zone breaching, and the fault geometries and displacement distributions which have resulted from breaching are described. Finally, a fault system from the North Sea is shown to have developed by a cycle of overlap zone formation and breaching, followed by formation of a second generation of overlap zones.

Terminology of overlap zones

Overlapping normal faults are defined in 2D as two sub-parallel fault traces arranged en echelon where both the overlapping lengths of the fault traces and the distance separating the traces are small relative to the full lengths of the constituent fault traces. Where the overlapping faults dip in the same direction (the 'synthetic overlapping' faults of Morley *et al.* 1990), the map view is a 2D cut through a 3D overlap structure; the volume of rock between the overlapping fault surfaces is the overlap zone. Where faults have opposed dips, the relative position of mapped fault traces changes according to the elevation of the mapped horizon and only at certain elevations does an overlap exist. We describe only synthetic overlap structures.

Peacock & Sanderson (1991) classified fault offsets (overlaps) into (i) displacement-parallel, in which the tip-lines of the overlapping faults parallel the slip direction, (ii) displacement-normal, in which the tip-lines are

From Holdsworth, R.E. & Turner, J.P. (compilers) *Extensional Tectonics: Faulting and Related Processes*. The Geological Society, Key Issues in Earth Sciences, **2 (2)**, 89–103.

First published in: "*Journal of the Geological Society*, **152**, 535–549".

perpendicular to the slip direction and (iii) displacement-oblique. In the case of normal faults, displacement-parallel and displacement-normal overlaps are best viewed on maps and on cross-sections, respectively. As the vertical extent of cross-section data is generally smaller than the lateral extent of map data, good data on displacement normal overlaps are less common. This study therefore concentrates on displacement parallel overlaps, i.e. those seen in map view.

Overlap zones may be subdivided into those which display displacement transfer and those which do not. Displacement transfer between two faults which overlap is identified by a decrease in displacement on one fault being matched by a complementary change on the other fault; within the transfer region displacement gradients on both faults are higher than elsewhere on the fault surfaces (Walsh & Watterson 1990; Peacock & Sanderson 1991). Faults characterized by displacement transfer are kinematically interdependent structures, the elastic strain fields of which overlapped during their growth causing the slip distribution on each to be influenced by the other. Kinematically independent overlapping faults did not interact either because they are too widely separated in space or because they grew at different times.

Kinematically interdependent faults may interact to the extent that the displacements on the individual fault surfaces may be aggregated to produce a displacement distribution which resembles that of a single fault. In this case the faults are referred to as geometrically coherent (Walsh & Watterson 1991).

Overlap zones which show displacement transfer have been variously referred to as transfer zones (Chadwick 1986), fault bridges (Ramsay & Huber 1987) and relay structures (Larsen 1988). We use the term relay zone to describe the rock volume between two overlapping faults which show displacement transfer. Displacement transfer across a relay zone causes bed rotation such that the orientation of bedding within the relay is significantly different from the regional orientation. We use the term relay ramp for the special case of a relay zone within a layered (bedded) sequence. Where the regional bedding is horizontal relay ramps are most easily identified as sites of increased bed dip.

Data

A study was undertaken of a number of interpreted 2D and 3D seismic reflection datasets, of varying quality and seismic resolution. As with all seismic interpretations there are uncertainties in measured throw values but the range of uncertainty is thought not to be critical to our conclusions.

Many, if not most, seismic interpretations are, to a greater or lesser extent, influenced by some preferred model of faulting (linked, listric, transfer etc.) and the preferred model varies between companies and between individuals. For the purpose of this study, the method of correlation of fault cuts from one shotline to another is of particular importance. Where there is a significant lateral offset in fault position between successive shotlines, one of three alternative interpretations can be made. The alternatives are (a) a fault bend, (b) a transfer fault, or (c) an overlap zone (Fig. 1). In order to reduce the effects of subjective interpretation, this study has concentrated mainly on faults having throws which are large relative to the seismic resolution and which, where possible, have overlap lengths which are large relative to the shotline spacing; the dangers of model driven interpretation and fault-cut correlation have therefore been minimized, but perhaps not excluded.

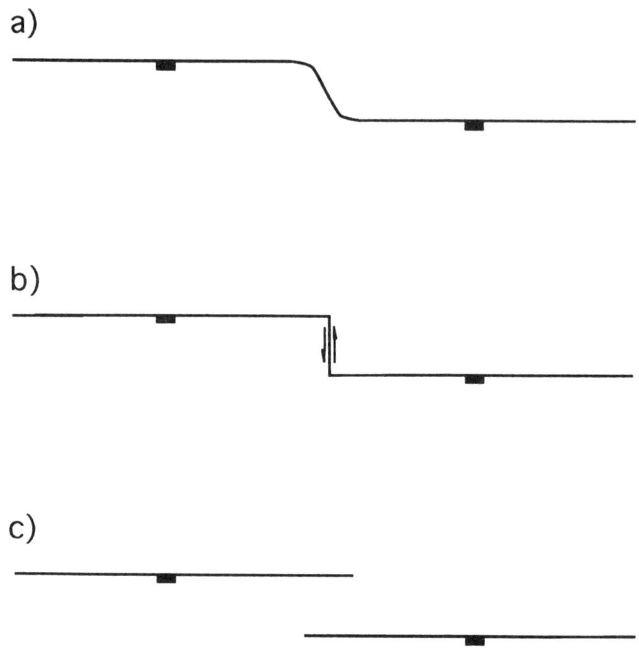

Fig. 1. Schematic diagram illustrating the possible structure at an offset in a fault trace; these are (**a**) a fault bend, (**b**) a transfer fault or (**c**) an overlap zone.

Additional data are presented from a sandbox analogue of faulting above a ductile layer; a detailed description of these data is given in Childs *et al.* (1993).

Overlap formation

Synthetic overlap zones as seen in 2D, i.e. in plan view or cross-section, may have one of three geometrical relationships when viewed in 3D (Fig. 2). The overlapping fault segments may be either:

(a) unconnected in 3D, or
(b) linked in the third dimension along branch-lines, or
(c) linked in the third dimension at branch-points.

The mode of initiation of the overlap zone is expected to be different in each case. Examples of each of these 3D relationships have been observed within the datasets examined, and in some cases, the modes of formation of the overlap zones can be established.

Unconnected segments

In general, it is more difficult to establish that faults are unconnected in 3D than to show that they are linked. To demonstrate that a fault does not intersect another requires the entire fault surface to be imaged, and this condition is rarely satisfied within seismic datasets.

Figure 3 shows an example of an overlap between two faults which are not connected and where one fault is contained entirely within the data volume. The traces of the two faults are approximately parallel (Fig. 3a) and the perpendicular distance between the two faults (the overlap separation) is approximately constant. The throw variations for these faults are represented on a strike-projection of

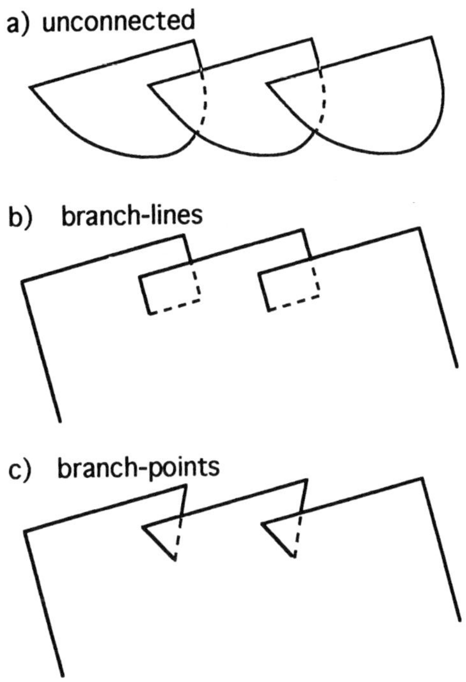

Fig. 2. Schematic diagram of the three possible structures underlying a segmented fault zone as seen in map view. The individual fault segments may be, (**a**) unconnected in three dimensions, (**b**) linked at depth along branch-lines or (**c**) linked at depth at branch-points.

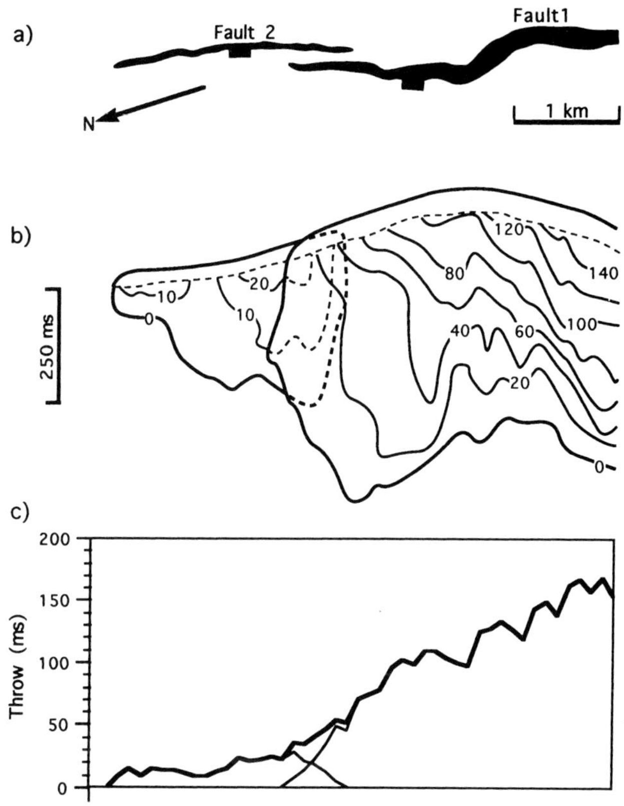

Fig. 3. (**a**) Map showing the traces of two synthetic overlapping faults from the Central North Sea; the faults are unconnected in 3D. The map is derived from a seismic survey with shotlines oriented perpendicular to the fault traces and spaced every 100 m. (**b**) Vertically exaggerated (×5) throw strike-projection of the two faults mapped in (a). The throw contours (ms), are derived from *c.* 200 datapoints: note that the contour interval is different for the two faults and that contours on the smaller fault are shown by broken lines where this fault lies behind the larger fault. The sub-horizontal light broken line shows the trace on the fault surfaces of the horizon mapped in (a). Heavy lines indicate the fault tip-lines; the tip-line of the smaller fault is shown as a broken line where it lies behind the larger fault. (**c**) Throw versus distance profiles for the two faults shown in (a). Throws are measured on the horizon which is mapped in (a) and indicated by the broken line in (b). The bold line shows the aggregate throw for the two faults.

throw (Fig. 3b) and a horizontal profile of throw (Fig. 3c). A throw strike-projection is constructed by horizontally projecting all the available throw data for a fault(s) surface onto a vertical plane parallel to the mean fault strike; the projected data points are then contoured for throw (Barnett *et al.* 1987; Badley *et al.* 1990). A throw profile shows the variation of throw of a horizon along the length of a fault trace. In Fig. 3 the strike-projection and the throw profile both show that displacement gradients within the overlap zone are higher than those outside the overlap and that displacement is transferred between the faults, i.e. they are separated by a relay zone. Furthermore, the aggregate throw profile for these faults has a regular form, resembling the throw profile for a single fault and this fault pair can be said to be geometrically coherent (Walsh & Watterson 1991; Roberts & Yielding 1994).

There are two ways in which overlap (and relay) zones between unconnected fault surfaces may develop (Fig. 4). In the first case (Fig. 4a) the individual faults may initiate as kinematically independent structures which overlap as they grow and increase in size. As fault growth continues, there is interaction between the elastic strain fields of the individual faults and they are no longer kinematically isolated. Such faults may either have nucleated at 'random' positions and have overlapped 'accidentally' or there may be some underlying mechanical control on their points of nucleation, e.g. along a basement shear zone or fault. Alternatively (Fig. 4b), overlap zones between unconnected fault surfaces may occur between faults which have been kinematically interdependent throughout their growth. This relationship requires that a growing single fault surface stepped out of plane as it propagated laterally, or with a significant lateral component, to form an unconnected fault segment at the propagating tip. This second model requires that the overlapping faults are geometrically coherent at all stages; this is not a requirement of the first model. Distinction between the two types can be uncertain as no fault propagates into a totally unfaulted rock volume and the distinction has meaning only when finite displacements are considered.

In the example described above (Fig. 3), there is no direct kinematic evidence concerning the mode of formation of the overlap zone. There is, however, indirect evidence to suggest that Fault 2 developed at the propagating tip of Fault 1. These faults are part of a fault system which developed in response to salt diapirism, with Fault 1 the largest fault within the array. The crest of the salt diapir lies to the south of the mapped region (Fig. 3a) and broadly coincides with the point of maximum displacement on Fault

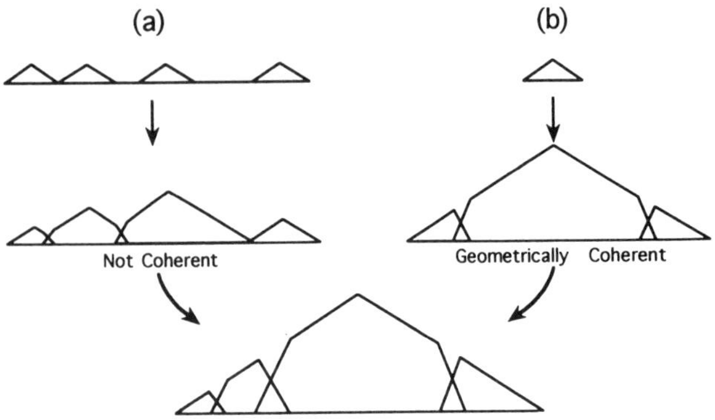

Fig. 4. Schematic horizontal throw profiles of four geometrically coherent fault traces illustrating the two ways in which such a fault system may develop. The individual faults may have developed in isolation and propagate towards one another, as in (a), or may develop sequentially and maintain geometric coherence throughout their growth, as in (b).

1. The fault array is believed to have propagated northwards in response to continued evolution of the salt diapir. This conclusion together with the geometric coherence between Faults 1 and 2 suggests that this overlap zone originated by a right-hand step in the northward propagation of Fault 1.

Linkage at a branch-line

Figures 5a–b show maps of a fault array on two horizons; the vertical interval between the two horizons varies between 300 and 400 ms (*c.* 375–500 m). The upper horizon (Horizon 2, Fig. 5a) is within the syn-rift sequence and shows a zone of overlap between two fault traces. The lower horizon (Horizon 1) marks the top of the pre-rift sequence. Horizon 1 shows a large fault (stippled in Fig. 5b) with a marked bend and an associated hanging-wall splay. A cross-section through these faults (Fig. 5c) shows that the hanging-wall splay does not intersect Horizon 2 and that the two overlapping faults on Horizon 2 link downwards at a branch-point. This branch-point can be traced on successive cross-sections as a sub-horizontal branch-line which is the same length as the overlap on Horizon 2. The overlap zone on Horizon 2 is clearly the result of upward propagation of the curved fault trace in the pre-rift sequence (Horizon 1). The fault bifurcated upwards, along a branch-line, into two fault surfaces such that the resulting splays were re-aligned parallel to the overall fault trend and perpendicular to the extension direction.

The aggregate throw profile for the overlapping faults on Horizon 2 (Fig. 5d) shows the two overlapping fault segments to be geometrically coherent. Outside the overlap zone there is no lateral variation in throw on either of the fault segments on Horizon 2 and the throw profile for each fault is flat topped. Within the overlap zone the lateral throw gradients increase. There is clear displacement transfer between the two splays which therefore bound a relay zone.

Linkage at a branch-point

The linking of laterally offset fault segments at a branch-point is difficult to establish from seismic data because of the restricted lateral resolution imposed by the

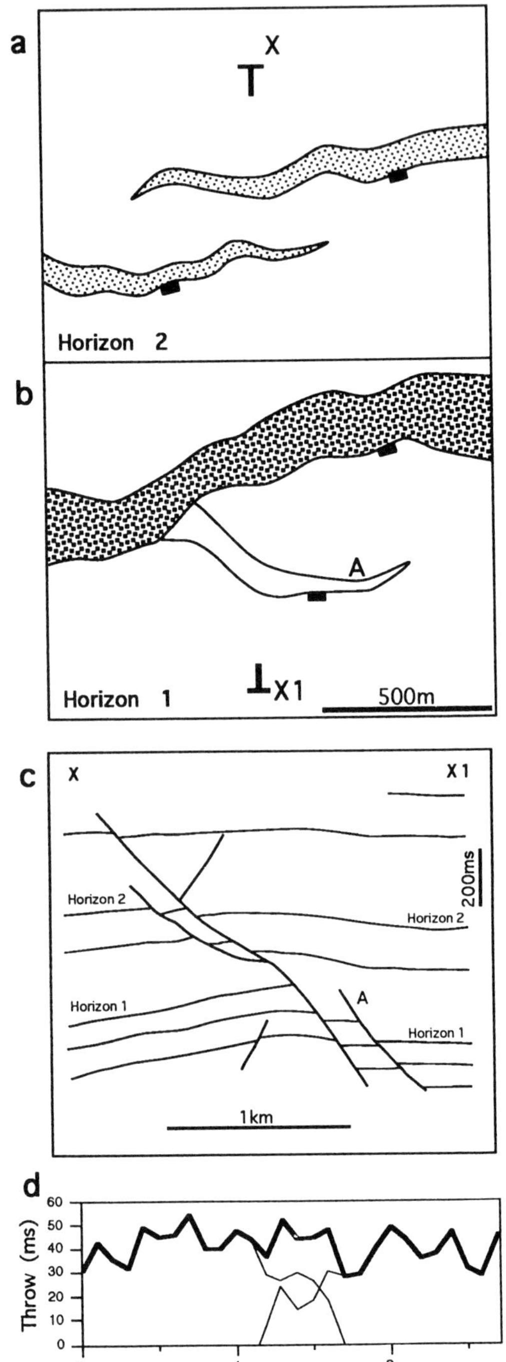

Fig. 5. (**a**) and (**b**) Two map views of the same fault array from the Northern North Sea. Horizon 2 is within the syn-rift sequence and Horizon 1 is at the top of the pre-rift sequence. Overlapping normal faults on Horizon 2 link at depth along a horizontal branch-line and connect to the large fault on Horizon 1 (stippled). Fault A on Horizon 1 does not intersect Horizon 2. The maps are constructed from seismic shotlines spaced every 100 m and oriented parallel to the line of section X–X1. (**c**) On a cross-section (X–X1) through the fault array shown in (a) and (b), the branch-line between the two overlapping faults on Horizon 2 is seen as a branch-point, near to the centre of the cross-section. (**d**) Throw profiles of the two overlapping faults on Horizon 2 shown in (a). The bold line is the aggregate throw for the two faults.

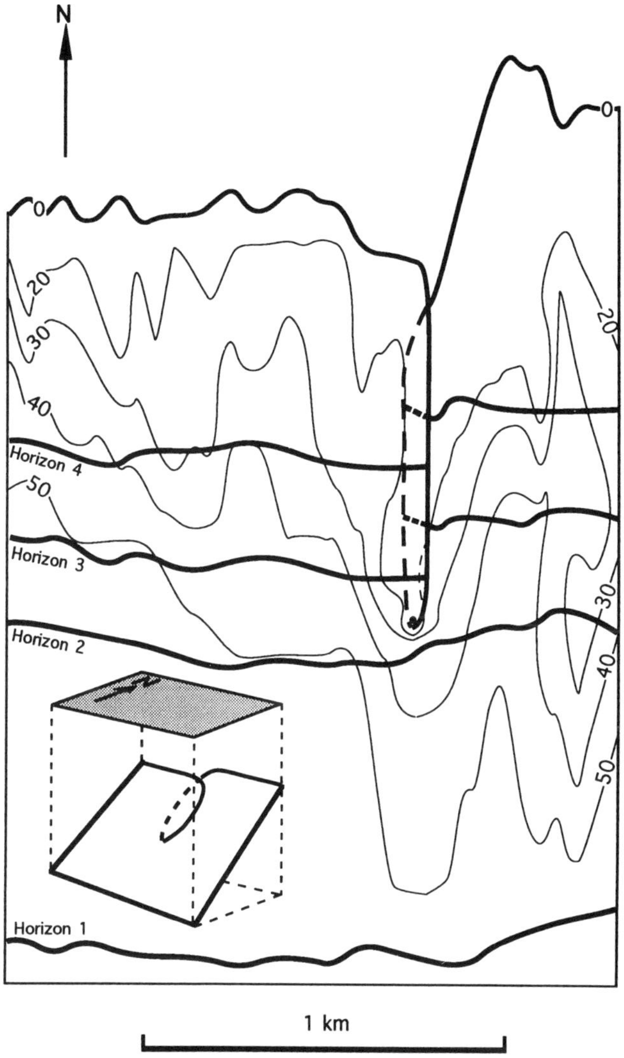

Fig. 6. A horizontal projection, with throw contours, of a fault surface in the Gulf of Mexico which divides upwards at a branch-point into two overlapping fault surfaces. The diagram is constructed by projecting the throw data and the traces of four horizons which intersect the fault surface vertically onto a horizontal plane (see inset); each horizon trace is equivalent to the map trace of the fault on that horizon. The data are from a seismic survey with shotlines oriented N–S and spaced every 50 m. The mean fault dip is 50° to the south, so that successively older and deeper horizons lie towards the bottom of the diagram. Throw contour values are in ms and the contours include the zero throw contour or tip-line (bold line). The 10 ms contour is omitted for clarity. Throw contours are based on 230 throw values derived from offsets of Horizons 1–4 and two other horizons which have been omitted. The vertical interval between the upper top tip-line and the lowermost mapped horizon is 2 s. The lateral offset of the upper tip-line across the overlap is due mainly to a change in fault dip across the overlap zone, but is also due partly to intersection with a N–S-trending, E-dipping, normal fault (not shown) which offsets the eastern segment of the overlap fault.

seismic shotline spacing. Figure 6 illustrates an example in which a single fault at depth bifurcates upwards into two overlapping fault segments. Bifurcation is believed to have occurred at a branch-point. The four lowermost horizons (Horizons 1 to 4) of the six intersected by this fault are shown in Fig. 6; on the lowest two (Horizons 1 and 2) the fault has a single continuous trace, but on the upper two (Horizons 3 to 4) there is a clearly defined 150 m sinistral offset of the fault trace with two overlapping fault lobes.

The overlapping fault lobes are both seen on two shotlines, on which they are interpreted as parallel fault traces. On a cross-section immediately to the west of the branch-point the hanging-wall lobe links downwards with the main fault where it offsets Horizons 1 and 2, but on a cross-section immediately to the east of the branch-point the footwall lobe links with the main fault. Therefore, within the limitations of the data resolution (50 m shotline spacing) fault bifurcation occurs at a point. The precise horizontal dimension of the overlap zone is difficult to establish. The occurrence of both fault lobes on two shotlines demonstrates the existence of an overlap. The overlap length may be less than 150 m (i.e. 3 shotline intervals) but it is likely to be larger if the faults extend beyond their interpreted tip-lines with throws too small to be resolved seismically. For these data the lower limit of seismically resolvable throw is *c.* 5 ms or 7 m.

Away from the overlap zone changes in displacement are gradual and the throw contours are widely spaced. The throws on Horizon 1 are approximately constant beneath the overlap zone, consistent with a single through-going fault surface at depth. Around the branch-point the throw pattern is complex, with throw gradients up to 0.25 and possibly higher immediately adjacent to the branch-point. On the two horizons above the branch-point high lateral throw gradients are not confined to the overlap zone but extend for more than 200 m on either side.

This fault occurs within a syn-sedimentary fault system. Within this system individual faults became active and died at different times. Growth indices suggest that the fault in question became active only after the deposition of Horizon 3. The branch-point between the two fault segments lies beneath Horizon 3, i.e. within the locally pre-rift sequence. The branch-point acted as a barrier or 'stick-point' to propagation of the fault tip-line, leading to high displacement gradients in the vicinity of the 'stick-point'. These high gradients are assumed to be accommodated by movement on sub-seismic faults. Above Horizon 3, i.e. within the syn-faulting sequence, the tip-line propagated upwards but the propagation direction within the pre-rift sequence cannot be determined directly. Huggins *et al.* (in press) suggest that in overlap zones resulting from bifurcation at a branch-point, lateral tip-lines of the overlapping lobes are approximately parallel to the propagation direction. If so the fault shown in Fig. 6 propagated upwards also within the pre-rift sequence.

An example of a laterally segmented normal fault, dextrally offset, which is believed to link downwards at a series of branch-points is shown in Fig. 7a. The mapped traces show little or no overlap, but it is likely that their lengths are underestimated by the seismic data (seismic resolution <10 ms or *c.* 12 m) and that they extend beyond the tip-points shown in Fig. 7a. While the lateral tip-points to fault traces cannot be mapped accurately, the shapes of the mappable tip-lines are assumed to reflect those of the actual tip-lines. The interpreted lateral tip-lines of each segment are approximately parallel to the fault dip direction and are seen as vertical lines on a strike-projection (Fig. 7b). Throws on each segment are approximately constant over most of the fault surface and significant displacement

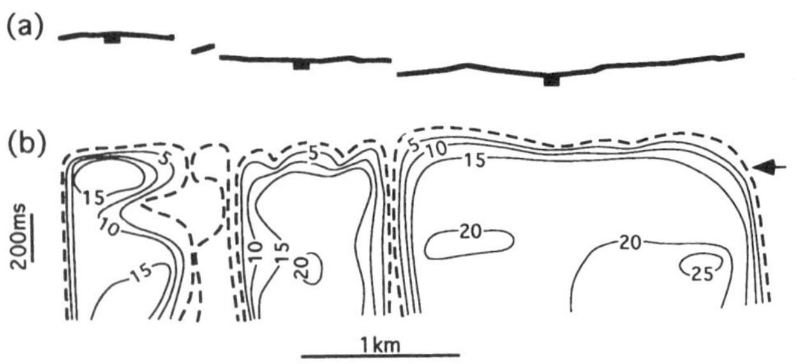

Fig. 7. (**a**) Fault trace map of a segmented fault zone from the Timor Sea derived from a seismic survey with shotlines oriented perpendicular to the fault traces and spaced every 50 m.
(**b**) Composite strike-projection of the faults shown in (a). The faults are interpreted on 13 horizons and the throw contours are defined by *c.* 1200 datapoints. Broken lines indicate fault tip-lines and throw contour values are in ms. The level of the mapped horizon in (a) is indicated by the arrow.

gradients occur only close to the tip-lines. This segmented fault overlies a larger older fault and was formed by reactivation of this structure. If the later extension direction was slightly oblique to the strike of the earlier structure, then upward realignment of the fault segments is to be expected (Woods 1992). Fault surface bifurcation has given rise to three fault segments separated by two overlap zones. The lateral tip-lines of the fault segments are approximately parallel to the fault propagation direction.

Conclusion: *overlap types*

Three simple geometrical relationships between overlapping faults have been recognized. Clearly the geometries described may occur in combination, e.g. fault segments which link upwards at a branch-point and downwards at a branch-line. Also, the geometrical relationship between overlapping faults may change during fault growth, e.g. branch-point linkage may develop into branch-line linkage (Huggins *et al.* in press).

Therefore displacement-parallel synthetic overlap zones may form by:
(i) Upward bifurcation from a fault bend;
(ii) Upward bifurcation from a reactivated fault;
(iii) Stepping during lateral propagation of a single fault;
(iv) Interference following lateral propagation of two initially independent faults.

It is to be expected that within a developing fault system overlap zones are continually being generated by these, and possibly other, processes.

The observations which may be of general application and used to constrain predictions of the three-dimensional forms of overlap zones are, (1) segmented faults which have flat-topped throw profiles may be characteristic of upward propagation from a single large fault (Figs 5d & 7b), and (2) fault segments which link at depth at a branch-point are likely to have shorter overlap lengths than either faults which are unconnected at depth or those which link at a branch-line.

Overlap interaction

Synthetic overlaps are differentiated on the basis of whether or not displacement is transferred across them. Displacement transfer across an overlap zone results in the formation of a relay zone. Fault overlap without displacement transfer is a simple superimposition of the deformation fields of two independent faults. The features associated with relay and 'non-relay' type overlaps are illustrated in Figs 8 and 9 respectively.

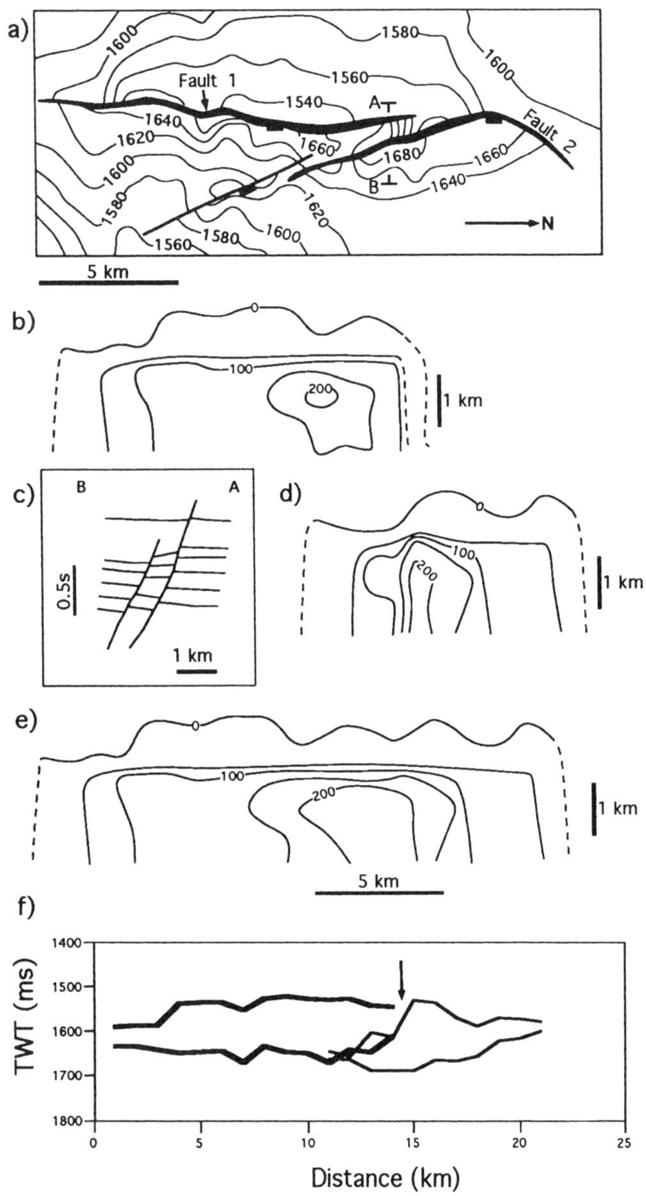

Fig. 8. (**a**) Structure contour map of a horizon deformed by two syn-sedimentary overlapping faults from the Northern North Sea (Faults 1 and 2) and an associated minor fault. TWT contours are in ms. The seismic lines are oriented E–W and are spaced every kilometre. The location of the cross-section shown in (**c**) A–B is indicated. (**b**), (**d**) and (**e**) are throw strike-projections of Fault 1, Fault 2 and their aggregate throw, respectively. Contours of throw are in metres (1 m = 0.75 ms). The strike-projection of Fault 1 is based on *c.* 100 throw readings and that of Fault 2 on *c.* 75 readings. (**f**) A horizon separation diagram of the horizon mapped in (a) on Faults 1 (heavy lines) and 2. The arrow indicates a possible point of linkage between these two faults. The correct relative lateral positions are maintained in the diagram for the fault traces on (a) and all three strike-projections.

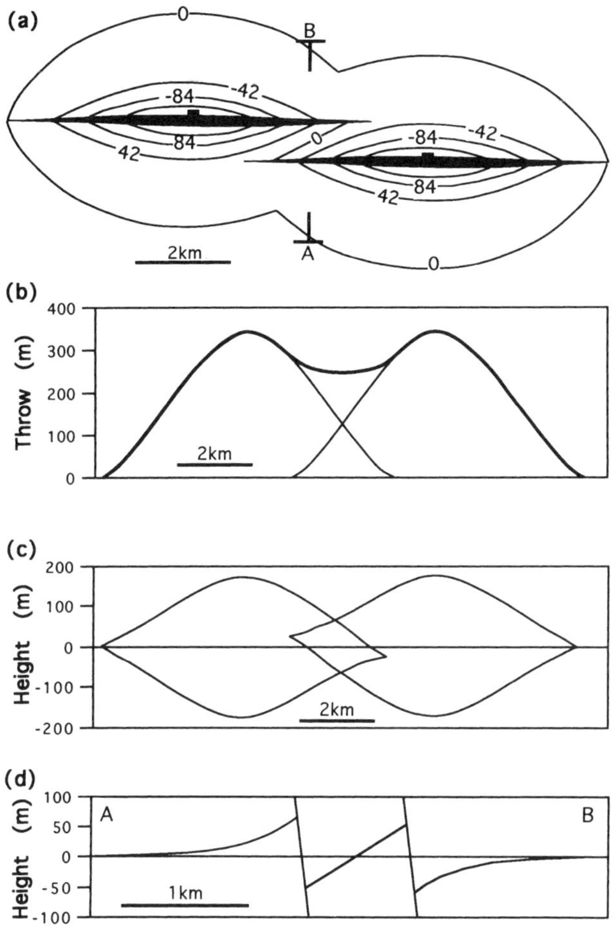

Fig. 9. The deformation field around two overlapping blind faults according to the fault model described by Gibson *et al.* (1989) is illustrated by (**a**) a structural contour map of a deformed horizon which was initially horizontal and passed through the fault centres, (**b**) throw profiles of the two faults and their aggregate throw profile measured on the horizon mapped in (a), (**c**) a composite horizon separation diagram of the two faults for the mapped horizon and (**d**) a cross-section (A–B). The faults both dip 60°, have widths (map trace lengths) of 8 km and maximum displacements of 400 m. The axial ratios of the fault surfaces (Gibson *et al.*) are 2:1 and the reverse drag radius is 3km. The hanging-wall fault pre-dates the footwall fault.

Figure 8 shows an example of two overlapping faults separated by a relay zone in which the entire horizontal extents (but not vertical extents) of both overlapping faults are contained within a seismic survey area. The two faults are distant from larger faults and the regional dips of horizons are low (Fig. 8a). Throw strike-projections of the two faults (Fig. 8b and d) show that each has an asymmetric throw distribution, with lateral displacement gradients within the relay zone three times higher than those outwith the relay. The aggregate throw strike-projection (Fig. 8e) of the two faults shows a symmetrical throw distribution characteristic of a single fault, demonstrating that the two faults form a geometrically coherent unit. The symmetry of the aggregate displacement pattern of these faults is also evident on a composite horizon separation diagram. A horizon separation diagram of a fault shows the traces of the footwall and hanging-wall cutoffs of a horizon on the fault on a strike-projection. The composite horizon separation diagram (Fig. 8f) shows that the footwall and hanging-wall cutoffs of the composite fault describe simple arcs, a pattern characteristic of single faults (Gibson *et al.* 1989). Those cutoffs which constitute the boundaries to the overlap zone, i.e. the hanging-wall cutoff of Fault 1 and the footwall cutoff of Fault 2, are parallel to each other and plunge steeply to the south, reflecting the relatively high bed dip within the overlap zone. In this example displacement transfer across the overlap is concentrated within the zone just to the south of the tip of Fault 1 (Fig. 8f) which has caused either high fault-parallel shear strain within this zone or linkage between the two faults; the shotline spacing is too great to distinguish between these two possibilities.

The characteristics of this example which are thought to be general features of relay zones include the increase in horizontal displacement gradient and high fault-parallel shear strain of the overlap zone. These features have been widely described (e.g. Chadwick 1986; Larsen 1988; Peacock & Sanderson 1991). The simple arcs of the cutoffs external to the relay zone are not general features of relays. However, where overlapping faults are geometrically coherent the same general form occurs of a regular overall footwall high and hanging-wall low across the overlap.

Kinematically independent or 'non-relay' overlaps may result either from spatial or from temporal separation of the constituent faults. In the former case, continued fault growth may cause the faults to interact to produce a relay zone whereas if faults grow at different times displacement transfer clearly is not possible.

The deformation associated with non-relay overlaps can be simply modelled as the superimposition of the deformation fields of two faults. Gibson *et al.* (1989) described the displacements in the rock volume containing an ideal single normal fault and proposed a method for modelling observed horizon geometries. The method can be applied only to faults which were kinematically independent during their growth. Figure 9a shows the structural contours predicted by the Gibson *et al.* (1989) model for two synthetic overlapping faults which are centred on an initially horizontal surface. The modelled faults are developed sequentially, i.e. in temporal isolation and therefore with no active interference between their deformations. The throw profiles of the two overlapping faults (Fig. 9b) are derived from a horizontal chord through the fault centres. Throw profiles of the model faults are symmetrical about their points of maximum displacement and contrast with the asymmetrical throw profiles of faults bounding relay zones (Fig. 8b & d). The aggregate throw profile for the modelled overlapping faults shows a low in the area of overlap, but the shape of the aggregate throw profile is entirely dependent on the length of overlap: an aggregate high would be produced if the overlap between the faults were to be increased.

The structural contours of the horizon deformed by the model faults (Fig. 9a) and the composite separation diagram (Fig. 9c) show independent footwall highs and hanging-wall lows associated with each fault in contrast to the pattern associated with the composite fault of Fig. 8.

Strain within the modelled overlap zone is due simply to the sum of the footwall uplift of the hanging-wall fault and hanging-wall rollover of the footwall fault, resulting in a net rotation of the horizon towards the mutual footwall. A cross-section through the model faults (Fig. 9d) shows

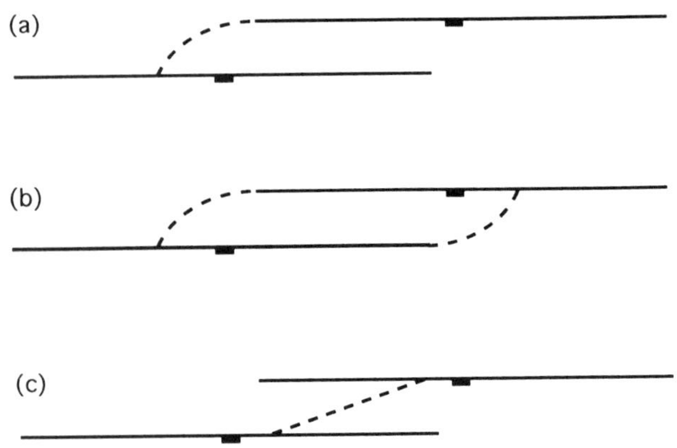

Fig. 10. Three possible ways in which a relay zone may be breached. (**a**) Propagation of either the hanging-wall or (as shown) the footwall fault. (**b**) Propagation of both the footwall and hanging-wall faults. (**c**) Development of a secondary link fault.

horizon dips in the overlap towards the mutual footwall of the overlapping faults. In contrast, fault-normal cross-sections through relay zones frequently show a horizon apparent dip towards the mutual hanging-wall of the overlapping faults (Peacock & Sanderson 1991, 1994; Huggins *et al.* in press) due to an element of fault normal shear within the relay.

Overlap zone breaching

An overlap zone is breached when the two overlapping fault segments are replaced by a through-going fault surface. Breaching may be achieved by (a) deflection of the tip of one of the overlapping faults, (b) deflections of both fault traces, or (c) development of a secondary 'link' fault as shown in Fig. 10. Breaching occurs when the shear strain across a relay zone can no longer be accommodated by continuous deformation. If the relay zone accommodates shear by flexure and bed-parallel slip then breaching could be a result of strain hardening within the relay zone. If shear is accommodated by movement on small faults within the relay zone breaching may occur as a result of strain being concentrated onto a single fault which grows to intersect the two overlapping faults.

Recognition of a breached overlap zone relies on identification of one of the fault patterns shown in Fig. 10 and demonstration that the observed pattern originated from an overlap. Demonstrating the previous existence of an overlap zone requires a knowledge of the kinematics of the fault development. In the examples described below kinematics are examined using a displacement backstripping method for syn-sedimentary faults (Chapman & Meneilly 1991; Petersen *et al.* 1992; Childs *et al.* 1993; Clausen & Korstgård 1994). Application of this method requires that sedimentation rates exceeded fault displacement rates and the assumption that in the time interval between two horizons the magnitude of fault slip was the same on all horizons on any single dip-section. The situations in which this assumption is valid are discussed in Childs *et al.* (1993). In general, constant slip within a vertical section can be assumed when the vertical extent of the syn-sedimentary sequence is small relative to the map trace lengths of the fault(s) being studied. If the assumption is valid then all up-dip displacement gradients on a dip-section are functions of sedimentation and fault slip rate and of sediment compaction. The displacements existing on a fault surface prior to deposition of a specified horizon can be approximated by subtracting the displacements on that horizon from those of underlying horizons. The implicit assumptions are discussed in the context of specific examples.

Example 1

The first example of a breached relay zone is from a sandbox model designed and constructed at the Bureau of Economic Geology at Austin, Texas, in order to study syn-sedimentary faulting above a ductile layer. A detailed analysis of the kinematics of faulting in this model, including a description of a breached overlap zone, is given in Childs *et al.* (1993). A second example of a breached overlap zone is described below.

The sandbox model was designed to examine the influence of sedimentation rate on structural style during gravity-driven extensional faulting of a brittle overburden (dry sand) overlying a ductile layer (silicone polymer). A total of eleven sand layers were added at regular intervals during the evolution of the model. The interfaces between layers are assigned horizon numbers, such that the interface between the silicone polymer and the lowermost layer is labelled Horizon 1 and the top of the uppermost layer is Horizon 12. The stages in model development which coincide with these horizons are numbered Time 1 to Time 12. In the interval between deposition of each layer the model was allowed to extend; faulting within the model initiated at Time 6.

A map of Horizon 7 (Fig. 11a) shows the same fault pattern as Fig. 10a with a through-going fault and a synthetic hanging-wall splay. The hanging-wall splay does not, however, intersect Horizons 10–12 (Fig. 11b) and is seen to terminate abruptly upwards between Horizons 9 and 10 (Fig. 11c), indicating that the splay died in the time interval between deposition of these two horizons. It is clear therefore that the geometry of the active faults changed through time. Displacement variations on the two faults are characterized by a region of reduced throw on the main fault which coincides with the position of the hanging-wall splay (Fig. 12b). Aggregating the throws on the main fault and on the splay produces regular throw variations (Fig. 12c) and demonstrates that the fault and splay are geometrically coherent.

Backstripping to the most recent time at which both the main fault and the hanging-wall splay were active (i.e. Time 9) causes the low displacements on the main fault on cross-section 6 to become zero. This exercise shows that the earlier fault geometry comprised two overlapping faults (Fig. 12d) and that breaching of the overlap zone was achieved by lateral propagation of the main fault through cross-section 6. This conclusion is confirmed by fault maps of the surface of the model drawn prior to addition of each sand layer. Backstripping also shows that each of the overlapping faults (Fig. 12e) had a regular displacement pattern. An aggregate displacement pattern for the two backstripped faults (Fig. 12f) shows a slight low in throw on cross-sections 7 and 8. This reduced fault throw in the zone of overlap occurs because some of the displacement across the zone is accommodated by continuous deformation either

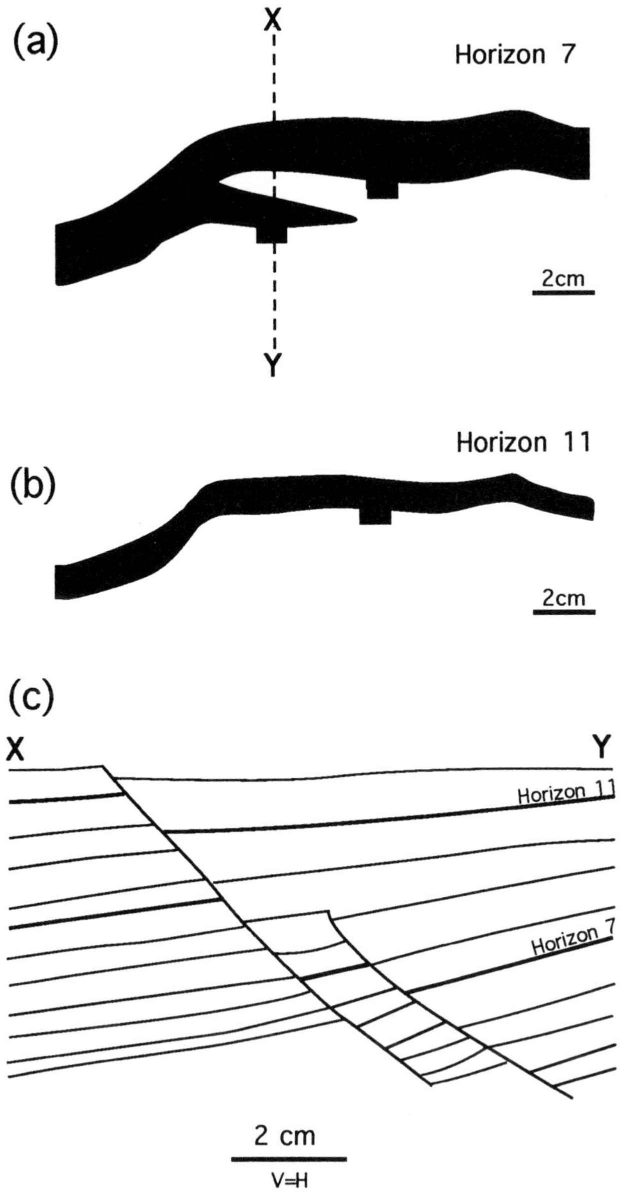

Fig. 11. Two maps and a cross-section (X–Y) illustrating the geometry of two faults in the sandbox model. A hanging-wall splay seen on the map of Horizon 7 (**a**) is not present on the map of Horizon 11 (**b**) and is seen on the cross-section (**c**) to terminate upwards immediately above Horizon 9. The lowermost horizon shown on the cross-section is the interface between the sand and the underlying silicon polymer.

as bed rotation between the overlapping faults or as normal drag.

Example 2

A second example of a breached overlap zone is from an area of normal faulting in the Northern North Sea. Seven seismic horizons have been interpreted (Fig. 13a) of which three (Horizons 1 to 3) are pre-rift, three are syn-rift and one is post-rift. There is no significant change in throw within the pre-rifting sequence on any dip-section, consistent with constant fault slip on a dip-section in the time interval between any two successive syn-rift horizons. Some consideration of the impact of compaction is required before application of the backstripping method. Compaction effects within the syn-rift sequence are evident in cross-section (Fig. 13a) as a culmination in the relatively thin syn-rift sequence overlying the footwall high in the pre-rift sequence. Compaction reduces fault offsets in the syn-rift sequence such that restored throw values in the previously compacted pre-rift sequence are maximum possible values. Although both the restored throws and the lateral extents of faults are likely to be slightly overestimated, the throw contour pattern on a restored fault is largely unaffected by the magnitudes of throws subtracted during backstripping; conclusions regarding the previous existence of an overlap zone are therefore not affected.

Comparison of the cross-section, map and the strike-projections of the faults within the overlap structure (Fig. 13a, b and c respectively) with corresponding diagrams for the sandbox model, shows the two structures to be similar. A bend in the main fault coincides with a synthetic hanging-wall splay branch-line (Fig. 13b). A region of low displacement on the main fault (Fig. 13c) coincides with the hanging-wall splay and is flanked by vertical displacement contours and overlain by horizontal displacement contours. The aggregate displacement contours for the main fault and the splay (Fig. 13e) indicate a regular pattern of displacement variation and geometric coherence between the faults. A cross-section shows the hanging-wall splay to terminate upwards within the syn-rift sequence indicating the interval within which this fault died. Displacement backstripping to zero displacement on the last horizon intersected by the hanging-wall splay (Horizon 4, Fig. 13a) provides the contour patterns shown in Fig. 13g. As in the previous example, the present day displacement low on the main fault restores to zero displacement and the reconstructed fault geometry comprises two overlapping faults. The displacement patterns on the restored overlapping faults (Fig. 13g) are again systematic as is the aggregate displacement pattern (Fig. 13i), indicating that the overlapping faults were geometrically coherent prior to overlap breaching.

A horizon separation diagram, showing the lines of intersection of Horizon 1 with both the main fault and the hanging-wall splay (Fig. 13d), shows that bed dips in the region between the main fault and the hanging-wall splay are higher than outside this region. The positions of these cutoff traces can be backstripped using the same method as for fault throws (Chapman & Meneilly 1991) with the additional assumption of partitioning of displacement equally between footwall uplift and hanging-wall subsidence. The resultant backstripped fault cutoff traces (Fig. 13h) show that high bed dips in the overlap zone existed prior to breaching, indicating displacement transfer across the overlap zone; the former overlap zone was therefore a relay zone. The restored horizon separation diagram indicates significant fault-parallel shear of the relay zone prior to breaching and approximately horizontal horizon dips outwith the relay zone. The restored relay zone is 700 m in length and *c*. 550 m wide with easterly horizon dips of *c*. 8°.

Example 3

This example is from a seismic dataset from offshore Australia in an area in which faulting is believed to be

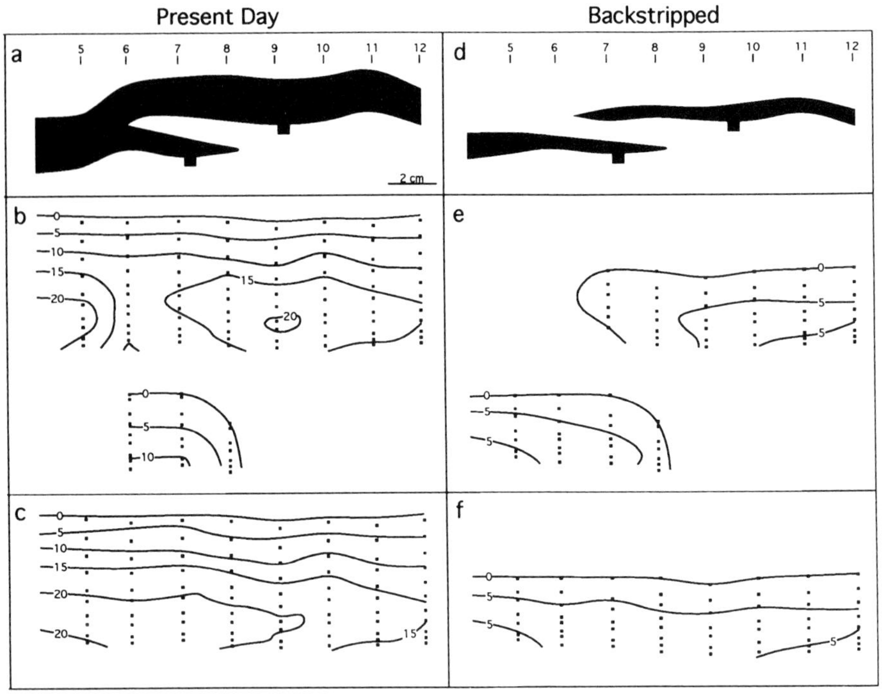

Fig. 12. Maps and throw strike-projections for the faults in Fig. 11 showing the structure in the final model (**a**–**c**) and at Time 9 (**d**–**f**). The fault throw readings are from eight cross-sections (numbered 5–12) and dots on the strike-projections indicate the positions of the datapoints. Throw contours are in millimetres. The horizontal scales on the strike-projections are the same as on the maps and the vertical and horizontal scales are equal. Throw data are not available to the left of Section 5 and the throw contours have been extrapolated into this region. Fault maps are available to the left of Section 5.

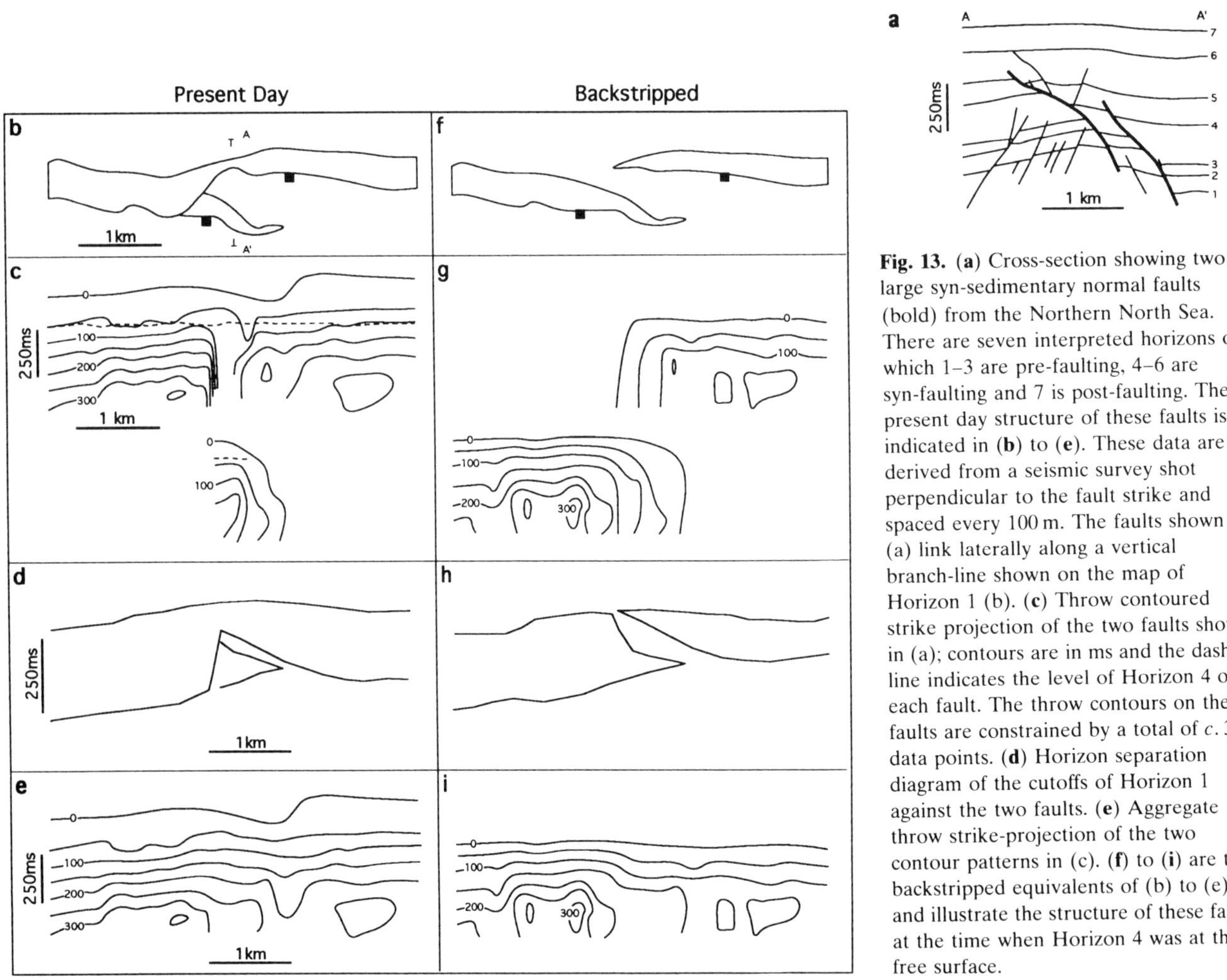

Fig. 13. (**a**) Cross-section showing two large syn-sedimentary normal faults (bold) from the Northern North Sea. There are seven interpreted horizons of which 1–3 are pre-faulting, 4–6 are syn-faulting and 7 is post-faulting. The present day structure of these faults is indicated in (**b**) to (**e**). These data are derived from a seismic survey shot perpendicular to the fault strike and spaced every 100 m. The faults shown in (a) link laterally along a vertical branch-line shown on the map of Horizon 1 (b). (**c**) Throw contoured strike projection of the two faults shown in (a); contours are in ms and the dashed line indicates the level of Horizon 4 on each fault. The throw contours on the faults are constrained by a total of *c.* 300 data points. (**d**) Horizon separation diagram of the cutoffs of Horizon 1 against the two faults. (**e**) Aggregate throw strike-projection of the two contour patterns in (c). (**f**) to (**i**) are the backstripped equivalents of (b) to (e) and illustrate the structure of these faults at the time when Horizon 4 was at the free surface.

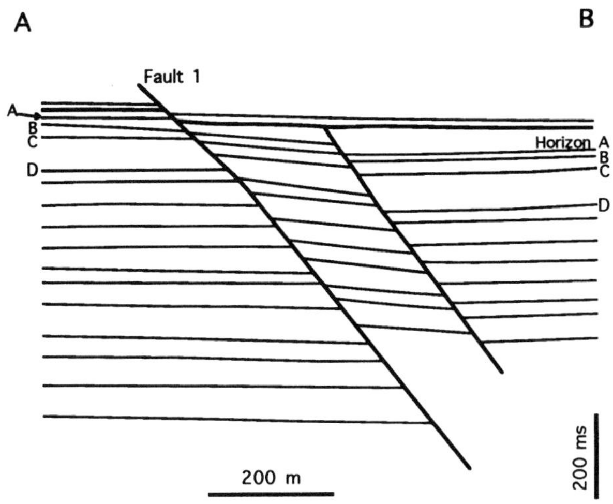

Fig. 14. Cross-section A–B (see Fig. 15a) through two normal syn-sedimentary faults from the Timor Sea. Horizon C is the uppermost pre-faulting horizon and Horizon D is the mapped horizon in Fig. 15a and c. Horizon A is the uppermost horizon intersected by the smaller of the two faults.

purely dip-slip. A total of 13 horizons are interpreted, five within the syn-rift sequence and the remainder pre-rift (Fig. 14). The faults formed both by reactivation of earlier faults at depth and by nucleation at the base of the syn-rift sequence (Horizon C, Fig. 14).

The overall fault geometry is similar to those of the previous examples. In map view (Fig. 15a), a bend in the large fault (Fault 1) coincides with a steeply dipping branch-line with a hanging-wall synthetic splay (Fig. 15b). A displacement low on the main fault is coincident with the hanging-wall splay, is flanked by vertical displacement contours and passes upwards into horizontal displacement contours. The synthetic splay terminates abruptly within the growth sequence (above Horizon A in Fig. 14) indicating that the splay died after deposition of Horizon A. Displacement backstripping to a time when the synthetic splay was last active (i.e. when Horizon A was the free surface) yields a zone of zero displacement immediately to the west of, and parallel to, the branch-line with the hanging-wall splay (Fig. 15d). The strike-projection of the restored hanging-wall splay has not been included in Fig. 15d but the restored fault map (of Horizon D at the time when Horizon A was the free surface) shows, as in previous examples, two overlapping faults (Fig. 15c).

Application of the backstripping method to this structure is not as straightforward as in the previous examples because vertical displacement gradients due to primary displacement variation may be significant. A displacement contour diagram for the main fault (Fig. 15b) can be divided into two parts along the trace of the branch-line with the synthetic splay. To the east of the branch-line, throws within the pre-rift sequence are approximately constant (90 ± 10 ms) and are consistent with upward propagation from a large fault at depth. To the west of the branch-line, fault throw is a maximum at the top of the pre-rift sequence and decreases radially, consistent with fault nucleation at the top of the pre-rift sequence. Since the observed downward decrease in throw is due to primary displacement variation, it is likely that the upward decreases in offset within the syn-rift sequence also incorporate these displacement variations. A maximum estimate of the strains due to primary displacement variation within the growth sequence can be obtained from the pre-rift sequence west of the branch-line, where vertical displacement gradients are up to one third of those within the overlying growth sequence. This value probably overestimates the strains due to primary displacement variation within the growth sequence, because displacement gradients within the pre-rift sequence record strains accumulated since the beginning of fault movement, while individual layers within the syn-rift sequence record only strains post-dating deposition of each layer. Application of the backstripping method to this example is further complicated by erosion of Horizon B and Horizon C. Although well data are not available for this area the seismic evidence suggests that erosion did not occur at other levels within the sequence. The effect on the backstripping method of local erosion within a syn-rift sequence is underestimation of the offsets of eroded or partially eroded syn-rift intervals which, in turn, results in overestimation of backstripped throw values.

The effects of primary displacement variations and of erosion on backstripping results are not insignificant, but have not been taken into account because the possible errors are too small to affect the main conclusions, i.e. that

Fig. 15. (**a**) Fault map of Horizon D showing a part of a large fault with a distinct bend (Fault 1) and an associated hanging-wall synthetic splay. A–B indicates the location of the cross-section shown in Fig. 14. (**b**) Throw strike-projection of Fault 1. Throw contours are in ms and they are based on *c*. 1000 data points derived from seismic lines spaced every 50 m and oriented N–S. The dashed line indicates the trace of the branch-line with the hanging-wall splay. (**c**) Fault map of Horizon D resulting from backstripping of the throw on Fault 1. (**d**) Backstripped throw strike-projection of Fault 1.

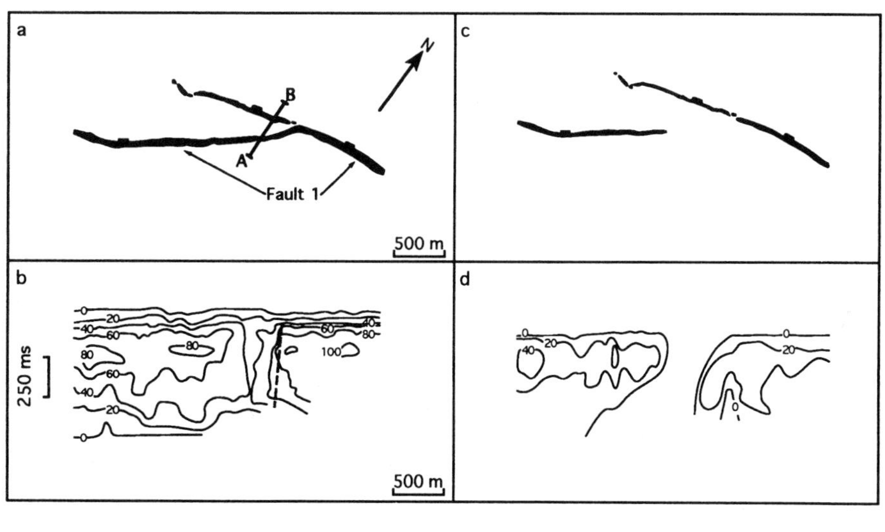

the main fault originated as two separate faults. However, the absolute values of the backstripped throws and the precise position of the restored fault tip-line are not accurate and these data could not be used, for instance, to determine accurately the fault propagation rate. The effect of erosion on the restored backstripped strike-projection is to underestimate the extent of the break in the main fault, rather than to change the conclusion that such a break existed.

On initial inspection, the main fault and the hanging-wall splay do not appear to be geometrically coherent but provide an aggregate displacement pattern with a low displacement region immediately west of the branch-line. This shortfall in measured fault throw is due to a large component of continuous deformation in the region between the main fault and the hanging-wall splay and is clearly seen in cross-section as a northward bed rotation (Fig. 14). It is not clear whether this deformation is a rigid rotation of bedding or ductile rotation due to sub-seismic faults. This bed rotation is absent above Horizon A and it appears that rotation ceased with termination of movement on the hanging-wall splay. The amount of rotation of the syn-rift horizons A–C does not decrease significantly upwards, suggesting there was no long term progressive rotation, but rotation between the main fault and hanging-wall splay occurred immediately prior to breaching of the overlap zone.

General characteristics of breached overlap zones

The common features of the breached relay zones described are: (a) a bend in the main fault coincident with the branch-line with a hanging-wall splay, (b) high horizon dips in the rock volume between the main fault and the splay, (c) displacement variations on the main fault which are characterized by vertical displacement contours at levels in the sequence at which the splay was active, giving way upwards to horizontal displacement contours, and (d) linkage synchronous with termination of movement on the hanging-wall splay. In each of the examples described, overlap zones were breached by propagation of the footwall fault; horizon and displacement geometries which would result from propagation of the hanging-wall fault are expected to be similar but no example is known to us. We expect the same process of relay zone breaching to occur on blind faults, with high bed dips in the volume between a main fault and an associated splay, but we have no relevant data for blind faults.

On the time scales resolvable from seismic and well data, the process of fault linkage and relay zone breaching is instantaneous. Fault systems with the relevant characteristics (as listed (a)–(c)) but with insufficient syn-rift stratigraphic control to perform backstripping are likely to exist. The best time resolution available to us is for the offshore Australia example; fault linkage is thought to occur within an interval of *c.* 1 Ma, during which time there was *c.* 20 m increase in fault throw. At this level of resolution the termination of movement on the synthetic splay and fault linkage were synchronous. However they would not be expected to be synchronous if shorter time intervals were examined. Since the data we present are time averaged over 10 s or 100 s of earthquake cycles they can provide little information on changes in fault geometries in time intervals corresponding to even large numbers of earthquake cycles, or their stable sliding time equivalents.

An example of the role of overlap zones in the development of a fault array

This section describes the role of overlap zones in the development of a normal fault system in the Northern North Sea. The fault maps (Fig. 16) are derived from interpretations of 100 seismic lines perpendicular to the overall fault trend. The faults, with throws of up to 500 m, are imaged within both the pre- and syn-rift sequences and are suited to analysis by throw backstripping. Two maps illustrating the existing fault pattern are shown in Fig. 16a and b. Horizon 1 is within the pre-rift sequence and Horizon 2 is syn-rift; the average interval between these two horizons is *c.* 400 ms (500 m). All faults shown on Horizon 2 link at depth to one or other of the faults on Horizon 1 and all the faults shown dip to the south.

The Horizon 1 map (Fig. 16a) shows overlap zones at locations A and B. That at A can be positively identified as

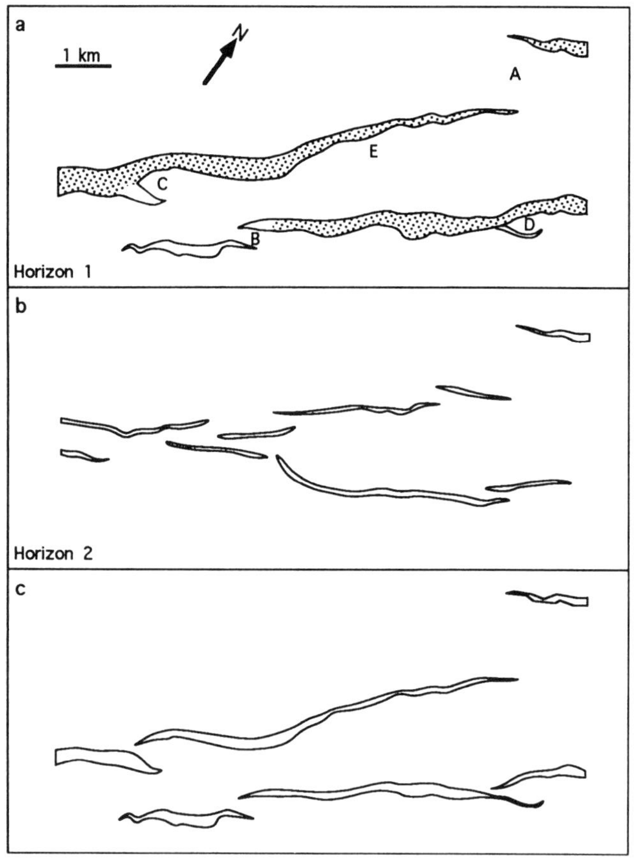

Fig. 16. Three maps of an array of syn-sedimentary normal faults from the Northern North Sea. The maps are derived from *c.* 100 seismic lines at 100 m spacing oriented perpendicular to the mean fault strike. The present day structure is shown by (**a**) and (**b**) which show the fault pattern within the pre-rift (Horizon 1, largest faults only) and the syn-rift (Horizon 2), respectively. The vertical interval between Horizons 1 and 2 averages *c.* 500 m. All the faults shown on Horizon 2 link at depth with one of the fault traces on Horizon 1. Those faults on Horizon 1 which intersect Horizon 2 are stippled. Locations A to E are discussed in the text. (**c**) Sketch map of Horizon 1 backstripped to the time of the earliest syn-rift horizon.

a relay zone as there is a large eastward bed rotation within the overlap and displacement transfer between the overlapping faults can be demonstrated. Displacement transfer across overlap B cannot be demonstrated. The fault patterns at locations C and D have been identified as breached relay zones using the displacement backstripping method. A complex pattern of small faults at E, in the hanging-wall of a large fault, may include a number of breached overlap zones, but the data are not adequate to resolve them.

Displacement backstripping to zero displacement on the lowermost syn-rift horizon (*c.* 250 ms above the top of the pre-rift sequence) provides a restored fault map for Horizon 1 (Fig. 16c), with four overlap zones dividing the faults into a series of sinistrally offset segments. The existing fault pattern developed by overlap zone breaching at C and D, leaving two dead hanging-wall splays. The overlap zones at A and B have not been breached. The large relay zone at A 'survived' by virtue of its size. From the restoration time to the present day, the increases in throw on faults bounding this relay zone have not been accompanied by significant lateral fault propagation, but were accompanied by bed rotation within the relay zone. The overlap zone at B 'survived' because both faults comprising the overlap zone died early in the development of the fault system, as shown by the fact that neither intersects Horizon 2.

Those faults on Horizon 1 which also intersect Horizon 2 are highlighted in Fig. 16a and are those in the pre-rift sequence which continued to be active until deposition of Horizon 2. At this time the three active faults within the pre-rift sequence propagated upwards as nine fault segments (Fig. 16b). Each of these segments links at depth along a sub-horizontal branch-line with one of the three active faults in the basement. The fault traces on Horizon 2 generally strike parallel to the overall structural trend but have local deviations on Horizon 1. Upward propagation of those parts of the basement faults which do not parallel the overall structural trend resulted in segmented faults. The strike of the fault at E in Fig. 16a is *c.* 15° from the general trend and upward propagation of this fault resulted in the formation of three sinistrally offset fault segments. Similarly, the bend in the large fault at D on Horizon 1, a result of breaching of a former overlap zone, propagated upwards as two sinistrally offset overlapping faults which parallel the overall fault trend.

This fault array achieved its present day geometry by a cycle of overlap zone formation, breaching and generation of new overlap zones by upward propagation into the syn-rift sequence. If the existing faults had continued to grow the overlap zones on Horizon 2 would have been breached, as eventually would the large ramp at A.

Discussion

The ephemeral nature of overlap zones has been demonstrated for syn-sedimentary normal faults but overlap formation and breaching are likely to occur throughout the development of any fault system. However, the details of overlap zone development and rupture are likely to vary with both the lithology and the type of overlap.

For a relay zone to develop, host rocks must be capable of accommodating shear strain across the relay without failure of the overlap zone, so a relay zone and displacement transfer are less likely to develop in rocks of very high shear modulus, e.g. granite. The overlap zones described occur within sedimentary sequences and separate faults with displacements of 10 m to several 100 m. At smaller scales, i.e. fault displacements <1 m, it is to be expected that factors such as the thicknesses of competent units and lithological anisotropy will exert a strong influence on both the size of overlaps which can develop and on their geometries. At larger scales i.e. fault displacements >1 km, it has been suggested that crustal thickness may determine the maximum lengths of overlapping fault segments (Jackson & White 1989).

Displacement transfer between overlapping fault segments which are unconnected in 2D has long been recognised (Dahlstrom 1969). This 2D relationship between faults has recently been termed soft-linkage (Walsh & Watterson 1991; Roberts & Yielding 1994). Soft-linkage contrasts with the hard-linked fault patterns described by Gibbs (1984) and others in which all faults are linked on any 2D map or cross-section and displacement transfer between faults occurs via transfer faults. While earlier workers recognized the existence of fault surfaces which are unconnected in 3D (Boyer & Elliot 1982) it was believed that faults which display 2D soft-linkage must be physically linked in 3D (Dahlstrom 1969). More recently it has been suggested that displacement transfer does not require 3D fault linkage (Walsh & Watterson 1991) and data presented here support this view. These data (Fig. 3) are from a seismic survey and fault throw resolution (<10 ms or 15 m) defines the scale on which these faults can be examined. Clearly, if the resolution of the data were improved, more faults would be imaged and it is possible, even likely, that faults not now seen as connected in 3D would be connected when examined on a finer scale. Similarly if the data resolution were coarsened these two faults would be seen as a single continuous fault trace. The observed topologies of fault surfaces are therefore dependent on both the scale of observation and on the resolution limit below which a discontinuous deformation appears to be continuous, or ductile.

All the data presented here are for displacement-parallel overlap zones with tip-lines of overlapping faults approximately parallel to the slip direction. Deformation within this type of overlap is therefore accommodated by simple shear with no volume strain. Displacement-normal overlap zones, with tip-lines of overlapping faults perpendicular to the fault slip direction, do require volume strain of the overlap zone. We believe that this type of overlap is fundamentally different from those described here, in terms of both development and breaching. Displacement-normal overlap zones have been widely discussed in the context of offset strike-slip faults (e.g. Aydin & Nur 1985).

Each example of relay zone breaching known to us, including examples not described here, occurred by propagation of the footwall fault but it is not known whether this is generally true of breaching of overlap zones close to the free surface. Breaching of overlap zones by both footwall and hanging-wall faults, resulting in fault bounded rhombs, is known to occur (Ramsay & Huber 1987; Peacock & Sanderson 1994; Huggins *et al.* in press) and also occurs with Mode I fractures.

Breached relay zones within both the seismic and analogue datasets are preserved as distinct fault bends with a synthetic hanging-wall splay. If the relay zone is well developed prior to breaching, i.e. there is significant

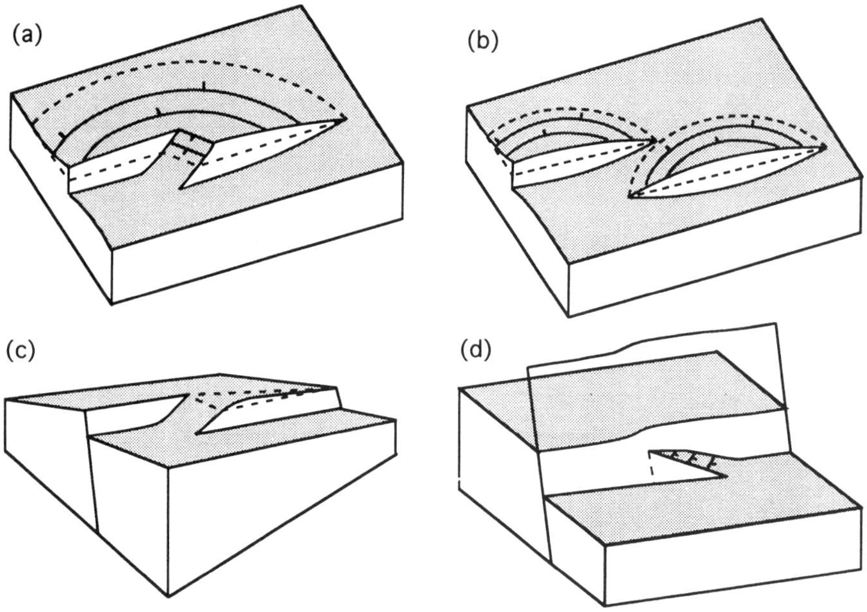

Fig. 17. Schematic isometric block diagrams of a faulted horizon illustrating the sites of potential closure associated with overlap zones. (**a**) High bed dip within a relay zone provides footwall closure. Heavy dashed line indicates oil–water contact and ticks on structure contours indicate the down-dip direction. (**b**) A non-relay overlap zone provides the same closure potential as two isolated faults. (**c**) Bed rotation within a relay zone reverses the regional dip (into page) to provide closure. (**d**) A curved fault trace is underlain by a relay zone breached by propagation of the footwall fault to provide potential hanging-wall closure.

displacement transfer between the overlapping faults, then after breaching the beds between the through-going fault and the hanging-wall splay will dip steeply away from the branch-line. If the overlap were to be breached by propagation of the hanging-wall fault then the beds within the breached overlap would dip steeply towards the branch-line with a footwall splay.

Relay and non-relay overlap zones have predictable and contrasting horizon geometries. High dips within a relay zone and the presence of a footwall high which is continuous across the relay (Anders & Schlische 1994), provide a site of enhanced hydrocarbon trapping potential (Fig. 17a). High bed dips within relay zones may be of particular importance in providing structural closure where the sense of rotation in the relay reverses a regional dip (Fig. 17c). In the case of non-relay overlap zones the associated culminations are no greater than if the faults were isolated and trapping potential is not increased (Fig. 17b). As the amplitude of a relay decreases so too does the height of the potential trapped oil column and in the case of small relay zones within a large, potentially sealing fault, the relay zone may act as a gap in the seal. Intra-reservoir relay zones increase reservoir connectivity. Growth of a breached relay zone results in the upward propagation of the kinked through-going fault to higher levels in the sequence. Therefore, fault trace bends on maps may typically be underlain by breached overlap zones and may indicate enhanced trapping potential at depth (Fig. 17d). Upward propagation of the kinked fault may also give rise to a secondary overlap zone at higher levels.

Where fault movement and hydrocarbon migration are synchronous the process and timing of relay zone breaching could be critical in determining migration paths and trapping. In such situations the backstripping method provides a powerful tool for studying the sealing potential of faults through time. Care should be taken, however, to ensure that the backstripping method is applied only to situations in which the methodological assumptions are valid.

Within the limits of the data resolution, the branch-lines which have been mapped appear to be either sub-parallel to the slip direction or normal to it. In common with branch-lines in other datasets we have examined, branch-lines in the data used for this study appear to be straight rather than curved. Although there is insufficient evidence on which to base a general conclusion, it is possible that there is a limited range of possible branch-line geometries and that they could be a key element in determining fault array kinematics.

It is generally true that the complexity of observed fault patterns increases concomitantly with improvement in data quality. With improved seismic data, more overlap zones are likely to be imaged, as in the case of the Flounder Field described by Sloan (1987). Analysis of the roles of overlap zones within fault systems requires detailed determination of 3D fault geometries but there are few published descriptions of such geometries. These detailed descriptions are scarce firstly, because good 3D seismic data from faulted areas have become available only in recent years and secondly because interpretations of 3D seismic data generally are concentrated on a particular reservoir interval with much of the data not interpreted. General practice is to represent the structure of an oilfield by reservoir horizon maps, but where fault branching and bifurcation have occurred a single map does not adequately describe the structure. Useful information regarding the development and internal structure of a reservoir may often be obtained by examination of the fault patterns on horizons above and below the reservoir.

Conclusions

(1) Overlap zones are ephemeral features within developing fault systems. Overlap zones are both created and destroyed during fault development but may be preserved when fault growth stops.

(2) Overlapping faults may be either unconnected in 3D, or link at either a branch-line or a branch-point. High lateral displacement gradients indicating displacement transfer may occur in both cases.

(3) Synthetic overlap zones between normal faults may develop by: (i) upward bifurcation from a fault bend; (ii)

upward bifurcation from a reactivated fault; (iii) stepping during lateral fault propagation; (iv) interference following lateral propagation of isolated faults.

(4) Overlap zones may be divided into those which show displacement transfer, i.e. relay zones, and those which do not. The pattern of deformation in the surrounding rock volume is different for each type.

(5) Overlap zone breaching results in a distinctive fault geometry. Former overlap zones can be identified, often by the presence of a bend in a map fault trace.

(6) The geometries at earlier growth stages of faults and fault arrays can be determined by throw backstripping but only where sedimentation rates in a syn-rift sequence were high relative to the fault displacement rate.

(7) The effects of a relay zone on reservoir integrity depend on the size of the overlap zone in relation to the height of the trapped hydrocarbon column.

We thank the companies which provided the seismic data on which this work is based, which include BHP Petroleum, BP Exploration Operating Company, Exxon Production Research Company, Maersk Olie og Gas, and Statoil; publication is with the permission of these companies. Thanks are also due to former staff of Britoil plc for providing access to data. We also thank B. Vendeville and colleagues at the Bureau of Economic Geology, University of Texas at Austin who provided results of their sandbox modelling. P. Bretan, G. Yielding and A. Roberts are thanked for their interpretations of some of the seismic datasets. We are grateful to D. Ellis and C. Bonson for preparation of diagrams and to P. Huggins and other colleagues in the Fault Analysis Group for much useful discussion. We thank A. Roberts, D. Sanderson and R. Schlische for their helpful reviews. This research was part funded by Amoco (UK) Production Company (contract RAD 59 (90)), by the OSO/NERC Hydrocarbon Reservoirs LINK Programme (project number 827/7053) and by the NERC Petroleum Earth Sciences Programme (grant D1/G1/189/03). Badley Earth Sciences FAPS software was used for projection and analysis of fault data.

References

ANDERS, M.H. & SCHLISCHE, R.W. 1994. Overlapping faults, intrabasin highs, and the growth of normal faults. *The Journal of Geology*, **102**, 165–180.

AYDIN, A. & NUR, A. 1985. The types and role of stepovers in strike-slip tectonics. *In*: BIDDLE, K.T. & CHRISTIE-BLICK, N. (eds) *Strike-slip deformation, basin formation and sedimentation*. Special Publication of the Society of Economic Paleontologists and Mineralogists, **37**, 35–44.

BADLEY, M.E., FREEMAN, B., ROBERTS, A.M., THATCHER, J.S., WALSH, J., WATTERSON, J. & YIELDING, G. 1990. Fault interpretation during seismic interpretation and reservoir evaluation. In: *The integration of geology, geophysics, petrophysics and petroleum engineering in reservoir delineation, description and managment*. Proceedings of the 1st Archie Conference, Houston, Texas. American Association of Petroleum Geologists, 224–241.

BARNETT, J.A.M., MORTIMER, J., RIPPON, J.H., WALSH, J.J. & WATTERSON, J. 1987. Displacement geometry in the volume containing a single normal fault. *American Association of Petroleum Geologists Bulletin*, **71**, 925–937.

BOYER, S.E. & ELLIOT, D. 1982. Thrust systems. *American Association of Petroleum Geologists Bulletin*, **66**, 1196–1230.

CHADWICK, R.A. 1986. Extensional tectonics in the Wessex Basin, Southern England. *Journal of the Geological Society, London*, **143**, 465–488.

CHAPMAN, T.J. & MENEILLY, A.W. 1991. The displacement patterns associated with a reverse reactivated, normal fault. *In*: ROBERTS, A.M., YIELDING, G. & FREEMAN, B. (eds) *The Geometry of Normal Faults*. Geological Society, London, Special Publications, **56**, 183–191.

CHILDS, C., EASTON, S.J., VENDEVILLE, B.C., JACKSON, M.P.A., LIN, S.T., WALSH, J.J. & WATTERSON, J. 1993. Kinematic analysis of faults in a physical model of growth faulting above a viscous salt analogue. *Tectonophysics*, **228**, 313–329.

CLAUSEN, O.R., & KORSTGÅRD, J.A. 1994. Displacement geometries along graben bounding faults in the Horn Graben, offshore Denmark. *First Break*, **12**, 305–315.

COWIE, P.A., & SCHOLZ, C.H. 1992. Displacement–length scaling relationship for faults: data synthesis and dicussion. *Journal of Structural Geology*, **14**, 1149–1156.

DAHLSTROM, C.D.A. 1969. Balanced cross-sections. *Canadian Journal of Earth Sciences*, **6**, 743–757.

ELLIOT, D. 1976. Energy balance and deformation mechanisms of thrust sheets. *Philosophical Transactions of the Royal Society of London*, **A283**, 289–312.

GIBBS, A.D. 1984. Structural evolution of extensional basin margins. *Journal of the Geological Society, London*, **141**, 609–620.

GIBSON, J.R., WALSH, J.J. & WATTERSON, J. 1989. Modelling of bed contours and cross-sections adjacent to planar normal faults. *Journal of Structural Geology*, **11**, 317–328.

GILLESPIE, P.A., WALSH, J.J. & WATTERSON, J. 1992. Limitations of dimension and displacement data from single faults and the consequences for data analysis and interpretation. *Journal of Structural Geology*, **14**, 1157–1172.

GRIFFITHS, P.S. 1980. Box-fault systems and ramps: atypical associations of structures from the eastern shoulder of the Kenya rift. *Geological Magazine*, **117**, 579–586.

HUGGINS, P., WATTERSON, J., WALSH, J.J. & CHILDS, C. (in press). Relay zone geometry and displacement transfer between normal faults recorded in coal-mine plans. *Journal of Structural Geology*.

JACKSON, J.A. & WHITE, N.J. 1989. Normal faulting in the upper continental crust: observations from regions of active extension. *Journal of Structural Geology*, **11**, 15–36.

LARSEN, P.H. 1988. Relay structures in a Lower Permian basement-involved extension system, East Greenland. *Journal of Structural Geology*, **10**, 3–8.

MARRETT, R. & ALLMENDINGER, R.W. 1991. Estimates of strain due to brittle faulting: sampling of fault populations. *Journal of Structural Geology*, **13**, 735–738.

MORLEY, C.K., NELSON, R.A., PATTON, T.L. & MUNN, S.G. 1990. Transfer zones in the East African Rift System and their relevance to hydrocarbon exploration in rifts. *American Association of Petroleum Geologists Bulletin*, **74**, 1234–1253.

PEACOCK, D.C.P. & SANDERSON, D.J. 1991. Displacements, segment linkage and relay ramps in normal fault zones. *Journal of Structural Geology*, **13**, 721–734.

—— & —— 1994. Geometry and development of relay ramps in normal fault systems. *American Association of Petroleum Geologists Bulletin*, **78**, 147–165.

PETERSEN, K., CLAUSEN, O.R. & KORSTGÅRD, J.A. 1992. Evolution of a salt-related listric growth fault near the D-1 well, block 5605, Danish North Sea: displacement history and salt kinematics. *Journal of Structural Geology*, **14**, 565–577.

RAMSAY, J.G. & HUBER, M.I. 1987. *The Techniques of Modern Structural Geology, Volume 2; Folds and Fractures*. Academic Press, London.

RANALLI, G. 1977. Correlation between length and offset in strike-slip faults, *Tectonophysics*, **37**, T1–T7.

ROBERTS, A. & YIELDING, G. 1994. Continental Extensional Tectonics. *In*: HANCOCK, P.L. (ed.) *Continental Deformation*. Pergammon Press, Oxford, 223–250.

SLOAN, M.W. 1987. Flounder—A complex intra-latrobe oil and gas field. *Australian Petroleum Exploration Association Journal*, **27**, 308–318.

STEWART, I.S. & HANCOCK, P.L. 1991. Scales of structural heterogeneity within neotectonic normal fault zones in the Aegean region. *Journal of Structural Geology*, **13**, 191–204.

WATTERSON, J. 1986. Fault dimensions, displacements and growth. *Pure and Applied Geophysics*, **124**, 365–373.

WALSH, J.J. & WATTERSON, J. 1990. New methods of fault projection for coal mine planning. *Proceedings of the Yorkshire Geological Society*, **48**, 209–219.

—— & —— 1991. Geometric and kinematic coherence and scale effects in normal fault systems. *In*: ROBERTS, A.M., YIELDING, G. & FREEMAN, B. (eds) *The Geometry of Normal Faults*. Geological Society, London, Special Publications, **56**, 193–203.

WOODS, E.P. 1992. Vulcan sub-basin fault styles—Implications for hydrocarbon migration and entrapment. *Australian Petroleum Exploration Association Journal*, **32**, 138–159.

Received 4 January 1994; revised typescript accepted 3 August 1994.

Active normal faulting and seismicity

Continental fault structure and the shallow earthquake source

Richard H. Sibson

SUMMARY: Plate boundaries in continental crust are generally less sharply defined than in the oceans, with seismicity spread over broad areas. Interplate displacements appear to be largely accommodated by networks of major fault zones. A simple 2-level model for these important structures accounts for the depth distribution of most continental earthquakes, and for the observed range of faulting styles and associated rock deformation textures. The model consists of a seismogenic frictional slip regime overlying quasi-plastic mylonite belts wherein shearing is largely accommodated aseismically, due mainly to the changing response of quartz to deformation with increasing temperature. Shear resistance increases with depth to a peak value in the vicinity of the frictional/quasi-plastic transition and then decreases rapidly. The depth to which microseismic activity extends appears inversely related to regional heat flow and can be satisfactorily modelled as the frictional/quasi-plastic transition for different geotherms using laboratory determined flow laws for quartz-bearing rocks. Larger earthquake ruptures ($M > 5.5$) tend to nucleate near the base of the seismogenic regime in the region inferred to have the highest shear resistance and concentration of distortional strain energy. Consideration is also given to the depression of isotherms and seismic activity in regions of thrusting, and to the question of the downward continuation of major fault zones through the lithosphere. Decoupling of the upper crust on flat-lying shear zones may accompany higher-level dip-slip (and perhaps in some circumstances, strike-slip) faulting, being favoured by above average continental heat flow and a high quartz content in the middle or deep crust. The average level of deviatoric stress within the seismogenic regime remains an outstanding problem.

Lithospheric plate boundaries are less sharply defined in continental crust than in the oceans. Active faulting occurs over broad areas and seismicity is generally diffuse (Barazangi & Dorman 1969). Furthermore, summation of seismic moments in such regions (admittedly over comparatively short sampling periods) suggests that sometimes only a minor fraction of the interplate motion is accommodated by earthquake faulting (North 1974; Chen & Molnar 1977). Nevertheless, while in some cases accepting the possibility of non-localized deformation within intervening blocks, the dominant role of major fault zone networks in absorbing intracontinental displacements, especially in collision zones, has been emphasized by a number of studies over the past decade (e.g. McKenzie 1972, 1978; Freund 1974; Molnar & Tapponier 1975, 1978; Tapponier & Molnar 1976, 1977, 1979; Dewey & Sengor 1979).

Thus a picture emerges of 'soft' intracontinental plate boundaries contrasting with 'hard' boundaries involving cold oceanic lithosphere where slip is comparatively localized and occurs largely by earthquake faulting (Davies & Brune 1971). Several explanations for the ill-defined nature of these continental boundaries have been put forward. McKenzie (1969, 1972) argued that the buoyancy of continental crust, inhibiting subduction, is a fundamental factor causing deformation to spread over broad areas in regions of continental collision. Irregularities along converging continental margins may also cause flaking or splintering of initially contacting sialic 'asperities' (Oxburgh 1972) to form swathes of jostling block motion, essentially cataclastic flow on a gross scale (Gallagher 1981). In addition, McKenzie pointed out that because of their generally ancient lineage compared with the oceans, continental regions usually contain zones of weakness persisting from earlier orogenic episodes, which are susceptible to reactivation (see also Watterson 1975; Sykes 1978). The presence of comparatively low melting point sialic material also renders continental crust inherently weaker than cold oceanic lithosphere.

This paper seeks to demonstrate how a simple conceptual model for continental fault zones [established previously on the basis of rock deformation textures (Sibson 1977), but now quantifiable from laboratory studies on the mechanical properties of quartz-bearing rock] can account for many observed characteristics of active continental faulting. Particular attention is paid to evidence for the downward continuation of fault zones, to the possibility of mid- to deep-crustal decoupling, and to inter-relationships between thermal structure and faulting style.

Fault activity

It is appropriate first to consider what is known about the kinematics of currently active faults, those with a history of Quaternary displacement. Matsuda (1975) presented a useful classification for the degree of fault activity based on time-averaged slip rate (Table 1). Plate boundaries generally consist of a dominant Class AA fault zone and/or a mesh of Class A structures. Intraplate seismic activity is usually restricted to Class B and C faults. With a time-averaged slip rate of

From Holdsworth, R.E. & Turner, J.P. (compilers) *Extensional Tectonics: Faulting and Related Processes*. The Geological Society, Key Issues in Earth Sciences, **2 (2)**, 107–133.
1476-315X/02/$15.00
First published in: "*Journal of the Geological Society*, **140**, 741–767".

TABLE 1: *Degrees of fault activity* (after Matsuda 1975)

	Time-averaged slip rate mm/yr	*Repeat time for 1m slip EQ yrs*
AA	10–100	10–100
A	1–10	10^2–10^3
B	0.1–1	10^3–10^4
C	0.01–0.1	10^4–10^5

~30 mm/yr (Dickinson *et al.* 1972), the San Andreas fault in California (Fig. 1) provides a good example of a dominant Class AA structure defining a plate boundary. Thus, although seismicity extends over a broad area of the western United States, Wallace (1981) demonstrated that at any one place along subsidiary active fault strands, the recurrence intervals between successive moderate to major earthquakes are typically 1–2 orders of magnitude greater than along the San Andreas fault itself.

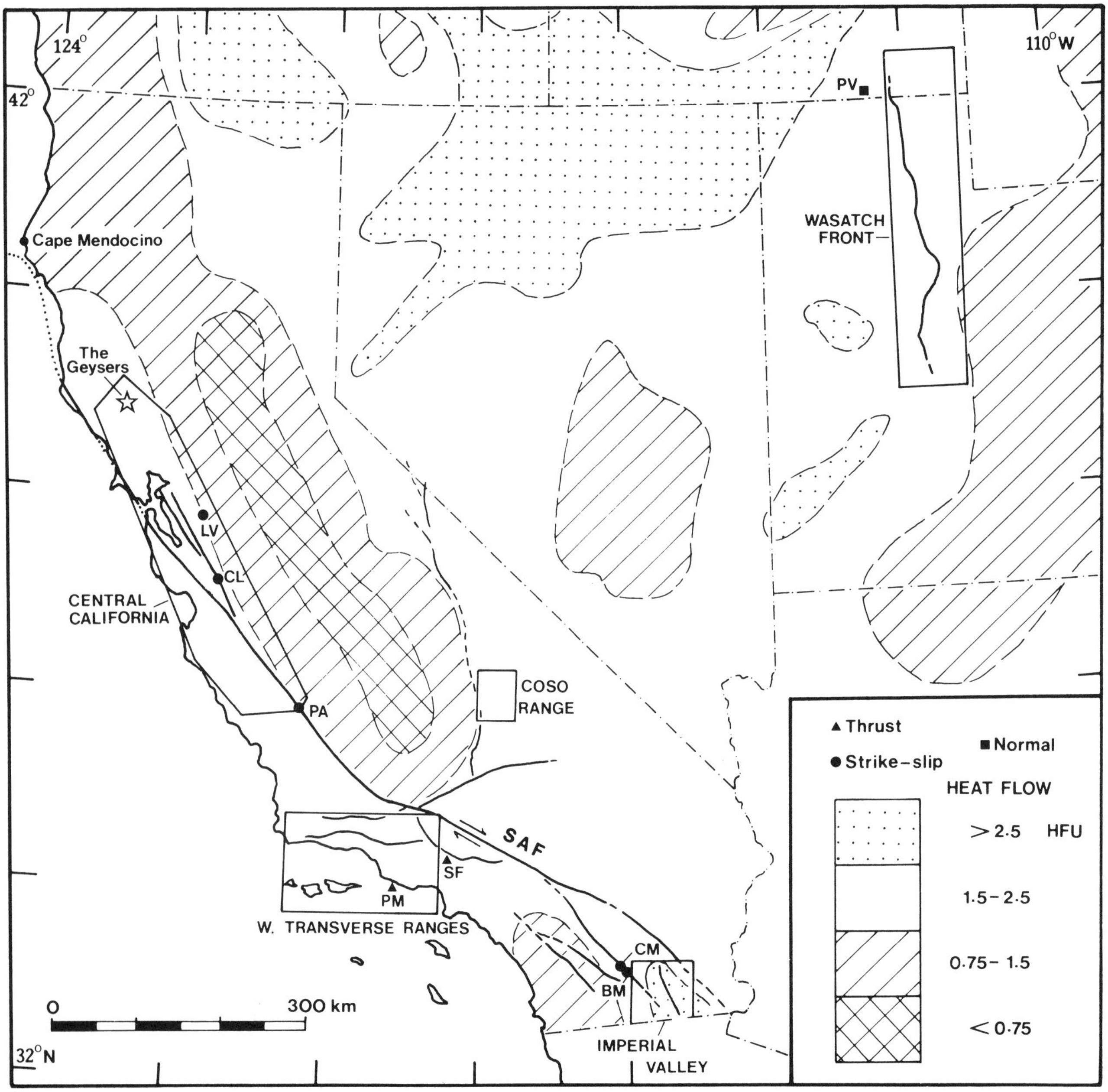

FIG. 1. Regional heat flow provinces of the western United States (after Lachenbruch & Sass 1980) in relation to major fault systems (SAF = San Andreas fault), sampling areas for background microseismicity, and epicentres of moderate earthquakes with well-determined focal depths and aftershock sequences (PA = M_L5.5 Parkfield, 1966; BM = M_L6.4 Borrego Mountain, 1968; CM = M_L5.9 Coyote Mountain, 1969; SF = M_L6.4 San Fernando, 1971; PM=M_L6.0 Point Mugu, 1973; PV=M_L6.0 Pocatello Valley, 1975; CL=M_L5.7 Coyote Lake, 1979; LV=M_L5.9 Livermore, 1980).

Modes of fault slip

Instrumental studies have shown that movement along active faults can occur by a variety of different slip modes whose chief properties are summarized in Table 2. Discrete slip episodes generally propagate over restricted areas of existing faults. Different types of slip event may be distinguished by means of their characteristic rise time, usefully defined as the time for 90% of the slip to be accomplished at a point on the fault surface (Kasahara 1981).

Both earthquakes and aseismic creep episodes may originate by Reid's (1910) mechanism of elastic rebound (Mavko 1981). Strain energy accumulates through differential plate motion to be released cyclically at seismic or lesser rates as the result of frictional or some other instability. Slip usually occurs on existing faults which may or may not have been fully locked before failure. Following Lensen (1970), the strain cycle for successive large earthquakes along the same fault may be divided into four phases. An *interseismic* α-phase with secular accumulation of elastic strain may lead into a *preseismic* β-phase of incipient strain release and anelastic behaviour (possibly with associated foreshock activity) immediately preceding the γ-phase of rapid *coseismic* slip. *Postseismic* adjustment during the δ-phase may involve decelerating afterslip and/or an aftershock sequence with aftershock frequency decaying inversely with time in accordance with the Omori Law. Much of the observed behaviour during the β, γ and δ phases and other time-dependent characteristics of earthquake rupture sequences have recently been explained in terms of a model coupling linear elastic fracture mechanics with subcritical crack growth by stress corrosion (Das & Scholz 1981).

Steady aseismic creep is generally taken to reflect the slip rate at depth, as seems to be the case for the central San Andreas fault where rigid block motion is taking place without significant accumulation of elastic strain, and the creep rate reaches ~30 mm/yr (Thatcher 1979; Burford & Harsh 1980).

Little is known about the partitioning of slip between the various slip modes for different regions, or about the factors governing that partitioning. Coseismic slip is often assumed to predominate, though estimates of seismic slip rate from summed seismic moments (see below) are, in general, unlikely to be accurate to better than perhaps a factor of two. Indeed, the potential importance of some of the other slip modes has only recently been realized. Recognition of slow earthquakes (which may or may not accompany 'normal' shocks) requires detection and interpretation of very long period strain signals; to date only a few have been described (Sacks *et al.* 1978; Kanamori & Stewart 1979). Significant post-seismic afterslip was generally dismissed as a localized near-surface propagation effect until the M_s 7.5 Guatemalan earthquake of 1976, when strike-slip rupturing occurred along 230 km of the Motagua fault. Slip increased by up to 40% in the months following the earthquake to average about 1.1 m along the fault trace (Bucknam *et al.* 1978). The coherent nature of observed afterslip over long (~50 km) segments of the rupture suggested that the process affected large areas of the fault to considerable depth.

Detailed studies of steady and episodic creep have largely been confined to California (King *et al.* 1973; Burford *et al.* 1983) though similar phenomena have been recognized along strike-slip faults in Turkey (Ambraseys 1970) and Iran (King *et al.* 1975), and on a thrust fault in the Himalaya (Sinvhal *et al.* 1973).

Earthquake source characteristics

The most physically meaningful measure of earthquake strength, seismic moment, comes from the dislocation theory of faulting (Aki 1966) and is defined as

$$M_0 = G\bar{u}A \tag{1}$$

where $\bar{u}$ is the mean coseismic slip occurring over a rupture area A (length L, and width W) on a fault in

TABLE 2: *Fault slip modes in continental crust*

Mode	*Slip*	*Characteristic Rise-time*	*Slip rate*	*Propagation rate*
Coseismic slip (may be preceded by accelerating preseismic creep)	<10 m	<10 s	0.1–1 m/s	~3 km/s
Slow earthquakes	<10 m?	10^2–10^3 s (minutes to hours)	1–100 mm/s	?
Postseismic afterslip	<1 m	up to months	0.01–10 μ/s	?
Episodic creep	<10 mm	10^2–10^5 s (minutes to days)	0.1–10 μ/s	1–10 km/day
Steady creep	–	∞	<3mμ/s = 100 mm/yr	–

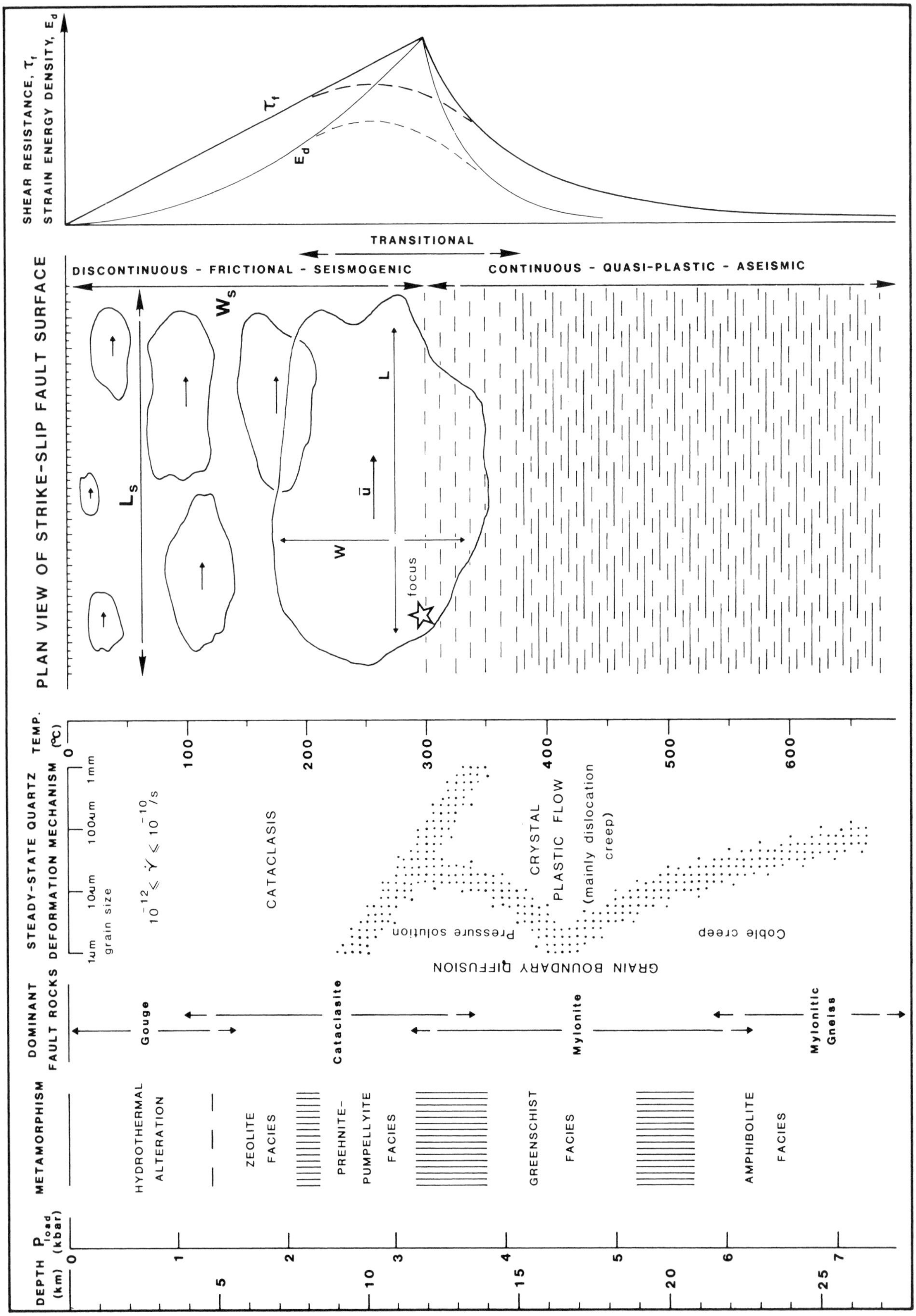
SHEAR RESISTANCE, τ_f
STRAIN ENERGY DENSITY, E_d
τ_f
E_d
TRANSITIONAL
DISCONTINUOUS - FRICTIONAL - SEISMOGENIC
CONTINUOUS - QUASI-PLASTIC - ASEISMIC
PLAN VIEW OF STRIKE-SLIP FAULT SURFACE
W_s
L_s
L
$\bar{u}$
W
focus
STEADY-STATE QUARTZ DEFORMATION MECHANISM
TEMP. (°C)
0
100
200
300
400
500
600
1μm
10μm
100μm
1mm
grain size
$10^{-12} \leq \dot{\gamma} \leq 10^{-10}$/s
CATACLASIS
CRYSTAL PLASTIC FLOW
(mainly dislocation creep)
Pressure solution
Coble creep
GRAIN BOUNDARY DIFFUSION
DOMINANT FAULT ROCKS
Gouge
Cataclasite
Mylonite
Mylonitic Gneiss
METAMORPHISM
HYDROTHERMAL ALTERATION
ZEOLITE FACIES
PREHNITE-PUMPELLYITE FACIES
GREENSCHIST FACIES
AMPHIBOLITE FACIES
DEPTH (km)
P_{load} (kbar)
0
1
2
3
4
5
6
7
5
10
15
20
25

crust with rigidity modulus G (Fig. 2). Earthquake strength may also be assessed by various magnitude scales based on amplitude measurements at selected frequencies within the spectrum of radiated waves. For crustal earthquakes, local magnitude, M_L (~1 s period) and surface wave magnitude, M_s (~20 s period) are widely used. The empirically derived Gutenberg-Richter equation

$$\log_{10} E_s = 1.5M_s + 4.8 \qquad \text{S.I.} \qquad (2)$$

relates M_s to the total radiated wave energy, E_s, (Richter 1958). Hanks & Kanamori (1979) recognized a moment-magnitude relationship

$$\log_{10} M_0 = 1.5M + 9.05 \qquad \text{S.I.} \qquad (3)$$

which is uniformly valid for $3 \leqslant M = M_L \leqslant 7$ and $5 \leqslant M = M_s \leqslant 7.5$, and may be used as the basis for a moment-derived magnitude scale which does not saturate at high magnitudes (Kanamori 1977). Different relationships may hold at lower magnitudes (Kanamori & Anderson 1975). Conventionally, following Hagiwara (1965), earthquakes may be ranked as great ($M > 8$), major ($8 > M > 7$), moderate ($7 > M > 5$), small ($5 > M > 3$), or as microearthquakes ($M < 3$).

Other useful relationships include that of Wyss (1979),

$$M = \log_{10} A - 1.85 \qquad \text{S.I.} \qquad (4)$$

empirically relating magnitude to rupture area for shocks with $M > 5.6$, and that of Ohnaka (1976), who found that the total radiated energy could be approximated directly from the source parameters by

$$E_s \sim GW\bar{u}^2 = M_0(\bar{u}/L) \qquad (5)$$

These empirical relationships are mutually consistent provided one condition is met. Thus the Gutenberg-Richter relationship (eqn. 2) can be derived from equations 3 and 5 if

$$\bar{u}/L = E_s/M_0 = 5.6 \times 10^{-5} \qquad (6)$$

Alternatively, equations 3 and 4 may be combined to give a rather similar expression

$$\bar{u}/\sqrt{A} = 5.7 \times 10^{-5} \qquad (7)$$

for a typical crustal rigidity, $G = 3.3 \times 10^{10}\ N/\mathrm{m}^2$. The expression $\bar{u}/L \sim \bar{u}/\sqrt{A}$ is a measure of the characteristic change in shear strain which accompanies earthquake faulting, and may be compared with geodetic estimates of earthquake strain changes, whose mean value is about 5×10^{-5} (Rikitake 1975). This leads directly to one of the most remarkable properties of crustal earthquakes. The drop in shear stress accompanying faulting, averaged over the rupture surface, is given by

$$\Delta\tau = CG(\bar{u}/\sqrt{A}) \qquad (8)$$

where the non-dimensional shape factor, C, is of the order of unity. For the great majority of shallow earthquakes, apparently regardless of source dimension, $\Delta\tau$ lies between 1 and 100 bars with a logarithmic mean of about 30 bars (Kanamori & Anderson 1975; Hanks 1977, 1979). In the case of a circular rupture with radius r, $C = 7\pi^{3/2}/16$ (Eshelby 1957) and equations 8 and 1 combine to give

$$M_0 = \frac{16}{7}\Delta\tau r^3 = \frac{49}{256}\frac{(\pi G\bar{u})^3}{\Delta\tau^2} \qquad (9)$$

Using this simple circular fault model, the near constancy of stress drop can be used to obtain estimates for the likely range of source dimensions and mean slip for earthquakes of different moment (or magnitude assuming eqn. (3) to hold) (Fig. 3). However, this scaling relationship tends to break down for larger non-circular crustal ruptures ($M > 6$), especially those with high aspect ratios (Scholz 1982).

Size-frequency distribution of earthquakes

Gutenberg & Richter (1954) demonstrated that in any seismically active region the annual number of earthquakes $N(M)$ with magnitudes $M \pm \delta M/2$ follows the exponential distribution

$$\log_{10} N(M) = a - bM \qquad (10)$$

where a and b are constants characteristic of that region, respectively related to the general level of activity and the proportion of large to small shocks. Though ranging from about 0.8 to 1.2, b generally has a value close to unity so that the frequency of earthquake occurrence increases by about a factor of 10 per unit decrease in magnitude. In any particular region there appears to be a maximum magnitude above which the distribution is truncated (Molnar 1979). As an example, we may consider a plate

FIG. 2. Conceptual model for a major fault zone in continental crust, schematically relating faulting mode, metamorphic environment, dominant steady-state quartz deformation mechanism and associated fault rocks. Profiles of shear resistance and distortional strain energy density versus depth are in arbitrary units. Metamorphic facies boundaries after Turner (1981) for a uniform crustal density of 2.80 g/cm^3 and thermal gradient of 25° C/km, on the assumption (unwarranted in the frictional regime) that $P_{load} = P_{H_2O}$. High conjectural deformation mechanism map for quartz is based on the work of Elliot (1973), Rutter (1976), White (1976), McClay (1977), Kerrich *et al.* (1977, 1980) and the author's own observations.

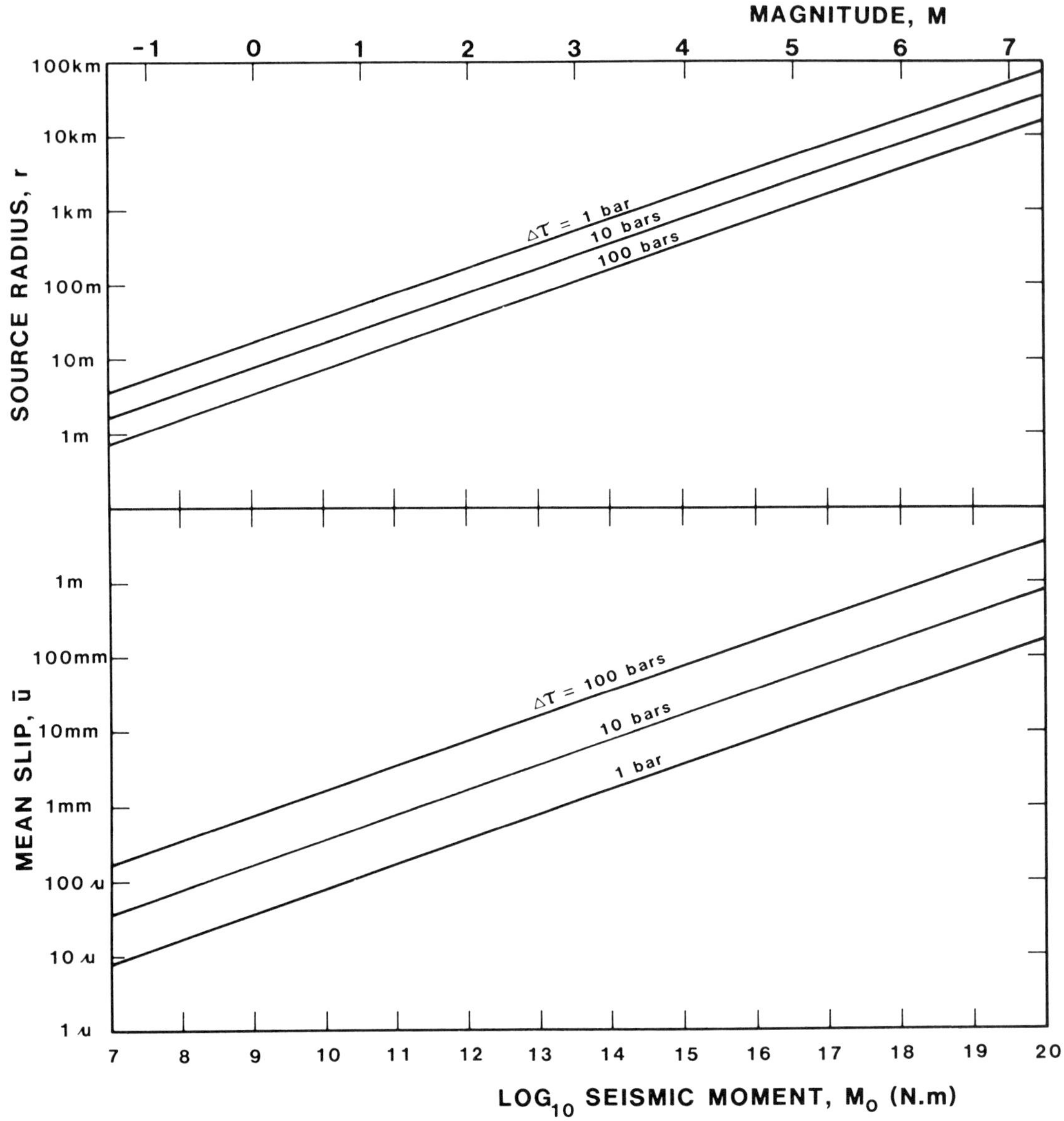

FIG. 3. Source radius and mean slip in relation to seismic moment for circular earthquake ruptures with $1 \leqslant \Delta\tau \leqslant 100$ bars (magnitude derived from the Hanks & Kanamori (1979) moment-magnitude relationship).

boundary extending for 1000 km along which seismicity follows a frequency-magnitude distribution with $a = 6$ and $b = 1$, roughly equivalent to the situation for crustal earthquakes in California or New Zealand. Table 3 illustrates the expected annual frequency of occurrence for earthquakes of varying magnitude down to $M = 0$, close to the lower limit of detection. Assuming uniform distribution of seismicity (not usually the case), the annual frequency of smaller events per kilometre of plate boundary is also given. Note the high level of microearthquake ($M < 3$) activity, even though seismicity may be distributed in a broad swathe without concentration along any one dominant fault.

Theory and experiment both suggest that the b-value is sensitive to the level of tectonic stress and/or the degree of stress heterogeneity within the fault zone. Low b-values (an abnormally high ratio of large

TABLE 3: *Earthquake frequency along a 1000 km plate boundary with a = 6 and b = 1.*

M	8	7	6	5	4	3	2	1	0
$N(M)$/yr	10^{-2}	10^{-1}	1	10	10^2	10^3	10^4	10^5	10^6
$N(M)$/km yr	—	—	—	—	—	1	10	10^2	10^3

to small shocks) correspond to comparatively high, long wavelength concentrations of stress and vice versa (Scholz 1968; Wyss 1973; Hanks 1979). In the usual case where $b \sim 1$, it follows from equations 4 and 10 that $N(M) \propto 1/A$, which may be justified on theoretical grounds (Hanks 1977; Andrews 1980).

Seismic slip rates from moment summation

Brune (1968) developed a technique for estimating the seismic slip rate along major fault zones or plate boundaries. A simple planar fault zone which extends for a distance, L_s, with seismic activity occurring to depth, W_s, possesses a total seismogenic area, $A_s = L_sW_s$ (Fig. 2). From equation 1, a single earthquake of rupture area A, and moment M_0, contributes a component of slip averaged over the seismogenic area given by

$$\langle \bar{u} \rangle = \bar{u}(A/A_s) = (M_0/GA_s) \qquad (11)$$

Over a time period, t, therefore, the seismic slip rate is

$$v_s = \frac{\sum_0^t \langle \bar{u} \rangle}{t} = \frac{1}{GL_sW_st} \sum_0^t M_0 \qquad (12)$$

In regions where seismicity is distributed over a broad area on faults of varying attitude and slip-sense (e.g. Central Asia), a more complete analysis involving summation of the full moment tensors may be used to establish the 'seismic volume flow' or volumetric strain rate throughout the seismogenic regime (Kostrov 1973; Anderson 1979; Molnar 1979).

In general, only the biggest earthquakes need be considered. This follows from consideration of the frequency-magnitude and moment-magnitude relationships. For $b = 1$, frequency of occurrence increases by a factor of 10 per unit drop in magnitude, but the sum moment contribution decreases by a factor $10 \times 10^{-1.5} = 10^{-0.5}$. Thus when magnitude drops by two units, the moment contribution decreases by a factor of 0.1. In a fuller treatment, equations 3 and 10 may be combined to give a moment-frequency relationship from which the total cumulative moment can be shown to be about twice the moment of the greatest earthquake in the region (Wyss 1973).

Because the moment contributions of the largest earthquakes are so dominant, the extent to which an estimate of v_s for a simple fault zone accurately reflects the long-term seismic slip rate clearly depends on the length of the sampling period with respect to the recurrence intervals for the biggest events. Inasmuch as reasonable estimates of seismic moment (obtained instrumentally, or from historical records of ground breakage or isoseismal areas) are at best available only for events of the past 100 years, the method can only work well for individual fault zones exhibiting Class AA or A activity (e.g. Thatcher *et al.* 1975). The possibility of long-term (several hundred year) fluctuations in seismicity, as demonstrated by Ambraseys (1971) for the Middle East, must also be borne in mind. Considerable uncertainties in the estimation of v_s may also arise through inadequate knowledge of the depth of the seismogenic zone (e.g. Chen & Molnar 1977).

Davies & Brune (1971) summed the moments of all shallow earthquakes with $M \geqslant 7$ which occurred over a 70 year period along the major plate boundaries. Despite the uncertainties, most of the seismic slip rates so obtained lay within a factor of two of the slip rates predicted from sea-floor spreading. However, North (1974) showed that along much of the fragmentary, partly intracontinental plate boundary extending from Gibraltar through the Middle East to Iran, the seismic slip rate from moment summation is deficient by over an order of magnitude.

Continental fault structure

A general conceptual model for continental fault structure was developed by Sibson (1977) from consideration of the diverse rock deformation textures found in ancient fault zones, and their relation to faulting style. In this model a seismogenic frictional regime, dominated by discontinuous pressure-sensitive deformation involving cataclasis and frictional sliding, gives way with increasing depth to a quasi-plastic regime where largely aseismic and continuous shearing is localized within mylonite belts. The passage from frictional to quasi-plastic behaviour in quartzo-feldspathic crust is determined primarily by the changing response of quartz to deviatoric stress with increasing temperature (Fig. 2). Intracrystalline plasticity involving dislocation climb and glide becomes the dominant deformation mechanism at temperatures greater than $300 \pm 50°$ C (Voll 1976; Tullis & Yund 1977; Kerrich *et al.* 1980), corresponding to the onset of a greenschist facies metamorphic environment. Other deformation mechanisms involving diffusive mass transfer (in particular, pressure solution) are also believed to operate in fault zones over the temperature range 200–400 °C, but their contribution is difficult to assess (Elliot 1976; McClay 1977; Rutter & Mainprice 1978).

Downward extension of larger earthquake ruptures may cause transient loss of continuity from time to time within the upper reaches of the quasi-plastic regime (Sibson 1980*a*). However, higher temperature at greater depths promotes strain-induced deformation and recrystallization of other important rock constituents. Voll (1976), for example, suggested that recrystallization of plagioclase feldspar becomes significant at $T > 450$ °C. In the deep crust, therefore, fully ductile continuous deformation is thought to predominate in the gneissose shear zones which develop under conditions of amphibolite facies metamorphism (Bak *et al.* 1975; Grocott & Watterson 1980). The frequent presence of retrograde mineral assemblages and/or hydrothermal veining in and around ancient crustal fault zones testifies to their important subsidiary role as conduits for aqueous fluid flow, often on a massive scale (Beach 1976; Sibson 1981).

Rheology of fault zones

On the basis of this conceptual model, shear resistance can be expected to increase approximately linearly with increasing depth and pressure through the frictional regime to reach a peak value in the vicinity of the frictional/quasi-plastic transition, beneath which it should decrease exponentially with increasing temperature. Sufficient knowledge now exists on the frictional and rheological properties of quartz-bearing rocks for this concept to be explored quantitatively for quartzo-feldspathic crust.

Frictional regime. Here it is assumed first that the simple law of effective stress holds (Hubbert & Rubey 1959), so that all effective normal stresses are given by

$$\sigma_n' = \sigma_n - P = \sigma_n - \lambda_v \sigma_v \quad (13)$$

where σ_n is the applied normal stress, P is the fluid pressure, λ_v is the pore fluid factor, and σ_v is the vertical stress, taken as equal to the lithostatic load. Under hydrostatic fluid pressures, $\lambda_v = 0.36$ for an average crustal density of 2.75 g/cm^3. Byerlee (1978) showed that almost irrespective of rock type, frictional failure on existing planes is governed by a composite linear relationship which, to mid-crustal depths, is adequately approximated by

$$\tau = \mu \sigma_n' \quad (14)$$

where τ and σ_n' are respectively the shear and effective normal stresses acting on the fault surface and the coefficient of static friction $\mu = 0.75$ (Fig. 4*a*).

Quasi-plastic regime. Dislocation creep in quartz is assumed dominant during quasi-plastic mylonitization, so that the constitutive flow law is of the form

$$\dot{\varepsilon} = A(\sigma_1 - \sigma_3)^n \exp(-Q/RT) \quad (15)$$

where $\dot{\varepsilon}$ is the strain rate, R is the gas constant, T is absolute temperature and A, n and Q are material constants. Experimental deformation of nominally 'dry' Westerly granite (average grain size ~0.75 mm; composition about 30% each of quartz, oligoclase and microcline, plus 10% biotite and accessories (Brace 1965)) at high temperatures and pressures has yielded $A = 8.83 \times 10^{-4}$ kbar^{-n} s^{-1}, $n = 2.9$ and $Q = 25.3$ k cal/mole for steady state creep. Microstructural studies show that quartz has accommodated most of the strain by crystal plastic flow combined with some microfracturing (Carter *et al.* 1981). This 'granitic' flow law may be compared with rather similar equations recently obtained for 'wet' and 'dry' quartzite by Koch *et al.* (1980) on the plot of flow shear resistance ($\tau = \frac{1}{2}(\sigma_1 - \sigma_3)$) versus temperature for the aseismic strain rates likely to be associated with crustal faulting (Fig. 4*b*). These range from geodetically determined values of 10^{-13} to 10^{-14}/s, related to the regional accumulation of elastic strain, to the strain rates of 10^{-10} to 10^{-12}/s inferred for mylonite belts at depth (Sibson 1977).

Some caution must be exercised in the application of such flow laws to the natural environment. Whereas no explicit pressure dependence appears in the constitutive equation, recent work has demonstrated that hydrolytic weakening in silicates is governed by the rate of water diffusion, which is affected by confining pressure (Tullis & Yund 1980; Blacic 1981). However, the capacity for ready flow of quartz-bearing rocks under moderate to high grade metamorphism at shear stresses below 1 kbar, as suggested by the plots in Fig. 4*b*, is in good accord both with field evidence for the onset of significant quartz plasticity at around 300° C, and with estimates of flow shear stress between 0.1 and 1 kbar obtained from palaeo-piezometric studies on quartz-bearing tectonites deformed under conditions of greenschist and amphibolite facies metamorphism (Kohlstedt & Weathers 1980). There are good grounds, therefore, for believing this granitic flow law appropriate to the mylonitization of quartz-bearing rocks, and it is used in the ensuing discussion.

Variations in shear resistance and strain energy concentration with depth

The general criterion for frictional failure on rock surfaces (eqn. 14) can be combined with the quasi-plastic granitic flow law to yield composite profiles showing the variation of shear resistance with depth for different geotherms, strain rates and faulting modes (Fig. 5). Somewhat similar procedures using different rheological parameters have been employed by Brace & Kohlstedt (1980), Meissner (1980), and Meissner & Strehlau (1982).

Within the frictional regime we consider the three dominant stress regimes and associated faulting modes recognized by Anderson (1951) on the basis of which of the three principal compressive stresses ($\sigma_1 > \sigma_2 > \sigma_3$) is vertical. Thrust, strike-slip and normal faults develop where σ_v is respectively σ_3, σ_2 and σ_1. From the frictional criterion, the minimum shear stress needed to induce slip at a given λ_v can be plotted against depth for the three faulting modes (Sibson 1974). In Fig. 6 these limiting conditions for slip with $\mu = 0.75$ are plotted for hydrostatic fluid pressures and on the assumption, in the case of strike-slip faulting,

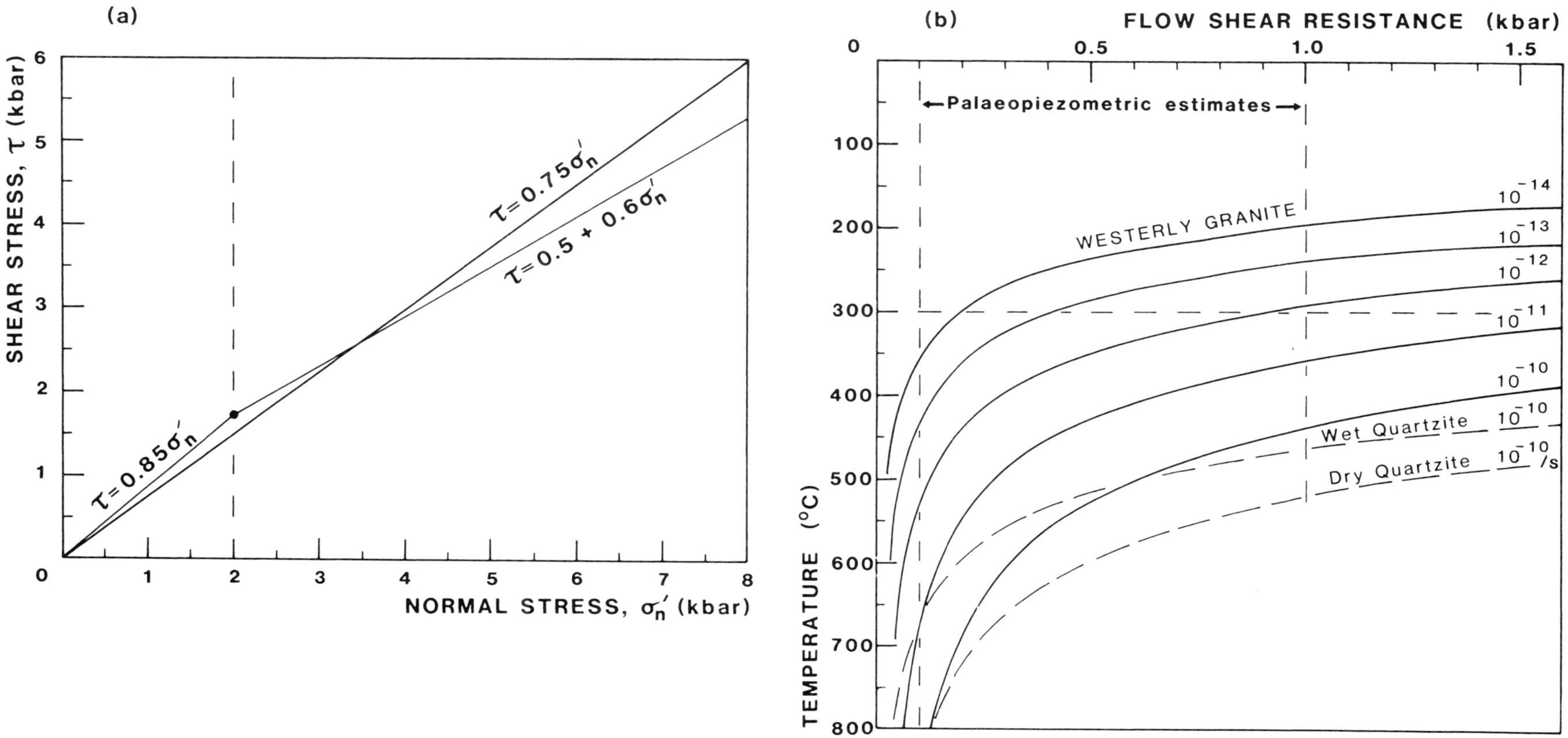

FIG. 4. (*a*) Byerlee's (1978) general composite relationship for rock friction with single, straight-line approximation; (*b*) Flow shear resistance versus temperature for Westerly Granite over the likely range of fault zone strain rates (from flow law of Carter *et al.* 1981); comparative quartzite flow laws from Koch *et al.* (1980)).

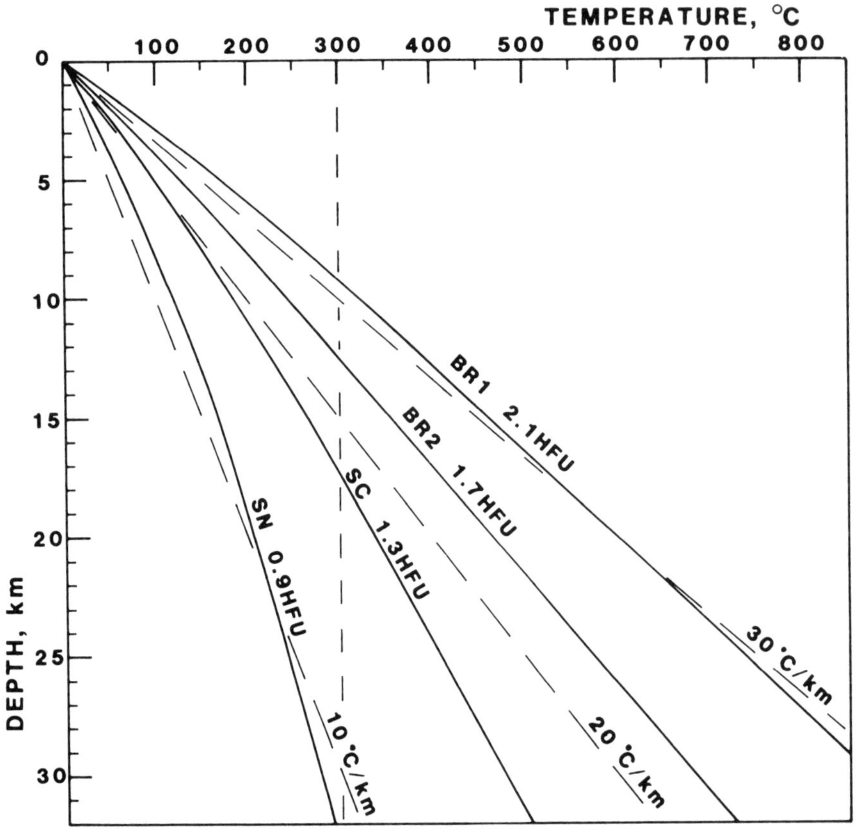

FIG. 5. Continental geotherms in the United States (after Lachenbruch & Sass 1977).

that $\sigma_2 = \frac{1}{2}(\sigma_2 + \sigma_3)$. Higher values of fluid pressure cause the frictional failure lines to rotate towards the depth axis, but at any given depth the stress levels around thrust, strike-slip and normal faults remain in the proportion 4:1.6 to 6:1. Note that as $\sigma_2 \rightarrow \sigma_3$ or σ_1, the failure condition for strike-slip moves towards those for thrust and normal faulting respectively.

Through the quasi-plastic regime, flow shear resistance is determined from the granitic law and plotted to intercept the frictional failure lines for the range of appropriate strain rates, and for each of the continental geotherms given in Fig. 5. Conductive geotherms BR1 and BR2 correspond respectively to surface heat flows of 2.1 and 1.7 HFU (1 HFU = 1 μcal/cm^2 ~ 42 mW/m^2), bracketing most of the values obtained from the Basin and Range province of active extensional tectonics, and from the region around the strike-slip San Andreas fault system (Lachenbruch & Sass 1977, 1980). Geotherms SC and SN correspond respectively to the surface heat flow characteristic of stable cratonic areas such as the eastern United States (1.3 HFU) and to the abnormally low heat flux of the Sierra Nevada batholithic terrain (0.9 HFU).

On the assumption that fault zones continue downward through the quasi-plastic regime as localized shear belts of finite width (this important question of downward continuity will be discussed further), some idea of the mylonitic strain rates likely to be associated with Class AA fault zones can be obtained directly from the San Andreas fault. At the surface, the entire fault zone is typically about 1 km in width (Allen 1981), possibly reflecting the width of a quasi-plastic shear zone at depth. An overall slip rate of 30 mm/yr (Dickinson *et al.* 1972) therefore yields a probable deep localized strain rate of at least 10^{-12}/s. However, seemingly continuous shear strain rates around 10^{-11}/s have been obtained directly from surface measurements at a number of localities along the creeping segment (Burford & Harsh 1980), so that a range of 10^{-11} to 10^{-12}/s seems the most appropriate (shaded sections of the strain rate fields in Fig. 6).

It is important to recognize the limitations inherent in these composite profiles of shear resistance versus depth (Fig. 6). First, it has been assumed that the middle and lower continental crust have a sufficiently high quartz content for the granitic flow law to be

FIG. 6. Shear resistance versus depth for varying strain rates under the different continental geotherms. Fluid pressure assumed hydrostatic ($\lambda_v = 0.36$) in the frictional regime (T = thrust, S = strike-slip, N = normal fault).

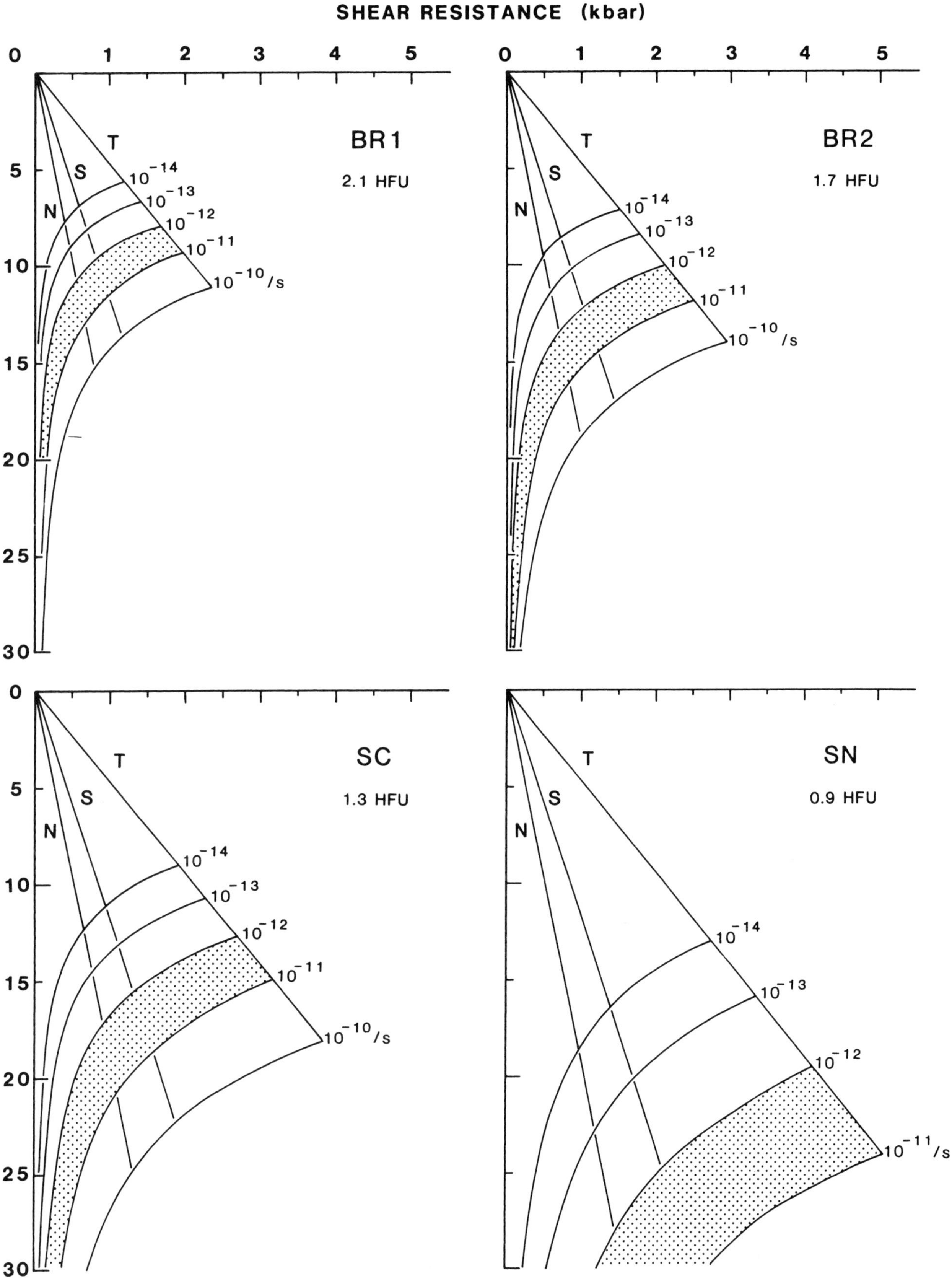
SHEAR RESISTANCE (kbar)
0 1 2 3 4 5
0 5 10 15 20 25 30
BR1
2.1 HFU
BR2
1.7 HFU
SC
1.3 HFU
SN
0.9 HFU
T
S
N
10^{-14}
10^{-13}
10^{-12}
10^{-11}
10^{-10}/s
10^{-11}/s

operative. In general, quartz content is expected to decrease with depth but recent interpretations have emphasized the compositional heterogeneity of the lower crust and the likelihood that it contains a significant fraction of acid and intermediate rocks metamorphosed to high amphibolite and granulite facies assemblages (Christensen & Fountain 1975; Smithson & Brown 1977; Fountain & Salisbury 1981). Moreover, increasing temperatures in the deep continental crust should counteract a diminishing quartz content by promoting plasticity of other important rock constituents such as feldspar. Secondly, the rather sharp peak in shear resistance inferred for the transition from frictional to quasi-plastic behaviour is unrealistic. It seems more likely that the transition must be spread over a broad interval in which a mixture of brittle and ductile shearing is accommodated (Sibson 1980*a*). The sawtooth stress cycles accompanying earthquake faulting in the frictional regime must also extend downward some distance into the quasi-plastic shear zones. Furthermore, no account has been taken of pressure solution and allied deformation mechanisms which are also likely to be operative in the transition region with the effect of smoothing and lowering the peak in shear resistance (Fig. 2). Natural strength profiles are also likely to possess a short wavelength ‘roughness’ arising from variations in lithology and fluid pressure, and from geometric irregularities in fault structure (Segall & Pollard 1980).

Despite these shortcomings, the profiles of shear resistance (Fig. 6) illustrate very clearly the strong controls exerted by geothermal gradient and quasi-plastic strain rate on the transition from frictional to quasi-plastic behaviour. Both the depth and the amplitude of the peak shear resistance increase for low geothermal gradients and/or fast strain rates, and vice versa. Distortional strain energy per unit volume varies as the square of shear stress, so that acute concentrations of strain energy, highly sensitive to the value of the peak shear resistance, are to be expected around the frictional/quasi-plastic transition once stress levels approach failure (Sibson 1977) (Fig. 2).

Thermal consequences of faulting

Long-term fault activity may profoundly affect the thermal regime of the crust. Two effects need to be considered: the disruption of existing conductive thermal structure by dip-slip faulting, and the temperature increases which may in some cases arise from shear heating within the fault zone.

Conduction versus convection. In tectonically active regions conductive heat transport through the crust may be severely perturbed or subordinated by the vertical transport of large crustal masses along faults, a form of convective heat transfer. The dominance of either conductive or convective heat transfer may be established from consideration of the Peclet number, P (Oxburgh & Turcotte 1974), defined by

$$P = \frac{v'l}{K} \qquad (16)$$

where v' is the time-averaged trans-isothermal component of slip rate, l is a characteristic dimension for the transported material, and K is the thermal diffusivity, usually about $10^{-6}\,m^2/s$. For $P < 1$, the conductive thermal structure is little affected by the faulting, but for $P > 1$ significant perturbations of the pre-existing thermal regime will occur. At high Peclet numbers isotherms are virtually transported as passive markers. In Fig. 7*a*, time-averaged slip rate for the various degrees of fault activity is plotted against characteristic dimension to define the conduction and convection dominated fields. Convective heat transport is clearly dominant in moderate to highly active regions of dip-slip faulting where individual fault blocks are, say, 10 km or greater in dimension. In regions of crustal extension and thinning involving normal faulting, geothermal gradient must increase in the vicinity of the faults, which should cause a shallowing of the frictional/quasi-plastic transition and a decrease in the peak shear resistance (cf. McKenzie 1978). In contrast, where active crustal thickening occurs by thrust stacking in regions of horizontal shortening, the geothermal gradient may become severely depressed at depth (Oxburgh & Turcotte 1974). This should lead to a deepening of the frictional/quasi-plastic transition and to an increase in the peak shear resistance, though an additional countering factor should be noted; thrust environments are also particularly favourable for the development of abnormally high fluid pressures (Hubbert & Rubey 1959; Sibson 1981). Note also that the near-surface gradient in the hanging wall block is likely to be enhanced as a consequence of erosion accompanying rapid uplift (Allis *et al.* 1979).

Shear heating. Here we are concerned only with long-term heating effects; the transient, sharply localized temperature increases that may arise from individual seismic slip episodes (McKenzie & Brune 1972; Sibson 1980*b*) are not under consideration. Then the rate of heat generation per unit area of the fault surface is given by

$$\dot{Q} = \tau_f v \qquad (17)$$

where v and τ_f are the time-averaged slip rate and shear resistance, respectively. Assuming uniform heat production along a discrete plane (or planar zone) within an infinite homogeneous medium, and adopting reasonable values for thermal diffusivity and conductivity (respectively, $10^{-6}\,m^2/s$ and $3\,W/m°K$), the temperature rise along the fault zone after time, t, is

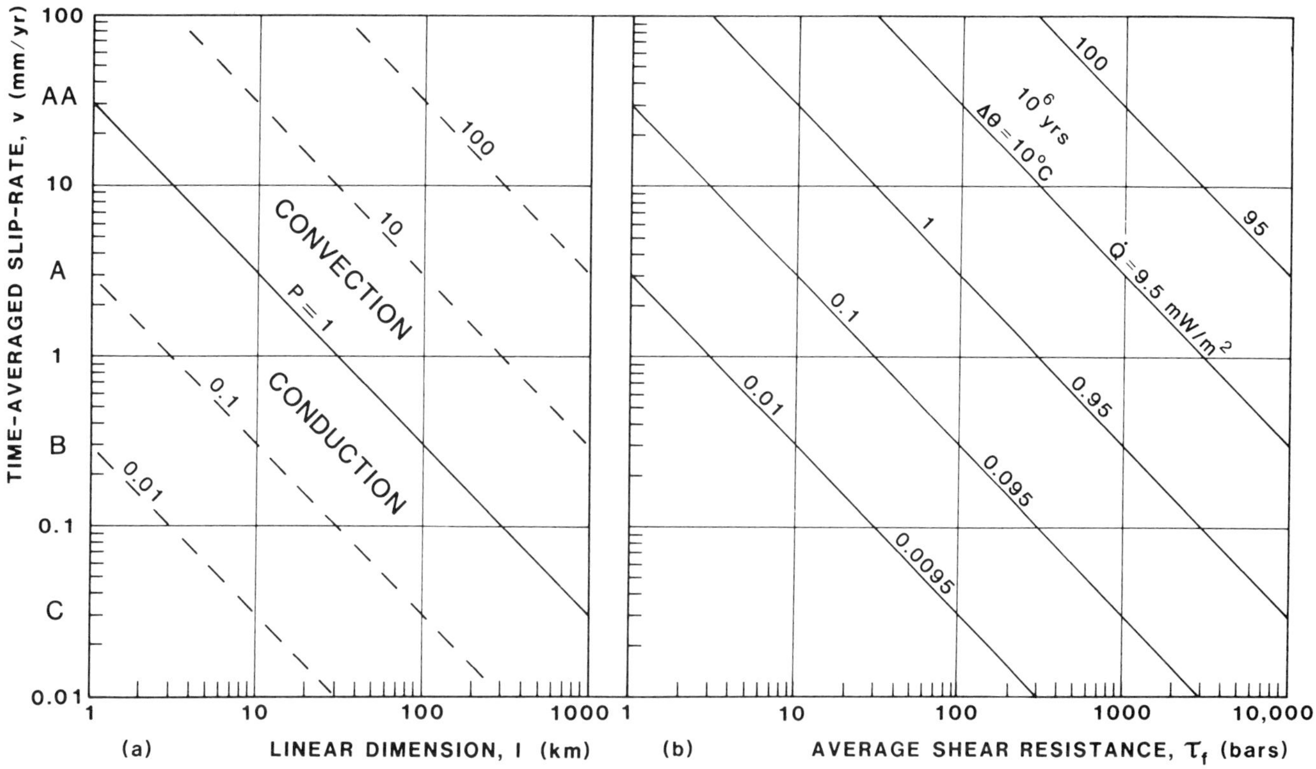

FIG. 7. Long-term thermal effects for the different classes of fault activity: (*a*) Convective and conductive fields in relation to fault block dimension, as defined by the Peclet Number, P; (*b*) Power dissipation, and temperature increase after 1 Ma of uniform shear heating, versus average shear resistance.

given by (Carslaw & Jaeger 1959, pp. 76–80)

$$\Delta\theta = 1.88 \times 10^{-4}\dot{Q}\sqrt{t} \quad \text{S.I.} \qquad (18)$$

In Fig. 7*b*, lines of equal power dissipation and temperature increase (calculated for a 1 Ma period) are drawn on a plot of time-averaged slip rate versus shear resistance. It is apparent that significant temperature increases approaching 100° C after 1 Ma are only likely to be achieved by shear heating along Class AA fault zones, and then only when the *average* crustal shear resistance is maintained in excess of 1 kbar (cf. Brun & Cobbold 1980; Scholz 1980). Consideration of the shear resistance profiles in Fig. 6 further suggests that this latter condition is most likely to be met along fast-moving thrust faults developed initially in cold, dry cratonic crust. The process is of course self-destructive through feedback thermal weakening, with the greatest likely effect concentrated around the peak in shear resistance and contributing to its smoothing-out.

It is evident from the considerations above that the thermal structure of Class AA thrust zones in particular may be exceedingly complicated and time-dependent. Note further that geochemical evidence suggests that substantial quantities of heat may also sometimes be transported through crustal fault zones by convective fluid transfer (Beach 1976). The combined effects of thrust stacking and shear heating have been considered for low-angle overthrusts lying sub-parallel to isotherms by Graham & England (1976) and Brewer (1981), while the general thermal structure of continental collision zones has been analysed for a range of boundary conditions by Toksoz & Bird (1977), and Bird (1978).

Average stress levels

Probably the most contentious and fundamental question in fault mechanics today is whether deviatoric stress levels associated with crustal faulting are of the order of 100 bars or 1 kbar on average (Hanks & Raleigh 1980). It has been argued convincingly (Brune *et al.* 1969; Lachenbruch & Sass 1980) that the absence of a detectable, localized heat flow anomaly arising from shear heating on the San Andreas fault prohibits a time-averaged shear resistance greater than perhaps 200 bars, roughly comparable with the larger stress drops accompanying earthquake faulting. For the San Andreas fault zone, the most appropriate profiles of shear resistance are those for strike-slip faulting with geotherms BR1 and BR2 and strain rate fields of 10^{-11}–10^{-12}/s (Fig. 6). Accepting these at face value, the average shear resistance through the frictional regime under conditions of hydrostatic fluid pressure is in the range 400–600 bars (this changes to 225–400 bars or 850–1250 bars as $\sigma_2 \rightarrow \sigma_1$ and σ_3 respectively).

There are several possible ways of reconciling these disparate stress estimates. Bear in mind first that no account has been taken of pressure solution and allied deformation mechanisms, whose effect would be to smooth out and lower the peak in shear resistance. The presence of abnormal fluid pressures or an abundance of low-strength clay minerals (Wang & Mao 1979) in the frictional regime would likewise reduce the average strength without greatly deepening the transition from frictional to quasi-plastic behaviour. On the other hand, *in situ* stress measurements to around 1 km depth adjacent to the San Andreas fault have yielded an average shear stress gradient of ~90 bars/km (Zoback *et al.* 1980), remarkably close to the frictional strike-slip gradients in Fig. 6. Furthermore, O'Neil & Hanks (1980) argued that high stresses remain a possibility because heat flow anomalies arising from shear heating may be 'washed out' by hydrothermal convection within the frictional regime, though there is a lack of direct evidence that this process occurs. Clearly, the controversy is far from being resolved.

As a final point we may recall the palaeostress estimates of 0.1–1 kbar obtained from quartz-bearing mylonites deformed (usually in thrust zones) under conditions of greenschist and amphibolite facies metamorphism (Kohlstedt & Weathers 1980), values which are compatible with most of the quasi-plastic portions of the shear resistance profiles for regions of average, or greater than average, continental heat flow.

Seismicity in relation to structure

Teleseismic location of moderate or greater earthquakes in continental crust, in particular estimation of their focal depths, is fraught with difficulties and the results are often unreliable (e.g. Jackson 1980). However, it is becoming increasingly evident from detailed studies of individual events by waveform modelling, analysis of *pP* phases, master event relocations, etc., that away from subduction margins intracontinental seismicity is concentrated largely in the upper half of the crust (Jackson & Fitch 1979, 1981; Chen & Molnar 1983). Similar conclusions have been reached following the installation over the past 15 years of densely instrumented local seismograph networks, which allow high resolution location of microearthquakes provided a satisfactory crustal velocity model has been established (Lee & Stewart 1982). The highest, A-quality locations have epicentral and focal depth uncertainties of ±1 km and ±2 km respectively, while for B-quality solutions the uncertainties increase to ±2.5 km and ±5 km.

These microearthquake studies are of two principal kinds. For moderate to highly active regions, the background microseismicity can be assessed over comparatively short time periods (weeks or months), yielding information on fault zone geometry, on the character of faulting through focal mechanism studies and on local frequency-magnitude relationships (e.g. Eaton *et al.* 1970; Pfluke & Steppe 1973; Prevot *et al.* 1980). Interest in the geometrical structure of fault zones at depth has been heightened recently by the development of fault 'barrier' theory to explain the unsteady propagation of larger ruptures (Das & Aki 1977; Aki 1979), and by a fresh appreciation of the mechanical implications of *en echelon* segmentation along faults (Segall & Pollard 1980). This latter concept, involving the recognition of pinned and unpinned patches at *en echelon* offsets, does much to explain the varied distribution of microearthquake activity in relation to the nucleation sites of larger ruptures along the segmented trace of the San Andreas fault in central California (Bakun *et al.* 1980; Lindh & Moore 1981). In the second kind of study, accurate location of microearthquakes in aftershock sequences following moderate or larger ruptures may yield information on the extent and geometry of the main rupture surface or surfaces, and on the migration of activity with time, allowing inferences to be drawn regarding stress redistribution following the main shock (e.g. Eaton *et al.* 1970; Hamilton 1972; Ellsworth 1975; Soufleris *et al.* 1982; Reasenberg & Ellsworth 1982).

Inasmuch as shallow seismic activity is thought to arise chiefly through frictional stick-slip on existing planes (Brace & Byerlee 1966), these high resolution microearthquake studies are of particular relevance to the underlying theme of this paper, providing direct information on the depth of frictional interaction around fault zones in different tectonic provinces.

Heat flow and the depth distribution of background microseismicity

Fig. 1 shows the location of sampling areas for background microearthquake activity in the western United States away from regions of active subduction, in relation to major active fault zones and heat flow provinces. Histograms illustrating the depth distributions for well-located events within these areas, normalized to percentages, are given in Fig. 8*a*. Dominant faulting modes, the sample size and data quality, and the depth above which 90% of the activity occurs are also indicated. Note that no aftershock sequences following moderate or larger events are included, and that microearthquakes in the geothermally active region occupying the Geysers and Clearlake Highlands quadrangles have been stripped from the central California catalogue for 1976 and are plotted separately. In all these areas, no consistent relationship is obvious between the depth of microseismic activity and major crustal layering, though clustering of activity sometimes occurs near velocity boundaries, either through an artefact of the location programme, or possibly as a result of fluctuations in strain energy concentration arising from the contrasting elastic properties of the layers (Sibson 1982*a*). This seems to rule out the possibility that the depth of the seismogenic zone is lithologically controlled, specifically by the passage from 'granitic' to 'gabbroic' crustal layers as suggested by Scholz *et al.* (1969).

All the regions considered here have background surface heat flows of 1.5–2.5 HFU, though much higher values are measured locally in the Coso Range, Imperial Valley, and of course around The Geysers geothermal area. Indeed, restriction of microearthquakes to the top few kilometres of the crust as at The Geysers, where vapour temperatures reach ~240° C (Majer & McEvilly 1979), has also been noted locally around the geothermal fields in the Imperial Valley where reservoir temperatures range up to 340° C (Gilpin & Lee 1978). With the exception of the western Transverse Ranges where activity extends significantly deeper, most of the microearthquakes occur within the top 10–12 km of the crust. Reference to the shear resistance profiles in Fig. 6 shows that the depth of this background microearthquake activity corresponds rather well with the modelled frictional/quasi-plastic transition for the appropriate strain rates and geotherms (BR1 and BR2). The abrupt decrease in shear resistance beneath the transition also does much to explain the sharp definition of the base of the seismogenic zone.

Insufficient data exist for statistical evaluation, but available information suggests that low intensity microearthquake activity extends to significantly greater depths in regions of lower heat flow. Focal depths extending to 20–25 km have been noted in the eastern United States and Canada (Sbar & Sykes 1977) where surface heat flow is generally 1.0–1.5 HFU (Lachenbruch & Sass 1977). Mixed mode microearthquakes have also been accurately located to depths of 35–40 km in the deep crust beneath the western foothills of the Sierra Nevada, where surface heat flow is less than 1.0 HFU (Marks & Lindh 1978; Wong & Savage 1983). While the strain rate fields appropriate to these regions are less certain, the extended depth distributions are broadly consistent with the deepening frictional regime which may be inferred from the shear resistance profiles for geotherms SC and SN (Fig. 6).

Together, these results support the hypothesis that temperature-dependent plasticity in quartz is the prime factor determining the depth of the seismogenic zone in continental crust.

Regions of active thrusting. The deeper activity in the western Transverse Ranges is of particular interest. This is a region of active thrust tectonics with late Quaternary uplift rates locally exceeding 5 mm/yr (Yerkes & Lee 1979) and a probable overall thrust slip rate of >10 mm/yr (Bird, in press). It seems likely that the deeper microseismic activity arises from depression of the thermal gradient at depth by the thrust stacking, as previously discussed. The same process may account for the deep microseismicity observed along the complex thrust front of the Himalayan arc where continental subduction may be taking place at rates in excess of 20 mm/yr (Chen & Molnar 1977). Although earthquakes on the main detachment thrust are largely confined to the top 20 km of the crust, microseismic activity is distributed within the cold footwall slab to depths approaching 60 km (*cf.* thermal modelling of continental collision zones by Toksoz & Bird 1977) (Seeber *et al.* 1981).

Aftershock sequences and the focal depths of larger shocks

Depth distributions of A-quality aftershock locations associated with moderate or larger earthquakes in the western United States whose focal depths are also reasonably well determined (to within at least ±5 km) are plotted in Fig. 8*b*. The following points emerge. Focal depths of the larger thrust events in the western Transverse Ranges again lie a few kilometres deeper than those found elsewhere. In general, the larger events ($M > 5.5$) appear to have nucleated towards the base of the seismogenic zone defined by background microseismicity within each region and to have ruptured upwards, so that their aftershocks are largely confined to the same zone. However, there is some evidence from aftershock studies that the very largest crustal ruptures, while still nucleating around the base of the background seismogenic zone, may propagate some distance downwards into the quasi-plastic regime as well as upwards (e.g. the *M* 7.5 Niigata, Japan, earthquake of 1964 (Aki 1966; Kayano 1968)).

Several other features of these aftershock sequences are worth noting. For both the Parkfield and Borrego Mountain aftershock sequences, the *b*-value in the frequency-magnitude distribution varied with depth and reached a minimum in the 8–10 km interval, indicating long wavelength stress concentrations at that depth (Eaton *et al.* 1970; Wyss 1973). At Parkfield, aftershocks along a subsidiary *en echelon* fault segment associated with the 'stopping' of the main rupture were also concentrated in this depth interval (Lindh & Boore 1981). One year after the Borrego Mountain earthquake, the Coyote Mountain event, interpreted by Thatcher & Hamilton (1973) as the product of '. . . relatively high stresses within a relatively small area . . .', nucleated in a bottom corner of the original aftershock zone and extended it along

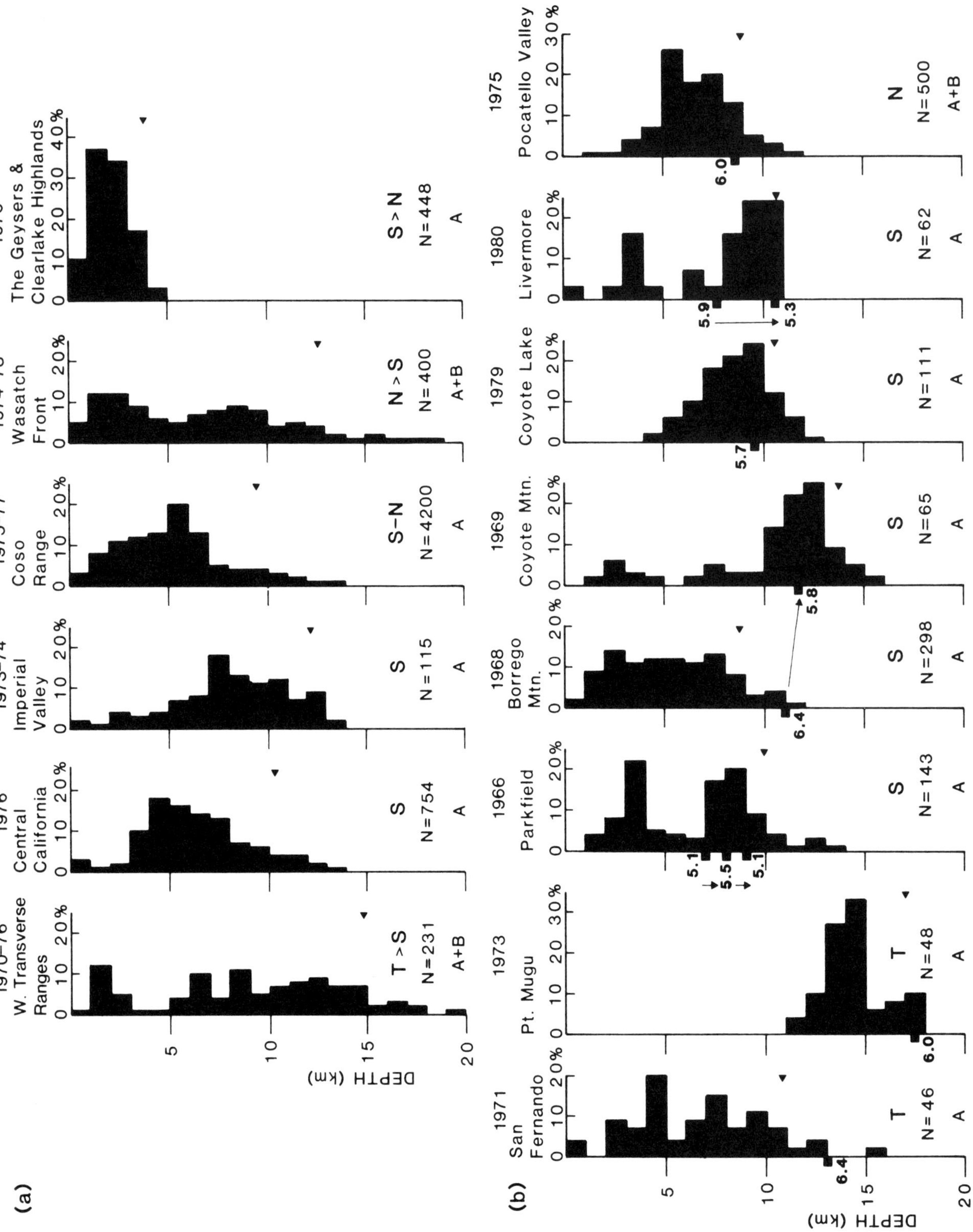
(a)
1970–76 W. Transverse Ranges
T > S
N = 231
A+B
1976 Central California
S
N = 754
A
1973–74 Imperial Valley
S
N = 115
A
1975–77 Coso Range
S–N
N = 4200
A
1974–78 Wasatch Front
N > S
N = 400
A+B
1976 The Geysers & Clearlake Highlands
S > N
N = 448
A
DEPTH (km)
(b)
1971 San Fernando
T
N = 46
A
1973 Pt. Mugu
T
N = 48
A
1966 Parkfield
S
N = 143
A
1968 Borrego Mtn.
S
N = 298
A
1969 Coyote Mtn.
S
N = 65
A
1979 Coyote Lake
S
N = 111
A
1980 Livermore
S
N = 62
A
1975 Pocatello Valley
N
N = 500
A+B
DEPTH (km)

strike. Concentrated stress release associated with rupture nucleation near the base of the seismogenic zone has also been invoked for the San Fernando (Hanks 1974; Bouchon 1978), Borrego Mountain (Heaton & Helmberger 1977; Ebel & Helmberger 1982), Pocatello Valley (Bache *et al.* 1980), Imperial Valley (Hartzell & Helmberger 1982), and Coyote Lake (Liu & Helmberger 1983) earthquakes.

The implication of all these observations is that the larger ($M > 5.5$) earthquakes nucleate around the peaks in shear resistance and distortional strain energy concentration associated with the frictional/quasi-plastic transition. Possibly, large ruptures can develop *only* if they nucleate in these areas of high strain energy concentration. There is considerable justification, therefore, for regarding the region around the base of the seismogenic zone as the prime asperity or sticking patch in crustal fault systems (*cf.* Wesson *et al.* 1973; Kanamori 1981). Variations in the depth and amplitude of peak shear resistance along strike (caused by changing rock type, fluid pressures, thermal gradients, etc.) may lead to substantial changes in strain energy density and affect the lateral extent of rupture propagation.

Thermal control of faulting style. Changes in the style of faulting appear to correlate with regional variations in heat flow along the San Andreas fault, and may be interpreted in terms of the geotherm-dependent profiles of shear resistance (Sibson 1982*b*). Intense microearthquake activity, aseismic creep and an absence of large shocks characterize the region of high heat flow (averaging ~2 HFU) in central California, with the very highest values of heat flow coinciding with the 50 km long segment of steady creep described by Burford & Harsh (1980). To the NW and SE along the ~400 km long segments that ruptured in the great *M*8 earthquakes of 1906 and 1857, the fault is currently locked with rather low levels of background microseismicity, and heat flow decreases to average ~1.7 HFU. Aseismic creep, occasional moderate shocks and more intense microseismic activity reappear SE of the 1857 rupture within the high heat flow region of the Imperial Valley (Goulty *et al.* 1978).

Although other geological factors may also play a role (Irwin & Barnes 1975; Allen 1981), these changes in behaviour are in accord with our expectations from the shear resistance profiles (Fig. 6), which predict that the depth and amplitude of peak shear resistance and strain energy concentration should be greatest in the colder parts of a fault zone and least in the hot. On this basis, the depth of the seismogenic zone should increase by several kilometres passing from the hotter to the colder segments of the San Andreas fault, and indeed a rather abrupt deepening of microseismic activity by several kilometres is apparent at both ends of the central Californian creeping segment (Moths *et al.* 1981).

Deep structure of fault zones: downward continuation or decoupling?

One of the outstanding problems affecting our understanding of crustal deformation by faulting is illustrated in Fig. 9 for the three dominant fault modes. Do major fault zones generally continue downward through the continental crust (and lithosphere?) or do they pass into flat-lying shear zones in the mid-to-deep crust so that décollement and decoupling of the upper crust occur on a regional scale? As both situations seem to occur, what factors favour one over the other?

In recent years these questions have been highlighted by a series of new structural interpretations and techniques. Listric normal faulting has been recognized on a hitherto unsuspected scale (De Charpal *et al.* 1978; McKenzie 1978; Bally 1981), and in western N America early to mid-Tertiary listric normal faulting in a cover sequence shows a consistent, if problematical relationship with flat-lying mylonite belts of regional extent, the so-called Cordilleran metamorphic core complexes (Coney 1980). In the region of the 'big bend' along the San Andreas fault (Fig. 1), the surface trace of the fault zone cuts across an anomalous strip of high velocity mantle beneath the Transverse Ranges, apparently without displacing it. This has prompted suggestions that the San Andreas fault system in the upper crust may be decoupled from the Pacific – N America lithospheric plate boundary at depth (Hadley & Kanamori 1977; Yeats 1981), although other possibilities exist (Bird, in press). Most intriguing of all, perhaps, is the extreme contrast between the Wind River thrust in Wyoming, apparently dipping fairly uniformly at about 35° through the

---

FIG. 8. Depth distributions of earthquakes in the western United States (see Fig. 1 for locations). Solid triangles indicate depth above which 90% of the activity occurs; the dominant faulting mode (T=thrust, S=strike-slip, N=normal), the sample size and the data quality are listed at the base of each histogram. (a) Background microearthquake activity (data from l. to r.): Lee *et al.* (1979*b*); McHugh & Lester (1978); Hill *et al.* (1975); Walter & Weaver (1980); Arabasz *et al.* (1980)). (b) Aftershock sequences following moderate earthquakes. Depth and local magnitude (M_L) of main shock (or shocks) are indicated by solid squares. Larger events in each sequence are linked by arrows. (Data from (l. to r.): Whitcomb *et al.* (1973), Hanks (1974), Bouchon (1978); Stierman & Ellsworth (1976); Eaton *et al.* (1970), Bakun (1972); Allen & Nordquist (1972), Hamilton (1972); Thatcher & Hamilton (1973); Lee *et al.* (1979*a*); Cockerham *et al.* (1980); Arabasz *et al.* (1980), Bache *et al.* (1980)).

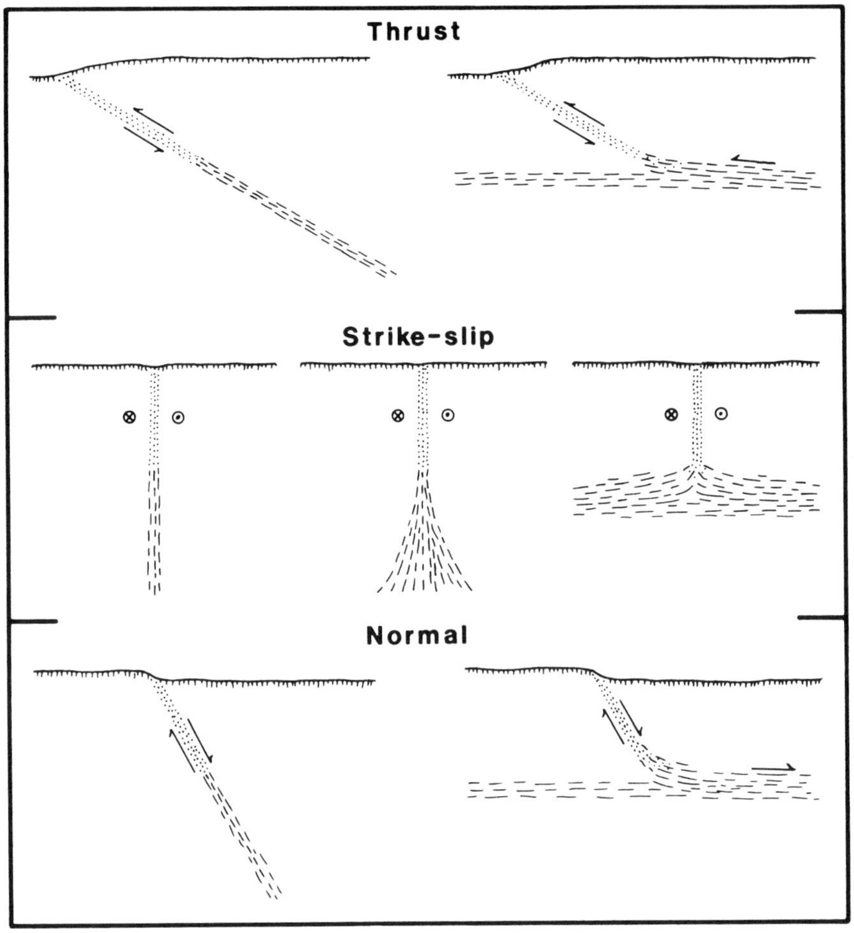

FIG. 9. Alternative models for the downward continuation of major fault zones (frictional regime dotted, quasi-plastic shear zones dashed).

entire crust, and the thin-skinned (6–15 km) crystalline allochthon overthrusting a sedimentary sequence for over 260 km in the southern Appalachians, as revealed by the COCORP programme of deep continental reflection profiling (Brewer *et al.* 1981).

Geological evidence

It is now widely recognized that thrust faults involving crystalline basement commonly flatten with depth into décollement zones developed under greenschist or greater grades of metamorphism in the middle or lower continental crust, to form partly or wholly crystalline thrust 'flakes' which are characteristically around 10–15 km in thickness (Oxburgh 1972; Armstrong & Dick 1974; Bally 1981). As discussed above, similar mid-crustal décollement regimes in crystalline quartzo-feldspathic rocks have also now been recognized in association with extensive cover sequences strongly disrupted by listric normal faulting.

The situation for transcurrent fault zones is rather more complicated (Fig. 9), and a series of possibilities exists ranging from downward continuation into quasi-plastic shear zones of constant width (as assumed up till now), through shear zones that widen with depth in the crust, to a fairly abrupt transition into a sub-horizontal decoupling shear zone (*cf.* Lachenbruch & Sass 1980; Prescott & Nur 1981). Coward (1980) discussed possible relationships between steep strike-slip and flat-lying shear zones developed under high grade metamorphic conditions in the Precambrian basement of southern Africa, but the field evidence as to which is the most appropriate general model is equivocal. In Brittany, NW France, the N and S Armorican Shear Zones of Hercynian (and possibly late Precambrian) age disrupt a crystalline basement complex, and provide a good example of a quasi-plastic, dominantly dextral strike-slip system formed in middle to lower continental crust under high greenschist to amphibolite facies metamorphic conditions (Cogne *et al.* 1974; Watts & Williams 1979; Jegouzo 1980; Gapais & Le Corre 1980). In scale and complexity, the whole transcurrent system of shear zones mimics quite remarkably the surficial patterns of

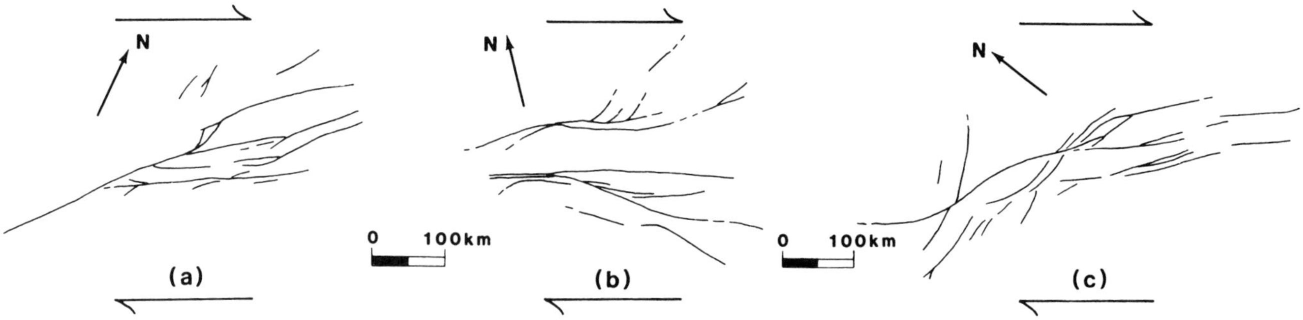

FIG. 10. Patterns of dextral strike-slip faulting at different crustal levels. Active surface faulting in the northern half of the South Island, New Zealand, and in southern California is shown in relation to regional slip vectors in (*a*) and (*c*) respectively (note that misaligned faults exhibit substantial dip-slip); the pattern of dominantly strike-slip, mid- to deep crustal quasi-plastic shear zones in Brittany is outlined in (*b*).

dominantly dextral strike-slip faulting now active in southern California and New Zealand (Fig. 10). Deformation is concentrated in sub-vertical high-strain mylonite belts, typically a few hundred metres wide, one of which can be traced for over 300 km. Percevault & Cobbold (1982) noted that substantial dextral shear strains ($\gamma \sim 1$) have also developed in the intervening regions between the major shear zones. The amount of strike-slip across individual mylonite belts is generally uncertain but is thought to reach tens or even hundreds of kilometres for the larger belts.

Sub-vertical transcurrent gneissose shear zones up to 40 km wide, with displacements possibly reaching 200 km in one instance, have also been described from crystalline basement in Greenland (Bak *et al.* 1975; Sorenson 1983). These structures developed in late Archaean – early Proterozoic time under conditions of high amphibolite to granulite facies metamorphism in the deep crust. The Nordre Stromfjord shear belt is of particular interest because an apparent deepening in the erosion level along strike is accompanied by a widening of the shear belt from 7 to 15 km. On this basis, the shear zone boundaries have been inferred to diverge downwards at an angle of about 40°, so that finite shear strain within the zone decreases with depth. Watterson (1979) also noted that shear strains are generally more intense in mid-crustal mylonite belts than in the gneissose shear zones characteristic of the deep crust.

Thus available geological evidence suggests that for strike-slip faults an abrupt transition to a sub-horizontal decoupling horizon immediately beneath the frictional regime is uncommon, but that in at least some instances transcurrent shear zones broaden considerably as they extend into the deep continental crust. Comparison of finite shear strain profiles obtained at different crustal levels across these ancient transcurrent structures (e.g. Grocott & Watterson 1980; Sorenson 1983) with geodetically determined profiles of shear strain rate across active strike-slip systems (e.g. Walcott 1979; Bibby 1981; Prescott *et al.* 1981) could be a fruitful exercise.

Seismological evidence

Tocher (1975) speculated on relationships between the base of the seismogenic zone in California, mid-crustal decoupling and the possible presence of an extensive crustal low velocity zone, anticipating much of the general theme under discussion. Subsequent work has not confirmed the widespread existence of such low velocity zones in California, although a good case can be made locally along the eastern margin of the Basin and Range province (Hill 1978; Smith 1978). Seismic evidence for thrust décollement at the base of the seismogenic zone comes from earthquake sequences in the vicinity of the 1971 *M*6.4 San Fernando rupture in the Transverse Ranges, California, and from aftershocks of the 1978 *M*7.4 Tabas, Iran, earthquake. In both cases the main thrust ruptures dipped through the seismogenic zone at moderate angles (30–55°), but focal mechanisms obtained for some of the deepest events were consistent with sub-horizontal shear surfaces (Hadley & Kanamori 1978; Berberian 1982). In the Himalayan foothills, Seeber *et al.* (1981) also inferred the existence of large thrust 'flakes', floored by sub-horizontal detachment faults, from concentrations of seismic activity at 10–12 km depth. McKenzie (1978) cited evidence from focal mechanism solutions for the shallowing of normal fault dips with depth.

Substantial dip-slip faulting may accompany or follow the propagation of major strike-slip ruptures, extending laterally from around the tips of the main earthquake faults, as in the case of the 1968 *M*7.5 Dasht-e-Bayaz (McKenzie 1972; Jackson & Fitch 1979), 1976 *M*7.5 Guatemala (Plafker 1976) and 1976 *M*7.7 Tangshan (Butler *et al.* 1979) events (Fig. 11). While consistent with the stress re-distributions expected at the ends of major strike-slip ruptures, such

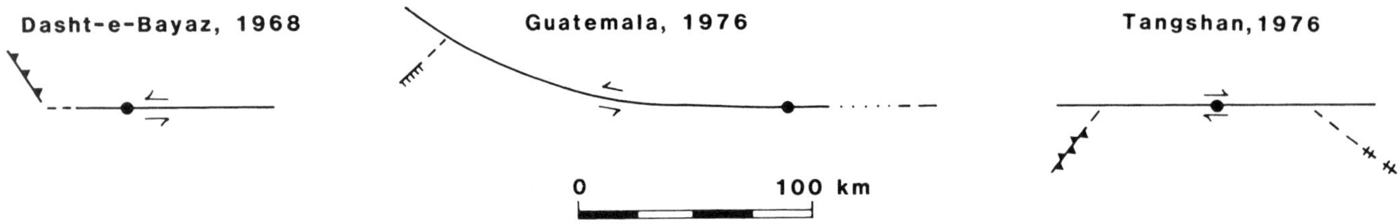

FIG. 11. Seismotectonic cartoons (only roughly to scale) indicating relationships between subsidiary dip-slip faulting (thrusts = saw-teeth, normal faults = barred; one-sided when dip direction known) and the ends of major strike-slip earthquake ruptures.

effects are also suggestive of upper crustal 'flaking' which would clearly be enhanced by a sub-horizontal decoupling zone.

Implications from the shear resistance profiles

In calculating the shear resistance profiles (Fig. 6), it was tacitly assumed that the width of the shear zone, and hence the strain rate, stays constant throughout the quasi-plastic regime. Although this seems to be a reasonable assumption in some instances (viz. the N and S Armorican Shear Zones), the evidence presented above suggests that major fault zones sometimes broaden with depth or pass into sub-horizontal décollement zones, in which case strain rate decreases and shear resistance drops off even more abruptly below the frictional/quasi-plastic transition. Under such circumstances, the shear resistance profiles imply that upper crustal decoupling along weak, flat-lying quasi-plastic shear zones is a distinct possibility, certainly for the steeper geothermal gradients (BR1 and BR2). Accepting that some widening of strike-slip fault zones with depth generally occurs, it is also easy to see how continental transforms may act as generators of upper crustal thrust 'flakes' in regions of transpression when the faults have become misaligned with respect to the interplate slip vector. This may have occurred in the Transverse Ranges, flanking the 'big bend' region of the San Andreas fault (Yeats 1981), and perhaps also in New Zealand where the Southern Alps are being ploughed up by oblique convergence along the Alpine Fault (Sibson *et al.* 1979; Wellman 1979).

Why should major decoupling horizons be located at mid-crustal levels? One possibility is that the general temperature-dependent weakening with depth, as expressed by the shear resistance profiles, is counteracted deeper down by diminishing quartz content, thereby creating a low-strength zone localized to the middle crust. Under such circumstances decoupling is clearly favoured by a quartz-rich middle crust and higher than average continental heat flow (cf. Armstrong & Dick 1974). Similar views have been expressed by Bird (1978) and Molnar & Gray (1979), who argued that delamination of low density upper crust is necessary for the subduction of continental lithosphere.

Discussion

Central to the theme of this paper is the proposition that the mechanical behaviour of continental fault zones at different crustal levels is governed largely by the temperature-dependent response of quartz to deviatoric stress. The modelled transition from frictional to quasi-plastic behaviour in quartz-bearing rocks, expressed as profiles of shear resistance versus depth for different geotherms, accounts fairly satisfactorily for the depth of background microearthquake activity in different heat flow provinces. Though reliable data are few, larger ruptures in continental crust also generally appear to be nucleating around the peaks in shear resistance and strain energy concentration inferred to occur at the frictional/quasi-plastic transition. Frictional behaviour and seismic activity may of course extend deeper in crust with a low quartz content. Progressive acquisition of high quality focal depth determinations for moderate to large shocks in different heat flow provinces will allow fuller evaluation of these conceptual fault models. Because of the current uncertainty over the average stress level in crustal fault zones, one key problem requiring solution is the extent to which the peak in shear resistance is smoothed out, and the manner in which this occurs. However, at the very least, the shear resistance profiles serve to emphasize the relationship between crustal strength and isothermal topography. On this basis, the depression of thermal gradients at depth in areas of fast thrust stacking (as evinced by the deeper microseismic activity characterizing such regions) may lead to the formation of a cold, high-strength 'keel', possibly a further factor causing deformation to spread out laterally in zones of continental convergence.

The shear resistance profiles also provide semi-quantitative support for many previously expressed hypotheses concerning the nature of intracontinental plate boundaries and fault systems (e.g. Oxburgh 1972; Armstrong & Dick 1974; Tocher 1975; Bird 1978). Formation of decoupling horizons in the middle to deep continental crust is favoured by a high quartz content and above average heat flow. Upper crustal 'flake' detachments extending over considerable areas may develop in association with regional dip-slip faulting or be spawned by misaligned transforms, contributing to the blurred seismic definition of intracontinental plate boundaries. Extensive rotation of crustal blocks floored by decoupling horizons may then occur (Luyendyk *et al.* 1980). It is apparent that plate interaction in such regions occurs on two different scales, that associated with conventional lithospheric thickness and that associated with crustal 'flaking'. Commonly, the effects of the latter tend to obscure the former, more fundamental interaction. In particular, high-level seismic activity may not be related in an obvious manner to the largely aseismic interactions at depth.

ACKNOWLEDGMENTS. Thanks are due to B. K. Atkinson, N. J. Price, E. H. Rutter and S. H. White, my former colleagues at Imperial College, London, for many stimulating discussions on materials science related to crustal rock mechanics, and also to the many scientists in the Office of Earthquake Studies, U.S. Geological Survey, Menlo Park, for teaching me the rudiments of seismology and the vagaries of focal depth determination. T. C. Hanks and A. G. Sylvester took pains to provide constructive criticism. Thanks also to Mrs Ellie Dzuro for her accurate typing. This work was partly supported by the U.S. Geological Survey during study leave from Imperial College, London.

References

AKI, K. 1966. Generation and propagation of G waves from the Niigata earthquake of June 16, 1964. 2, Estimation of earthquake moment, released energy, and stress-strain drop from G wave spectrum. *Bull. Earthquake Res. Inst. Tokyo Univ.* **44,** 73–88.

—— 1979. Characterization of barriers on an earthquake fault. *J. geophys. Res.* **84,** 6140–8.

ALLEN, C. R. 1981. The modern San Andreas fault. *In:* ERNST, W. G. (ed.). *The Geotectonic Development of California.* Prentice-Hall Inc., Englewood Cliffs, N.J., 511–34.

—— & NORDQUIST, J. M. 1972. Foreshock, main shock, and larger aftershocks of the Borrego Mountain earthquake. *Prof. Pap. U.S. geol. Surv.* **787,** 16–23.

ALLIS, R. G., HENLEY, R. W. & CARMAN, A. F. 1979. The thermal regime beneath the southern Alps. *In*: WALCOTT, R. I. & CRESSWELL, M. M. (eds.). *The Origin of the Southern Alps.* Bull. R. Soc. N.Z. **18,** 79–85.

AMBRASEYS, N. N. 1970. Some characteristic features of the Anatolian fault zone. *Tectonophysics,* **9,** 143–65.

—— 1971. Value of historical records of earthquakes. *Nature, London,* **232,** 375–9.

ANDERSON, E. M. 1951. *The Dynamics of Faulting.* 2nd ed., Oliver & Boyd, Edinburgh, 206 pp.

ANDERSON, J. G. 1979. Estimating the seismicity from geological structure for seismic risk studies. *Bull. seism. Soc. Am.* **69,** 135–58.

ANDREWS, D. J. 1980. A stochastic fault model. I, Static case. *J. geophys. Res.* **85,** 3867–77.

ARABASZ, W. J., SMITH, R. B. & RICHINS, W. D. 1980. Earthquake studies along the Wasatch Front, Utah: network monitoring, seismicity and seismic hazards. *Bull. seism. Soc. Am.* **70,** 1479–99.

ARMSTRONG, R. L. & DICK, H. J. B. 1974. A model for the development of thin overthrust sheets of crystalline rock. *Geology,* **2,** 35–40.

BACHE, T. C., LAMBERT, D. G. & BARKER, T. G. 1980. A source model for the March 28, 1975, Pocatello Valley earthquake from time-domain modelling of teleseismic P-waves. *Bull. seism. Soc. Am.* **70,** 405–18.

BAK, J., GROCOTT, J., KORSTGARD, J. A., SORENSON, K., NASH, D. F. & WATTERSON, J. 1975. Tectonic implications of Precambrian shear belts in western Greenland. *Nature, London,* **254,** 566–9.

BAKUN, W. H. 1972. Focal depths of the 1966 Parkfield, California, earthquakes. *J. geophys. Res.* **77,** 3816–22.

——, STEWART, R. M., BUFE, C. G. & MARKS, S. M. 1980. Implication of seismicity for failure of a section of the San Andreas fault. *Bull. seism. Soc. Am.* **70,** 185–201.

BALLY, A. W. 1981. Thoughts on the tectonics of folded belts. *In*: MCCLAY, K. A. & PRICE, N. J. (eds.). *Thrust and Nappe Tectonics.* Spec. Publ. geol. Soc. London, **9,** 13–32.

BARAZANGI, M. & DORMAN, J. 1969. World seismicity map of ESSA Coast and Geodetic Survey epicenter data for 1961–1967. *Bull. seism. Soc. Am.* **59,** 369–80.

BEACH, A. 1976. The interrelations of fluid transport, deformation, geochemistry and heat flow in early Proterozoic shear zones in the Lewision complex. *Philos. Trans. R. Soc. London,* **A280,** 569–604.

BERBERIAN, M. 1982. Aftershock tectonics of the 1978 Tabas-e-Golshan (Iran) earthquake sequence: a documented active 'thin- and thick-skinned tectonic' case. *Geophys. J. R. astron. Soc.* **68,** 499–530.

BIBBY, H. M. 1981. Geodetically determined strain across the southern end of the Tonga-Kermadec-Hikurangi subduction zone. *Geophys. J. R. astron. Soc.* **66,** 513–33.

BIRD, P. 1978. Initiation of intracontinental subduction in the Himalaya. *J. geophys. Res.* **83,** 4975–87.

—— (in press). Kinematics of present crust and mantle flow in southern California. *Bull. geol. Soc. Am.*

BLACIC, J. D. 1981. Water diffusion in quartz at high pressure: tectonic implications. *Geophys. Res. Lett.* **8,** 721–3.

BOUCHON, M. 1978. A dynamic source model for the San Fernando earthquake. *Bull. seism. Soc. Am.* **68,** 1555–76.

BRACE, W. F. 1965. Some new measurements of linear compressibility of rocks. *J. geophys. Res.* **70,** 391–8.

—— & BYERLEE, J. D. 1966. Stick-slip as a mechanism for earthquakes. *Science,* **153,** 990–2.

—— & KOHLSTEDT, D. L. 1980. Limits on lithospheric stress imposed by experiments. *J. geophys. Res.* **85,** 6248–52.

BREWER, J. A. 1981. Thermal effects of thrusting. *Earth planet. Sci. Lett.* **56,** 231–44.

——, COOK, F. A., BROWN, L. D., OLIVER, J. E., KAUFMAN, S. & ALBAUGH, D. S. 1981. COCORP seismic reflection profiling across thrust faults. *In*:

McClay, K. A. & Price, N. J. (eds.). *Thrust and Nappe Tectonics*. Spec. Publ. geol. Soc. London, **9**, 501–11.

Brun, J. P. & Cobbold, P. R. 1980. Strain heating and thermal softening in continental shear zones: a review. *J. struct. Geol.* **2,** 149–58.

Brune, J. N. 1968. Seismic moment, seismicity, and rates of slip along major fault zones. *J. geophys. Res.* **73,** 777–84.

——, Henyey, T. L. & Roy, R. F. 1969. Heat flow, stress and rate of slip along the San Andreas fault, California. *J. geophys. Res.* **74,** 3821–7.

Bucknam, R. C., Plafker, G. & Sharp, R. V. 1978. Fault movement (afterslip) following the Guatemala earthquake of February 4, 1976. *Geology,* **6,** 170–3.

Burford, R. O. & Harsh, P. W. 1980. Slip on the San Andreas fault in central California from alinement array surveys. *Bull. seism. Soc. Am.* **70,** 1233–61.

——, Nason, R. D. & Harsh, P. W. 1983. Results of fault-creep studies in central California. *Earthquake Info. Bull.*

Butler, R., Stewart, G. S. & Kanamori, H. 1979. The July 27, 1976, Tangshan, China earthquake—a complex sequence of interplate events. *Bull seism. Soc. Am.* **69,** 207–20.

Byerlee, J. D. 1978. Friction of rocks. *Pure appl. Geophys.* **116,** 615–26.

Carslaw, H. S. & Jaeger, J. C. 1959. *Conduction of Heat in Solids*. 2nd ed. Clarendon Press, Oxford.

Carter, N. L., Anderson, D. A., Hansen, F. D., & Kranz, R. L. 1981. Creep and creep rupture of granitic rocks. *Mon. Am. Geophys. Union,* **24,** 61–82.

Chen, W. P. & Molnar, P. 1977. Seismic moments of major earthquakes and the average rate of slip in central Asia. *J. geophys. Res.* **82,** 2945–69.

—— & —— 1983. Focal depths of intracontinental and intraplate earthquakes and their implications for the thermal and mechanical properties of the lithosphere. *J. geophys Res.* **88,** 4183–214.

Christensen, N. I. & Fountain, D. M. 1975. Constitution of the lower continental crust based on experimental studies of seismic velocities in granulite. *Bull. geol. Soc. Am.* **86,** 227–36.

Cockerham, R. S., Lester, F. W. & Ellsworth, W. L. 1980. A preliminary report on the Livermore Valley earthquake sequence January 24 – February 26, 1980. *U.S. Geol. Surv. Open-File Rept.* 80–714, 45 pp.

Cogne, J., Durand, S. & Estoule-Choux, J. 1974. Le Massif Armoricain. *In*: Debelmas, J. (ed.). *Geologie de la France*, Doin, Paris, **1,** 105–61.

Coney, P. J. 1980. Cordilleran metamorphic core complexes: an overview. *In*: Crittenden, M. D., Coney, P. J. & Davis, G. H. (eds.). *Cordilleran Metamorphic Core Complexes*. Mem. geol. Soc. Am. **153,** 7–31.

Coward, M. P. 1980. Shear zones in the Precambrian crust of southern Africa. *J. struct. Geol.* **2,** 19–27.

Das, S. & Aki, K. 1977. Fault plane with barriers: a versatile earthquake model. *J. geophys. Res.* **82,** 5658–70.

—— & Scholz, C. H. 1981. Theory of time-dependent rupture in the earth. *J. geophys. Res.* **86,** 6039–51.

Davies, G. F. & Brune, J. N. 1971. Regional and global fault slip rates from seismicity. *Nature, Phys. Sci.* **229,** 101–7.

De Charpal, O., Guennoc, P., Montadert, L. & Roberts, D. G. 1978. Rifting, crustal attenuation and subsidence in the Bay of Biscay. *Nature, London,* **275,** 706–11.

Dewey, J. F. & Sengor, A. M. C. 1979. Aegean and surrounding regions: complex multiplate and continuum tectonics in a convergent zone. *Bull. geol. Soc. Am.* **90,** 84–92.

Dickinson, W. R., Cowan, D. S. & Schweikert, R. A. 1972. Test of new global tectonics—discussion. *Bull. Am. Assoc. Petrol. Geol.* **56,** 375–84.

Eaton, J. P., Lee, W. H. K. & Pakiser, L. C. 1970. Use of micro-earthquakes in the study of the mechanics of earthquake generation along the San Andreas fault in central California. *Tectonophysics,* **9,** 259–82.

Ebel, J. E. & Helmberger, D. V. 1982. P-wave complexity and fault asperities: the Borrego Mountain, California, earthquake of 1968. *Bull. seism. Soc. Am.* **72,** 413–37.

Elliot, D. 1973. Diffusion flow laws in metamorphic rocks. *Bull. geol. Soc. Am.* **84,** 2645–64.

—— 1976. The energy balance and deformation mechanisms of thrust sheets. *Philos. Trans. R. Soc. London,* **A283,** 289–312.

Ellsworth, W. L. 1975. Bear Valley, California, earthquake sequence of February–March 1972. *Bull. seism. Soc. Am.* **65,** 483–506.

Eshelby, J. D. 1957. The determination of the elastic field of an ellipsoidal inclusion and related problems. *Proc. R. Soc. London,* **A241,** 376–96.

Fountain, D. M. & Salisbury, M. H. 1981. Exposed cross-sections through the continental crust: implications for crustal structure, petrology and evolution. *Earth planet. Sci. Lett.* **56,** 263–77.

Freund, R. 1974. Kinematics of transform and transcurrent faults. *Tectonophysics,* **21,** 93–134.

Gallagher, J. J. 1981. Tectonics of China: continental scale cataclastic flow. *Mon. Am. Geophys. Union,* **24,** 259–73.

Gapais, D. & Le Corre, C. 1980. Is the Hercynian belt of Brittany a major shear zone? *Nature, London,* **288,** 574–6.

Gilpin, B. & Lee, T-C. 1978. A microearthquake study in the Salton Sea geothermal area, California. *Bull. seism. Soc. Am.* **68,** 441–50.

Goulty, N. R., Burford, R. O., Allen, C. R., Gilman, R., Johnson, C. E. & Keller, R. P. 1978. Large creep events on the Imperial Fault, California. *Bull. seism. Soc. Am.* **68,** 517–21.

Graham, C. M. & England, P. C. 1976. Thermal regimes and regional metamorphism in the vicinity of overthrust faults: an example of shear heating and inverted metamorphic zonation from southern California. *Earth planet. Sci. Lett.* **31,** 142–52.

Grocott, J. & Watterson, J. 1980. Strain profile of a boundary within a large ductile shear zone. *J. struct. Geol.* **2,** 111–17.

Gutenberg, B. & Richter, C. F. 1954. *Seismicity of the Earth and Associated Phenomena*, 2nd ed., Princeton University Press, Princeton, New Jersey, 310 pp.

Hadley, D. & Kanamori, H. 1977. Seismic structure of the Transverse Ranges, California. *Bull. geol. Soc. Am.* **88,** 1469–78.

—— & —— 1978. Recent seismicity in the San Fernando region and tectonics in the west-central Transverse Ranges, California. *Bull. seism. Soc. Am.* **68,** 1449–57.

Hagiwara, T. 1965. Brief description of the project proposed by the earthquake prediction research group of

Japan. *Proc. U.S.–Japan Conf. on Research related to Earthquake Prediction Problems*, 10–12.

HAMILTON, R. M. 1972. Aftershocks of the Borrego Mountain earthquake from April 12 to June 12, 1968. *Prof. Pap. U.S. geol. Surv.* **787,** 31–54.

HANKS, T. C. 1974. The faulting mechanism of the San Fernando earthquake. *J. geophys. Res.* **79,** 1215–29.

—— 1977. Earthquake stress drops, ambient tectonic stresses, and stresses that drive plate motions. *Pure appl. Geophys.* **115,** 441–58.

—— 1979. *b* values and $w^{-\gamma}$ seismic source models: implications for tectonic stress variations along active crustal fault zones and the estimation of high-frequency strong ground motion. *J. geophys. Res.* **84,** 2235–42.

—— & KANAMORI, H. 1979. A moment magnitude scale. *J. geophys. Res.* **84,** 2348–50.

—— & RALEIGH, C. B. 1980. The conference on magnitude of deviatoric stresses in the earth's crust and upper mantle. *J. geophys. Res.* **85,** 6083–5.

HARTZELL, S. & HELMBERGER, D. V. 1982. Strong motion modeling of the Imperial Valley earthquake of 1979. *Bull. seism. Soc. Am.* **72,** 571–96.

HEATON, T. H. & HELMBERGER, D. V. 1977. A study of the strong ground motion of the Borrego Mountain, California, earthquake. *Bull. seism. Soc. Am.* **67,** 315–30.

HILL, D. P. 1978. Seismic evidence for the structure and Cenozoic tectonics of the Pacific Coast States. *In*: SMITH, R. B. & EATON, G. P. (eds.). *Cenozoic Tectonics and Regional Geophysics of the Western Cordillera.* Mem. geol. Soc. Am. **152,** 145–74.

——, MOWINCKEL, P. & LAHR, K. M. 1975. Catalog of earthquakes in the Imperial Valley, California, June 1973 - May 1974. *U.S. Geol. Surv. Open-File Rept.* 75-041, 29 pp.

HUBBERT, M. K. & RUBEY, W. W. 1959. Role of fluid pressure in the mechanics of overthrust faulting. *Bull. geol. Soc. Am.* **70,** 115–205.

IRWIN, W. P. & BARNES, I. 1975. Effect of geologic structure and metamorphic fluids on seismic behavior of the San Andreas fault system in central and northern California. *Geology*, **3,** 713–6.

JACKSON, J. 1980. Errors in focal depth determination and the depth of seismicity in Iran and Turkey. *Geophys. J. R. astron. Soc.* **61,** 285–301.

—— & FITCH, T. J. 1979. Seismotectonic implications of relocated aftershock sequences in Iran and Turkey. *Geophys. J. R. astron. Soc.* **57,** 209–29.

—— & —— 1981. Basement faulting and the focal depths of the larger earthquakes in the Zagros mountains (Iran). *Geophys. J. R. astron. Soc.* **64,** 561–86.

JEGOUZO, P. 1980. The South Armorican Shear Zone. *J. struct. Geol.* **2,** 39–47.

KANAMORI, H. 1977. The energy release in great earthquakes. *J. geophys. Res.* **82,** 2981–7.

—— 1981. The nature of seismicity patterns before large earthquakes. *In*: SIMPSON, D. W. & RICHARDS, P. G. (eds.). *Earthquake Prediction: an International Review. Maurice Ewing Series* **4,** Am. Geophys. Union, Washington D.C., 1–19.

—— & ANDERSON, D. L. 1975. Theoretical basis of some empirical relations in seismology. *Bull. seism. Soc. Am.* **65,** 1073–95.

—— & STEWART, G. S. 1979. A slow earthquake. *Phys. Earth Planet. Inter.* **18,** 167–75.

KASAHARA, K. 1981. *Earthquake Mechanics*. Cambridge University Press, Cambridge, 248 pp.

KAYANO, I. 1968. Determination of origin times, epicentres and focal depths of aftershocks of the Niigata earthquake of June 16, 1964: a preliminary report of the cooperative study of aftershocks of the Niigata earthquake. *Bull. Earthquake Res. Inst. Tokyo Univ.* **46,** 223–69.

KERRICH, R., ALLISON, I., BARNETT, R. L., MOSS, S. & STARKEY, J. 1980. Microstructural and chemical transformations accompanying deformation of granite in a shear zone at Mieville, Switzerland; with implications for stress corrosion cracking and superplastic flow. *Contrib. Mineral. Petrol.* **73,** 221–42.

——, BECKINSALE, R. D. & DURHAM, J. J. 1977. The transition between regimes dominated by intercrystalline diffusion and intracrystalline creep evaluated by oxygen isotope thermometry. *Tectonophysics,* **38,** 241–57.

KING, C.-Y., NASON, R. D. & TOCHER, D. 1973. Kinematics of fault creep. *Philos. Trans. R. Soc. London,* **A274,** 355–360.

KING, G. C. P., BILHAM, R. G., CAMPBELL, J. W., MCKENZIE, D. P. & NIAZI, M. 1975. Detection of elastic strainfields caused by fault creep events in Iran. *Nature, London*, **253,** 420–3.

KOCH, P. S., CHRISTIE, J. M. & GEORGE, R. P. 1980. Flow law of "wet" quartzite in the α-quartz field (abstract). *EOS, Trans. Am. Geophys. Union,* **61,** 376.

KOHLSTEDT, D. L. & WEATHERS, M. S. 1980. Deformation-induced microstructures, paleopiezometers and differential stresses in deeply eroded fault zones. *J. geophys. Res.* **85,** 6269–85.

KOSTROV, B. V. 1973. Seismic moment, earthquake energy and seismic flow of rocks. *Publs. Inst. Geophys. Pol. Acad. Sci.* **62,** 25–47.

LACHENBRUCH, A. H. & SASS, J. H. 1977. Heat flow in the United States and the thermal regime of the crust. *Mon. Am. Geophys. Union,* **20,** 626–75.

—— & —— 1980. Heat flow and energetics of the San Andreas fault zone. *J. geophys. Res.* **85,** 6185–222.

LEE, W. H. K. & STEWART, S. V. 1982. *Principles and Applications of Microearthquake Networks.* Academic Press, Princeton, N.J., 193 pp.

——, JOHNSON, C. E., HENYEY, T. L. & YERKES, R. L. 1978. A preliminary study of the Santa Barbara, California, earthquake of August 13, 1978, and its major aftershocks. *U.S. Geol. Surv. Circular* 797, 11 pp.

——, HERD, D. G., CAGNETTI, V., BAKUN, W. H. & RAPPORT, A. 1979*a*. A preliminary study of the Coyote Lake earthquake of August 6, 1979 and its major aftershocks. *U.S. Geol. Surv. Open-File Rept.* 79-1621, 43 pp.

——, YERKES, R. F. & SIMIRENKO, M. 1979*b*. Earthquake activity and focal mechanisms in the western Transverse Ranges, California. *U.S. Geol. Surv. Circular* 799-A, 26 pp.

LENSEN, G. J. 1970. Elastic and non-elastic surface deformation in New Zealand. *Bull. N.Z. Soc. Earthquake Eng.* **3,** 131–43.

LINDH, A. G. & BOORE, D. M. 1981. Control of rupture by fault geometry during the 1966 Parkfield earthquake. *Bull. seism. Soc. Am.* **71,** 95–116.

LIU, H. L. & HELMBERGER, D. V. 1983. The near-source ground motion of the 6 August 1979 Coyote Lake,

California, earthquake. *Bull. seism. Soc. Am.* **73,** 201–18.

LUYENDYK, B. P., KAMERLING, M. J. & TERRES, R. 1980. Geometric model for Neogene crustal rotations in southern California. *Bull. geol. Soc. Am.* **91,** 211–17.

MCCLAY, K. 1977. Pressure solution and Coble creep in rocks and minerals: a review. *J. geol. Soc. London*, **134,** 57–70.

MCHUGH, C. A. & LESTER, F. W. 1978. Catalog of earthquakes along the San Andreas fault system in central California for the year 1976. *U.S. Geol. Surv. Open-File Rept.* 78-1051, 91 pp.

MCKENZIE, D. P. 1969. Speculations on the consequences and causes of plate motion. *Geophys. J. R. astron. Soc.* **18,** 1–32.

—— 1972. Active tectonics of the Mediterranean region. *Geophys. J. R. astron. Soc.* **30,** 109–85.

—— 1978. Active tectonics of the Alpine-Himalaya belt: the Aegean Sea and surrounding regions. *Geophys. J. R. astron. Soc.* **55,** 217–54.

—— & BRUNE, J. N. 1972. Melting on fault planes during large earthquakes. *Geophys. J. R. astron. Soc.* **29,** 65–78.

MAJER, E. L. & MCEVILLY, T. V. 1979. Seismological investigations at The Geysers geothermal field. *Geophysics*, **44,** 246–69.

MARKS, S. M. & LINDH, A. G. 1978. Regional seismicity of the Sierra foothills in the vicinity of Oroville, California. *Bull. seism. Soc. Am.* **68,** 1103–15.

MATSUDA, T. 1975. Magnitude and recurrence intervals of earthquakes from a fault. *Zisin. J. seism. Soc. Japan*, **28,** 269–83.

MAVKO, G. M. 1981. Mechanics of motion on major faults. *Ann. Rev. Earth Planet. Sci.* **9,** 81–111.

MEISSNER, R. 1980. Bruch- und kriechprozesse in kontinentaler kruste. *Z. dt. geol. Ges.* **131,** 591–603.

—— & STREHLAU, J. 1982. Limits of stresses in continental crusts and their relation to the depth-frequency distribution of shallow earthquakes. *Tectonics,* **1,** 73–89.

MOLNAR, P. 1979. Earthquake recurrence intervals and plate tectonics. *Bull. seism. Soc. Am.* **69,** 115–33.

—— & GRAY, D. 1979. Subduction of continental lithosphere: some constraints and uncertainties. *Geology,* **7,** 58–62.

—— & TAPPONIER, P. 1975. Cenozoic tectonics of Asia: Effects of a continental collision. *Science*, **189,** 419–26.

—— & —— 1978. Active tectonics of Tibet. *J. geophys. Res.* **83,** 5361–75.

MOTHS, B. L., LINDH, A. G., ELLSWORTH, W. L. & FLUTY, L. 1981. Comparison between the seismicity of the San Juan Bautista and Parkfield regions, California (abstr.). *EOS, Trans. Am. Geophys. Union*, **62,** 958.

NORTH, R. G. 1974. Seismic slip rates in the Mediterranean and Middle East. *Nature, London*, **252,** 560–3.

OHNAKA, M. 1976. A physical basis for earthquakes based on the elastic rebound model. *Bull. seism. Soc. Am.* **66,** 433–51.

O'NEIL, J. R. & HANKS, T. C. 1980. Geochemical evidence for water-rock interaction along the San Andreas and Garlock faults of California. *J. geophys. Res.* **85,** 6286–92.

OXBURGH, E. R. 1972. Flake tectonics and continental collision. *Nature, London*, **239,** 202–4.

—— & TURCOTTE, D. L. 1974. Thermal gradients and regional metamorphism in overthrust terrains with special reference to the Eastern Alps. *Schweiz. Mineral. Petrogr. Mitt.* **54,** 641–62.

PERCEVAULT, M. N. & COBBOLD, P. R. 1982. Mathematical removal of regional ductile strains in central Brittany: evidence for wrench tectonics. *Tectonophysics,* **82,** 317–28.

PFLUKE, J. H. & STEPPE, J. A. 1973. Magnitude-frequency statistics of small earthquakes from San Francisco Bay to Parkfield. *In*: KOVACH, R. L. & NUR, A. (eds.). *Proceedings of the Conference on Tectonic Problems of the San Andreas Fault System, Stanford Univ. Publ. Geol. Sci.* **13,** 13–23.

PLAFKER, G. 1976. Tectonic aspects of the Guatemala earthquake of 4 February, 1976. *Science*, **193,** 1201–8.

PRESCOTT, W. H. & NUR, A. 1981. Accommodation of relative motion at depth on the San Andreas fault system in California. *J. geophys. Res.* **86,** 999–1004.

——, LISOWSKI, M. & SAVAGE, J. C. 1981. Geodetic measurement of crustal deformation on the San Andreas, Hayward, and Calaveras faults near San Francisco, California. *J. geophys. Res.* **86,** 10853–69.

PREVOT, R., HATZFELD, D., ROECKER, S. W. & MOLNAR, P. 1980. Shallow earthquakes and active tectonics in eastern Afghanistan. *J. geophys. Res.* **85,** 1347–57.

REASENBERG, P. & ELLSWORTH, W. L. 1982. Aftershocks of the Coyote Lake, California earthquake of August 6, 1979: a detailed study. *J. geophys. Res.* **87,** 637–55.

REID, H. F. 1910. The mechanics of the earthquake. *In*: *The California Earthquake of April 18, 1906, Rep. State Earthquake Invest. Comm.*, Washington D.C., Carnegie Inst.

RICHTER, C. 1958. *Elementary Seismology*. W. H. Freeman, San Francisco, 768 pp.

RIKITAKE, T. 1975. Statistics of ultimate strain of the earth's crust and probability of earthquake occurrence. *Tectonophysics,* **26,** 1–21.

RUTTER, E. H. 1976. The kinetics of rock deformation by pressure solution. *Philos. Trans. R. Soc. London*, **A283,** 203–19.

—— & MAINPRICE, D. H. 1978. The effect of water on stress relaxation of faulted and unfaulted sandstone. *Pure appl. Geophys.* **116,** 634–54.

SACKS, I. S., SUYEHIRO, S., LINDE, A. T. & SNOKE, J. A. 1978. Slow earthquakes and stress redistribution. *Nature, London*, **275,** 599–602.

SBAR, M. L. & SYKES, L. R. 1977. Seismicity and lithospheric stress in New York and adjacent areas. *J. geophys. Res.* **82,** 5771–86.

SCHOLZ, C. H. 1968. The frequency-magnitude relation of microfracturing in rock and its relation to earthquakes. *Bull. seism. Soc. Am.* **58,** 399–415.

—— 1980. Shear heating and the state of stress on faults. *J. geophys. Res.* **85,** 6174–84.

—— 1982. Scaling laws for large earthquakes: consequences for physical models. *Bull. seism. Soc. Am.* **72,** 1–14.

——, WYSS, M. & SMITH, S. W. 1969. Seismic and aseismic slip on the San Andreas fault. *J. geophys. Res.* **74,** 2049–69.

SEEBER, L., ARMBRUSTER, J. G. & QUITTMEYER, R. C. 1981. Seismicity and continental subduction in the Himalayan Arc. *In*: GUPTA, H. K. & DELANY, F. M. (eds.). *Zagros – Hindu Kush – Himalaya: Geodynamic Evolution.* Am. Geophys. Union Geodynamics Ser. **3,** 215–42.

SEGALL, P. & POLLARD, D. D. 1980. Mechanics of discontinuous faults. *J. geophys. Res.* **85,** 4337–50.
SIBSON, R. H. 1974. Frictional constraints on thrust, wrench and normal faults. *Nature, Phys. Sci.* **249,** 542–3.
—— 1977. Fault rocks and fault mechanisms. *J. geol. Soc. London*, **133,** 191–214.
—— 1980*a*. Transient discontinuities in ductile shear zones. *J. struct. Geol.* **2,** 165–71.
—— 1980*b*. Power dissipation and stress levels on faults in the upper crust. *J. geophys. Res.* **85,** 6239–47.
—— 1981. Fluid flow accompanying faulting: field evidence and models. *In*: SIMPSON, D. W. & RICHARDS, P. G. (eds.). *Earthquake Prediction: an International Review*. Maurice Ewing Series **4,** Am. Geophys. Union, Washington D.C., 593–603.
—— 1982*a*. Fault zone models, heat flow, and the depth distribution of earthquakes in the continental crust of the United States. *Bull. seism. Soc. Am.* **72,** 151–63.
—— 1982*b*. Seismic style in relation to heat flow along the San Andreas fault system. *Proc. 23rd U.S. Symposium on Rock Mechanics, Berkeley, California.* Am. Inst. Min. Metall. Petrol. Eng., 361–71.
——, WHITE, S. H. & ATKINSON, B. K. 1979. Fault rock distribution and structure within the Alpine Fault Zone: a preliminary account. *In*: WALCOTT, R. I. & CRESSWELL, M. M. (eds.). *The Origin of the Southern Alps*. Bull. R. Soc. N.Z. **18,** 55–56.
SINVHAL, H., AGRAWAL, P. N., KING, G. C. P. & GAUR, V. K. 1973. Interpretation of measured movement at a Himalayan (Nahan) thrust. *Geophys. J. R. astron. Soc.* **34,** 203–10.
SMITH, R. B. 1978. Seismicity, crustal structure and intraplate tectonics of the interior of the western Cordillera. *In*: SMITH, R. B. & EATON, G. P. (eds.). *Cenozoic Tectonics and Regional Geophysics of the Western Cordillera*. Mem. geol. Soc. Am. **152,** 111–44.
SMITHSON, S. B. & BROWN, S. K. 1977. A model for lower continental crust. *Earth planet. Sci. Lett.* **35,** 134–44.
SORENSON, K. 1983. Growth and dynamics of the Nordre Stromfjord shear zone. *J. geophys. Res.* **88,** 3419–37.
SOUFLERIS, C., JACKSON, J. A., KING, G. C. P. & SPENCER, C. P. 1982. The 1978 earthquake sequence near Thessaloniki (northern Greece). *Geophys. J. R. astron. Soc.* **68,** 429–58.
STIERMAN, D. J. & ELLSWORTH, W. L. 1976. Aftershocks of the February 21, 1973, Point Mugu, California, earthquake. *Bull. seism. Soc. Am.* **66,** 1931–52.
SYKES, L. R. 1978. Intraplate seismicity, reactivation of pre-existing zones of weakness, alkaline magmatism and other tectonism postdating continental fragmentation. *Rev. geophys. Space Phys.* **16,** 621–88.
TAPPONNIER, P. & MOLNAR, P. 1976. Slip-line field theory and large-scale continental tectonics. *Nature, London*, **264,** 319–24.
—— & —— 1977. Active faulting and Cenozoic tectonics of China. *J. geophys. Res.* **82,** 2905–30.
—— & —— 1979. Active faulting and Cenozoic tectonics of the Tien Shan, Mongolia, and Baykal regions. *J. geophys. Res.* **84,** 3425–59.
THATCHER, W. 1979. Systematic inversion of geodetic data in central California. *J. geophys. Res.* **84,** 2283–95.
—— & HAMILTON, R. M. 1973. Aftershock and source characteristics of the 1969 Coyote Mountain earthquake, San Jacinto fault zone, California. *Bull. seism. Soc. Am.* **63,** 647–61.
——, HILEMAN, J. A. & HANKS, T. C. 1975. Seismic slip distribution along the San Jacinto fault zone, southern California, and its implications. *Bull. geol. Soc. Am.* **86,** 1140–6.
TOCHER, D. 1975. On crustal plates. *Bull. seism. Soc. Am.* **65,** 1495–1500.
TOKSOZ, M. N. & BIRD, P. 1977. Modelling of temperatures in continental convergence zones. *Tectonophysics*, **41,** 181–93.
TULLIS, J. & YUND, R. A. 1977. Experimental deformation of dry Westerly Granite. *J. geophys. Res.* **82,** 5705–18.
—— & —— 1980. Hydrolytic weakening of experimentally deformed Westerly granite and Hale albite rock. *J. struct. Geol.* **2,** 439–51.
TURNER, F. J. 1981. *Metamorphic Petrology* (2nd ed.). McGraw Hill, New York, 524 pp.
VOLL, G. 1976. Recrystallisation of quartz, biotite and feldspars from Erstfeld to the Levantina Nappe, Swiss Alps, and its geological significance. *Schweiz. Miner. Petrogr. Mitt.* **56,** 641–7.
WALCOTT, R. I. 1979. Plate motion and shear strain rates in the vicinity of the Southern Alps. *In*: WALCOTT, R. I. & CRESSWELL, M. M. (eds.). *The Origin of the Southern Alps*. Bull. R. Soc. N.Z. **18,** 5–12.
WALLACE, R. E. 1981. Active faults, palaeoseismology and earthquake hazards in the western United States. *In*: SIMPSON, D. W. & RICHARDS, P. G. (eds.). *Earthquake Prediction: an International Review.* Maurice Ewing Ser. **4,** Am. Geophys. Union, Washington D.C., 209–16.
WALTER, A. W. & WEAVER, C. S. 1980. Seismicity of the Coso Range, California. *J. geophys. Res.* **85,** 2441–58.
WANG, C-Y. & MAO, N. 1979. Shearing of saturated clays in rock joints at high confining pressures. *Geophys. Res. Lett.* **6,** 825–8.
WATTERSON, J. 1975. Mechanism for the persistence of tectonic lineaments. *Nature, Phys. Sci.* **253,** 520–2.
—— 1979. Strain and strain-rate gradients at the ductile levels of fault displacements. *In*: *Proc. Conf. VIII, Analysis of actual fault zones in bedrock,* U.S. Geol. Surv. Open-file Rept. 79-1239, 235–57.
WATTS, M. J. & WILLIAMS, G. D. 1979. Fault rocks as indicators of progressive shear deformation in the Guincamp region, Brittany. *J. struct. Geol.* **1,** 323–32.
WELLMAN, H. W. 1979. An uplift map for the South Island of New Zealand, and a model for uplift of the Southern Alps. *In*: WALCOTT, R. I. & CRESWELL, M. M. (eds.). *The Origin of the Southern Alps*. Bull. R. Soc. N.Z. **18,** 13–20.
WESSON, R. L., BURFORD, R. O. & ELLSWORTH, W. L. 1973. Relationship between seismicity, fault creep and crustal loading along the central San Andreas fault. *In*: KOVACH, R. L. & NUR, A. (eds.). *Proceedings of the Conference on Tectonic Problems of the San Andreas Fault System,* Stanford Univ. Publ. Geol. Sci. **13,** 303–21.
WHITCOMB, J. H., ALLEN, C. R., GARMANY, J. D. & HILEMAN, J. A. 1973. The 1971 San Fernando earthquake series: focal mechanisms and tectonics. *Rev. geophys. Space Phys.* **11,** 693–730.
WHITE, S. H. 1976. The effects of strain on the microstructures, fabrics and deformation mechanisms in quartzites. *Philos. Trans. R. Soc. London*, **A283,** 69–86.
WONG, I. G. & SAVAGE, W. U. 1983. Deep intraplate

seismicity in the western Sierra Nevada, central California. *Bull. seism. Soc. Am.* **73,** 797–812.

WYSS, M. 1973. Towards a physical understanding of the earthquake frequency distribution. *Geophys. J. R. astron. Soc.* **31,** 341–59.

—— 1979. Estimating maximum expectable magnitude of earthquakes from fault dimensions. *Geology*, **7,** 336–40.

YEATS, R. S. 1981. Quaternary flake tectonics of the California Transverse Ranges. *Geology*, **9,** 16–20.

YERKES, R. F. & LEE, W. H. K. 1979. Late Quaternary deformation in the western Transverse Ranges, California. *U.S. Geol. Surv. Circular* 799-B, 12 pp.

ZOBACK, M. D., TSUKUHARA, H. & HICKMAN, S. 1980. Stress measurements at depth in the vicinity of the San Andreas fault: implications for the magnitude of shear stress at depth. *J. geophys. Res.* **85,** 6113–56.

Received 28 April 1982; revised typescript received 5 May 1983.

RICHARD H. SIBSON, Department of Geological Sciences, University of California, Santa Barbara, California 93106.

Discussion

MR RICHARD W. MURPHY said: In your analysis of the San Andreas fault system please do not omit the tectonic underpinning of observed variations in seismic character. You have identified four segments of the San Andreas: two without microearthquakes (the '1906 segment' and the '1857 segment') and two with microearthquakes (central California and the Imperial Valley). Comparison of surface trace of the fault system with slip directions as revealed by plate tectonic analysis shows that the two segments without microearthquakes are wrench-compressional zones, whereas the segments with microearthquakes are wrench-extensional or wrench-parallel. The obvious inference is that the compressional zones are locked, at least on a short-term basis, whereas the active zones are more freely releasing heat and strain energy.

These observations are intended to supplement, not detract from, a fine paper in which you have brought order and meaning to a host of rather loose and chaotic data.

In reply, the AUTHOR thanked Mr Murphy for his comments and observations. Lack of time had prevented a full analysis of the several hypotheses put forward to explain changes in seismic behaviour along the San Andreas fault, but a more lengthy discussion appears in Sibson (1982*b*). It does indeed seem probable that the thermo-mechanical properties of fault systems are intimately linked to fault geometry, with changes in seismic style reflecting their interplay.

DR DAVID C. RUBIE said: I wish to comment on Dr Sibson's discussion of the depression of isotherms by underthrusting. If the thermal structure of a thrust sheet is temporarily preserved and is carried down to deep crustal levels during thrusting, then metamorphism in the blueschist/eclogite facies is likely to result. The Sesia–Lanzo Zone of the Western Alps is an example of continental crust affected by such a process (Rubie *et al.* 1981). Current work in the Sesia–Lanzo Zone indicates that the reaction of albite→quartz + jadeite, in rocks of granitic composition, created a significant increase in ductility during underthrusting. If metamorphism occurs, the mechanical properties of thrust sheets may change, with rocks becoming weaker at moderate temperatures (200–400° C) than suggested by flow laws derived for granite. This process of 'reaction-enhanced ductility' would tend to decrease the depth of the frictional/quasi-plastic transition within active thrust belts.

In reply, the AUTHOR said that while he recognized the additional complications introduced by such high pressure – low temperature metamorphic reactions, their occurrence is in fact generally cited as *evidence* for the process described: isotherm depression at depth as a result of under thrusting. Consequently, while the temperature of the frictional/quasi-plastic transition may be somewhat diminished, a deepening of microseismic activity is still to be expected in regions where such processes are operative.

DR ROSEMARY A. WILLIAMS stated: At first glance the 12 km depth cut off and horizontal scattering in seismic activity shown by Dr Sibson lend themselves admirably to interpretation in terms of a horizontal décollement at around 12 km. This strongly suggests that the San Andreas fault is a shallow adjustment to the deeper movement between the Pacific and American plates. The western part of California would then represent a crustal wedge riding on the deeper oceanic plate and the real plate margin would lie to the E as Dr Sibson mentioned for the model he dismissed. If I was looking for data to support this model, I would be happy to use the data he has presented, and I should like to know the technical grounds for his rejection of this model.

In reply, the AUTHOR agreed on the possibility of such a décollement, but considered the evidence far from definitive. His reservations were based in part on a wish to keep structures as simple as possible for as long as possible, but mainly on the lack of a good 'ancient' example where horizontal decoupling in association with major transcurrent faults could be convincingly demonstrated. In contrast, the evidence for downward continuation of many ancient strike-slip systems through the crust is strong, as documented in the text. The restriction of microseismic activity to the upper crust is common to most tectonically active continental regions and is not a unique characteristic of the San Andreas system.

Reference

RUBIE, D. C., DANTHINE, R., KOONS, P., RAMSAY, J. G. & THOMPSON, A. B. 1981. High pressure deformation and metamorphism in the Sesia–Lanzo Zone, Western Alps. *J. struct. Geol.* **3,** 193 (abstr.).

Active normal faulting and crustal extension

J.A. Jackson

SUMMARY: A world-wide review of fault-plane solutions and focal depths for large normal-faulting earthquakes on the continents shows that the overwhelming majority of such earthquakes nucleate in the depth range 6–15 km on faults dipping between 30 and 60°. In the few cases where levelling or seismic data are good enough, these normal faults are shown to be approximately planar from the surface to their nucleation depth at the base of the brittle crust. There is evidence that, in some cases, as a result of the enormous transitory increase in strain rate during large earthquakes, rupture continues into the normally 'ductile' lower crust on surfaces with substantially gentler dips. These low-angle surfaces may be analogous to some of the 'detachments' seen in metamorphic core complexes of the western USA, but the nature of the motion on them depends on strain rate as well as on rheological contrasts between the 'detachments' and the blocks on either side. Such contrasts, however, are unlikely to introduce substantial curvature to an originally planar shear zone.

If the observed spread in active normal-fault dips is caused by rotation of the faults and the blocks they bound during extension, then a maximum β value of 1.7 can be accommodated by *seismic* activity on a single generation of normal faults. With continued extension, either a new generation of steeper faults, cutting the rotated first faults, is likely to form, or the deformation will continue aseismically on faults dipping at less than 30°.

Large earthquakes have not been observed to nucleate on very low-angle (<20°) normal faults within the continental crust anywhere in the world. Such faults can of course move aseismically, but are unlikely to do so on a large scale within the upper crust in areas where steep normal faults are seismically active. Thus extensional models that require concentrated simple shear on large sub-horizontal faults within the brittle upper crust will also require a spatial separation between aseismic, very low-angle faulting and seismic high-angle faulting. Since seismogenic high-angle faults dominate the topography of extending regions, upper-crustal very-low-angle faulting, if it occurs on a large scale, presumably does so in areas that are relatively flat.

There is now little doubt that the formation of many continental sedimentary basins and passive margins has involved a considerable amount of both lithospheric and crustal stretching. At shallow levels, at least some, and perhaps the majority, of the crustal extension in these basins is accommodated by normal faults. Attempts to reconcile the observed normal faulting with estimates of overall crustal extension deduced from heat flow, subsidence or crustal thickness have proved difficult (see e.g. de Charpal *et al.* 1978; Le Pichon & Sibuet 1981; Wood & Barton 1983; Ziegler 1983), and there are profound disagreements about the nature of the normal faulting itself, particularly in the Basin and Range Province (e.g. Wernicke 1981; Gans & Miller 1983).

There are several continental areas where seismically active normal faulting is occurring today. This paper is an attempt to review the information available from seismological (and to some extent geomorphological) data, which throws light on the geometry and kinematics of the normal faults in these actively deforming regions. There are several advantages to be gained from studying large earthquakes.

1 Earthquakes larger than magnitude (M_s) 5.5 have source (fault) dimensions of at least 10–15 km, which are comparable with the thickness of the brittle upper crust. The faulting associated with such earthquakes usually dominates the topography, and thus there is no doubt that we are dealing with the major structures in a region and are not concerned with the minor accommodation structures that are always present.

2 By studying active faults we can see their geometry and orientation at the time of movement: information which may not be easy to deduce from the older geological record where original orientations may have been changed by tilting or uplift.

3 The interaction between adjacent faults is much more obvious in actively deforming areas than in the older geological record, where the resolution of timing creates its own difficulties.

4 Vertical movements associated with faulting can only be detected in the presence of a horizontal reference level. In the case of active faults this may either be sea-level (e.g. in the Aegean) or surveyed levelling lines. The absence of such reference levels in older geological records can

From Holdsworth, R.E. & Turner, J.P. (compilers) *Extensional Tectonics: Faulting and Related Processes*. The Geological Society, Key Issues in Earth Sciences, **2 (2)**, 135–149.
1476-315X/02/$15.00
First published in: "Coward, M.P., Dewey. J.F. & Hancock, P.L. (eds) *Continental Extensional Tectonics*, The Geological Society, London, Special Publications **28**, 3–17".

make absolute vertical movements almost impossible to deduce.

A basic tenet of this paper is that seismically active normal faulting looks much the same wherever it is occurring in the continents today. This appears to be true in spite of the fact that the actively extending regions differ dramatically in both scale and altitude (e.g. Tibet and the Aegean), and almost certainly differ also in the nature and magnitude of the forces responsible for their extension. Thus the kinematics of normal faulting appear to be largely independent of dynamics, and, for this reason, older, now static extended areas are likely to have been stretched by normal faults in the same way as currently active regions. This paper first reviews the geometry of active normal faults, and then goes on to discuss plausible models for crustal extension that are compatible with these observations. Finally, it examines the problems faced when trying to reconcile some interpretations of older, now static, structures, especially in the Basin and Range Province, with the data from seismically active normal faults. This paper does not intend to imply that any such interpretations of static structures are wrong: it merely attempts to highlight the apparent incompatibility between some proposed mechanisms of crustal extension and active-fault geometries, in the belief that eventual understanding of this incompatibility will also increase our understanding of extensional processes.

Active normal faults

The nature of the data

Figure 1 shows cross-sections through two active normal faults in Greece, both of which broke the surface in earthquakes of $M_s > 6.0$. In both cases the dip of the surface slip vector was measured in the field, the dip of the fault at the hypocentre (nucleation point of the earthquake) was well constrained by the fault-plane solution, the focal depth was estimated from waveform modelling, and the epicentre (projection of the hypocentre to the surface) was determined by a relative location technique using a well-located aftershock as a reference. The data used in Fig. 1 are described fully by Soufleris & Stewart (1981), Soufleris *et al.* (1982), Mercier *et al.* (1979) and Jackson *et al.* (1982). The point of Fig. 1 is simply this: in spite of the uncertainties involved, the seismogenic faults (both of which were major structures extending 12–15 km along strike) cannot have had dips of less than about 40–50° at depths shallower than about 6–8 km. The simplest interpretation of the data is that the faults are, to a first approximation, planar with a dip of 40–50° from the surface to a depth of at least 6–8 km. It is, of course, possible that the faults have minor curvature (the most extreme is shown by the dashed lines) and they could even have small changes of curvature, inflexions, or ramp-and-flat geometries, all of which are certainly sufficient to cause minor accommodation structures, antithetic faults and some curvature to bedding in the hanging wall (see e.g. Gibbs 1984; White *et al.* 1986). However, the degree of overall curvature is strongly limited by the seismic data: if the fault dip becomes horizontal, it does so *below* the nucleation point of the earthquake.

It is clear that the main constraints on the geometry shown in Fig. 1 are the dip of the fault at the hypocentre and the focal depth. Fortunately both are easily obtainable from the seismic waveforms generated by the earthquakes, which are recorded, and are thus available for study, world-wide.

Figures 2 & 3 present focal depths and dips for normal-faulting earthquakes in Greece, Turkey, western USA, the Gulf of Suez, Tibet, Baykal, NE China, Yunnan and East Africa. Much of this data is from published sources, the main ones being Chen & Molnar (1983a), Tapponnier & Molnar (1977, 1979), McKenzie (1972, 1978b), Eyidogan & Jackson (1985) and Shudofsky (in press), but much is as yet unpublished and will be reported in greater detail with fuller references elsewhere. All of the events used in these figures have magnitudes (M_s or m_b) greater than 5.2.

Focal depths

Focal depths reported by international agencies (e.g. USGS, ISC) are usually based on arrival-time data and are notoriously unreliable. Since about 1975, a different technique has become available, which models the shape of the observed seismic waveform by generating synthetic seismograms on a computer. This method is usually capable of estimating focal depths to within about 2 km (see e.g. Langston & Helmberger 1975; Chen & Molnar 1983a).

Figure 2 shows a histogram of focal depths determined from waveform modelling for 43 normal-faulting earthquakes in the continental regions listed above. The depths range from about 6–18 km and thus occur in the upper part of the continental crust. Moreover, where local seismic networks have been installed to locate aftershocks accurately, it is clear that large

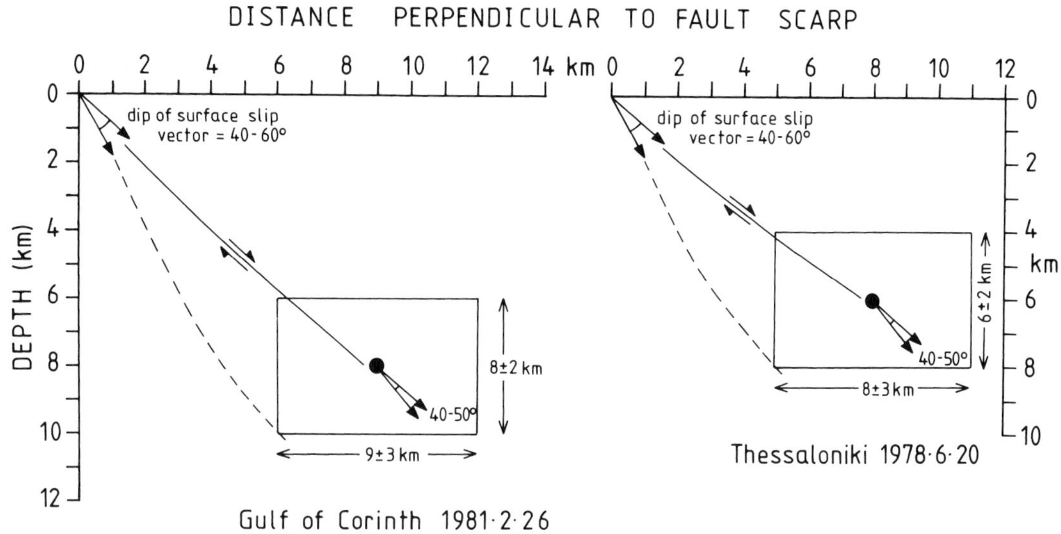

FIG. 1. Cross-sections through two active normal faults in Greece. References to the data used in this figure are given in the text. The hypocentre (focus) of each earthquake is marked by a filled circle surrounded by a rectangle indicating the uncertainty in its depth and epicentre. The uncertainties in the fault dip at the surface and at the hypocentre are marked by arrows. The position of the hypocentre as marked is consistent with a roughly planar geometry, but the hypocentre could lie anywhere within the rectangle indicated. The fault must pass through its outcrop and the focus and have correct dip at each of these positions. The most listric (concave upward) geometry allowable within these constraints is shown by a dashed line. From Eyidogan & Jackson (1985).

normal-faulting earthquakes nucleate at or near the base of the aftershock zone (e.g., Soufleris *et al.* 1982; King *et al.* 1984; Smith & Bruhn 1984; Lahr *et al.* 1976; Doser & Smith 1985). Below the depth to which aftershocks occur, deformation evidently takes place aseismically, presumably by some form of creep. There is mounting evidence that the depth of this transition from apparently 'brittle' (and seismic) to apparently 'ductile' (and aseismic) behaviour is controlled principally by the temperature gradient in the crust (e.g. Sibson 1982, 1984; Meissner & Strehlau 1982; Chen & Molnar 1983b; Wiens & Stein 1983). Variations in temperature gradient are probably responsible for variations in the focal depths from region to region; for example events in Greece and western Turkey appear to nucleate at depths as shallow as 6–8 km, whereas those in the Basin and Range Province appear to have focal depths of up to 15 km (see e.g. Doser 1985; Doser & Smith 1985). Thus Fig. 2 confirms the general view that the lower crust is largely aseismic and that the largest earthquakes nucleate at or near the base of the 'brittle' upper crust. Figure 2 does not include five significantly deeper normal-faulting earthquakes: two in southern Tibet (at 85 and 90 km; Chen & Molnar 1983a) and three in East Africa (at 25, 27 and 29 km; Shudofsky, 1985). All of these earthquakes nucleated in or near the uppermost upper mantle beneath continental crust and their interpretation is not clear. Under some thermal gradients it is possible that the uppermost mantle may deform in a brittle (seismic) fashion even though the lower crust deforms aseismically (see e.g. Chen & Molnar 1983b). Such deep events do

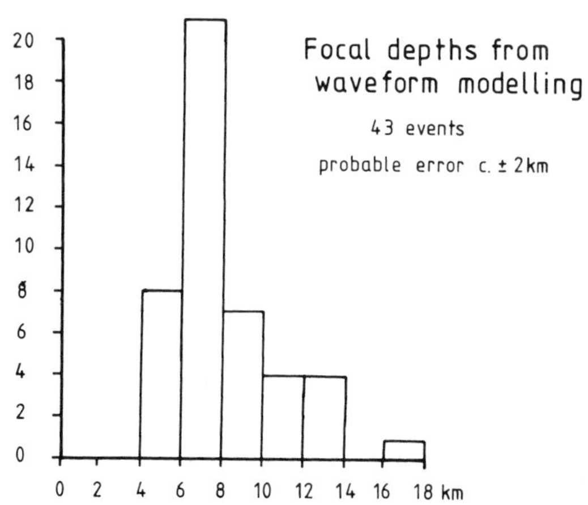

FIG. 2. Histogram of focal depths for normal faulting earthquakes. Detailed tabulation of the data will be published elsewhere.

not appear to be common, and their source dimensions (about 15 km) are small compared with their focal depths, making it unlikely that they will correlate directly with any structures visible at the surface.

Fault dips from fault-plane solutions

It is well known that either of the two nodal planes in a first-motion fault-plane solution for an earthquake could be the actual fault plane. For this reason, in the histogram of Fig. 3(a) both nodal planes are plotted for all the 82 normal-faulting earthquakes whose fault-plane solutions have been collected. All these fault-plane solutions have been constructed using a crustal *P*-wave velocity (of either 6.5 or 6.8 km sec^{-1}) at the focus. Although half the dips shown in Fig. 3(a) must be from auxiliary rather than fault planes, it is clear that both nodal planes in normal-faulting earthquakes commonly dip in the range 30–70°. Nodal planes with dips less than 20° or more than 70° are very rare. Uncertainties in nodal-plane dips are variable and difficult to estimate. In places with good station coverage, such as the Mediterranean and western USA, both the dip and strike of the nodal planes is usually well constrained, whereas in remote parts of central Asia the dips may be much better constrained than the strikes. In particular, the uncertainty in dip usually depends on the dip itself: very low-angle faulting requires an almost vertical (auxiliary) nodal plane to pass near the centre of the focal sphere, which is the part of the radiation pattern best sampled by teleseismic observations. In almost all the earthquakes shown in Fig. 3(a), dilatational first motions dominate the teleseismic distance range (30–80°), thereby excluding the possibility of an almost-vertical nodal plane, and thus also excluding the possibility of a very low-angle fault plane. Molnar & Chen (1982) give a useful discussion of these uncertainties for those unfamiliar with these types of data. Fault-plane solutions the dips of which are very poorly constrained are not included in Fig. 3, where the majority of the dips are probably correct to within 10–15°.

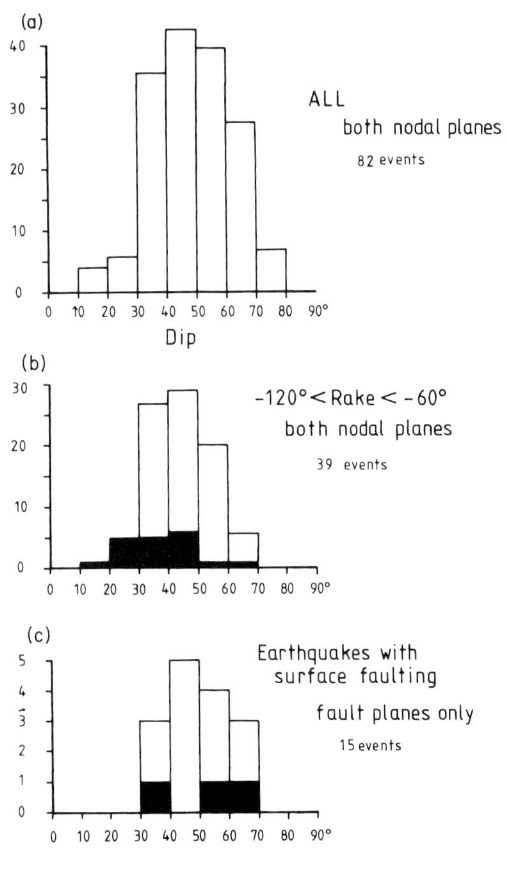

FIG. 3. Dips of nodal planes for normal faulting earthquakes measured from fault-plane solutions. Note that these dips indicate the orientation of the normal fault at the focus of the earthquake. The black regions of histograms (b) and (c) are explained in the text.

Figure 3(b) shows a sub-set of Fig. 3(a) in which again the dips of both nodal planes are plotted, but this time only earthquakes for which the strike-slip component is constrained by the data to be small (i.e. with rake, as defined by Aki & Richards 1980, in the range −60 to −120°) are included. The dark parts of each column represent nodal planes the slip vectors of which are well determined unlike the dips. These have been drawn (and plotted) as almost pure normal faults but they could include a slightly greater component of strike-slip motion, i.e. they could have steeper but not gentler dips. Figure 3(b) shows that for normal-faulting earthquakes whose strike-slip motion is small, nodal planes (and hence fault planes) with dips less than 20° or more than 70° are virtually unknown, and that the great majority dip in the range 30–60°.

Finally, Fig. 3(c) shows dips for those earthquakes whose faulting was observed to break the ground surface, allowing one of the nodal planes of the fault-plane solution to be unambiguously chosen as the fault plane. In a few cases (shaded black) this choice was made from the very obvious pattern of the aftershock distribution. Note that in Fig. 3(c), unlike in Fig. 3(a) & (b), *only fault-plane dips* at the focus (and not auxiliary planes) are plotted. Figure 3(c) confirms what is suggested by Fig. 3(a) & (b): that, above m_b 5.5, seismic activity on normal faults with dips gentler than about 30° is almost unknown.

Conclusions

The data from active normal faults is quite unequivocal: in Greece, Turkey, the western USA, central Asia and Africa active normal faults nucleate at or near the base of the 'brittle' (seismogenic) upper crust with relatively steep dips (in the range 30–60°). The earthquakes discussed in this paper have source dimensions comparable to or bigger than their focal depths and thus break through most of the brittle upper continental crust. From the 82 normal-faulting earthquakes studied in this paper (as well as the five deeper events in Tibet and East Africa) there is *not one single example* of seismic activity on a sub-horizontal fault. From Fig. 3 we must conclude that normal faults with dips much less than 30° do not move seismically.

There are few examples where the fault geometry is as well constrained as in Fig. 1: generally because the earthquake epicentre is not sufficiently well determined. Other examples include the Dixie Valley (1954) and Borah Peak (1983) earthquakes in the western USA where levelling as well as seismic data support an approximately planar fault geometry from the surface to the nucleation depth (Eddington & Smith, this volume; Stein & Barrientos, in press; Doser & Smith 1985), though once again, minor departures from a planar geometry, sufficient to cause internal deformation of the hanging wall block, cannot be excluded by the available data. Thus, although the data in Fig. 3 permit the faults to steepen as they approach the surface, the evidence from the few earthquakes in Greece and the western USA for which the observations are good enough suggests that they do not steepen: i.e. that the faults are approximately planar. This does not, of course, deny the existence of faults that flatten abruptly with depth and even become horizontal; but it does put constraints on the nature of the movement on them, as such faults have never been seen to move *seismically* in the upper continental crust. A number of points can now be made.

1 There are many well-documented examples of large normal faults that are strongly curved in cross-section and become almost flat at depths within the upper crust. The best known are those within thick sedimentary sequences such as the Gulf of Mexico and Niger Delta. Most of these do not contribute to extension of the entire crustal thickness as the material beneath the fault surface is generally undeformed, and the extension in the up-dip direction is usually accommodated by shortening in the down-dip direction; achieved by either thrusting, folding or lateral compaction. Such faults represent a readjustment of the sediment load above the basement. However, the presence of such 'superficial' faults in places where the basement has genuinely been extended by other deeper faults can lead to confusion: especially where the sedimentary carapace is decoupled from the basement by salt (see e.g. Gibbs 1984).

2 It is, of course, possible that the steep upper-crustal seismogenic faults (of the type shown in Fig. 1) flatten abruptly below the hypocentre into the lower crust, where motion on them is aseismic. This possibility is discussed in a later section.

3 The data shown in Figs 2 & 3 cannot exclude the possibility of substantial *aseismic* motion on large extensional faults with dips less than 30° in the upper crust.

4 In the geological literature the term 'high-angle normal fault' is associated with a dip greater than 45° and 'low-angle normal fault' with a dip of less than 45° (e.g. Hobbs *et al.* 1976). This is unfortunate in the sense that seismically active normal faults are common throughout the dip range 30–60°. In this paper I will use the term 'very low-angle normal fault' when referring to a dip of less than 20° and 'steep normal fault' when referring to the dip range 30–60°.

Rotation and crustal extension

There is no doubt that substantial movement on roughly planar faults dipping at 30–60° throughout the upper crust occurs in extensional terranes. A direct and inevitable consequence of this is the rotation of both the faults and the blocks they bound about a horizontal axis. There are two main reasons for such rotation, the first being strain compatibility. Movement on planar faults represents simple shear yet the overall result of such movement must be pure shear: i.e. extending the crust in the horizontal direction by thinning it in the vertical. Overall pure shear can be achieved by a combination of simple shear and rotation (Fig. 4). The second reason is that substantial movement on crustal-scale, planar non-rotational faults over the width of an entire sedimentary basin would produce a massive isostatic disequilibrium and a long-wavelength gravity anomaly much larger than is ever observed. Where good gravity data exist, such as in the North Sea (Barton & Wood 1984), there is positive evidence that the

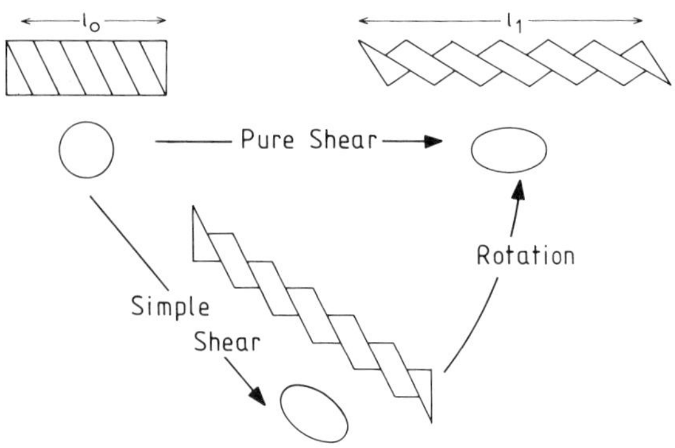

FIG. 4. Cartoon to illustrate the accomplishment of pure shear by simple shear and rotation, and to illustrate how planar non-rotational faulting on a crustal scale will inevitably lead to a large gravity anomaly. The relative extension is $\beta = l_1/l_0$ (as defined by McKenzie 1978a).

continental crust in sedimentary basins has no significant strength at all on wavelengths larger than the separation of major fault-bounded blocks; which may be 25 km or more in central Greece and the Basin and Range Province (Jackson & McKenzie 1983; Fletcher & Hallett 1983). In addition to a large gravity anomaly, planar non-rotational faults that cut the entire crust would lead to a depression of the Moho depth beneath the thickest sediments in the basin, rather than the elevation that is usually observed. Reconciliation with the gravity and Moho data is possible if the planar non-rotational faults are restricted to the upper crust: but then only if considerable flow is allowed in the lower crust.

It is therefore unlikely that planar non-rotational faults contribute significantly to major crustal extension processes, though on a small scale, especially in sediment cover, they certainly exist. Instead, faults and fault-bounded blocks rotate as they move, producing the tilted terranes characteristic of extended regions. This is the basis of the rotating block or 'domino' model for crustal extension, described by Ransome *et al.* (1910) and subsequently by many others (e.g. Morton & Black 1975; Le Pichon & Sibuet 1981; Wernicke & Burchfiel 1982; Jackson & McKenzie 1983). This model does not need elaboration here, though it is worth noting that the rotation of blocks in the extending terrane requires either a curved fault or substantial internal deformation of the hanging wall block where the extending terrane adjoins a non-extending stable region (see Wernicke & Burchfiel 1982).

Of interest here is whether the data in Fig. 3 put any quantitative constraints on the block-rotation process that accommodates stretching in the upper crust. Let us suppose that the spread in fault dips of 30–60°, implied by Fig. 3, is caused by rotation of the faults as they move. This would mean that the maximum rotation (and hence stretching) is achieved by a fault which starts with a dip of 60° and rotates to 30°. Such a rotation implies a relative extension, β, of 1.7 (see Fig. 4 and Jackson & McKenzie 1983 for relevant algebraic expressions). Indeed a value for β of 1.7 is the maximum extension that can be achieved with a single set of faults if they are to be seismically active. If stretching is to continue beyond $\beta = 1.7$ there are three possibilities:

1 A second generation of steep faults becomes active, cutting the first generation faults, which have rotated to a dip of 30°, making them inactive.

2 The faults continue to move at dips of less than 30°, but do so aseismically.

3 The faults become inactive once their dips reach 30°, and extension, with new faults, continues somewhere else nearby.

Note that $\beta = 1.7$ is the *maximum* value beyond which some change in the nature of extension must occur; if the faults start moving with dips of less than 60°, such a change will occur at smaller values of β.

The first of the three possibilities listed above is well documented as a process: there is good reason to think that as normal faults become flatter gravity will impede, rather than help, movement on them, and this is generally suggested as the reason why a second generation of steep faults may initiate. It is encouraging to note that, where values for stretching factors exceeding $\beta = 1.7$ on land have been suggested, two generations of normal faults have usually been described: e.g. in Afar (Morton & Black 1975); Yerrington, and the Snake Range in Nevada (Proffett 1977; Gans & Miller 1983); and parts of Mexico (Angelier & Colletta 1983). However, most sedimentary basins in which β exceeds 1.7 are in deep water and relatively unexplored. It is not certain whether an earlier generation of faults within the blocks bounded by the youngest faults would be revealed by seismic reflection, and resolution of fault structure in places like the Bay of Biscay (de Charpal *et al.* 1978) may have to await drilling. By contrast, in places where β is less than 1.7, such as the North Sea, there is no reason to expect to find anything other than a single generation of tilted fault blocks. If a rotation from an initial fault dip of 60° to one of 30° is all that can be expected from a single generation of faults, this in turn implies a maximum sedimentary tilt of 30° and a maximum throw on each fault equal to the initial spacing between adjacent faults.

It is perhaps worth noting that 'Andersonian' fault mechanics (Anderson 1951) predicts that normal faults should initiate in intact crust with dips of about 60°, which might explain the upper boundary to the observed range of seismically active normal fault dips. Furthermore, Sibson (1985 & pers. comm.) points out that, for a typical static rock-friction coefficient of 0.75, provided the maximum compressive stress (σ_1), remains vertical, *frictional* normal-slip failure cannot occur on planes dipping less than 37° unless the least principal compressive stress (σ_3) becomes effectively tensile, perhaps through abnormally high fluid pressures. This may go some way to explaining the lower observed boundary of seismically active normal fault dips: though in fact dips as gentle as 30° are known.

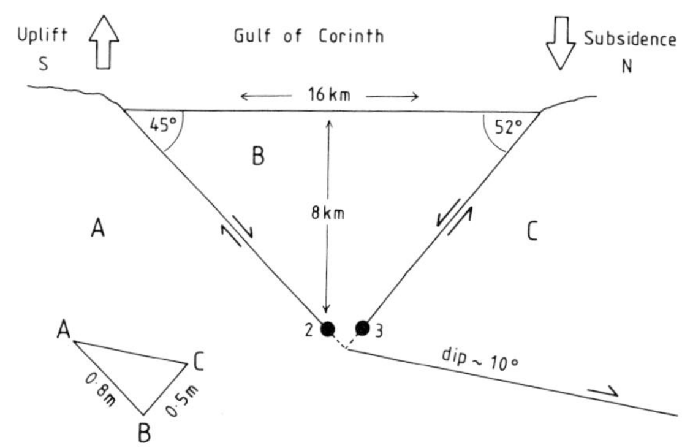

FIG. 5 A cross-section through the eastern Gulf of Corinth (Greece) constructed from seismic and surface data obtained from the earthquakes in 1981. Note the asymmetry of the vertical movement of the coastline. The dip of the very low-angle fault below the hypocentres of events 2 (1981.2.26) and 3 (1981.3.4) was estimated from the vector triangle on the left, showing the directions and amounts of the relative motions between blocks A, B and C. (see King *et al.* 1984).

Faults in the lower crust

While it is clear that extension in the upper 'brittle' (i.e. seismogenic) crust occurs by rotating fault blocks it is less certain what happens in the lower crust. Several deep seismic reflection profiles have suggested the presence of discontinuities in the lower crust (e.g. Brewer & Smythe 1984), and two observations from actively extending areas strongly support the presence of faults below the seismogenic zone, at least in some cases.

The asymmetry of vertical movements in normal faulting

Figure 5 shows a cross-section of the eastern Gulf of Corinth: an active graben in Greece. In 1981 normal faults on both sides of the Gulf moved in earthquakes of M_s>6.4, creating observed surface breaks about 12 km long in each event. Figure 4 is contructed from data reported in Jackson *et al.* (1982) and Kim *et al.* (1984). The section is approximately perpendicular to fault strike and the hypocentre of each event lay about 5 km to the side of the line of section.

The faults on both sides of the Gulf can be projected downwards to their respective hypocentres, which are both in the middle of the Gulf at roughly the same depth. The asymmetry of the vertical motions, however, clearly indicates that the southern fault projects northwards under the northern fault at a depth below that of the deepest aftershocks (see Jackson *et al.* 1982; King *et al.* 1984). Only this can explain the emergent nature of the southern shore (with raised beaches and uplifted solution notches in limestone) and the rapidly sinking northern shore. An estimate of the dip of the lower part of the southern fault can be made by vector addition, assuming that all the motion is transferred, presumably aseismically, on to this fault at depth (see King *et al.* 1984). This estimate suggests a dip of 10°, which is significantly less than the 45° dip observed in the upper crust. The change in dip of the slip vector is presumably responsible for the internal deformation of the hanging wall block, manifested by numerous small aftershocks and the presence of the main antithetic fault itself.

Earthquake faulting with a very similar geometry occurred in Bulgaria in 1928, where again the asymmetry of the vertical movements, this time revealed by levelling, clearly implies that the graben is asymmetrical at depth, with one fault continuing beneath the other. This work by Jankhof (1945: in Bulgarian) is reported by Richter (1958) and Jackson & McKenzie (1983).

Low-angle normal faulting in earthquakes

In a recent study, Eyidogon & Jackson (1985) investigated the normal faulting associated with several large earthquakes in western Turkey. All the events studied nucleated at depths of 6–10 km and broke through the upper crust on faults dipping between 30 and 45°. However, the seismograms of two of these events, at Gediz (1970.3.28) and Alasehir (1969.3.28), contain large-amplitude, long-period signals which originated *after* the faulting that broke through the upper crust. The extremely directional nature of these signals allows resolution of the fault orientation responsible for them. Eyidogan & Jackson (1985) concluded that these signals were caused by movement on almost flat (less than 10°) normal faults nucleating 15–25 sec after the start of the main shock and at the base of the seismogenic upper crust. As shown in Fig. 3, such faults are never seen in *first-motion* fault-plane solutions of earthquakes, and yet in Turkey they are apparently seen to move seismically during or immediately after earthquakes on steep normal faults in the upper crust. The explanation suggested by Eyidogan & Jackson (1985) for these observations is as follows. During the periods between large earthquakes the lower crust deforms by aseismic creep at strain rates in the region of 10^{-11} to 10^{-14} sec^{-1}. Meanwhile the upper crust deforms only by microearthquakes, whose source dimensions are small compared to the 'brittle' thickness of the crust, and therefore do not affect the lower crust beneath. When a large earthquake occurs, on a fault that breaks right through the 'brittle' upper crust, the top part of the lower crust suddenly experiences strain rates of about 10^{-4} sec^{-1}. With an increase in strain rate of 10 orders of magnitude, the lower crust is unlikely to respond by creep alone and will probably fail in a brittle fashion, but rupture will be attenuated as it propagates downwards into the 'ductile' layer. This behaviour is similar to that of toffee, which at slow strain rates will flow, and when hit with a hammer will shatter. In short, because of the extreme increase in strain rates during earthquakes, it is likely that rupture propagates down as well as up, and, under these special circumstances, the uppermost lower crust behaves in a brittle rather than a ductile fashion: but only for a short time. This brief display of brittle behaviour is sufficient to radiate seismic energy and thus reveal the geometry of the faults in the lower crust.

This behaviour is illustrated in Fig. 6, taken from Eyidogan & Jackson (1985). The fault geometry is clearly very similar to that proposed for the Gulf of Corinth (Fig. 5). It is not clear how widespread such behaviour is. The later part of the *P*-wave seismograms have hitherto attracted less attention than the first cycle, which is sufficient to estimate the fault depth and seismic moment. Moreover, the *P*-wave radiation from flat faults is much more directional than that from faults dipping at 45°, and helps greatly in their detection. If the upper-crustal faults propagated downwards at a constant dip, they would be much more difficult to detect in the lower crust. Yielding (1985) and Nabelek (1985) both suggest a similar flattening of a fault, from a steep dip in the upper crust to a gentler dip in the uppermost lower crust, during the 1980 El Asnam earthquake in Algeria: though it should be noted that this was a thrust (reverse fault).

Rheology and the nature of faults in the lower crust

Both examples in the preceding section suggest that, at least for the short time during or immediately after large earthquakes, faults exist as discrete surfaces of concentrated simple shear in the lower crust beneath the depth at which earthquakes nucleate. Such faults are easier to detect if they decrease their dip with depth (because of the directional radiation they emit and the accompanying internal deformation of the hanging wall block); but there is no particular reason why they should do so.

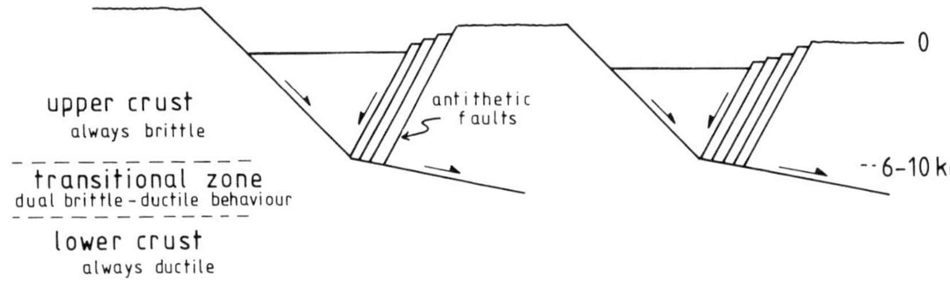

FIG. 6. Summary cross-section to illustrate the gross geometry of the normal faults investigated by Eyidogan & Jackson (1985). Note the change of dip at the base of the 'always brittle' crust. This section is designed to show the crude geometry of the faults and basement only. The effects of tilting, sediment fill and footwall uplift are not shown. Although the faults are shown as planar segments, they may in reality have limited curvature. Note that the very low-angle surface in the 'transitional zone' may only be a 'fault' (*sensu stricto*) at high strain rates.

The behaviour suggested in Figs 5 & 6 raises problems of terminology as well as rheology. The 'ductile' lower crust is only aseismic at low strain rates, and there will clearly be a transition zone in which the deformation is brittle (i.e. seismic) during big earthquakes and ductile (i.e. aseismic) otherwise. In this region, pseudotachylites are likely to intermix and be reworked by mylonites, as described by Sibson (1980). Some ultramylonites may consist almost entirely of reworked pseudotachylite created by this process. But what is the nature of the aseismic creep that occurs in the lower crust, and how is it distributed between the shear zones (or 'faults') and the blocks either side? This is a question of some importance, because if the deformation is partly distributed within the blocks either side of the shear zones (e.g. by pure shear) it will not be possible to restore a cross-section to its original configuration by simply reversing movement on the lower-crustal 'faults'. In short, are lower-crustal shear zones 'faults' in the same sense as is used in the upper crust?

A combination of reduced grain size, retrogressive mineral assemblages and crystallographic fabrics is likely to make a lower-crustal mylonite zone weaker than the surrounding rock (see e.g. White 1976; Etheridge, in press). The magnitude of this difference in strength will presumably govern how much the deformation in the lower crust is distributed between simple shear on the mylonite zone and pure shear within the blocks on either side. As discussed by Jackson & McKenzie (1983), both pure and simple shear produce rotations, but, for given values of total extension, β_{TOT}, and the initial dip of a fault or passive marker, the amount of rotation is different for each shear type. For a model in which the faults in the upper 'brittle' crust rotate by toppling (like dominoes) and the faults in the lower crust rotate as passive markers it is easy to calculate the difference in dip between the fault above and the passive marker below (see Fig. 7 and Jackson & McKenzie 1983). Such a model corresponds to an extremely weak lower crust, with no simple shear concentrated on the lower crustal 'fault zone'. What Jackson & McKenzie (1983) omitted to point out was the extremely small magnitude of this effect: for faults with initial dips in the range 30–60° the difference in dip (or 'curvature') between top and bottom layers is less than 8° for all values of β_{TOT} less than 3 (Fig. 7). Moreover, if the the dip of the fault in the brittle layer is restricted to be more than 30° (as implied by Fig. 3) the maximum 'curvature' is less than 4°. If the amount of simple shear, β_B, decreases gradually with depth, and the amount of pure shear, β_D, correspondingly increases to maintain the same overall extension at all depths (such that $\beta_{TOT} = \beta_B \times \beta_D$), then the change of dip is gradual rather than abrupt, but is still the same amount, in total, between top and bottom.

A number of conclusions can be drawn from the discussion in this section.

1 A change in rheology with depth, leading to either an abrupt or a gradual change in the ratio of pure to simple shear, is not an efficient way to produce fault curvature, if the overall extension, β_{TOT}, is to stay the same at all levels. The change of dip produced in this way is always very small, though is probably sufficient to require internal deformation, antithetic faults, and curved sedimentary horizons within hanging wall blocks. Geometries such as those shown in Figs 5 & 6 will not be produced unless a second-generation fault in the upper crust continues to use an original first-generation fault in the lower crust. For example, a fault initially dipping at 60° throughout the crust will rotate to 30° in both upper and lower crust if $\beta_{TOT} = 1.7$ (using the expressions in Fig. 7). If a new generation of faults in the upper crust, again rotating from 60 to 30°, now contributes a further β of 1.7 (so that the total

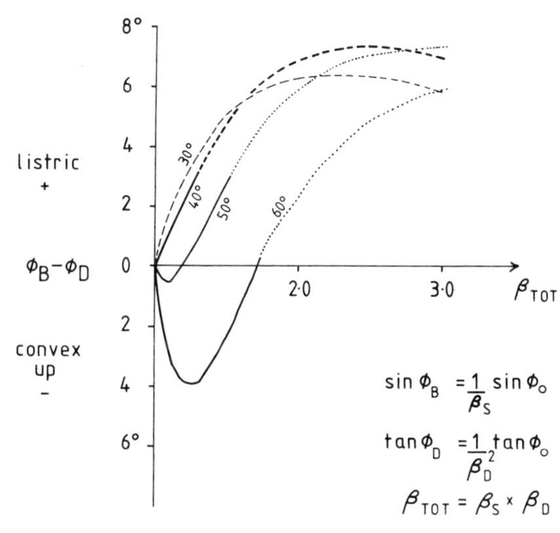

FIG. 7. Graph to show the maximum difference between the dip in the brittle and ductile portions of an originally planar surface, as a function of starting dip (ϕ_0) and extension (β_{TOT}). ϕ_B and ϕ_D are the final dips in the brittle and ductile layers respectively. β_B and β_D are the extensions achieved by simple shear (and rotation) and pure shear, respectively. The curves correspond to the case where $\beta_{TOT} = \beta_B$ and $\beta_D = 1$ in the brittle layer, and $\beta_{TOT} = \beta_D$ and $\beta_B = 1$ in the ductile layer. The expressions used are from Jackson & McKenzie (1983). The solid parts of the curves indicate regions where the dip in the brittle layer is greater than 30° (i.e. the fault may be seismically active).

stretching factor is $\beta_{TOT} = 1.7 \times 1.7 = 2.9$), then the original fault in the lower crust will have rotated by pure shear to a dip of 12°. Such a geometry would resemble that shown in Figs 5 & 6, though this process remains hypothetical. There is no evidence for this amount of extension in either central Greece or Western Turkey.

2 A planar fault in the crust is likely to stay approximately planar as deformation proceeds if β_{TOT} is to be the same at all levels. As a consequence, the shape of the 'fault' or shear zone is very little guide as to how the deformation is distributed between the 'faults' and the blocks either side. This is particularly relevant to the interpretation and restoration of deep seismic reflection sections from the lower crust.

3 If a weak layer exists at the base of the seismogenic crust, then a sub-horizontal decoupling of the layers above and below this layer is possible. Under these circumstances an abrupt change in dip is possible between the weak layer and the steep faults in the upper crust. Clearly, the nature of the weak layer itself is important in this case. If it marks an abrupt change from simple shear on rotating faults to penetrative pure shear below, as suggested by Gans & Miller (1983) for the Snake Range Decollement, then β_{TOT} may remain the same at all levels and the weak zone may have little or no simple shear concentrated on it. If, on the other hand, significant simple shear is concentrated on the weak layer, this movement will make the shear zone rotate towards the horizontal in all cases except where it is horizontal already.

In summary, gradual changes in rheology with depth are likely to control the distribution of deformation between simple shear on faults or mylonite zones and pure shear within the blocks in between. But this distribution will hardly affect the curvature of a planar shear zone at all: planar features, once formed, are there to stay.

Very low-angle normal faulting, detachments and decollements

In the last few years there has been considerable interest in the 'Cordilleran Metamorphic Core Complexes' of the western USA, and in particular in surfaces or zones of mylonitization, referred to variously as 'low-angle normal faults', 'detachments' or 'decollements', which are currently gently dipping or sub-horizontal, and are thought to have played an important role in crustal extension (see e.g. Crittenden *et al.* 1980; Frost & Martin 1982; Armstrong 1982). Of interest here is: (i) the nature of movement on or across these zones; (ii) how they were lifted from the depths of 10–15 km needed to form mylonites to be exposed at the surface today; and (iii) whether the present-day active faulting in extensional regions throws any light on their origin. The examples cited below are chosen because they illustrate different possible extension mechanisms. This paper does not imply any view as to whether they are appropriate to the particular area they were designed to explain.

Second-generation normal faults

Any normal faulting on discrete fault zones, if it is to achieve major crustal extension, will involve rotation of the faults and the blocks they bound. There clearly exists a class of faults in the Basin and Range Province, which, though now almost horizontal, were much steeper when they were active, and have subsequently been passively rotated by a later generation of steep faults. Such faulting has been described in Nevada by Proffett (1977) and Gans & Miller (1983), who claim that the faults, when active, had relatively steep dips of 30–60° and were approximately planar from the surface to depths of about 10 km. This interpretation is clearly compatible with observations of seismically active normal faulting, in that it requires no large fault to be active outside the dip range 30–60°.

Brittle-ductile decoupling zones

In the northern Snake Range of Nevada, Gans & Miller (1983) and Gans *et al.* (1985) suggested that beneath the rotating fault blocks in the upper 'brittle' plate was a weak layer (the northern Snake Range Decollement), originally at a depth of about 7 km, which completely decoupled the fault movement above from penetrative ductile flow (pure shear) below. In this model, the upper-plate faults end abruptly and at steep angles against the decollement, but because the total extension in the upper and lower plates is approximately the same there need be little or no discrete simple shear on the decollement itself. A similar model was suggested by Rehrig & Reynolds (1980) for some of the core complexes in Arizona. In such interpretations no active very low-angle faulting is expected, except conceivably of the type described by Eyidogan & Jackson (1985) (Fig. 6), in which rupture during earthquakes on steep upper-plate faults propagates momentarily downwards, directed along the decollement surface by the strong *L–S* mylonite fabrics.

Large earthquakes are not seen to nucleate at depths shallower than about 5 km, presumably because the rocks at such levels are not strong enough to store the necessary elastic strain. Thus, as stretching proceeds and (in the absence of sedimentation) the upper plate thins, the decollement approaches the surface, and at some stage the upper-plate faults presumably become aseismic as their vertical extent is reduced. This type of model is not in conflict with observations of seismically active normal faulting.

Low-angle 'brittle' detachment faults

More problematic and interesting are the types of structure described by Wernicke (1981), who suggested that large-scale extension may occur by discrete simple shear on very low-angle fault zones that penetrate the whole crust (or even lithosphere). If such structures move only below the depth at which large earthquakes nucleate there is no difficulty reconciling them with the data shown in Fig. 3, and they may even represent the low-angle faults detected by Eyidogan & Jackson (1985) (Fig. 6). Possible origins for such structures in the lower crust may be either: (i) through rotation—where second-generation upper-crustal faults continue to transfer motion on to old first-generation faults in the lower crust; or (ii) as weak layers. However, weak layers that are controlled by temperature are likely to be roughly horizontal, and are unlikely to cut downwards through the entire lower crust unless they follow, or are influenced by, older pre-existing weak layers such as thrust faults. Alternatively, some *a priori* appeal to rupture propagation must be made, requiring a fault to form *ab initio* at a very gentle dip.

But what if such very low-angle faults do indeed move at extremely gentle dips right through the brittle upper crust? A fault dipping at 10° will have an across-strike width of 85 km from the surface to a depth of 15 km. Faults with dips less than 10° will occupy an even greater across-strike width in the upper crust. Eighty-five kilometres is considerably larger than the average spacing between the active high-angle normal faults that bound range fronts in the Basin and Range Province (Fletcher & Hallett 1983) or the Aegean (Jackson & McKenzie 1983). Since all large seismically active normal faults appear to be steep and to cut right through the brittle upper crust, they would cut, and render inactive, any very low-angle faults in the same vicinity (as described by Proffett 1977 and Gans & Miller 1983). Thus, not only are the very low-angle faults likely to be aseismic (Fig. 3), but they must also be in an area of almost no seismicity at all. The inevitable conclusion is that, if large-scale very low-angle normal faulting in the brittle upper crust really does occur *and contributes to significant crustal extension* (unlike, for instance, in the Gulf of Mexico), then extending regions may be expected to contain separate provinces in each of which one of two quite different types of upper-crustal deformation occurs: one in which very low-angle (<30°) normal faulting is accompanied by an almost total lack of sizeable earthquakes, and the other in which large earthquakes accompany movement on steep (30–60°) normal faults. The scale on which this spatial separation must occur clearly depends on the dip of the very low-angle faults and the thickness of the upper brittle layer. Note that both areas, because simple shear is involved, will exhibit rotations about the horizontal axis, though in the very low-angle terrains, because the faults are flatter, the tilting will be less.

The conclusion that upper-crustal extension by aseismic or creep processes does not occur in the same place as seismic faulting has other implications. In particular, steep normal faults produce high topography that dominates seismically active extending regions. Large-scale very low-angle normal faulting in the upper crust will presumably occur in regions that are relatively flat. The vertical movements in such areas will depend on the nature of the very low-angle 'faulting' itself, and whether the footwall or hanging wall (or both) deform internally: either uplift or subsidence could result.

In regions of continental shortening, a spatial separation of seismic and aseismic deformation is most unlikely. In Iran for example, it is known that only about 10% of the convergence between Arabia and Eurasia is manifest by seismic faulting (North 1977; Ambraseys & Melville 1982). Because the regions of active seismicity (mostly thrusting) totally dominate the topography, and relatively aseismic regions like the Southern Caspian Sea, central Iran and Western Afghanistan are almost flat, it is usually assumed that the aseismic creep occurs in the same geographical regions as the seismicity. In short, it is unlikely that 10% of the shortening produces all of the topography while 90% produces no elevation at all. The same remarks apply further E in Asia, where aseismic regions within the deforming belt (such as the Tarim basin and Ordos plateau) have very little topography.

Uplifting very low-angle detachments and decollements to the surface

While aseismic simple shear on flat or gently dipping shear zones below the nucleation depth of earthquakes is compatible with observed siesmic faulting, it is not clear how these shear zones are uplifted from depths of about 10–15 km to the surface, where they are now observed. Both progressive rotation (Davis 1983) and unroofing of the footwall under a gently dipping fault (Spencer 1984) have been suggested; but both require faults dipping less than 30° to be active in the upper crust. They will presumably move aseismically, and so this mechanism leads inevitably to the spatial separation of the very low- and steep-angle faulting discussed earlier.

A second possibility involves extreme progressive thinning of the upper crust by several generations of rotating faults. This will, in the absence of sedimentation, bring a weak decoupling horizon at the base of the brittle crust closer to the surface, as envisaged by Rehrig & Reynolds (1980) and Gans *et al.* (1985). A combination of the resultant high heat flow and small vertical extent of the upper-crustal faults is likely to make this process aseismic as well.

A third possibility, alluded to by Davis (this volume) is that very low-angle faults that were originally active below the seismogenic depth, perhaps having been partially uplifted by one of the above-mentioned processes, are finally raised to the surface by structural uplift and rotation in the footwalls of later, steep (seismically active), range-bounding faults. This mechanism has the merit of not requiring aseismic movement on very low-angle faults at shallow crustal levels.

A note on terminology

The processes described in this paper merit a short plea for a less casual use of various common terms. In particular, the terms 'brittle' and 'ductile' are frequently used to mean 'seismic' and 'aseismic' and to imply 'strong' and 'weak'. They are often generalized to apply to the whole of a crustal layer even when the data on which they are based is from the seismicity (or lack of it) on faults or shear zones: where, as for example White (1976) and Etheridge (in press) point out, grain size, mineralogy and fabrics are likely to make rheological behaviour quite different from that in the blocks either side of the fault zone. Moreover, it is clear that some fault zones which are usually 'ductile' (and aseismic) may deform in a 'brittle' (and seismic) fashion during the high strain rates associated with earthquakes.

Similar difficulties surround the word 'listric' which is applied sometimes to faults whose dips become horizontal with depth and at other times to those with any concave-up curvature at all. Are the faults shown in Figs 1, 5 & 6 listric? This is more than a semantic question—a certain amount of curvature above the hypocentre in Fig. 1 is certainly allowed by the data, but if the surfaces become flat (as in Figs 5 & 6) they do so below the hypocentre, in a region where deformation is assumed to have a 'ductile' character. Here it is questionable whether they should be called faults at all, because if *any* deformation occurs by penetrative flow in the blocks either side, it will not be possible to restore the section to its predeformation state by reversing simple shear on these surfaces alone. Thus lower-crustal 'shear zones', 'detachments' and 'decollements' may or may not behave like 'faults' in the usual sense, depending on the nature and strain rate of the deformation on them and in their surrounding blocks.

Conclusions

1 Normal faults that move in large earthquakes on the continents break through the entire upper (seismically active) crust with dips apparently restricted to the range 30–60°. It is, of course, possible to take the view that 20 years' data (on which Fig. 3 is based) is not long enough to obtain a representative picture of normal faulting in earthquakes. This may be so, but does not explain why 20 years is long enough to reveal seismogenic faults in the dip range 30–60° and not a single example (to date) where large-scale seismic motion on a normal fault dipping less than 20° in the upper crust can be proven. Nor does it explain why, seismically, the Basin and Range Province looks no different from any other actively extending terrain on the continents. Very low-angle normal faults, if they contribute to extension of the upper crust, are clearly something special.

2 Although limited curvature of large seismogenic faults (in cross-section) is permitted by the seismic observations, the few examples where the data are sufficiently good suggest that the faults are approximately planar from the surface to the base of the seismogenic upper crust.

3 Such limited curvature is expected if changing rheology with depth simply varies the distribution of extension between simple shear on the faults and pure shear in the blocks either side.

4 If the observed spread of active fault dips (30–60°) is caused by rotation, this suggests that either, (i) a second generation of steeper normal faults is initiated when (or before) β reaches a value of 1.7, or (ii) when β exceeds 1.7, extension occurs aseismically by motion on faults dipping less than 30°.

5 To avoid intersecting faults being simultaneously active, the second alternative (above) requires a spatial separation, within extending regions, of seismically active areas where steep (30–60°) faulting occurs and almost totally aseismic areas where very low-angle faulting occurs. The scale on which this separation must take place depends on the dip of the very low-angle faults and the thickness of the upper seismogenic layer.

6 There is evidence from both vertical movements at the surface and seismic waveforms for very low-angle normal faulting in the uppermost lower crust, beneath the depth at which earthquakes nucleate: at least during the momentarily high strain rates associated with earthquakes in the upper crust. The long-term effects of such lower-crustal motions depend on whether, at slow strain rates, deformation concentrates on these 'faults' or is spread penetratively throughout the lower crust. If the latter occurs it will not be possible to restore cross-sections to their pre-deformation states by reversing simple-shear movement on these lower-crustal 'faults' alone.

7 A remaining problem is that of lifting the low-angle 'shear zones', decollements' or 'detachments', found in the metamorphic core complexes of the western USA, from the depths of 10–15 km to the surface. Mechanisms that require them to be active at gentle ($<30°$) dips in the upper crust as they rise will also require them to be aseismic: leading to the difficulties expressed in (5) above.

ACKNOWLEDGMENTS: R. Sibson, B. Wernicke and P. Gans provided helpful reviews of this paper for which I am grateful. I have also benefitted from many useful discussions of this subject with D. McKenzie and G. Yielding. D. Doser, R. Smith and G. Shudofsky kindly sent me preprints of their work prior to publication. This work was supported by a grant from NERC and Cambridge University Department of Earth Sciences Contribution No. 791.

References

AKI, K. & RICHARDS, P. 1980. Quantitative Seismology, 932 pp. W. H. Freeman & Co., San Francisco.

AMBRASEYS, N.N. & MELVILLE, C.P. 1982. *A History of Persian Earthquakes* 219 pp. Cambridge University Press.

ANDERSON, E.M. 1951. *The Dynamics of Faulting*, 2nd edn, 206 pp. Oliver & Boyd, Edinburgh.

ANGELIER, J. & COLLETTA, B. 1983. Tensional fractures and extensional tectonics. *Nature*, **301**, 49–51.

ARMSTRONG, R.L. 1982. Cordilleran Metamorphic Core Complexes. *Ann. Rev. Earth planet. Sci.* **10**, 129–54.

BARTON, P. & WOOD, R. 1984. Tectonic evolution of the North Sea Basin: crustal stretching and subsidence. *Geophys. J. R. astron. Soc.* **79**, 987–1022.

BREWER, J.A. & SMYTHE, D.K. 1984. MOIST and the continuity of reflector geometry along the Caledonian–Appalachian orogen. *J. geol. Soc. London*, **141**, 105–20.

DE CHARPAL, O., GUENNOC, P., MONTADERT, L. & ROBERTS, D.G. 1978. Rifting, crustal attenuation and subsidence in the Bay of Biscay. *Nature*, **275**, 706–11.

CHEN, W-P. & MOLNAR, P. 1983a. Focal depths and fault plane solutions of earthquakes under the Tibetan plateau. *J. geophys. Res.* **88**, 1180–96.

—— & —— 1983b. The depth distribution of intracontinental and intraplate earthquakes and its implications for the thermal and mechanical properties of the lithosphere. *J. geophys. Res.* **88**, 4183–214.

CRITTENDEN, M.D., JR, CONEY, P.J. & DAVIS, G.H. (eds) 1980. Cordilleran Metamorphic Core Complexes. *Mem. geol. Soc. Am.* **153**, 490 pp.

DAVIS, G.H. 1983. Shear zone model for the origin of Metamorphic Core Complexes. *Geology*, **11**, 342–7.

—— This volume. A shear-zone model for the structural evolution of metamorphic core complexes in southeastern Arizona, 247–66.

DOSER, D.I. 1985. Source parameters and faulting processes of the 1959 Hebgen Lake, Montana, earthquake sequence. *J. geophys. Res.* **90**, 4537–55.

—— & SMITH, R.B. 1985. Source parameters of the 28 October 1983 Borah Peak, Idaho, earthquake from body wave analysis. *Bull. seism. Soc. Am.* **75**, 1041–51.

EDDINGTON, P.K. & SMITH, R.B. This volume. Kinematics of Basin and Range intraplate extension, 371–92.

ETHERIDGE, M.A. In press. On the reactivation of extensional fault systems. *Phil. Trans. R. Soc. London.*

EYIDOGAN, H. & JACKSON, J.A. 1985. A seismological study of normal faulting in the Demirci, Alasehir and Gediz earthquakes of 1969–70 in western Turkey: implications for the nature and geometry of deformation in the continental crust. *Geophys. J. R. astron. Soc.* **81**, 569–607.

FLETCHER, R.C. & HALLETT, B. 1983. Unstable extension of the lithosphere: a mechanical model for Basin-and-Range structure. *J. geophys. Res.* **88**, 7457–66.

FROST, E.G. & MARTIN, D.L. (eds) 1982. *Mesozoic–Cenozoic evolution of the Colorado River region, California, Arizona and Utah.* Cordilleran Publishers, San Diego, California.

GANS, P.B. & MILLER, E.L. 1983. Style of mid-Tertiary extension in east-central Nevada. *In:*

Guidebook Part 1, Geol. Soc. Am. Rocky Mt. & Cordilleran Sections meeting: Utah Geol. & Mining Surv. Spec. Studies, **59**, 107-60.
—— , —— ,McCarthy, J. & Ouldcott, M.L. 1985. Tertiary extensional faulting and evolving ductile-brittle transition zones in the northern Snake Range and vicinity: new insights from seismic data. *Geology*, **13**, 189-93.
Gibbs, A.D. 1984. Structural evolution of extensional basin margins. *J. geol. Soc. London*, **141**, 609-20.
Hobbs, B.E., Means, W.D. & Williams, P.F. 1976. *An outline of structural geology*, 571 pp. John Wiley & Sons.
Jackson, J.A. & McKenzie, D.P. 1983. The geometrical evolution of normal fault systems. *J. struct. Geol.* **5**, 471-82.
——, Gagnepain, J., Houseman, G., King, G., Papadimitriou, P., Soufleris, C. & Virieux, J. 1982. Seismicity, normal faulting and the geomorphological evolution of the Gulf of Corinth, (Greece): the Corinth earthquakes of February and March 1981. *Earth planet. Sci. Lett.* **57**, 377-97.
Jankhof, K. 1945. Changes in ground level produced by the earthquakes of April 14 to 18 1928 in southern Bulgaria. *In: Tremblements de Terre en Bulgarie, Nos 29-31* (in Bulgarian), pp. 131-136. Institute meteorologique central de Bulgarie, Sofia.
Kim, W-Y, Kulhanek, O. & Meyer, K. 1984. Source processes of the 1981 Gulf of Corinth earthquake sequence from body wave analysis. *Bull. seism. Soc. Am.* **74**, 459-77.
King, G., Ouyang, Z., Papadimitriou, P., Deschamps, A., Gagnepain, J., Houseman, G., Jackson, J., Soufleris, C. & Virieux, J. 1984. The evolution of the Gulf of Corinth (Greece): an aftershock study of the 1981 earthquake. *Geophys. J.R. astron. Soc.* **80**, 677-93.
Lahr, K.M., Lahr, J.C., Lindh, A.G., Bufe, C.G. & Lester, F.W. 1976. The August 1975 Oroville earthquakes. *Bull. seism. Soc. Am.* **66**, 1085-99.
Langston, C.A. & Helmberger, D.V. 1975. A procedure for modelling shallow dislocation sources. *Geophys J.R. astron. Soc.* **42**, 117-30.
Le Pichon, X. & Sibuet, J.C. 1981. Passive margins: a model of formation. *J. geophys. Res.* **86**, 3708-21.
McKenzie, D.P. 1972. Active tectonics of the Mediterranean region. *Geophys. J.R. astron. Soc.* **30**, 109-85.
—— 1978a. Some remarks on the development of sedimentary basins. *Earth planet. Sci. Lett.* **40**, 25-32.
—— 1978b. Active tectonics of the Alpine-Himalayan belt: the Aegean Sea and surrounding regions. *Geophys. J. R. astron. Soc.* **55**, 217-54.
Meissner, R. & Strehlau, J. 1982. Limits of stress in continental crusts and their relation to depth—frequency distribution of shallow earthquakes. *Tectonics*, **1**, 73-89.
Mercier, J-L., Mouyaris, N., Simeakis, C., Roundoyannis, T. & Angelidhis, C. 1979. Intraplate deformation: a quantitative study of the faults activated by the 1978 Thessaloniki earthquakes. *Nature*, **278**, 45-8.
Molnar, P. & Chen, W-P. 1982. Seismicity and mountain Building. *In*: Hsu, K. J. (ed.) *Mountain Building Processes*, pp. 41-57. Academic Press.
Morton, W.H. & Black, R. 1975. Crustal attenuation in Afar. *In*: Pilger, A. & Rosler, A. (eds) *Afar depression of Ethiopia, Inter-union comm. Geodyn. Sci. Rep. No. 14.* pp. 55-65. E. Schweizerbart'sche Verlagsbuchhandlung, Stuttgart.
Nabelek, J. 1985. Geometry and mechanism of faulting of the 1980 El Asnam, Algeria, earthquake from inversion of teleseismic body waves and comparison with field observations. *J. geophys. Res.* **90**, 12,713-28.
North, R.G. 1974. Seismic slip rates in the Mediterranean and Middle East. *Nature*, **252**, 560-3.
Proffett, J.M. 1977. Cenozoic geology of the Yerington district, Nevada, and implications for the nature of Basin and Range faulting. *Bull. geol. Soc. Am.* **88**, 247-66.
Ransome, F.L. Emmons, W.H. & Garrey, G.H. 1910. Geology and ore deposits of the Bullfrog district, Nevada. *Bull. U. S. geol. Surv.* **407**, 130 pp.
Rehrig, W.A. & Reynolds, S.J. 1980. Geologic reconnaissance of a northwest trending zone of metamorphic core complexes in southern and western Arizona. *In*: Crittenden, M.D. jr, Coney, P.J. & Davis, G.H. (eds) *Cordilleran Metamorphic Core Complexes, Mem. geol. Soc. Am.* **153**, 131-57.
Richter, C.F. 1958. *Elementary Seismology*. W.H. Freeman, San Francisco.
Shudofsky, G.N. 1985. Source mechanisms and focal depths of east African earthquakes using Rayleigh wave inversion and body wave modelling. *Geophys. J. R. astron. Soc.* **83**, 563-614.
Sibson, R. H. 1980. Transient discontinuities in ductile shear zones. *J. struct. Geol.* **2**, 165-71.
——1982. Fault zone models, heat flow and depth distribution of earthquakes in the continental crust of the United States. *Bull. seism. Soc. Am.* **72**, 151-63.
—— 1984. Roughness at the base of the seismogenic zone: contributing factors. *J. geophys. Res.* **89**, 5791-800
—— 1985. A note on fault reactivation. *J. struct. Geol.* **7**, 751-4.
Smith, R.B. & Bruhn, R.L. 1984. Intraplate extensional tectonics of the eastern Basin-Range: inferences on structural style from seismic reflection data, regional tectonics and thermal-mechanical models of brittle-ductile deformation. *J. geophys. Res.* **89**. 5733-62.
Soufleris, C. & Stewart, G.S. 1981. A source study of the Thessaloniki (northern Greece) earthquake sequence. *Geophys. J. R. astron. Soc.* **67**, 343-58.
Soufleris, C., Jackson, J., King, G., Spencer, C. & Scholz, C. 1982. The 1978 earthquake sequence near Thessaloniki (northern Greece). *Geophys. J. R. astron. Soc.* **68**, 429-58.
Spencer, J.E. 1984. Role of tectonic denudation in warping and uplift of low angle normal faults. *Geology*, **12**, 95-8.
Stein, R.S. & Barrientos, S.E. In press. The 1983 Borah Peak, Idaho, earthquake: geodetic evidence for deep rupture on a planar fault. *In*: Stein, R. &

BUCKNAM, R. (eds) *Workshop XXVIII on the Borah Peak Earthquake*, US. Geol. Surv. Open-File Rep.
TAPPONIER, P. & MOLNAR, P. 1977. Active faulting and tectonics in China. *J. geophys. Res.* **82**, 2905-30.
—— & —— 1979. Active faulting and Cenozoic tectonics of the Tien Shan, Mongolia and Baykal regions. *J. geophys. Res.* **84**, 3425-59.
WERNICKE, B. 1981. Low angle normal faults in the Basin and Range province: nappe tectonics in an extending orogen. *Nature*, **291**, 645-8.
—— & BURCHFIEL, B.C. 1982. Modes of extensional tectonics. *J. struct. Geol.* **4**, 105-15.
WHITE, N.J., JACKSON, J.A. & MCKENZIE, D.P. In press. The relationship between the geometry of normal faults and that of the sedimentary layers in their hanging walls. *J. struct. Geol.*
WHITE, S. 1976. The effects of strain on the microstructures, fabrics and deformation mechanisms in quartzites. *Phil. Trans. R. Soc. London,* **A283**, 69-86.
WIENS, D.A. & STEIN, S. 1983. Age dependence of oceanic intraplate seismicity and implications for lithospheric evolution. *J. geophys. Res.* **88**, 6455-68.
WOOD, R. & BARTON, P. 1983. Crustal thinning and subsidence in the North Sea. *Nature*, **302**, 134-6.
YIELDING, G. 1985. Control of rupture by fault geometry during the 1980 El Asnam (Algeria) earthquake. *Geophys. J. R. astron. Soc.* **81**, 641-70.
ZIEGLER, P. 1983. Crustal thinning and subsidence in the North Sea. *Nature*, **304**, 561.

J.A. JACKSON, Department of Earth Sciences, Bullard Laboratories, Madingley Rise, Madingley Road, Cambridge CB3 0EZ, UK.

Active normal faulting in central Greece: an overview

STEVEN ROBERTS & JAMES JACKSON

Bullard Laboratories, Madingley Road, Cambridge CB3 0EZ, UK

Abstract: Rapid extension in central Greece is accommodated by large active normal faults that largely control the geomorphology. The faults bound graben whose asymmetry is evident in the topography, bathymetry, vertical movements of the coastline and tilting of Neogene sediments adjacent to them. Graben up to 20 km in width appear to be simple half-graben, whereas the hangingwalls in wider graben are generally broken by faults antithetic to the major bounding faults. The graben form basins up to 100 km long, but the faults bounding these basins are segmented, with individual segments no longer than about 20 km. The fault segments step en echelon or change polarity along strike of the large basins, but are not linked by simple strike-slip ('transfer') faults: the regions between the main fault segments appear to be pervasively deformed by numerous small faults. Drainage in central Greece is controlled by fault segmentation and footwall lithology. Where footwalls consist of Mesozoic limestone the drainage is often axial, breaking through the fault system into the main basin by flowing round the ends of fault segments or through gaps where the fault segments step. In such places, drainage directly off the fault scarps is relatively minor, and the fault segmentation controls the spacing of major fans along the basin margin. Where footwalls consist of Neogene sediments, drainage may flow directly across the fault scarp, though in such cases the catchment areas are relatively small and linear. The origin of the fault segmentation is unclear, but may be related to rotations of the fault-bounded blocks about horizontal and vertical axes.

The Aegean Sea, and its surrounding regions of Greece, Albania, western Turkey, southern Yugoslavia and Bulgaria, is one of the most rapidly extending areas on the continents today. Its present day activity is manifest by its high seismicity and by the dramatic influence of large active normal faults on the topography, geomorphology and movement of the coastline relative to sea level. The region has attracted much interest, both in order to examine active extension in general, and to use the Aegean as a modern analogue to help interpret the geological record of older, now inactive, stretched sedimentary basins. Much of the published work in the Aegean has been either very general (covering the whole region) or specifically concerned with individual faults, earthquakes or basins. This paper is concerned with generalizations that can be made about the geometry of the normal faulting, and the way in which it influences geomorphology and drainage, in a sub-region of the Aegean extensional province in central Greece, from the northern Peloponnese to the island of Evvia (Fig. 1). This is one of the most active regions within the extensional province, and one of the most accessible: many of the largest active faults crop out near the coast, and are easier to study than those under the Aegean Sea itself, where greater extension has led to total submergence. For comparison, the across-strike width of the region in Fig. 1 (130 km) is very similar to that of the East Shetland Terrace in the northern North Sea.

Background

Regional considerations suggest that the N–S rate of extension across the whole province (i.e. between Crete and Bulgaria) is 40–60 mm a^{-1} (McKenzie 1978; Le Pichon & Angelier 1979; Jackson & McKenzie 1988b). An examination of this century's seismicity reveals that most of this extension is likely to be achieved by the motion of large faults during earthquakes of magnitude 6.0 and larger, and that the contribution of smaller faults (and earthquakes) and aseismic creep to the regional deformation of the upper crust is probably not significant (Jackson & McKenzie 1988a, b). This is an important deduction, affecting the perspective with which this paper discusses normal faulting. In an earthquake of magnitude 6.0, slip occurs on a fault with a length of about 10–15 km, similar in dimension to the thickness of the upper seismogenic layer of the continental crust, which in Greece is in the region of 10–15 km, based on accurate microearthquake surveys (Soufleris *et al.* 1982; King *et al.* 1985; Hatzfield *et al.* 1987; Lyon-Caen *et al.* 1988). Throughout this paper we will distinguish such faults, which we may refer to as 'large' faults, from those

From Holdsworth, R.E. & Turner, J.P. (compilers) *Extensional Tectonics: Faulting and Related Processes*. The Geological Society, Key Issues in Earth Sciences, **2 (2)**, 151–168.
1476-315X/02/$15.00
First published in: "Roberts, A.M., Yielding, G. & Freeman, B. (eds) *The Geometry of Normal Faults*, The Geological Society, London, Special Publications **56**, 125–142".

with lengths significantly smaller than the thickness of the seismogenic layer, and which slip in earthquakes smaller than about magnitude 5.0. (For example, an earthquake of magnitude 4.0 involves movement on a fault of only about 100 m length.) Since it is known that large faults take up most of the deformation, a knowledge of the geometry and kinematics of the large faults is a prerequisite for an understanding of how the extension is achieved: in other words, any model of the active deformation that is inconsistent with the motion on the large faults cannot be correct. This is not to say that small faults are unimportant, but simply that they do not account for much of the overall strain. In Greece, as elsewhere, most small earthquakes occur away from the large faults and represent internal deformation of the blocks bounded by major faults. They may also provide information on the transfer of deformation from one large fault to another, which we discuss later.

The region of Fig. 1 is one of the most seismically active in the Aegean extensional province. The present rate of N–S extension across this part of central Greece is likely to be in the range 10–20 mm a^{-1} (Jackson & McKenzie 1988, 1989): a rate in agreement with a comparison of two geodetic surveys carried out in 1890–1900 and 1988 (Billiris *et al.* 1989). However, the overall motion across the region of Fig. 1 is NE–SW (see inset to Fig. 1b), with roughly equal amounts of N–S extension and E–W right-lateral shear, the shear component being caused by the motion of Turkey relative to Eurasia (McKenzie 1972, 1978; McKenzie & Jackson 1983, 1986).

Geological studies of the Neogene basins in central Greece, summarized by Mercier *et al.*

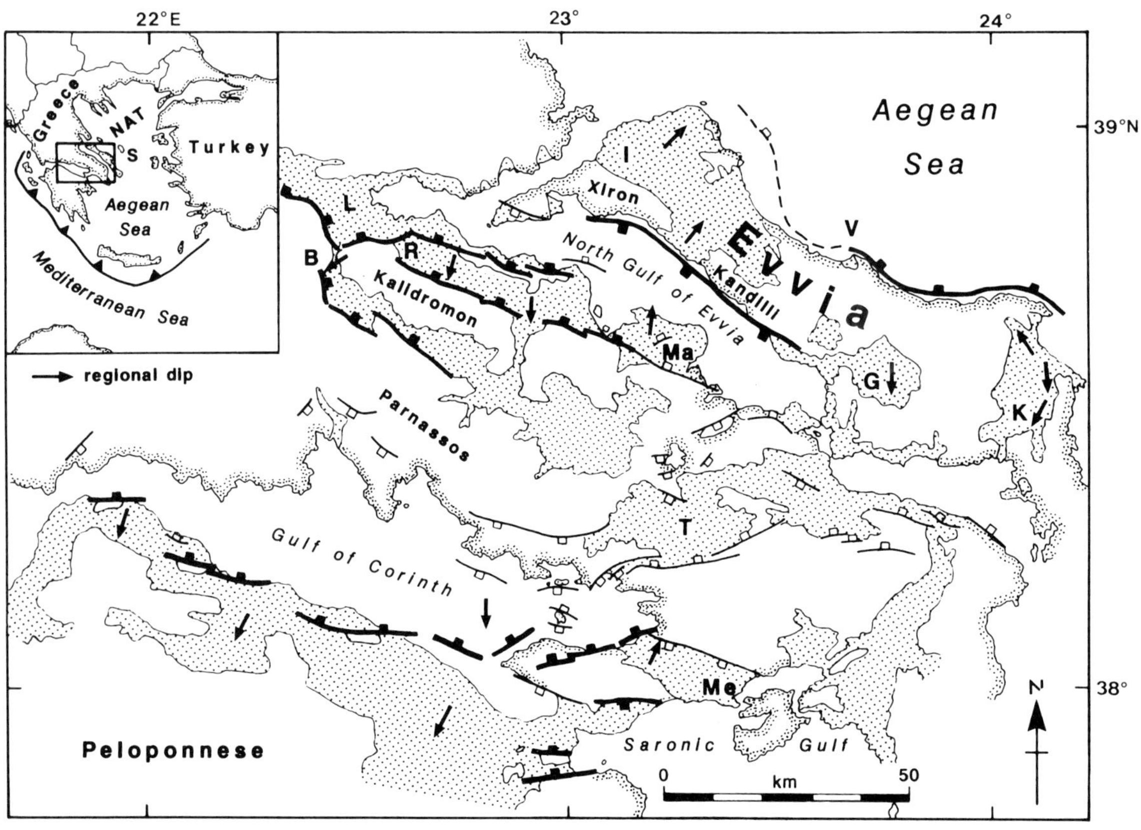

Fig. 1.(a) General fault map of central Greece, showing the principal areas (stippled) where Miocene–Recent sediments are preserved. The region is identified in the inset map, where NAT is the North Aegean Trough and S is the island of Skyros. Major range-bounding faults are shown in heavy lines with filled blocks on their downthrown side. Other 'large' faults, capable of generating earthquakes of magnitude 6.0 or greater, but with apparently less displacement or topographic expression, are shown in thinner lines with open blocks. The distinction between these two fault types is, to some extent, subjective. Some faults, such as those on the NE flanks of Kalidromon and Parnassos and on the NE side of the Megara (Me) basin; may no longer be active. Basins identified by letters are: Lamia (L), Bralos (B), Renginion (R), Istea (I), Vlachia (V), Gides, Kymi (K), Malessina (Ma), Thiva (T), Megara (Me). Arrows show approximate regional directions of dip within the Neogene basins, from Roberts (1988) and Brooks & Ferentinos (1984).

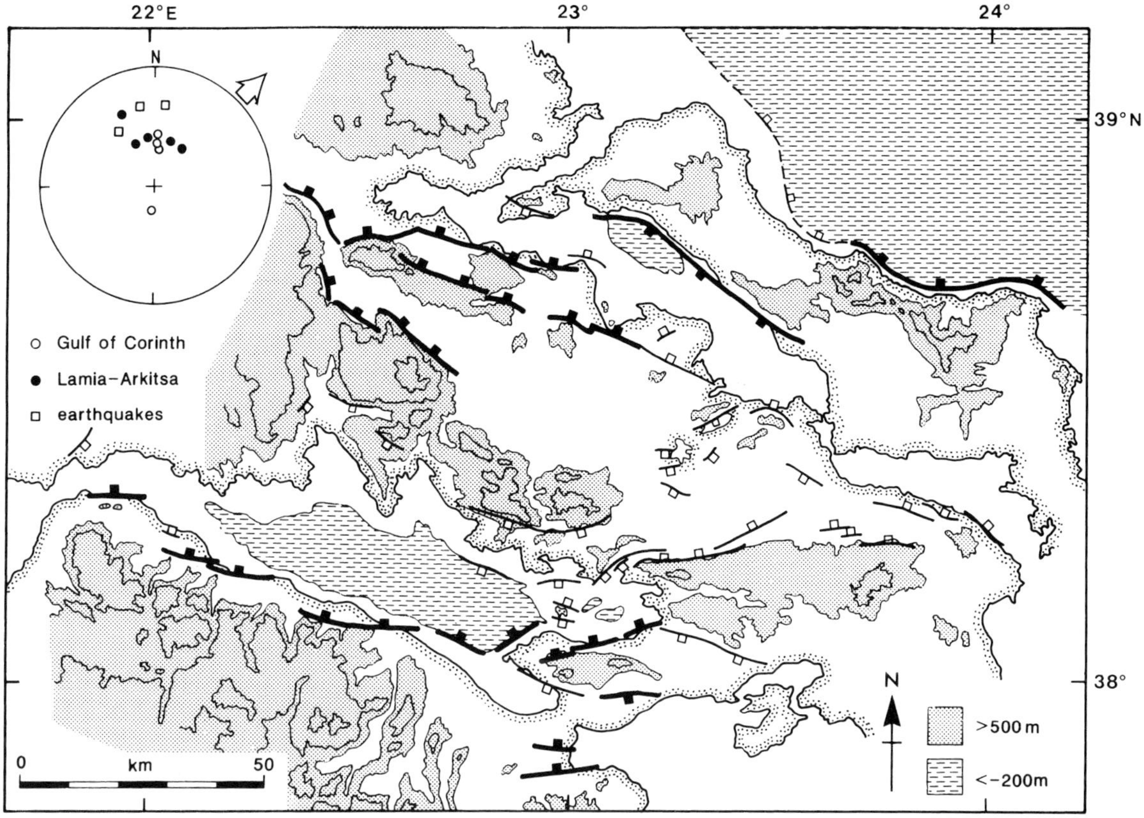

Fig. 1.(b) Topography (500 m and 1000 m contours, stippled above 500 m) and bathymetry (shaded below 200 m) in central Greece. Faults are as in Fig. 1(a). The inset shows the a stereographic projection of the slip vectors on major faults, measured at the surface in the Lamia-Arkitsa and Gulf of Corinth systems, or from fault plane solutions for earthquakes larger than magnitude 6.0. The open arrow is the direction of motion in the North Aegean Trough, and is the expected motion between the top of the figure and the bottom (see McKenzie 1978; McKenzie & Jackson 1983, 1986).

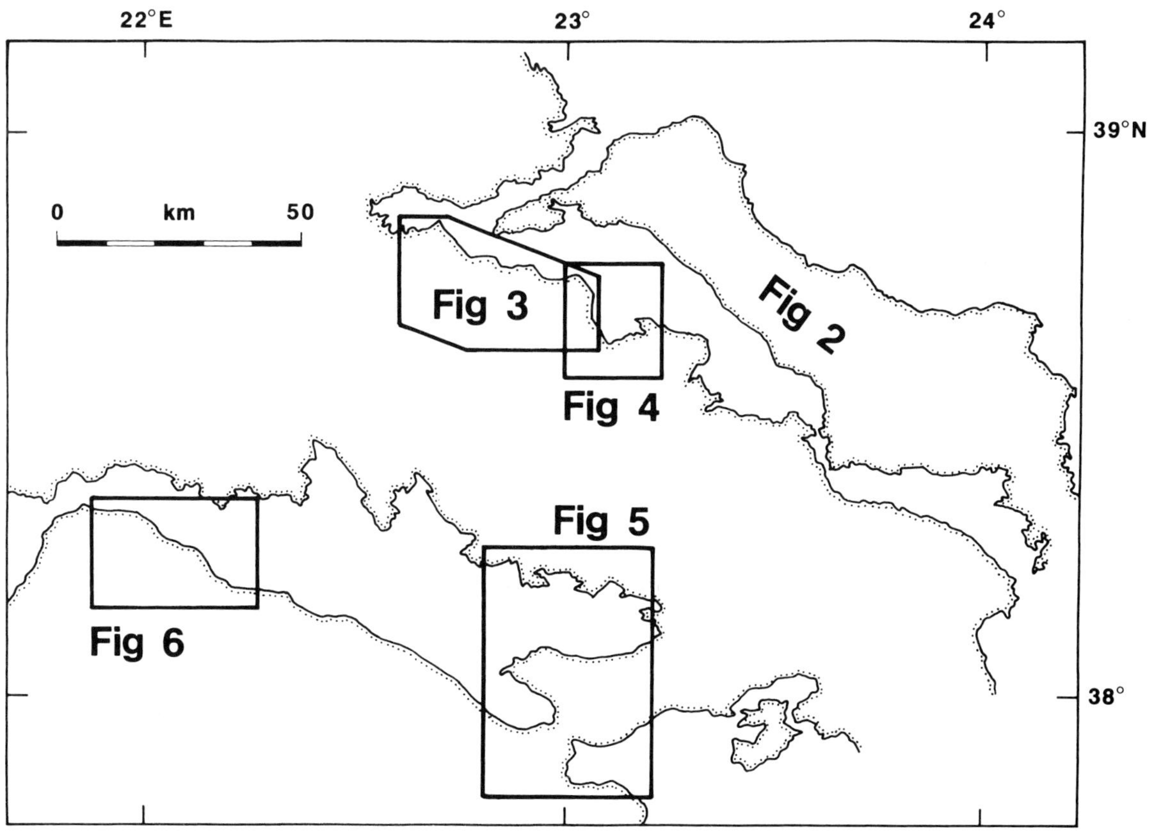

(1979, 1987, 1989), indicate that extension has dominated the deformation of the region since at least the early Pliocene (*c*. 5 Ma BP), though it may not always have been constant in orientation or rate. Within the Aegean region there is evidence that, in places, extension began during the Middle Miocene (Mercier *et al*. 1976; Sengor *et al*. 1985); though most of it is likely to have occurred in the last 5 Ma (Kissel & Laj 1988; Mercier *et al*. 1989). Many authors regard the extension in the Aegean province as being linked dynamically with the subduction of the Mediterranean Sea floor in the Hellenic Trench (e.g. McKenzie 1978; Le Pichon 1982; Mercier *et al*. 1987).

Normal faulting

General structure

The structure of central Greece is dominated by a series of large normal faults with an E–W to NW–SE strike (Fig. 1). These faults are very clearly associated with the topography, and are responsible for the development of large basins, such as the Gulfs of Corinth and northern Evvia, and the basins NE of Mt Parnassos and Evvia. All the major, range-bounding normal faults are thought to have been active in the Pleistocene, but the relative timing of their activity is less certain: some, such as that bounding the NE flank of Kalidromon (Figs 1a & 3), may no longer be active (e.g. Philip 1974; Lemeille 1977; Mercier *et al*. 1979). In spite of the high historical seismicity, surface faulting accompanying specific earthquakes has been reported only from the Gulf of Corinth and northern Gulf of Evvia basins.

All the basins are asymmetric, most of them with the dominant faulting on the SW side. This asymmetry is often clear from the topography, but is also seen in the tilting of young sediments and vertical movements of the coastline (discussed later). However, only the smaller basins, up to 15–20 km in width, such as those of Renginion (Fig. 3) and NE of Parnassos (Fig. 1a), look like simple 'domino-style' rotated half-graben. The larger basins, such as the Gulfs of Corinth and northern Evvia, though obviously asymmetric, are bounded by faults that are antithetic to the dominant ones. Thus on the north side of the Gulf of Corinth are a number of recent normal fault scarps that dip south. Several of these are 'large' in the seismological sense of being long enough to generate earthquakes of magnitude 6.0, but individually appear to have much less total displacement than the large north-dipping faults on the south side of the Gulf. The fault responsible for the earthquake of 3 March 1981 near Kaparelli (Fig. 5) is an example (Jackson *et al*. 1982a).

Previous impressions that all the dominant faulting dips NE (e.g. Jackson & McKenzie 1983) are incorrect. A major SW-dipping normal fault system occurs along the SW side of northern Evvia, dominating the tilting, drainage and topography of the central part of the northern Gulf of Evvia and its borders (Figs 1 & 2). The structure of the northern Gulf of Evvia and of the island of Evvia itself varies dramatically along strike. At the western end of the Gulf the structure is clearly dominated by a fault system on the south side, dipping north. Farther east, at about the longitude of Arkitsa (Figs 1 & 3), the polarity of the half-graben changes, and is controlled by the fault on the north side of the Gulf, dipping south. This fault appears to die out east of Mt. Kandilli (Figs 1a & 2), where the morphology of central Evvia is dominated by a fault along the north coast of the island, dipping north and bounding a basin offshore. Thus northern Evvia resembles a twisted horst block, that cannot be realistically assessed in two-dimensions (see also Stiros 1985).

It is unclear how structures in the central part of the N Gulf of Evvia accommodate the change in polarity part of the half graben, but there is no evidence for significant strike-slip faulting in the N to NE direction. Antithetic faults are seen both east and west of the transition in polarity, on the western tip of Evvia and in the Martinon region (Fig. 4). The Atalanti–Martinon fault system (Figs 3 & 4), part of which moved in two earthquakes of magnitude 6.5 and 6.7 in 1984, is sited in the region of the transition, and changes in nature along strike. Its eastern part, from Tragana to the coastline, is opposite the large south-dipping fault flanking Mt. Kandilli on Evvia and resembles the antithetic faults on the north side of the Gulf of Corinth, in that, though 'large' in the seismological sense, it has relatively minor displacement, indicated by both the topography (a few tens of metres) and the occurrence of Miocene–Pliocene marls on both sides of the fault (Fig. 4). Its western part, from Tragana to Atalanti (Figs 3 & 4), is closer to the north-dipping fault system running west from Arkitsa, and is associated with greater topography (a few hundred metres) and a substantial basin in its hanging-wall. Changes in half graben polarity have attracted attention elsewhere, notably in the Gulf of Suez (e.g. Garfunkel & Bartov 1977), the Basin and Range (Stewart 1980) and East Africa (Rosendahl *et al*. 1986). In the absence of off-

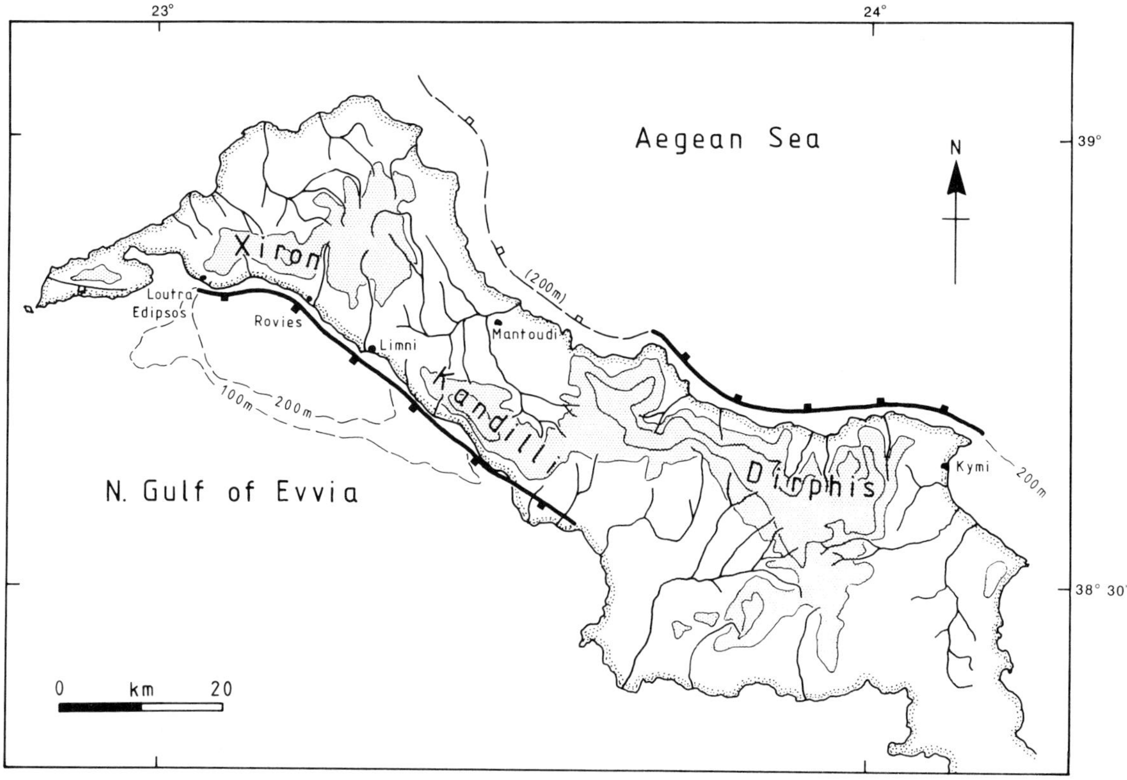

Fig. 2. Topographic map of north and central Evvia, showing the principal drainage pattern. Contours are at 400 m and 800 m, stippled above 400 m. The fault along the NE side of Dirphis is drawn to follow the 200 m bathymetric contour: the coastal morphology suggests that it is most active, or has largest displacement, in its eastern part (heavy line).

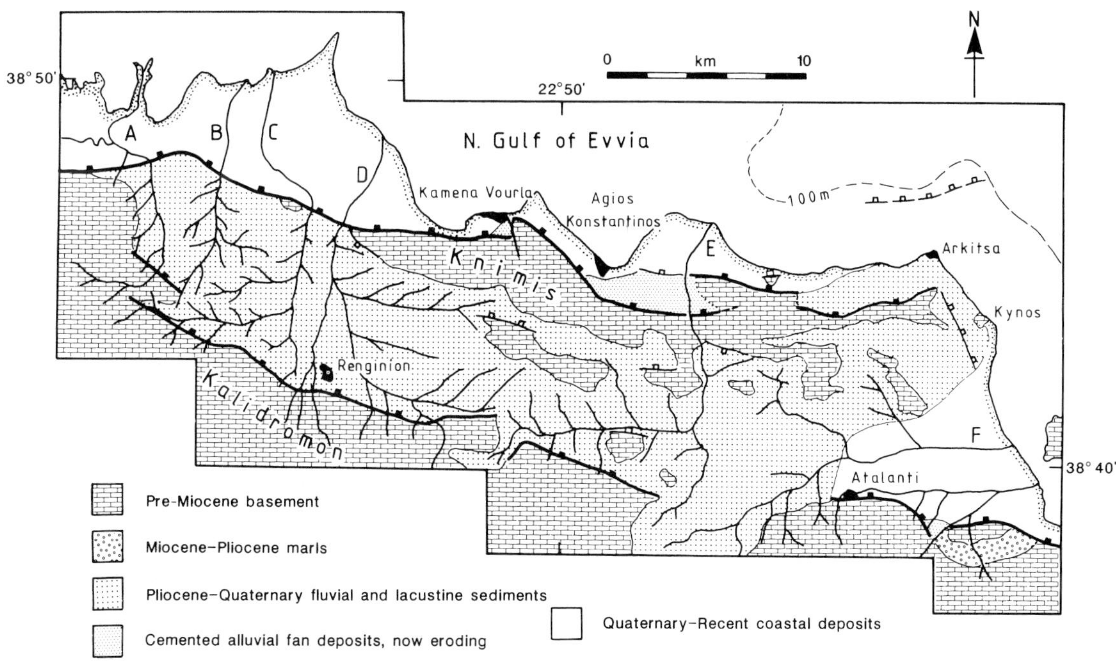

Fig. 3. Summary geological map of the Renginion graben region, illustrating the faulting and principal drainage. The geology is modified from Philip (1974). Major range-bounding faults are shown in thick lines with filled blocks on their downthrown side. Thinner lines with open blocks are smaller faults.

shore data from the N Gulf of Evvia, it is premature to compare this region with other areas.

Two possible explanations have been suggested for the presence of antithetic faulting in the half-graben of central Greece. The first attributes the antithetic faults to an abrupt decrease in the dip of the dominant faulting at depth, probably below the seismic-aseismic transition (e.g. King *et al.* 1985; Eyidogan & Jackson 1985; Jackson 1987). In this scheme, antithetic faulting would migrate in time towards the centre of the basin as the hanging wall block moves away from the point of intersection of the dominant fault with the flatter, deeper portion of the fault. This would lead to a staircase morphology of antithetic faults (see Fig. 25a of Eyidogan & Jackson 1985), each with relatively minor displacement. This resembles the situation in the Gulf of Corinth to some extent, particularly at the eastern end (Jackson *et al.* 1982a). A second explanation, advanced by Jackson & White (1989), is that the spacing of the dominant faults across strike (50 km from the southern Gulf of Corinth to the NE flank of Parnassos) is too great for the intervening block to rotate coherently about a horizontal axis, as the stresses required for it to do so would be

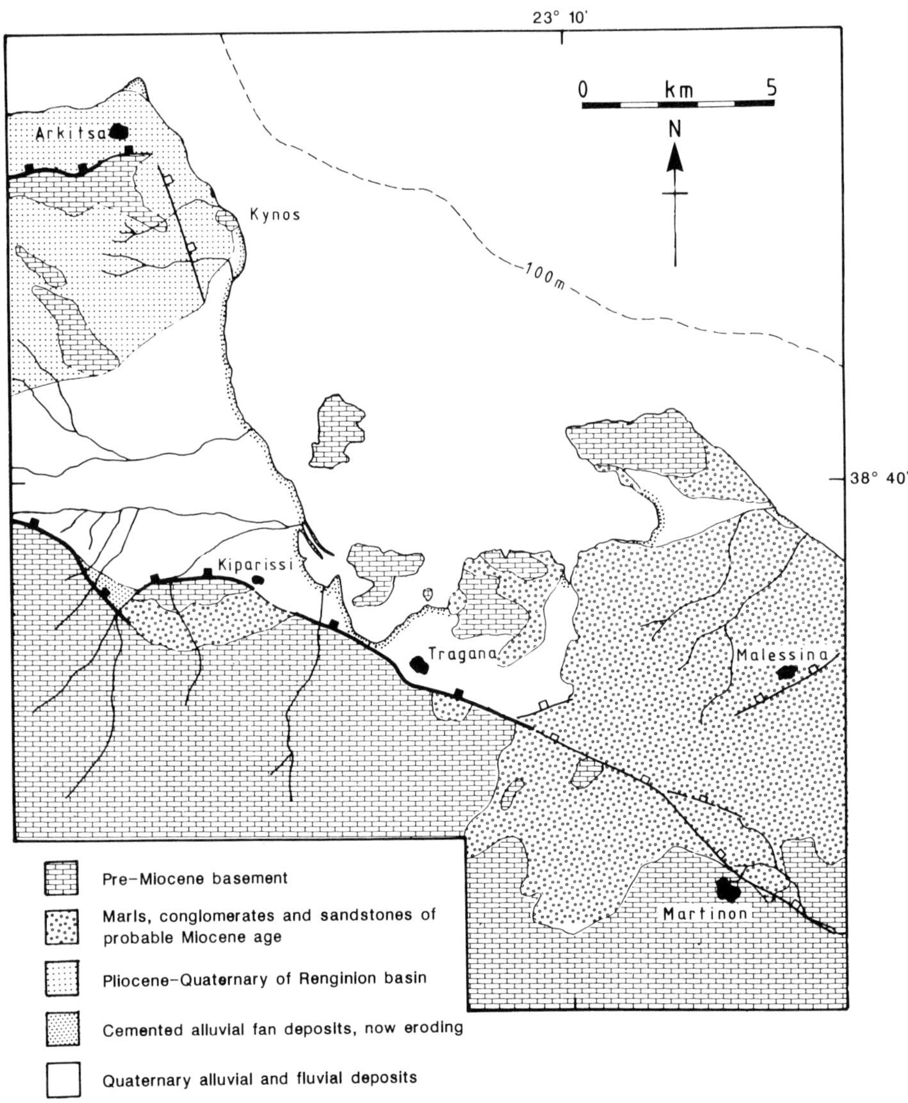

Fig. 4. Summary geological map of the Kiparissi-Martinon region, based on Philip (1974) and Rondoyanni (1984). The earthquakes of 1894 are known to have ruptured the fault from Martinon to Kiparissi, and also the segment west of Kiparissi through Atalanti (see Fig. 3). In its eastern part this fault has very subdued morphology (thin line with open blocks). Its central and western parts (heavy lines with filled blocks) follow topographic escarpments 500–1000 m high.

large (several hundred bars). Instead it breaks up by antithetic faulting.

The major fault systems that bound large topographic depressions are easy to identify. However, central Greece contains many 'large' faults with relatively small displacement, particularly around the north side of the Gulf of Corinth and the Thiva basin (Fig. 1a). We are sure that many such faults are omitted from Fig. 1, particularly because there are historical records of large earthquakes in regions where no faulting is marked (e.g. Sieberg 1932; Galanopoulos 1955; Bousquet & Pechoux 1977; Ambraseys & Jackson 1990). For this reason Fig. 1 is not a reliable guide to seismic risk.

Fault morphology, dips and slip vectors

Much of the most dramatic fault morphology in central Greece occurs where the footwall blocks are composed of Mesozoic limestone. Polished surfaces cutting through cemented limestone breccia are occasionally exposed, and sometimes revealed by earthquake movement or quarrying of scree resting against the scarp. The details of such surfaces are discussed by Hancock & Barka (1987) and Stewart & Hancock (1989). They often contain corrugations with wavelengths of cm to tens of metres and with axes aligned parallel to the slip vector. Some of the surfaces contain smaller and less continuous grooves and slickensides indicating different directions of movement on the same fault plane (e.g. Pechoux *et al.* 1973; Mercier 1976). The dips of these polished surfaces are generally steep (40–70°) and the axes of the major corrugations, when projected upwards from their exposure, generally do not intersect the hillside. In many places the slope of the preserved limestone hillside appears to be a reasonable approximation to the dip of the fault. The slip vector azimuths, based on the major corrugations or grooves, are generally N–S (± 20°), similar to those from earthquake focal mechanisms and conspicuously different from the NE–SW motion across the region as a whole (Fig. 1b, inset); an observation that led McKenzie & Jackson (1983, 1986) to suggest that the fault blocks rotate clockwise about a vertical axis. The dips seen in earthquake focal mechanisms are also steep (30–70°), and are consistent with the large faults having approximately planar cross sections from the surface to the base of the seismogenic upper crust (Jackson & McKenzie 1983; Eyidogan & Jackson 1985; Jackson 1987; Jackson & White 1989), though the available data cannot rule out changes of dip of up to about 15°.

Continuity along strike

Although basins such as the Gulf of Corinth and northern Gulf of Evvia are more than 100 km long, the faulting that flanks them is discontinuous, even when the polarity of the half-graben is unchanged (Fig. 1). Where faults crop out on land they occur in segments no longer than about 20 km. Even where the faulting is confined entirely offshore, like that flanking northern Evvia, changes in the strike, morphology, and vertical movements along the coastline suggest that the fault segments are not continuous for more than about 20 km. This segmentation of the faulting has a profound influence on the drainage and sedimentation, which is discussed later. The maximum observed segment length of 20 km potentially limits the maximum size of earthquakes in central Greece to about magnitude (M_s) 6.7, though larger earthquakes involving movement of more than one segment are possible, as in the 1970 Gediz earthquake in western Turkey (Eyidogan & Jackson 1985). Of interest here is how the fault segments are linked to each other along the basin margins.

It is clear that the segments are not linked by single strike-slip ('transfer') faults in the simple manner sometimes envisaged elsewhere (e.g. Bott 1976; Gibbs 1984; Etheridge 1986). Instead, the region between the fault segments appears to be intensively deformed by widespread minor faulting, probably accompanied by some folding. The evidence for this is both morphological and seismological.

The morphological evidence is illustrated by the example in Fig. 5, which is a map of the eastern part of the Gulf of Corinth. The Perakora peninsula occupies the region between two major normal faults dipping north: one following the NW flank of Mt Gerania, through Pisia (and which moved in the earthquake of 25 February 1981), and the other lying offshore along the NW coast of the peninsula and continuing to the west (the western part of this second fault may have moved in the earthquake of 24 February 1981). Both faults are marked by major topographic (or bathymetric) escarpments about 1000 m high. The high ridge in the footwall of the Pisia fault dies out sharply at its western end, precisely where the fault displacement itself died out in the 1981 earthquake. At its eastern end, the higher of the two faults bounding the NE flank of Gerania also loses its clear topographic expression, terminating in a canyon that allows drainage to enter the Gulf of Corinth round the end of the fault segment, and which also marked the eastern limit of surface ruptures in 1981. The topographic change associated with the eastern termination of this fault is partly influenced by footwall lithology, which changes from limestone (near Pisia) to more easily eroded igneous rocks farther east: but the 1981 surface ruptures continued at least 2 km east of the change in lithology to reach the canyon draining north. The western part of the Perakora peninsula, between the Gerania fault and the fault offshore, is severely deformed by many smaller normal faults, up to a few km in length, some of which were reactivated in 1981. The topography associated with these smaller normal faults is generally less than 100–200 m, giving a hummocky appearance to the peninsula. It is clear that no strike-slip fault crosses the peninsula to link the major faults.

A similar geometry on a smaller scale is seen on the north side of the eastern Gulf of Corinth (Fig. 5) in the Kaparelli fault system, which also moved in 1981 (Jackson *et al.* 1982a). Here two fault segments, each about 6 km long, were offset by 3 km. The region between the two segments was deformed by many discontinuous cracks and fissures consistent with the warping expected when passing along strike from a footwall to a hanging-wall block (see Fig. 12 of Jackson *et al.* 1982a). In both these examples the gradual decrease in seismic displacement along strike as the ends of the segments are reached, mirrored also in the topography, makes it inconceivable that the fault motion is transferred elsewhere by a single strike-slip fault, in the way transform faults are believed to link ridge segments in the oceans.

The seismological evidence for the nature of

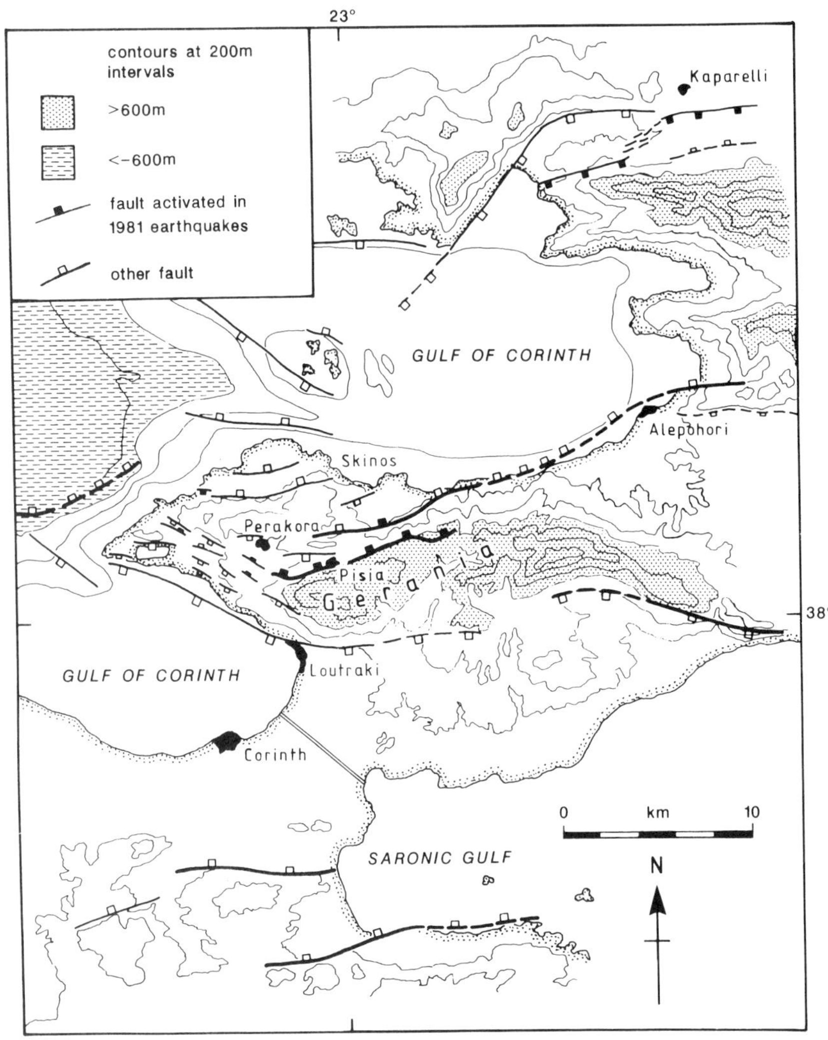

Fig. 5. Faulting, bathymetry and topography in the eastern Gulf of Corinth, after Jackson *et al.* (1982a), Perissoratis *et al.* (1986) and Collier (1988). Major range-bounding faults are shown in thick lines. Faults active at the surface in the 1981 earthquakes have filled blocks in their hanging-walls.

the linkage between fault segments comes from detailed aftershock studies of large normal faulting earthquakes. Here the data are not just from Greece (Soufleris *et al.* 1982; King *et al.* 1985; Lyon-Caen *et al.* 1988), but also from elsewhere (e.g. Idaho, USA.: Richens *et al.* 1987; Italy: Deschamps & King 1984 and Westaway & Jackson 1987). The regions where normal fault segments end or are offset are usually the sites of intense aftershock activity dispersed over a volume, with a great variety of focal mechanisms that may even include thrusting. These observations suggest a widespread fracturing by small faults and not linkage by a single fault.

Jackson & White (1989) suggest that the maximum fault segment length of about 20 km is a worldwide characteristic of active normal faults on the continents. There is no doubt that normal faults are also discontinuous on a scale smaller than this. However, it is this maximum segment length that imposes a scale on the structure, morphology and drainage in regions of active extension, particularly in central Greece. The pervasive fracturing and minor faulting between offset fault segments may contribute to the erosion and establishment of drainage systems in these locations (discussed later; see also Leeder & Gawthorpe 1987).

Uplift and subsidence

Vertical movements in central Greece are evidently related to three factors: (1) a general subsidence caused by stretching (2) a general uplift of the Peloponnese; (3) block motions caused by normal faulting that are superimposed on the first two. Most vertical movements are estimated relative to sea level. Over the last 2 Ma the movement of sea level is reasonably well known, but is best determined over the last 20 000 years, during which time sea level has not been higher than present (van Andel & Shackleton 1982). Earlier in the Pleistocene sea levels were probably also never more than 10 m above the present level (Inman 1983). A reliable indicator of uplift relative to sea level is the occurrence above the highest tide mark of borings, usually in crystalline limestone, containing the bivalve *Lithophaga lithophaga (sp)* (see e.g. Lyell 1867). Bored sea cliffs erode more quickly (probably over several thousand years: Inamura 1926; Schneider 1976; Matsuda *et al.* 1978; Pirazzoli & Kawana 1986) than the intact limestone, to produce a characteristic 'notched' profile.

Subsidence in central Greece is greatest in the Aegean Sea, NE and E of Evvia, where the crustal thickness deduced from gravity data constrained by seismic refraction is less than on the mainland (Makris & Vees 1977; Makris & Stobbe 1984), indicating greater stretching offshore. Fifty km east of central Evvia, the island of Skyros (marked S on the inset to Fig. 1a) is the crest of a tilted block bounded by a normal fault with a deep basin on its NE side (see Jackson *et al.* 1982b, Jackson & McKenzie 1983). This structure is clear in the morphology of Skyros, which has a steep, rugged and uplifting NE coast and gently sloping and embayed west side (Guernet 1971; Melentis 1973; Lemeille 1977; Roberts 1988).

Several authors have remarked on the uplift of the Peloponnese, particularly along its northern edge, adjacent to the Gulf of Corinth (e.g. Dufaure *et al.* 1975; Sebrier 1977; Keraudren & Sorel 1987; Mariolakos & Stiros 1987; Collier 1988, 1990; Ori 1989). The magnitude of this uplift is uncertain, but clearly large, with Gilbert fan delta deposits now preserved more than 1000 m above sea level (Ori 1989). The timing of the uplift is also uncertain, because the uplifted sediments have proved difficult to date. Much of the uplift may be Pleistocene in age, as most of the outer part of the Hellenic Arc was subsiding through the Pliocene (Mercier *et al.* 1987). The cause of the uplift may well not be solely footwall uplift associated with the normal fault system along the southern edge of the Gulf of Corinth.

Two lines of evidence suggest that some of the uplift is causd by underplating the Peloponnese with sediments overthrust in the Hellenic Trench system further west, in the manner described by Angelier *et al.* (1982) to explain the uplift of Crete. Firstly, the thick sediments in the Ionian basin west of Greece do not appear to influence the isotopic composition of volcanic rocks above the subducting slab (Barton *et al.* 1983; Briqueu *et al.* 1986). Secondly, the seismicity indicates that the subducting slab is almost flat under the Peloponnese, steepening to about 45° only in the far NE of the Peloponnese (Hatzfeld *et al.* 1989), where the first volcanic rocks are found. The highest elevations are in the northern Peloponnese, and it may be that the general uplift caused by underplating is accentuated here by uplift in the footwall of the Gulf of Corinth normal fault system. The southern side of the Gulf of Corinth itself contains a complicated record of the interaction between uplift and sea-level change, preserved in both the exposed sediments and in terrace morphology (Freyberg 1973; Keraudren & Sorel 1987; Collier 1988, 1990). However, the hangingwalls of the active faults along the southern coast of the Gulf, and the whole of the northern coast, are subsiding, indicating that fault movement is more than able to keep pace with any regional uplift.

Rugged coastal morphology and exposed 'notches' containing recent *L. lithophaga* borings 1–2 m above sea level indicate uplift in the footwalls of the normal faults bounding the NE and SW coasts of Evvia. The footwall of the north-dipping Kamena Vourla–Arkitsa fault system contains exposed *L. lithophaga* borings at its eastern end near Kynos (Fig. 3) and river terraces in its central part, again indicating uplift relative to sea level.

Where the hangingwall blocks of the large faults are exposed on land, it is not always straightforward to distinguish tectonic vertical motions from sedimentation in rapidly prograding alluvial fan systems or from subsidence caused by settling of the foundations of archaeological sites built on unconsolidated alluvium (which is accentuated by the frequent vibrations associated with earthquakes: e.g. Ambraseys & Sarma 1969; Blackman 1971; Owen 1987). One case of hangingwall subsidence is clear: movement of about one metre on the Kiparissi–Martinon fault (Fig. 4) in the earthquakes of 1894 caused flooding of the hangingwall between Kiparissi and Tragana, making islands of Mesozoic limestone outcrops that were previously attached to land (Skuphos 1894).

Tilting

The clearest indication of relative vertical movements associated with faulting is seen in the regional dips of sediments in the Neogene basins (Fig. 1a). Sections of these sediments are fine-grained lacustrine deposits that are likely to have been deposited close to the horizontal. Most are Miocene to Pleistocene in age (Lemeille 1977; Steininger & Rogl 1984; Mettos *et al.* 1986), and were deposited in basins that were not necessarily bounded by the currently active faults. Nevertheless, in nearly all the basins the regional dip directions are those expected from the motions of the nearby major active fault systems (and are reflected in the geomorphology and drainage, discussed later). Note particularly the change in dip between northern and central Evvia (Fig. 1a), reflecting the change in polarity of the fault system along the coast.

However, there are some basins exceptional to this generalization. In the Kymi basin (Fig. 1a) dip directions are very variable and much steeper (locally up to 85°) than elsewhere (generally up to 15–25°). Sediments in the Kymi basin are early Miocene in age (Lemeille 1977) and have probably been deformed by more than one episode of faulting (e.g. Guernet 1971; Lemeille 1977; Katsikatsos *et al* 1981). The present drainage and morphology of the Kymi basin is controlled by the large normal fault offshore, with uplift, deep erosion and incision occurring along the eastern coastal margin of the basin and decreasing southwestwards. The main rivers draining the basin flow towards the NE, with deeply incised valleys close to the coast.

In the Megara basin (Fig. 1) the regional dip is to the NE, towards a degraded normal fault scarp with a WNW–ESE strike that bounds the basin. However, the present-day geomorphology of the basin is clearly not controlled by this fault, but by a probably more recently-active fault offshore along its NW end, which has caused uplift and incision of the drainage in the NW part of the basin (Jackson *et al.* 1982a).

The sediments exposed on land in the Malessina basin (Figs 1 & 4) dip N or NE in the northern part, towards the major fault bounding Mt. Kandilli along the SW coast of Evvia, and only dip S in the southernmost part of the basin adjacent to the Martinon fault. The structure of this basin does not appear to be greatly influenced by the Martinon fault, which has a relatively small topographic expression in its eastern part.

Relations between drainage and faulting

Relationships between drainage and faulting in central Greece are illustrated using three examples.

The Renginion–Atalanti region

The Renginion basin (Fig. 3) contains fluvial and lacustrine sediments of Pliocene–Pleistocene age (Philip 1974; Ioakim & Rondoyanni 1988), with an average dip of about 10–20° towards the south, and bounded on the south side by a fault system along the limestone mountain of Kalidromon. Sediments on the north side of the basin are mostly resting unconformably on the limestone mountain of Knimis, which is the footwall of the Kamena Vourla–Arkitsa fault system. The average elevation of the Renginion basin is 300–400 m above sea level and it is currently being eroded rapidly. The segmentation of the Kamena Vourla–Arkitsa fault system is clearly expressed in the topography by the resistant Mesozoic limestone of Knimis, and exposures of fault planes strongly suggest it is still active. The morphology of the Kalidromon fault is less well developed, and it may no longer be active, though there is evidence for movement in the Pleistocene (Philip 1974; Mercier 1976). In the SE part of the basin is the western end of the Atalanti–Martinon fault, which last moved in earthquakes in 1894 (Skuphos 1984).

Only in the western part of the Renginion basin is the main drainage perpendicular to the strike of the faults (streams A, B and C on Fig. 3). These streams have relatively narrow drainage basins and cut through the Kamena Vourla fault system west of Knimis where the footwall is no longer the resistant Mesozoic basement.

In the eastern part of the basin, south of the Knimis massif, the main drainage is axial, eventually reaching the northern Gulf of Evvia either by flowing round the end of Mt. Knimis (streams D and F on Fig. 3), or by cutting through Mt. Knimis at an obvious step in the Kamena Vourla–Arkitsa fault system (stream E on Fig. 3). The drainage basins for streams D–F are far greater than those of A–C, though the combined sediment output of streams A–C is able to create a large fan system in the Gulf. The sediment output and drainage into the Gulf directly off the limestone footwall of the Kamena Vourla–Arkitsa fault system is relatively insignificant: though even this is greatest at a step in the faulting by Kamena Vourla itself.

Evvia

The topography and position of the drainage divide in north and central Evvia (Fig. 2) closely follow the two major fault systems: along the SW coast in the north and along the NE coast in the central part of the island. The deepest water in the N Gulf of Evvia is in the hangingwall of the fault system running from near Loutra Edipsos to Mt Kandilli. The saddle in the footwall escarpment of this system is located in relatively soft Neogene sediments (Figs 1a & 1b), whereas the high mountains to either side are of basement limestones and serpentinites. Much deeper water exists off the NE coast of Evvia, with the highest and most rugged topography associated with the basement limestones of Mt Dirphis.

The largest drainage basins are established in Neogene sediments and flow away from the large coastal fault systems down the dip slopes of the footwall blocks. Only small linear basins drain directly off the fault scarps into the hangingwalls, even where the Neogene sediments in the saddle of the Loutra Edipsos–Kandilli escarpment extend to the coast (Figs 1a & 2).

The western Gulf of Corinth: Egion region

The NW–SE trend of the western Gulf of Corinth is controlled by a set of almost E–W fault segments stepping to the right along the coast (Fig. 1) and offshore (Higgs 1988). Between the coastal set of faults and other, probably older or less active, normal faults to the south are Neogene sediments with regional dips of up to 30° towards the south. The sediments between the stepping coastal faults are mostly alluvial conglomerates, well exposed near Egion (Fig. 6). Older, more consolidated marls, sandstones and conglomerates are exposed in the southern part of the basin (Ori 1989). Two basement ridges, Mts Bura and Panahaiko, occur in the footwalls of the coastal faults, the fault bounding Mt Bura having probably moved in the earthquake of 1861 (Montessus de Ballore 1924; Bousquet *et al.* 1983).

Drainage enters the Gulf of Corinth either directly, in relatively straight channels cutting through Neogene footwall blocks, or by axial channels that reach the Gulf through steps in the fault segments. The drainage south of Mt Panahaiko flows almost E–W, and so does the River Selinous as it enters the Gulf between the Egion and Mt Bura faults, where the land surface is tilted about 3° to the south. The drainage breaks through the Mt Bura fault at both ends of the resistant basement ridge, using a step in the faulting at the eastern end. The streams entering the Gulf of Corinth in this region form large submarine fans, which are prone to slumping, and are described by Ferentinos *et al.* (1988). A discussion of the sedimentary facies likely to be associated with such a drainage pattern is given by Leeder & Gawthorpe (1987).

General principles

It is clear from these examples that the pattern of drainage entering large depressions such as the Gulfs of Corinth and Evvia is related to both the faulting and the footwall lithology. The resistance of footwall ridges in basement, especially basement carbonate, leads to much axial drainage, but the size of the drainage basins depends on the continuity of the footwall ridges, which in turn is mostly determined by the continuity of the faulting itself. Much sediment is derived from basins in the dip slopes of footwall blocks and reaches the sea by rivers

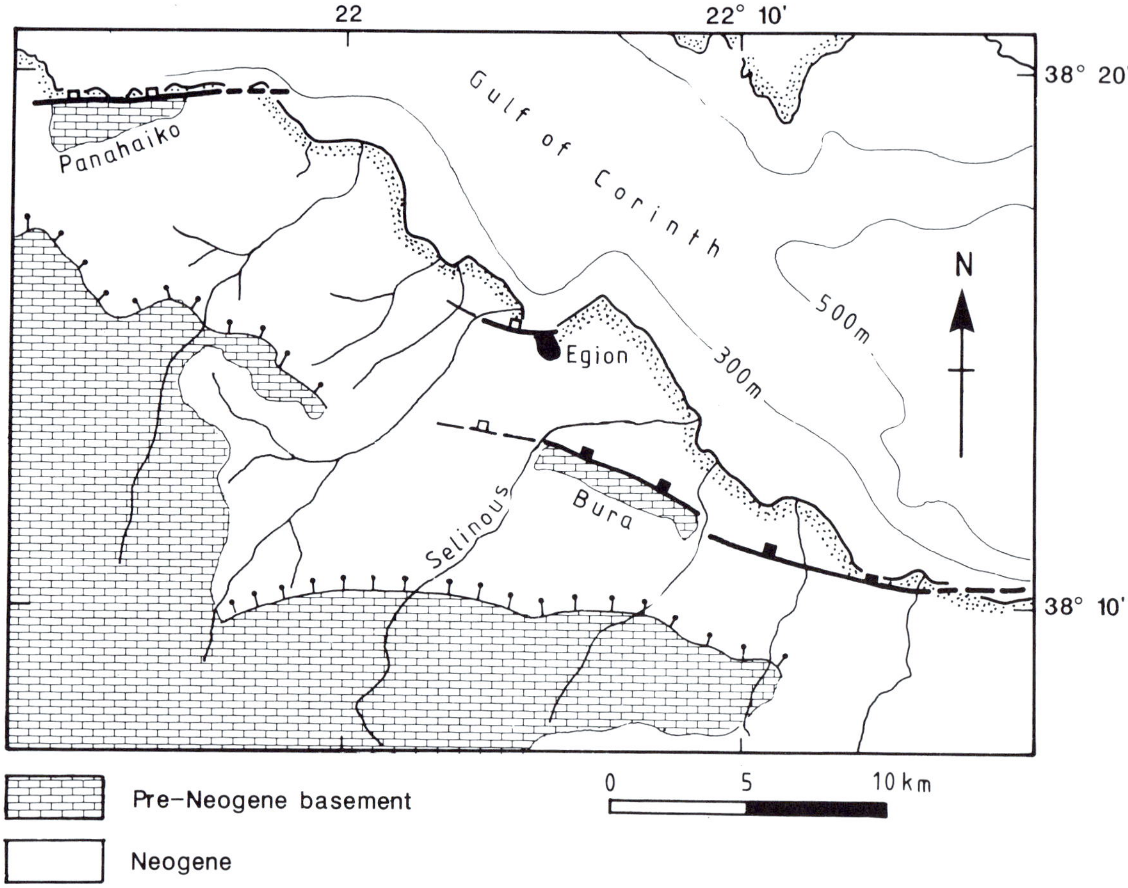

Fig. 6. Drainage and faulting near Egion, in the western Gulf of Corinth. Geology is modified from Dufaure (1978). The fault thought to be responsible for the 1861 earthquake is identified by filled blocks in its hanging-wall. The barbed basement-Neogene contacts are faults that are probably less active than the coastal fault system farther north (see also Ori 1989).

that break through gaps caused by steps between fault segments. Such basins often have larger catchment areas than the relatively narrow drainage systems that flow roughly perpendicular to strike, cutting through footwalls made of soft sediment. The smallest fans and drainage systems of all are those flowing directly off the fault controlled escarpments into the sea (see also Sabot & Maroukian 1982).

Perhaps the most important general principle is that the drainage in central Greece only starts to make sense when viewed on a scale comparable with the length of the fault segmentation, which may be up to 20 km. This may be a useful lesson when examining palaeocurrent data in tectonically older regions.

Perched basins

The segmentation of fault scarps may also affect the preservation of sediments along the escarpment systems bounding the major graben in central Greece. Near Kiparissi (Fig. 4) is a left step separating the Martinon and Atalanti (Fig. 3) segments of a fault system known to have moved in earthquakes in 1894. Southwest of Kiparissi is an area about 4×2 km^2 containing sediments with the same lithology (marls) and fossil fauna and flora as those in the Malessinna basin farther east (Fig. 4), which are thought to be Miocene in age (Rondoyanni–Tsiambaou 1984; J. Mercier, *pers. comm.*). The Kiparissi sediments are about 100 m above the level of the coastal plain to the north, and dip south by up to 16°. It appears that the Kiparissi sediments are part of a much larger basin, probably predating the Atalanti–Martinon faulting, that is exposed in the step between major fault segments, but probably buried beneath the coastal plain either side of the step. (It was pointed out earlier that the 600 m high topography across the fault south of Tragana dies out progressively eastwards: near Martinon the elevation across the fault is only tens of metres, with sediments of the Malessina basin exposed on both sides: see Fig. 4.)

A second example is seen in the left step in the fault system near Agios Konstantinos, where the stream marked E in Fig. 3 breaks through Mt Knimis. Here, alluvial fan sediments dipping south up to 15° are found with an elevation up to 300 m above sea level. These are lithologically similar to those of the Renginion basin (Philip 1974), but are unfossiliferous. They occupy an area similar in size to those near Kiparissi.

It is unclear why the 'perched basins' of Kiparissi and Agios Konstantinos are exposed, though their location is clearly related to the fault segmentation. One possibility is that they are preserved at a relatively high elevation because the displacement on both fault segments dies out in the region where they meet or overlap. Another is that one fault segment is growing along strike at the expense of the other, which is dying out (in the manner of propagating rifts in the oceans), thus uplifting in its footwall sediments previously lowered in the hangingwall of the 'dying' fault. There are few constraints on the dating and magnitude of displacement on these faults, and further speculation is not justified. Other possible examples of 'perched basins' have been described by Johnson (1930) in the Basin and Range Province of the western USA. A lesson to be learned from the occurrence of perched basins is the way in which local fault control can influence the extent and distribution of erosional unconformities or sediment preservation within extended basins. When working with sub-surface data such control must be understood first, before any discussion of regional sea-level variations becomes relevant.

Discussion

Two obvious questions arise from the observations summarized above. (1) What is the overall extension across the region of Fig. 1? (2) What is the origin of the fault segmentation, that so dramatically influences the geomorphology and sedimentation in central Greece?

No seismic refraction line exists across central Greece, though the limited data from offshore are sufficient to show that the crustal thickness is less in the Aegean than on the mainland (Makris & Vees 1977; Makris & Stobbe 1984). Recent studies in the northern North Sea (White 1990) and South China Sea (Su *et al.* 1990) have found that, in regions of block tilting, the overall extension estimated from the simple domino model (i.e. $\beta=\sin\phi_0/\sin\phi_1$, where ϕ_0 and ϕ_1 are the initial and final dips of the faults) was, within error, compatible with extension estimated from both the change in crustal thickness and the observed syn- and post-rifting subsidence. These studies give us confidence that the same approach to estimating extension from the faulting in central Greece is not unreasonable. Seismogenic normal faults in central Greece typically dip 40–45° (Jackson & White 1989) and associated tilting of Miocene–Pleistocene sediments is typically up to 15–20° (Roberts 1988), suggesting that the initial fault dips were about 60°. These observations are consistent with β values of about 1.2–1.3, or extensions across the region in Fig. 1 of 20–30 km. We expect these to be minimum values: they assume that the blocks rotate coherently, which, as mentioned above, they probably do not, since they are observed to break up by antithetic faulting. Nonetheless, we believe that this estimate of extension is the right order of magnitude, though it is probably uncertain by 50%. It would require fault offsets of 5–7 km on blocks 20–25 km wide, which, in view of the topography and bathymetry associated with some of the larger faults, we do not consider to be impossible. If this extension occurred in the last 5 Ma, the average rate of stretching across central Greece would be 4–6 mm/yr, which is rather lower than the probable rate averaged over the last 100 years, thought to be in the region of 10–20 mm/yr (Jackson & McKenzie 1988a, b; Billiris *et al.* 1989).

The segmentation of the major normal faults is a fundamental structural characteristic of central Greece, and of other extensional provinces (e.g. Moore 1960, Garfunkel & Bartov 1977, Armijo *et al.* 1986). Various authors have considered possible origins of such discontinuous faulting, including: the elastic interaction of en echelon shear cracks (Segall & Pollard 1980); rotation of the stress field with depth (Mandl 1987); and the tendency of growing faults to develop arcuate strike lines, commonly concave towards the downthrown side (Mandl 1988). In the case of central Greece, there is a kinematic consideration that, in our opinion, makes it almost inevitable that the faulting should be discontinuous. We mentioned earlier that the slip vectors on the major faults were approximately N–S, whereas the overall direction of motion across the region of Fig. 1 is NE–SW (see inset to Fig. 1b). McKenzie & Jackson (1983 & 1986) pointed out that faulting with N–S slip vectors could take up the overall deformation if the faults rotate clockwise about a vertical axis as they move, which we now know occurs, from paleomagnetic studies on Evvia and Skyros (Kissel *et al.* 1986).

A component of right-lateral shear is thus distributed across central Greece by the normal faulting (even though movement on the normal faults has a left-lateral component) and leads to a clockwise sense of rotation. However, it is most unlikely that the shear gradient is uniform across the deforming zone (cf. New Zealand: Walcott 1984), and the rate of clockwise rotation is therefore likely to vary. This would inevitably break up any continuous linear marker (or fault) that spanned the zone. We are aware of no theoretical argument that is able to predict the maximum length of the fault segments that form; though, following Jackson & White (1989), we note that the observed maximum length of about 15–20 km is similar to the thickness of the seismogenic upper crust, and that it seems intuitively reasonable that the thickness of the deforming layer should influence the length and continuity of the structures that form within it.

Conclusions

Central Greece is one of the most actively extending regions on the continents, with a present-day motion of 10–20 mm a^{-1} across an area 150 km wide, and a minimum overall extension in the region of 20–30 km, most of which probably occurred during the last 5 Ma. The region consists of a number of fault-bounded half-graben, with most (but not all) the major faults dipping N to NE. Only the smaller graben, with widths up to about 20 km, appear to be simple half-graben that are bounded by significant faults on only one side and rotate coherently about a horizontal axis. The wider graben, such as the Gulf of Corinth and Northern Gulf of Evvia, while clearly asymmetric, are bounded by antithetic faults, with apparently relatively minor displacement, facing the major faults. The asymmetry of the graben is apparent in the topography, bathymetry, vertical movements of the coastline and in the regional tilting of exposed Neogene sediments adjacent to the major faults, though some dips in the Neogene sediments may be related to movement on faults that are no longer active.

The graben systems form basins up to 100 km long, such as the Gulf of Corinth and the Northern Gulf of Evvia, but the faulting bounding such basins is discontinuous, with fault segments reaching a maximum of about 20 km in length. In the Northern Gulf of Evvia the polarity of the half graben changes along strike. The island of Evvia itself is a twisted horst, with large bounding faults en echelon, rather than back-to-back. There is no seismological, geomorphological or structural evidence for fault segments being linked by simple strike-slip ('transfer') faults: instead, zones between steps or overlaps in major fault segments appear to be severely deformed by minor faulting. Zones where fault segments overlap or step are sometimes locations where sediments that would normally have been dropped to considerable depth in the hanging walls are exposed at the surface as 'perched basins'.

The segmentation of the normal fault systems is perhaps the most dramatic structural characteristic of central Greece and along with the lithology of the footwalls, profoundly influences the geomorphology and drainage. Where the footwalls consist of Mesozoic limestone, the drainage systems are mostly axial, flowing into the basins round the ends of fault segments or where fault segments step. Drainage directly off the fault scarps into the basins is relatively minor. In such regions the fault segmentation imposes a scale on the drainage, controlling the spacing of major fans along the margin of the basin. Where the footwalls consist of Neogene sediment, drainage directly across the fault scarps into the basins is more important, though the catchment areas of these streams are often relatively small and linear, compared with those in the limestone footwall terrains.

Thus the two main controls on the morphology and drainage of central Greece are the fault segmentation and the lithology of the footwalls. The origin of the segmentation is unclear, though it is a common feature of other extensional terranes, where, as in Greece, the maximum segment length is in the region of 15–20 km; similar to the thickness of the seismogenic upper crust. In central Greece the segmentation may be related to the rotation of blocks about a vertical, as a well as a horizontal, axis.

We are grateful to Professor J.-L. Mercier and his colleagues for many discussions and for their work in central Greece, which has constantly guided us. We thank I.G.M.E. for permission to work in Greece, and Stathis Stiros for much logistical help. J.-L. Mercier, M. Leeder, C. Stark and A. Roberts provided thoughtful reviews that improved the original manuscript. This work was funded by N.E.R.C. We thank North Sea Sun Oil for general support of our work on active normal faulting. Cambridge Earth Sciences contribution No. 1529.

References

AMBRASEYS, N. N. & JACKSON, J. A. 1990. Seismicity and strain in central Greece between 1890 and 1988. *Geophysical Journal International*, **101**, 663–708

—— & SARMA, S. 1969. Liquefaction of soils by earthquakes. *Bulletin of the Seismological Society of America*. **59**, 61–64.

ANGELIER, J., LYBERIS, N. LE PICHON, X., BARRIER, E. & HUCHON, P. 1982. The neotectonic development of the Hellenic Arc and the Sea of Crete: a synthesis. *Tectonophysics*, **86**, 159–196.

ARMIJO, R., TAPPONNIER, P., MERCIER, J. L. & HAN TONG-LIN. 1986. Quaternary extension in southern Tibet: field observations and tectonic implications. *Journal of Geophysical Research*, **91**, 13, 803–13, 872.

BARTON, M., SALTERS, V. & HUIJSMANS, J. 1983. Sr-isotope and trace element evidence for the role of continental crust in calc-alkaline volcanism on Santorini and Milos, Aegean Sea, Greece, *Earth & Planetary Science Letters*, **63**, 272–291.

BILLIRIS, H. PARADISSIS, D., VEIS, G. ENGLAND, P., PARSONS, B., CROSS, P., RANDS, P., RAYSON, M., SELLERS, P., ASHKENAZI, V., DAVISON, M. & JACKSON, J. 1989. Geodetic determination of the strain in Greece in the interval 1900–1988. (Abstr.) *EOS: Transactions of the American Geophysical Union*, **70**, 719.

BLACKMAN, D. J. 1971. Evidence of sea level change in ancient harbours and coastal instalations. *In*: BLACKMAN, D. J. (ed.) *Marine Archaeology*, Butterworths, London, 115–137.

BOUSQUET, N., DUFAURE, J. J. & PECHOUX, P.Y. 1983. Temps historique et évolution des paysages égéens. *Mediterranée*. **2**, 3–10.

—— & PECHOUX, P.Y. 1977. La Seismicité du Basin Egéen pendant L'Antiquité. Méthodologie et premiers resultats. *Bulletin de la Sociéte Géologique de France*, **19**, 679–684.

BOTT, M.H.P. 1976. Formation of sedimentary basins of graben type by extension of the continental crust. *Tectonophysics*, **36**, 77–86.

BRIQUEU, L., JAVOY, M., LANCELOT, J. R. & TASUMOTO, M. 1986. Isotope geochemistry of recent magmatism in the Aegean arc: Sr, Nd, Hf and O isotopic ratios in the lavas of Milos and Santorini – geodynamic implications. *Earth & Planetary Science Letters*, **80**, 41–54.

BROOKS, M. & FERENTINOS, G. 1984. Tectonics and sedimentation in the Gulf of Corinth and the Zakynthos and Kefallania channels, Western Greece. *Tectonophysics*, **101**, 25–54.

COLLIER, R. 1988. *Sedimentary facies evolution in continental fault-bounded basins formed by crustal extension: the Corinth basin, Greece*. Ph.D. Thesis, University of Leeds.

COLLIER, R. E. 1990. Eustatic and tectonic controls upon Quaternary coastal sedimentation in the Corinth basin, Greece. *Journal of the Geological Society, London*, **147**, 301–314.

DESCHAMPS, A. & KING, G. C. P. 1984. Aftershocks of the Campania-Lucania (Italy) earthquake of 23 November 1980. *Bulletin of the Seismological Society of America*, **74**, 2483–2517.

DUFAURE, J.J. 1978. Le Peloponnese, Carte Géomorphologique. Institute Géographie, Paris.

——, KERAUDREN, B. & SEBRIER, M. 1975. Les terrasses de Corinthe (Grèce): Chronologie et déformations. *Comptes Rendues, Academie de Science, Paris*, **281**, 1943–1945.

ETHERIDGE, M. A. 1986. On the reactivation of extensional fault systems. *Philosophical Transactions of the Royal Society, London*, **317A**, 179–194.

EYIDOGAN, H. & JACKSON, J. A. 1985. A seismological study of normal faulting in the Demirci, Alasehir and Gediz earthquakes of 1969–70 in W. Turkey: implications of the nature and geometry of deformation in the continental crust. *Geophysical Journal of the Royal Astronomical Society*, **81**, 569–607.

FERENTINOS, G., PAPATHEODOROU, G. & COLLINS, M. B. 1988. Sediment transport processes on an active submarine fault escarpment: gulf of Corinth, Greece. *Marine Geology*, **83**, 43–61.

FREYBURG, B. von 1973. Geologie des Isthmus von Korinth. *Erlanger Geologische Abhandlungen*, **95**, 12–154.

GALANOPOULOS, A. 1955. 'Seismiki geographia tis Hellados'. *Annales Géologique de Pays Helleniques*. **6**, 83–121.

GARFUNKEL, Z. & BARTOV, Y. 1977. The tectonics of the Suez rift. *Bulletin of the Geological Survey of Israel*. **71**.

GIBBS, A. D. 1984. Structural evolution of extensional basin margins. *Journal of the Geological Society, London* **141**, 609–620.

GUERNET, C. 1971. *Contribution a l'étude géologique de l' Eubée et des régions voisines*. Thesis, Université de Paris.

HANCOCK, P. L. & BARKA, A. A. 1987. Kinematic indicators on active normal faults in W. Turkey. *Journal of Structural Geology* **9**, 573–584.

HATZFELD, D., CHRISTODOULOU, A., SCORDILIS, E. PANAGIOTOPOULOS, D. & HATZIDMITRIOU, P. 1987. A microearthquake study of the Mygdonian graben (N. Greece). *Earth & Planetary Science Letters*, **81**, 379–396.

HATZfIELD, D., PEDOTTI, G., HATZIDIMITRIOU, P., PANAGIOTOPOULOS, D., SCORDILIS, M., DRAKOPOULOS, J., MAKROPOULOS, K., DELIBASSIS, Hellenic subduction beneath the Peloponnese: first results of a microearthquake study. *Earth & Planetary Science Letters*.

HIGGS, B. 1988. Syn-sedimentary structural controls on basin formation in the Gulf of Corinth, Greece. *Basin Research*, **1**, 155–165.

INAMURA, I. 1926. On the recurrence of destructive earthquakes. *Proceedings of the Imperial Academy, Tokyo* **12**, 264–267.

INMAN, D. L. 1983. Application of coastal dynamics to the reconstruction of paleocoastlines in the vicinity of La Jolla California. *In*: MASTERS, P. M. & FLEMMING, N. C. (eds): *Quaternary coastlines and Marine Archeology*. Academic, New York.

IOAKIM, C. & RONDOYANNI, T. 1988. Contribution to the geological study of Zeli region, Locris (central Greece). *Revue de Micropaléontologie*, **31**, 129–136.

JACKSON, J. A. 1987. Active normal faulting and continental extension. *In*: COWARD, M. P., DEWEY, J. F. & HANCOCK, P. L. (eds) *Continental Extensional Tectonics*, Geological Society, London, Special Publication **28**, 3–17.

——, GAGNEPAIN, J., HOUSEMAN, G., KING, G. C. P., PAPADIMITRIOU, P., SOUFLERIS, C. & VIREUX, J. 1982a. Seismicity, normal faulting, and the geomorphological development of the Gulf of Corinth (Greece): the Corinth earthquakes of February & March 1981. *Earth & Planetary Science Letters*, **57**, 377–397.

——, KING, G. & VITA-FINZI, C. 1982b. The neotectonics of the Aegean: an alternative view. *Earth & Planetary Science Letters*, **61**, 303–318.

—— & MCKENZIE, D. P. 1983. The geometrical evolution of normal fault systems. *Journal of Structural Geology*, **5**, 471–482.

—— & —— 1984. Active tectonics of the Alpine-Himalayan Belt between W. Turkey and Pakistan. *Geophysical Journal of the Royal Astronomical Society*, **77**, 185–264.

—— & —— 1988a. The relationship between plate motions and seismic moment tensors, and the rates of active deformation in the Mediterranean and Middle East. *Geophysical Journal*, **93**, 45–73.

—— & —— 1988b. Rates of active deformation in the Aegean Sea and surrounding regions. *Basin Research*, **1**, 121–128.

—— & WHITE, N. J. 1989. Normal faulting in the upper continental crust: observations from regions of active extension. *Journal of Structural Geology*, **11**, 15–36.

JOHNSON, D. 1930. Geomorphological aspects of Rift Valleys. *Proceedings of the 15th International Geological Congress*, **2**, 354–373.

KATSIKATSOS, G., BRUIJN, DE H., VAN DER MEULEN, A. J. 1981. The Neogene of the Island of Euboea (Evia): A Review. *Geologie en Mijnbouw*, **16**, 509–516.

KERAUDREN, B, & SOREL, D. 1987. The terraces of Corinth (Greece): a detailed record of eustatic variations of the sea level during the last 500,000 years. *Marine Geology*, **77**, 99–107.

KING, G. C. P., OUYANG, Z. X., PAPADIMITRIOU, P., DESCHAMPS, A., GAGNEPAIN, J., HOUSEMAN, G., JACKSON, J. A., SOUFLERIS, C. & VIRIEUX, J. 1985. The evolution of the Gulf of Corinth (Greece): an aftershock study of the 1981 earthquakes. *Geophysical Journal of the Royal Astronomical Society*, **80**, 677–693.

KISSEL, C. & LAJ, C. 1988. The Tertiary geodynamical evolution of the Aegean arc: a paleomagnetic reconstruction. *Tectonophysics*, **146**, 183–201.

——, —— & MAZAUD, A. 1986. First paleomagnetic results from Neogene formations in Evia, Skyros, and the Volos region, and the deformation of Central Aegea. *Geophysical Research Letters*, **13**, 1446–1449.

LEEDER, M. R. & GAWTHORPE, R. L. 1987. Sedimentary models for extensional tilt-block/half-graben basins. *In*: COWARD, M. P., DEWEY, J. F. & HANCOCK, P. L. (eds.): *Continental Extensional tectonics*. Geological Society, London, Special Publication, **28**, 139–152.

LEMILLE, F. 1977. *Etudes néotectoniques en Grèce centrale nordorientale: Eubée centrale, Attique, Béotie, Locride et dans les Sporades du Nord, (Skiros)*. Thesis, Université de Paris-Sud, Orsay.

LE PICHON, X. 1982. Land-locked oceanic basins and continental collision: the eastern Mediterranean as a case example. *In*: HSU K. (ed.) *Mountain Building Processes*, Academic, New York 201–211.

—— & ANGELIER, J. 1979. The Hellenic arc and trench system: a key to the neotectonic evolution of the eastern Mediterranean area. *Tectonophysics*, **60**, 1–42.

LYELL, C. 1867. *Principles of Geology*. 10th Edn., Murray, London.

LYON-CAEN, M., ARMIJO, R., DRAKOPOULOS, J., BASKOUTAS, J., DELIBASSIS, N., GAULON, R., KOUSKOUNA, V., LATOUSSAKIS, J., MAKROPOULOS, K., PAPADIMITRIOU, P., PAPANASTASSIOU, D. & PEDOTTI, G. 1988. The 1986 Kalamata (S. Peloponessos) earthquake: Detailed study of a normal fault and tectonic implications. *Journal of Geophysical Research* **93**, 14,967–15,000.

MAKRIS, J. & STOBBE, C. 1984. Physical properties and state of the crust and upper mantle of the easten Mediterranean Sea deduced from geophysical data. *Marine Geology*, **55**, 347–363.

—— & VEES, R. 1977. Crustal structure of the Central Aegean sea and the islands of Evvia and Crete, Greece, obtained by refraction seismic experiments. *Journal of Geophysics*, **42**, 329–341.

MANDL, G. 1987. Discontinuous fault zones. *Journal of Structural Geology*, **9**, 105–110.

—— 1989. *Mechanics of Tectonic Faulting*. Elsevier, Amsterdam.

MARIOLAKOS, R. & STIROS, S. 1987. Quaternary deformation of Isthmus and gulf of Corinthos (Greece), *Geology*, **15**, 225–228.

MATSUDA, T. OTA, Y., ANDO, M. & YONEKURA, N. 1978. Fault mechanism and recurrance time of major earthquakes in S. Kanto district, Japan, as deduced from coastal terrace data. *Bulletin of the Geological Society of America*, **89**, 1610–1618.

MCKENZIE, D. P. 1972. Active tectonics of the Mediterranean region. *Geophysical Journal of the Royal Astronomical Society*, **30**, 109–185.

—— 1978. Active tectonics of the Alpine-Himalayan belt: the Aegean Sea and surrounding regions. *Geophysical Journal of the Royal astronomical Society*, **55**, 217–254.

—— & JACKSON, J. A. 1983. The relationship between strain rates, crustal thickening, paleomagnetism, finite strain and fault movements within a deforming zone. *Earth & Planetary Science Letters*, **65**, 182–202, erratum 1984, **70**, 444.

—— & —— 1986. A block model of distributed deformation by faulting. *Journal of the Geological Society, London*, **143**, 249–253.

MELENTIS, J. 1973. La géologie de l'ile de Skiros. *Bulletin de la Sociéte Géologique de Grèce* **5**,

108–114.

MERCIER, J. L. 1976. La néotectonique, ses méthods et ses buts. Un example: l'arc égéen (Mediterranée orientale). *Revue de Géologie Dynamique et de Géographie Physique*, **18**, 323–346.

MERCIER, J. L., CAREY, E., PHILIP, H. R. & SOREL, D. 1976. La neotectonique plio-quaternaire de l'arc égéen externe et de la Mer égéen et ses relations avec séismicité. *Bulletin de la Sociéte Géologique de France*, **18**, 159–176.

——, DELIBASSIS, N., GAUTIER, A., JARRIGE, J. J., LEMILLE, F., PHILIP, H., SEBRIER, M. & SOREL, D. 1979. La néotectonique de l'Arc Egéen. *Revue de Géologie Dynamique et de Géographie Physique*, **21**, 67–92.

——, SOREL, D. & SIMEAKIS, K. 1987. Changes of the state of stress in the overiding plate of a subduction zone: The Aegean Arc from the Pliocene to the Present. *Annales Tectonicae*, **1**, 20–39.

——, ——, VERGELY, P. & SIMEAKIS, K. 1989. Extensional tectonic regimes in the Aegean basins during the Cenozoic. *Basin Research*, **2**, 49–71.

METTOS, A., IOAKIM, C. & RONDOYANNI, T. 1986. Les formations neogenes lacustres de l'Attique bu Nord-Béotie: stratigraphie palynologie et tectonique. *Geologie Mediterranéenne*, **12–13**, 167–174.

MONTESSUS de BALLORE, F.: La Geologie sismologique. Armand Colin Paris 1924.

MOORE, J. G. 1960. Curvature of normal faults in the Basin and Range province of the western United States. *Professional Papers of the U.S. Geological Survey*, **400**, 409–411.

ORI, G. G. 1989. Geologic history of the extensional basin of the Gulf of Corinth (?Miocene-Pleistocene), Greece. *Geology*, **17**, 918–921.

OWEN, G. 1987. Deformation processes in unconsolidated sands. 11–24. *in*: JONES, M. E. & PRESTON, R. M. F. (eds) *Deformation of Sediments and Sedimentary Rocks*. Geological Society, London, Special Publication, **29**, 11–24.

PECHOUX, P., PEGORARO, P., PHILIP, H. & MERCIER, J. 1973. Déformations Mio-Pliocene et Quaternaires en extension et en compression sur les rivages du golfe Malaique et du canal d'Atalanti (Egée, Grèce). *Comptes Rendues, Academie de Science, Paris*, **276**, 1813–1816.

PERISSORATIS, C. MITROPOULOS, D. ANGELOPOULOS, I. 1986. Marine geological research at the E. Korinthiakos Gulf. *I.G.M.E. Special issue of Geology and Geophysical Research*, Institute of Geology and Mineral exploration, Athens. 381–401.

PHILIP, H. 1974. *Etude néotectonique des rivages Egéens en Locride et Eubée nord-occidentale (Gréce)*. Thesis, Université des Sciences et Techniques du Languedoc, Montpellier.

PIRAZZOLI, P. A. & KAWANA, R. 1986. Determination de movements crustaux quaternaires d'apres la deformation des anciens rivages dans les iles Ryukyu, Japan. *Revue de Geologie Dynamique et de Geographie Physique*, **27**, 269–278.

RICHENS, W. D., PECHMANN, J. C., SMITH, R. B., LANGER, C.J., GOTER, S. K., ZOLLWEG, J. E. & KING, J. J. 1987. The Borah Peak, Idaho, earthquake and its aftershocks. *Bulletin of the Seismological Society of America*, **77**, 694.

ROBERTS, S. C. 1988. *Active normal faulting in central Greece and western Turkey. PhD. Thesis*, University of Cambridge.

RONDOYANNI-TSIAMBAOU, T. 1984. *Etude néotectonique des rivages occidentaux du Canal D'Atalanti, (Gréce-Central)*. Thesis, Université de Paris-Sud, Orsay.

ROSENDAHL, B. R., REYNOLDS, D. J., LORBER, P. M., BURGESS, C. F., MCGILL, J., SCOTT, D., LAMBIASE, J. J. & DERKEN, S. J. 1986. Structural expressions of rifting: lessons from Lake Tanganika, Africa, *In*: FROSTICK, L. E., RENAUT, R.W., REID, I. & TIERCELIN, J. J. (eds.) *Sedimentation in the African Rifts*. Geological Society, London, Special Publication, **25**, 29–43.

SABOT, V. & MAROUKIAN, H. 1982. Geomorphology and tectonics in and around the Gulf of Corinth, Greece, *Proceedings of the Hellenic Arc and Trench System*, Athens, 174–182.

SCHNEIDER, J. 1976. Biological and inorganic factors in the destruction of limestone coasts. *Contributions to Sedimentology*, **6**.

SEBRIER, M. 1977. *Tectonique récente d'une transversale a l'arc egéen: le Golfe de Corinthe et ses régions périphériques*. Thesis, Université de Paris-Sud, Orsay.

SEGALL, P. & POLLARD, D. D. 1980. Mechanics of discontinuous faults. *Journal of Geophysical Research*, **85**, 4337–4350.

SENGOR, A. M. C., GORUR, N. & SAROGLU, F. 1985. Strike slip faulting and related basin formation in zones of tectonic escape: Turkey as a case study. *In*: Strike slip deformation, basin formation, and sedimentation. *Society of Economic Palaeontologists and Mineralogists, Tulsa*, **37**, 227–264.

SIEBERG, A. 1932. *Untersuchungen uber Erdbeben*. Aachen.

SKUPHOS, T. G. 1894. Die zwei grossen Erdbeben in Lokris. *Zeitschrift der Gesellschaft für Erdkunde. XXIX*. Berlin.

SOUFLERIS, C., JACKSON, J. A. KING, G., SPENCER, C. & SCHOLZ, C. 1982. The 1978 earthquake sequence near Thessaloniki (N. Greece). *Geophysical Journal of the Royal Astronomical Society* **68**, 429–458.

STEININGER, F. F. & ROGL, F. 1984. Paleogeography and palinspastic reconstruction of the Neogene of the Mediterranean and Paratethys. *In*: DIXON, J. E. & ROBERTSON, A. H. F. (eds): *The Geological Evolution of the Eastern Mediterranean*. Geological Society, London, Special Publication, **17**, 659–668.

STEWART, I. S. & HANCOCK, P. L. 1989. Normal fault zone evolution and fault scarp degradation in the Aegean region *Basin Research*, **1**, 139–153.

STEWART, J. H. 1980. Regional tilt patterns of the late Cenozoic basin-range fault blocks, western United States. *Bulletin of the Geological Society of America*, **91**, 46–464.

STIROS, S. C. 1985. *Archeological and geomorphic*

evidence of late Holocene vertical motions in the N. Euboean Gulf (Greece) and its Tectonic implications. Institute of Geology and Mineral exploration, Athens, internal report.

SU DAQUAN, WHITE, N., & MCKENZIE, D. 1990. Extension and subsidence of the Pearl River Mouth basin, northern South China Sea. *Basin Research*, in press.

VAN-ANDEL, T. H. & SHACKLETON, J. C. 1982. Late Paleolithic and Mesolithic coastlines of Greece and the Aegean. *Journal of Field Archeology*, **9**, 445.

WALCOTT, R. I. 1984. The kinematics of the plate boundary zone through New Zealand: a comparison of short- and long-term deformations. *Geophysical Journal of the Royal Astronomical Society*, **79**, 613–633.

WESTAWAY, R. & JACKSON, J. A. 1987. The earthquake of 1980 November 23 in Campania-Basilicata (southern Italy). *Geophysical Journal of the Royal Astronomical Society*, **90**, 375–443.

WHITE, N. 1990. Resolving the extension discrepancy in the North Sea. *In*: BLUNDELL, D. J. & GIBBS, A. (eds) *Tectonic Evolution of the North Sea Rifts*, Oxford University Press.

Analogue Modelling

Physical models of extensional tectonics at various scales

B. Vendeville, P.R. Cobbold, P. Davy, J.P. Brun & P. Choukroune

SUMMARY: In a preliminary series of experiments, using physical models mechanical processes of extensional tectonics have been investigated at various scales. By a suitable choice of model materials, experiments were performed at low cost in a natural gravity field. Upper layers of the lithosphere were modelled using sand; lower layers, using silicone putties of two different densities; the mantle asthenosphere was modelled using honey. The models deformed under their own weight or under absolute horizontal tension. Rates of extension were controlled using a stepper motor. Surface deformation and faulting were monitored using 35 mm time-lapse photography. Lower lithosphere topography was photographed through the transparent asthenosphere. Fault patterns in models with lithosphere only, were observed by serial sectioning. Otherwise, the brittle–ductile interface was observed after suctioning off the sand.

Simple experiments with uniformly extended sand layers only show that; (i) spacing of normal faults is a measure of the layer thickness; (ii) the length of fault trace increases with the amount of downthrow; and (iii) faults tend to form domino domains. Some experiments with a brittle layer on a ductile substrate show a mechanism of passive rifting where; (i) major faults occur in conjugate pairs, defining rift valleys; (ii) minor faults localize additional extension in rift-valley floors; and (iii) isostatic uplift of the viscous substrate causes uplift and tilting of rift rims. In freely floating continents, gravitational spreading leads to: (i) highly localized extension and thinning at continental margins and (ii) internal rifting.

Physical and numerical models of tectonic processes can provide insights into the mechanical and thermal processes that operate in nature. In favourable circumstances, they can even produce structures which exist in nature but have escaped detection.

For the last two years, the experimental tectonics group at Rennes has been engaged in pilot studies of physical modelling within a normal gravity field and at the scale of the Earth's lithosphere (Cobbold *et al.* unpublished manuscript). Since the pioneering work of Hubbert (1937) and Ramberg (1967), it has been known that this technique places severe restrictions on the strength of model materials used. For example, if a model has linear dimensions 10 million times smaller than its natural counterpart (so that 1 m represents 10,000 km), then the model material must also be 10 million times weaker than rock, if gravity is to have an equivalent effect on the model to that it has in nature. Thus for a brittle crust, obeying a simple frictional (Navier–Coulomb) failure law with no time dependence, the cohesive strength must be 10 million times smaller in the model. A lithified sedimentary rock must therefore be represented by a material such as dry sand. Similarly, if a laboratory experiment is to take place in a reasonable time (say, a day) and this is to represent a typical tectonic period (say, 20 Ma), then the viscosity of a ductile lower crust must be reduced by a factor of about 10^{16} and the corresponding model material must have the consistency of pitch. It has often been argued (e.g. Ramberg 1967, p. 44) that such consistencies make it difficult for the experimenter to construct and dissect his models, so that he would be better advised to use a centrifuge. We maintain that a centrifuge is neither necessary, nor even always desirable, but leave more detailed discussion of this topic to a companion paper (Cobbold *et al.* unpublished manuscript).

The purpose of what follows is to describe a number of pilot experiments covering the field of extensional tectonics at various scales. We illustrate the contrasting results of using models that (a) are purely brittle or (b) have a brittle layer underlain by a viscous substrate. In a broad sense, we are therefore studying extensional tectonics at the scale of (a) the upper crust only, or (b) the entire crust, if not the lithosphere. Because of current uncertainties regarding the rheology of the lithosphere (see, e.g. Kusznir & Park, this volume), it is not always easy to be definitive about what one is modelling. We have constructed simple (some would say oversimplified) models with materials of known rheology and have subjected them to simple conditions of boundary motion. No attempt has been made to introduce or study thermal effects. Our experiments are similar in this respect to those of Oertel (1965), Cloos (1955, 1968), Hoeppener *et al.* (1969) and to other more recent ones in the field of extensional tectonics

From Holdsworth, R.E. & Turner, J.P. (compilers) *Extensional Tectonics: Faulting and Related Processes*. The Geological Society, Key Issues in Earth Sciences, **2** (**2**), 171–183.
1476-315X/02/$15.00
First published in: “Coward, M.P., Dewey, J.F. & Hancock, P.L. (eds) *Continental Extensional Tectonics*, The Geological Society, London, Special Publications **28**, 95–108”.

(Horsfield 1980; Elmohandes 1981; Mulugeta 1985).

In this paper, we focus on the distribution of faults obtained in the brittle layer, including such geologically important aspects as (i) fault location and localization, (ii) fault spacing, (iii) fault vergence and (iv) relationships between fault length and amount of throw. Not enough experiments have been done to allow us to develop unequivocal mechanical interpretations in all cases. Nevertheless our preliminary experiments draw attention to some factors which control the geometry of fault systems during extensional tectonics.

Experimental procedure

Materials

Three kinds of material have been used: dry quartz sand, silicone putty and honey.

1 Dry quartz sand (Fontainbleau sand, well sorted, with grains of 200–300 μm, well rounded) was used to simulate the brittle behaviour of lithified rock in the upper crust. The sand fails according to a Navier–Coulomb law, with negligible cohesion and with an angle of internal friction, $\Phi = 30°$ (Mandl *et al.* 1977). Failure leads to formation of narrow planar shear bands or faults, orientated at about $\pm$ 30° to the compression direction. Bands form by pore-space dilatation, with accompanying strain-softening (Mandl *et al.* 1977). Once formed, the bands remain weaker than the surrounding non-dilatated sand. The density of the starting material ($\varrho = 1.3$ g cm^{-3}) can be reduced by mixing with ethyl-cellulose powder ($\varrho = 0.3$ g cm^{-3}) which has similar mechanical properties. The cohesion can be increased by mixing with dry cement powder. Finally, the colour can be changed by dyeing with methylene blue or other organic pigments.

2 Silicone putty (pink opaque gomme GS1R, manufactured by Rhone–Poulenc, France) was used to model the lower crust, or the lower lithosphere, depending on the scale of the problem considered. Rheological tests, performed by the manufacturer or by us using various viscometers, show that this silicone putty is an almost perfectly Newtonian fluid in the range of strain rates 10^{-6} sec^{-1} $< \dot{\epsilon} <$ 10^{-2} sec^{-1}, with a typical viscosity (batch dependent) of about 10^4 Pa s at 20°C. This behaviour contrasts with that of other makes of silicone putty (see Dixon & Summers 1985), which are non-Newtonian. The density of 1.16 g cm^{-3} can be increased to 1.37 g cm^{-3} by charging with finely powdered galena. This also increases the viscosity sevenfold, and imparts a grey colour to the silicone putty.

3 Natural clear honey ($\varrho = 1.40$ g cm^{-3}) was used to model the asthenosphere. The honey is nearly Newtonian, with a viscosity of about 10 Pa s at 20°C. It is transparent over a distance of about 5 cm and translucent over greater distances. This property is very convenient for the observation of an overlying layer of silicone putty representing the lithosphere.

The densities of sand, silicone and honey were chosen and adjusted so as to give a gravitationally stable density stratification representative of the crust, mantle lithosphere and asthenosphere. Similarly, the viscosity ratio of silicone and honey (greater than 100:1) is assumed to be representative of that between lithosphere and asthenosphere. For fuller discussion of these matters, see Cobbold *et al.* (in prep.)

Apparatus

Models were constructed and then deformed in two kinds of housings. The wide housing (Fig. 1a) is also shallow ($50 \times 50 \times 10$ cm^3) and suitable for one-layer or two-layer models; whereas the deep housing (Fig. 1b; $20 \times 20 \times 20$ cm^3) with its transparent base is suitable for three-layer models.

Model construction

Viscous layers were poured into the chosen housing and allowed to attain a horizontal upper surface under the spreading action of gravity. Because higher layers were successively less dense, no difficulties were encountered in this procedure.

Uppermost sand units were layered to reveal fault throws. Thin layers of uniform thickness and contrasting colours were poured sequentially by hand.

Model deformation

In most experiments, horizontal extension was either imposed or allowed by displacing one side wall of the housing at a constant velocity, in the range $1 < v < 10$ cm hr^{-1}, provided by a stepper motor and a screw-jack. In some experiments, basal stretching was rendered uniform by the use of a thin rubber sheet (1 mm), stretched between two moving walls. The other horizontal dimension of the rubber sheet was kept constant by the use of hooks along two stationary curtain rails.

Model observation

The motion of passive markers and the development of faults at the upper surface of the model were monitored using time-lapse photography (35 mm reflex Minolta camera with programmable back). For three-layer models, the topography of the silicone–honey interface was photographed occasionally through the honey and transparent base. No attempt was made to photograph the model through the transparent lateral glass walls, because boundary friction was judged to be too severe and the result not representative of the internal deformation. At the end of experiments with sand or sand/silicone models, the sand was wetted by carefully spraying the upper surface with water. This introduced extra cohesion and enabled vertical sections to be cut through both the sand and silicone.

The sections were then photographed. Some of the sectioned models were further hardened by cooling to $-30°C$ in a commercial freezer. With three-layer models, no practical cutting or cooling techniques were discovered. Instead the geometry of the sand layers was observed by siphoning off the sand at selected sites using a vacuum pipe fitted with a thin nozzle (1 mm aperture).

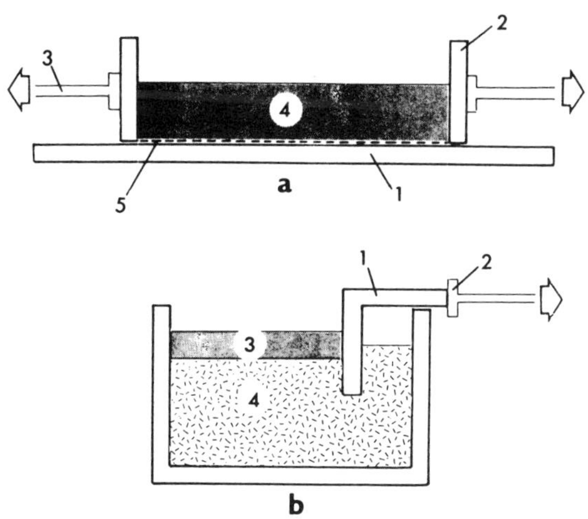

FIG. 1. Model housings and applied boundary motions. (a) Wide housing has rigid base (1) and mobile end walls (2) attached to pistons (3). Uniform basal extension of model (4) can be provided by rubber sheet (5). (b) Deep housing has one mobile wall (1) attached to a piston (2). Sand/silicone (3) floats on honey (4).

Uniform sand models

Seven experiments were performed using models made of sand alone, to investigate the effect of a uniform basal stretching on the development of faults. The models may represent the upper crust, or part of it, in areas subject to uniform ductile extension, without isostatic response. Uniform basal extension was achieved using a rubber sheet in the wide housing. Parameters that differed from one model to another were (i) the inclination of the stretched base with respect to the horizontal, and (ii) the thickness of the model.

Horizontal models

Cross-sections and a plan view (Fig. 2) illustrate typical patterns of normal faults obtained (in models initially 2.5 and 5 cm thick) after a stretch of $\lambda = 1.5$. Our principal observations are listed below.

(1) Fault spacing is strongly dependent on model thickness (cf. Figs 2b & c). (2) In a given model there are approximately equal numbers of faults with opposite senses of dip (vergence). (3) Faults with the same vergence are often but not always concentrated in domains, a result previously obtained by Oertel (1965), Hoeppener *et al.* (1969), Freund (1974), and Faugère & Brun (1984). (4) In fault domains bedding dips tend to be largest where fault dips are gentlest. We interpret this as evidence for block tilting, domino-fashion, as demonstrated in detail by Faugère & Brun (1984). (5) Domains are largest and more clearly defined in thin models (Fig. 2c), whereas thick models show more cross cutting faults as well as horst and rift blocks (Fig. 2b). The reason for this is not yet clear: it may be, for example, that the thinner models with their larger number of faults, are less constrained by the vertical end walls of the housing. (6) Major faults tend to be planar within domino domains; more sigmoidal (initially planar?) where they bound horsts or rifts. (7) At their widest parts, horst and rift blocks contain minor faults (Fig. 2b) also arranged with alternating vergences. These minor faults clearly are responsible for horizontal extension within the blocks. There is a possibility that extension is not constant with depth and this may explain the sigmoidal shape of the bounding major faults. (8) In plan view (Fig. 2d), deformed passive markers show that surface strain is remarkably homogeneous on a bulk scale, but heterogeneous at fault-block scale. (9) Fault traces appear longest where the throw is greatest. Fault

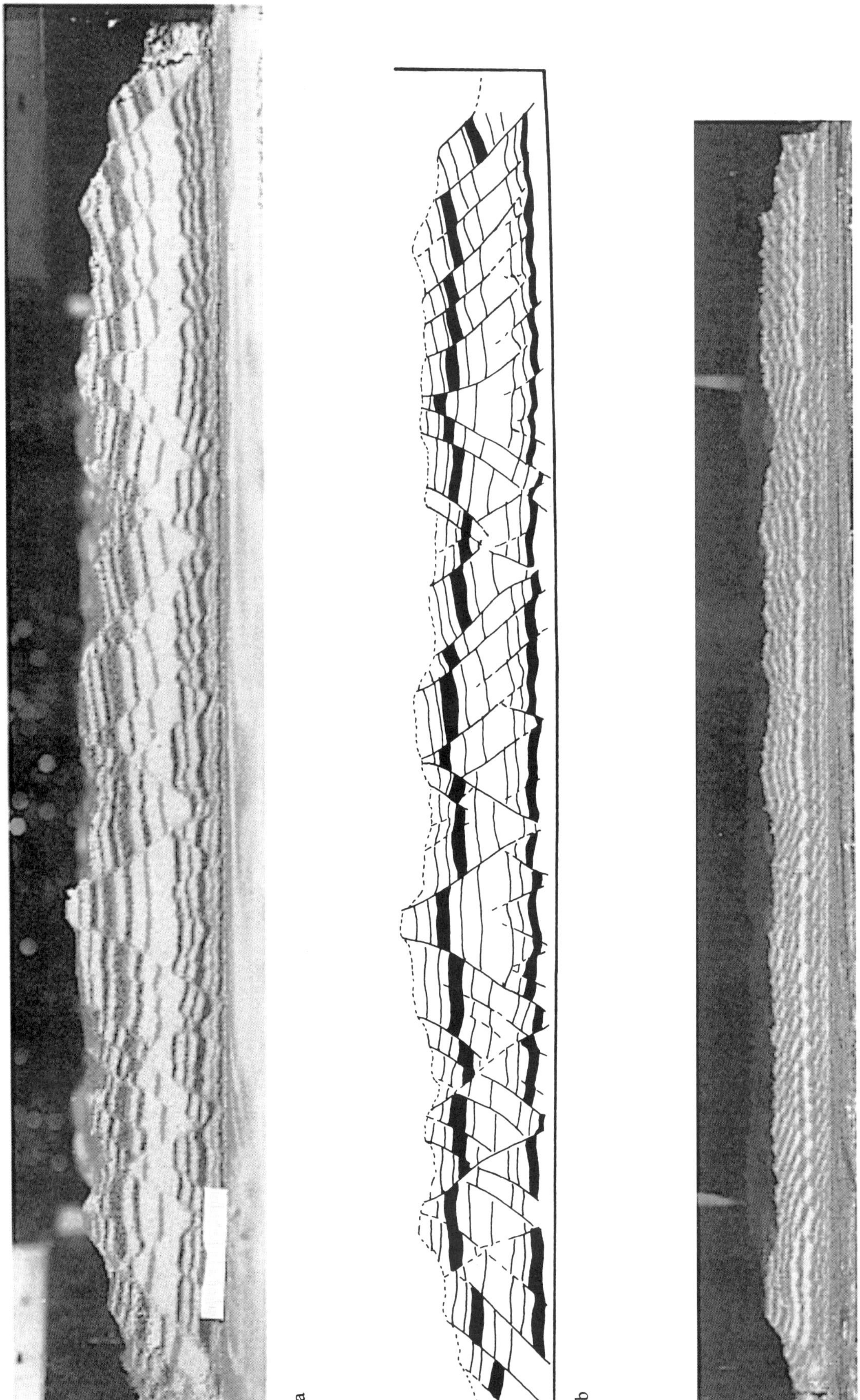

2a

2b

2c

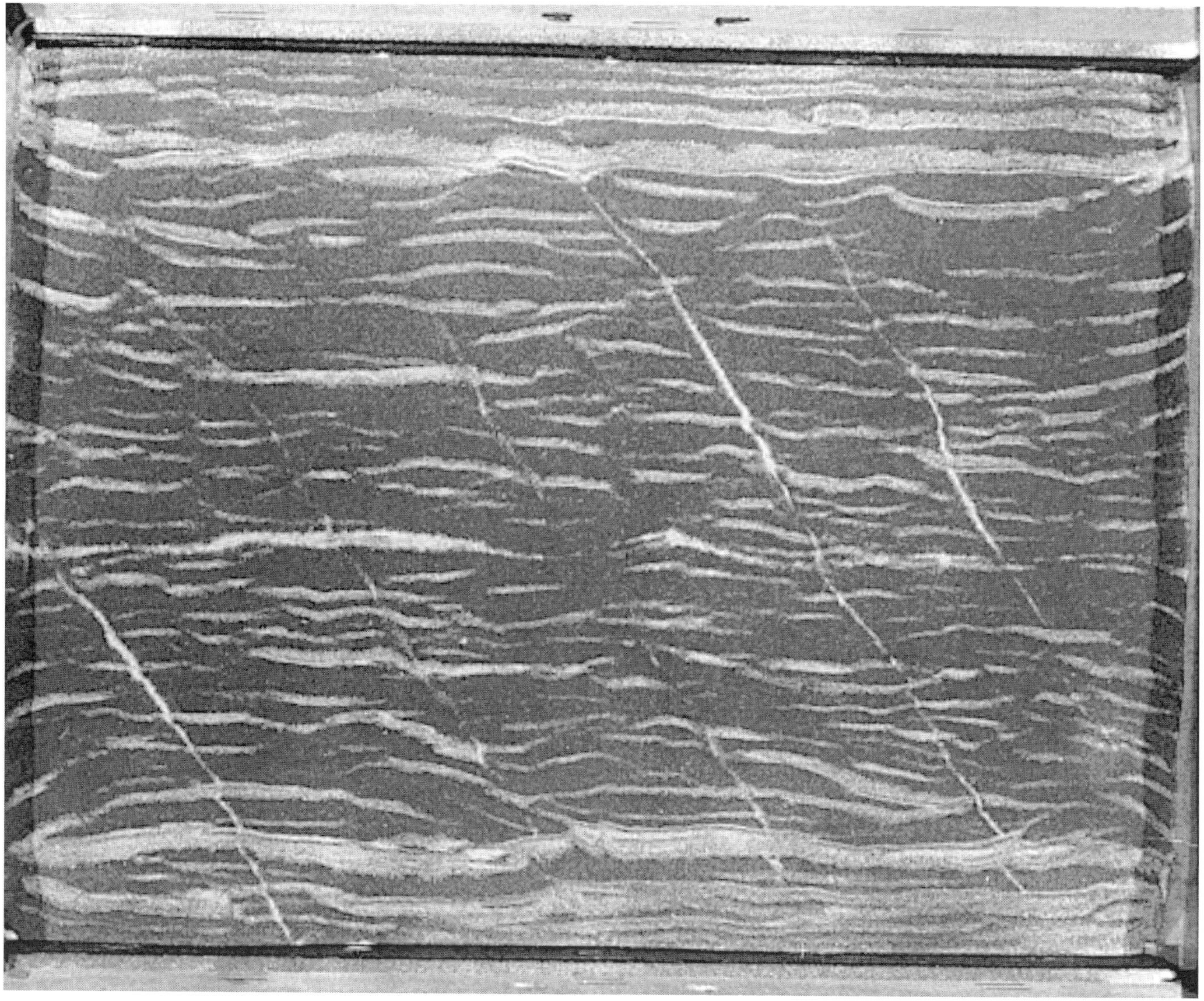

2d

FIG. 2. Sand models uniformly and horizontally stretched at base ($\lambda = 1.5$), showing (a) longitudinal section through model initially 5 cm thick, (b) line drawing of same, (c) longitudinal section through model initially 2.5 cm thick, and (d) surface view of model initially 5 cm thick. Diagonal lines in (d) are passive markers (surface scratches).

traces are locally slightly oblique to the bulk kinematic axes or slightly wavy.

The last observation is more readily understood if one examines a series of time-lapse photographs of the upper surface of a model. These show that the trace length and the throw of an individual fault increase together with time, so that a fault scarp maintains a characteristic pea-pod shape, unless it coalesces with another fault scarp.

Models with inclined bases

Two experiments were performed by making models as before (2.5 cm thick) and then inclining the housing at 7° or 14° to the horizontal *before* applying a basal extension. Sections through both models, after a stretch of $\lambda = 1.4$, show that practically all faults are nearly planar and dip downslope, forming a single domino domain (Fig. 3). Using the occasional conjugate pair, we find that the acute bisector is vertical to within a degree or so. This suggests that faults formed in response to a vertical principal stress, which itself resulted from gravitational loading. Although at the free surface the principal compressive stress must in theory have acted in a direction normal to that surface and not vertically, this stress state appears to have had no control on fault orientation. In fact, stress magnitudes at the sand surface were probably all close to zero. We infer that faults nucleated lower in the model or at its base and maintained their orientation as they propagated upwards. Thus we attribute the dominance of one fault set over its conjugate to the non-parallelism of stress with the stretching base.

A third experiment was performed in the same way as the other two, except that whereas the base was inclined at 14° previously, the top was made horizontal. The model was thus of variable thickness (triangular in cross-section). As before, a dominant family of faults was obtained at $\lambda = 1.3$, but fault spacing varied directly with local thickness of the model. This confirms the thickness/spacing relationship established earlier.

Sand model with internal silicone layer

One experiment was performed in the wide housing to investigate the effect of having ductile horizons at the base and within a brittle system. The layers from bottom to top were pure silicone (3 mm), sand (18 mm), pure silicone (3 mm) and sand (20 mm). End-wall velocity was 3.5 cm hr^{-1} and total stretch was $\lambda = 1.25$. Uniform basal extension was achieved via a rubber sheet. A cross-section (Fig. 4) reveals notable differences with pure sand models. (1) The number of faults is smaller. (2) There is only one domino domain, affecting the lower sand layer. (3) Horst and rift blocks are common. They contain few minor faults. (4) The fault pattern is markedly different from one sand layer to

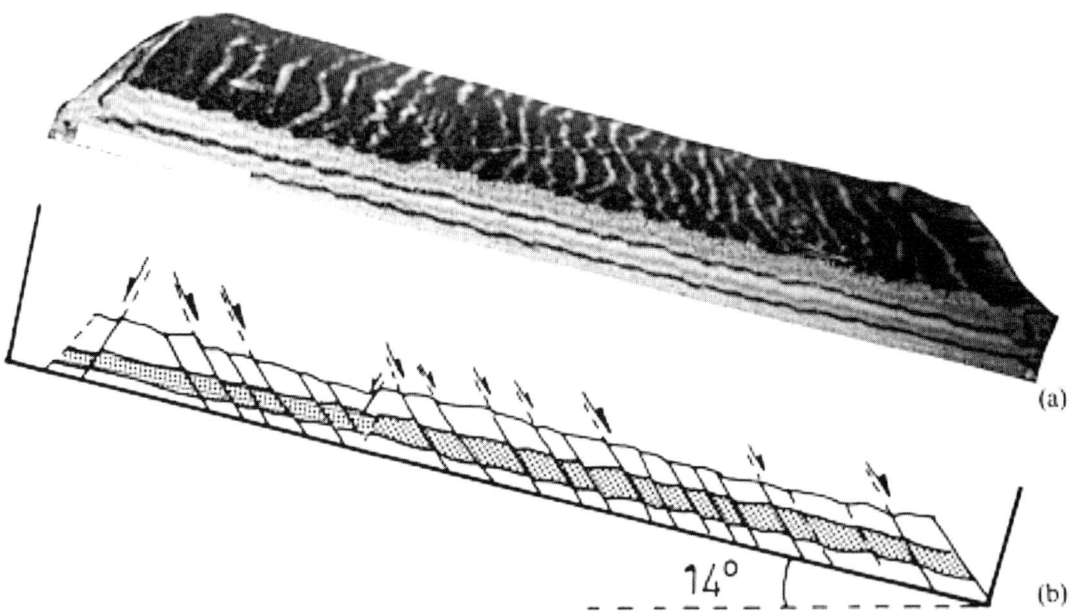

FIG. 3. (a) Sand model uniformly stretched to $\lambda = 1.4$ downslope inclined at 14° to the horizontal. (b) Line drawing of same.

FIG. 4. Longitudinal section through sand model with internal silicone layer (thin, at mid-height), for $\lambda = 1.26$.

the other, as one crosses the intervening silicone layer. Indeed, large unfaulted blocks on one side of the silicone appear to face domains with smaller blocks on the other side. Some major fault traces appear to cut through the silicone without any offset; whereas others terminate at the silicone.

We infer that the ductile silicone layers acted as decoupling and accommodating horizons. They thus reduced the number of faults, facilitating horsts and rifts, reduced the need for domino domains and allowed subsurface fault patterns to be different from surface patterns. In spite of the presence of ductile horizons, we observed little tendency for faults to be curved (listric).

Two-layer models (sand/silicone)

Models were constructed with an upper layer of sand and a lower layer of charged silicone putty, in order to simulate the behaviour of an upper brittle crust above a lower ductile crust. In the absence of lateral constraint, the silicone spreads under its own weight and the overlying sand layer is extended.

In an early set of experiments (Faugère & Brun 1984; Faugère 1985; Brun & Choukroune 1983), the spreading rate was in no way constrained at the end of the model. Basal friction was rendered negligible by floating on mercury. In horizontal models the silicone tended to extrude out from under the sand, imparting a basal shear stress to the sand layer, revealed by observation of passive markers embedded in the silicone. In the central part of the sand layer, where basal shear stress was greatest, a domino domain appeared (Fig. 5a). If a model was tilted before spreading was allowed, thus reversing the sense of basal shear stress, a domino pattern was again obtained, but facing the other way, that is downslope (Fig. 5b).

More recently, we performed four new experiments in which the spreading rate was controlled via screw-jacks and a stepper motor. Velocities chosen were 1 to 1.8 cm hr^{-1}. Models were constructed and deformed in the wide housing. Basal friction was reduced using liquid soap. At these low spreading rates, we infer that shear stresses at the sand/silicone interface were smaller than in the experiments of Faugère & Brun (1984). We illustrate a section through one model at a stretch of $\lambda = 1.21$ (Fig. 6). It shows two major rifts, containing minor faults and limiting large blocks of almost unfaulted sand. Time-lapse photographs of the top surface revealed the following history. (1) The earliest

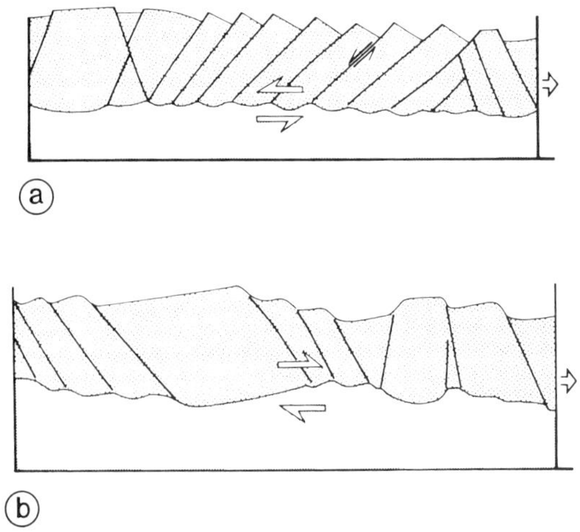

FIG. 5. Line drawings of longitudinal sections through two-layer models of sand (stippled) upon silicone (grey) extended in horizontal position (a), or inclined at about 10° to the right (b). Large arrows show slip sense of basal shear stress, small arrows show slip sense on faults (after Faugère & Brun 1984).

faults to form were a conjugate pair with initial dips of 60° and an intersection point at the sand/silicone interface, 5 cm from the mobile wall. They define a major rift valley. (2) Minor faults formed in the rift floor and accommodated localized extension. The rift widened. (3) At a bulk stretch of 1.2, the second rift developed 5 cm from the first one. (4) Both rifts widened.

The cross-section (Fig. 6) shows that rift development, in its late stages (left-hand side), resembles boudinage: the sand layer necking almost symmetrically and the silicone rising into the neck. We infer that this is the result of isostatic response beneath the rift. Other features to be noticed are the uplifted rims of the rift and the associated curvature of the intervening blocks. Notice the minor normal faults at the outer arcs. Neither major nor minor faults are listric.

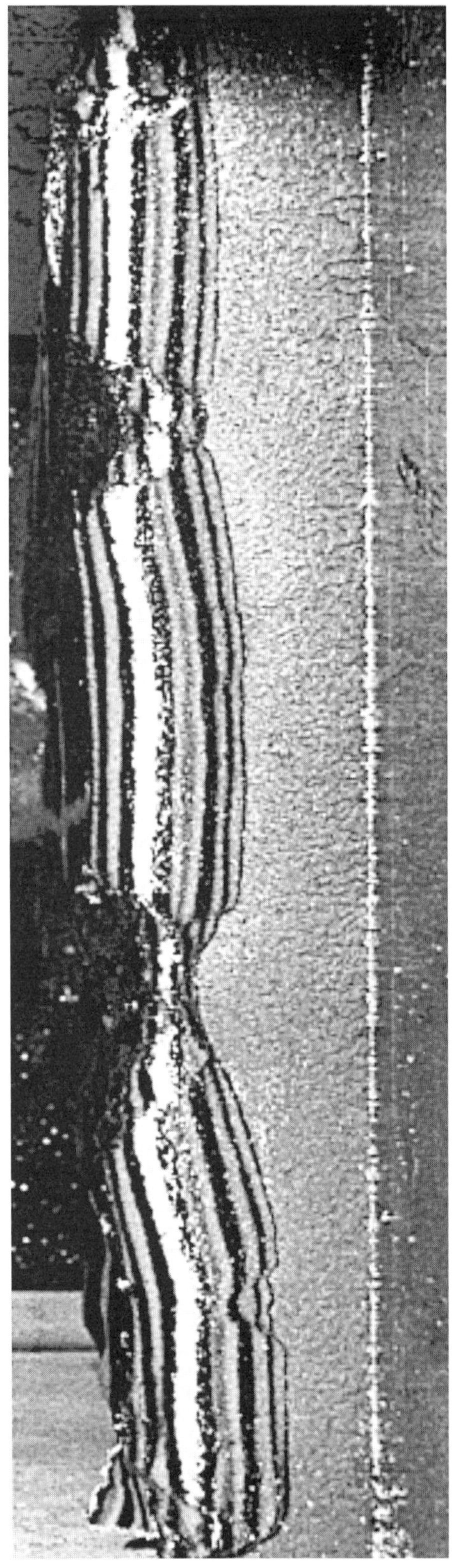

FIG. 6. Longitudinal section through two-layer model (sand upon silicone) at a stretch of $\lambda = 1.21$ and for an end-wall velocity of $v = 1.18$ cm hr^{-1}.

Three-layer models (sand/silicone/honey)

Four experiments have so far been conducted in the deep housing to explore mechanisms of extension at lithospheric scale. In all experiments, honey was used for modelling the mantle asthenosphere; sand and silicone for the lithosphere. We distinguish the experiments on the basis of boundary motions and lithospheric structure. These conditions lead to passive rifting (Sengör & Burke 1978).

Uniaxially stretched simple lithosphere

The lithosphere was modelled simply as a rigid crust (1.8 cm of lightened sand) resting on a heavier ductile mantle (2.1 cm of pink silicone). It was subjected to a uniaxial end-wall velocity of 4.6 cm hr^{-1}. This was greater than the spreading rate due to gravity alone and was obtained because of high adherence between the silicone and the end plate. Time-lapse photography (Fig. 7a & b) and a balanced cross-section (Fig. 7c) show that a single major rift formed near the moving end wall. Most of the extension then became localized within the rift floor. Notice the isostatic upwelling of the honey and the typical rift geometry with uplifted and tilted rims.

Uniaxially stretched complex lithosphere

The lithosphere was modelled as a brittle upper crust (0.7 cm of sand), ductile lower crust (1 cm of pink silicone) and ductile mantle (3.6 cm of heavy stiff grey silicone). It was subjected to a uniaxial end-wall velocity of 1.46 cm hr^{-1}, greater than the spreading rate due to gravity alone. Time-lapse photography (not illustrated) showed marginal major faults at the moving wall and also a zone of rifting in the central part of the model.

Uniaxially spreading simple lithosphere

The lithosphere was modelled simply (1.1 cm of lightened sand, 1.5 cm of pink silicone) and subjected to a low end-wall velocity (1.15 cm hr^{-1}), almost equal to the gravitational spreading velocity. In contrast with the uniaxially stretched simple lithosphere experiment, two major rifts appeared, affecting almost half the model length (Fig. 7d). This may have been a result of the low velocity, or it may have been due to other factors, such as unintentional variations in initial sand thickness. Further experiments are required to clarify these uncertainties. Notice, however, the clear correlation between depressions in the upper surface of the sand and upwelling of the honey; this contrasts with the undeflected nature of the sand/silicone interface. As in the first experiment, the rift floors localized most of the model extension. Notice again the typical rift geometry with tilted rims.

Biaxially spreading rectangular continent

Continental lithosphere within a rectangular continent (26×15 cm^2, equivalent to 1300×750 km^2) was modelled as an upper brittle layer (0.8 cm of sand) and a lower ductile layer (1.5 cm of pink silicone) freely floating on an asthenosphere (honey) without any lateral constraint (Fig. 8a). There was no equivalent to an oceanic lithosphere.

The continent spread biaxially under its own weight. The initial velocity was about 1 cm hr^{-1}, diminishing rapidly. Most of the resulting extension occurred in marginal zones initially about 1–1.3 cm wide (equivalent to 50–65 km wide) and separated from the unextended interior by sharp breakaway faults (Fig. 8b). The margins attained stretches of about $\lambda = 3.0$ after 10 hours (10 Ma in nature), whereas the bulk stretch was only $\lambda = 1.3$. Notice that the margins and breakaway faults followed the initial rectangular shape, but corners became rounded. More strikingly perhaps, two corners became progressively rifted off, one completely, the other partially. The propagating rifts had stubby open tips, reminiscent of natural rift valleys. Rift width at initiation was about 0.5 cm (25 km), increasing to 1.5 cm (75 km).

Certain similarities can be seen between the pattern of extension and surface topography in this model and extension in the Basin and Range Province of the western United States. Thus the marginal zone, oblique rift valley and rifted-off corner in the model resemble the Great Basin, Rio Grande rift and Colorado plateau. This raises the question as to how much the pattern of extension in the western US may be due to faulting in a crust spreading under its own weight, with isostatic readjustment below.

Conclusions

Mechanical processes of lithospheric extension can be modelled at various scales and in three dimensions using physical models with brittle and ductile layers, deformed in a natural gravity field. From our preliminary range of experiments, we draw the following tentative conclusions. (i) In a strain-softening brittle upper crust, faults have a tendency to persist and if necessary, rotate or deform to accommodate finite deformation. (ii) Spacing of normal faults in a brittle layer is a measure of layer thickness. Faults propagate laterally as downthrow increases. (iii) Uniformly extended brittle layers with straight horizontal bases tend to develop domino domains. The sense of basal shear stress exerts considerable control on the sense of vergence. (iv) An underlying viscous substrate can favour the development in the upper brittle layer of localized rifts bounded by major conjugate faults. (v) Tilted rift margins can result

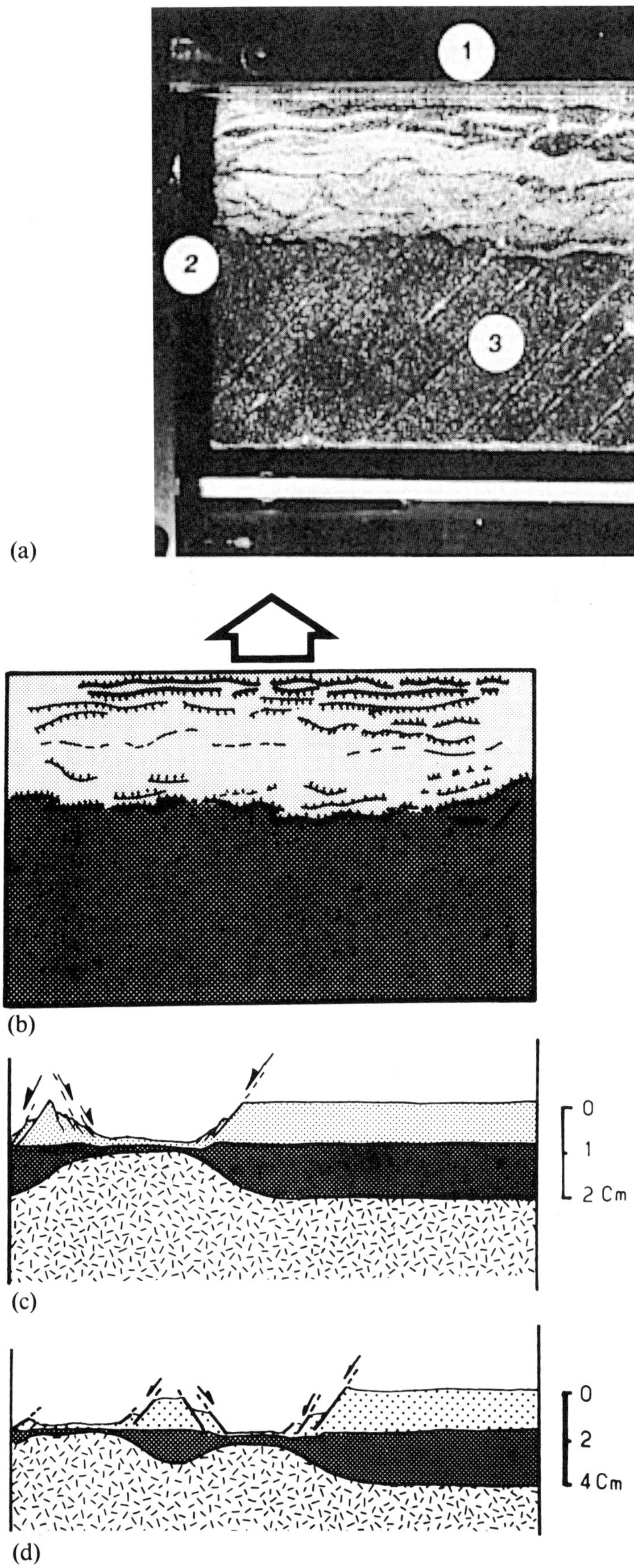

FIG. 7. Uniaxially extending three layer models. Photograph of top surface of model uniaxially stretched at $v = 4.6$ cm hr^{-1} (a), shows moving and wall (1), stationary side wall (2) and lightened sand (3). Line drawing of above (b) shows fault scarps. Longitudinal balanced cross-section through same (c), shows sand (light stipple), silicone (heavy stipple), honey (ornament) and one major rift. Longitudinal balanced cross-section through another model (d), allowed to spread at 1.15 cm hr^{-1}, shows two major rifts.

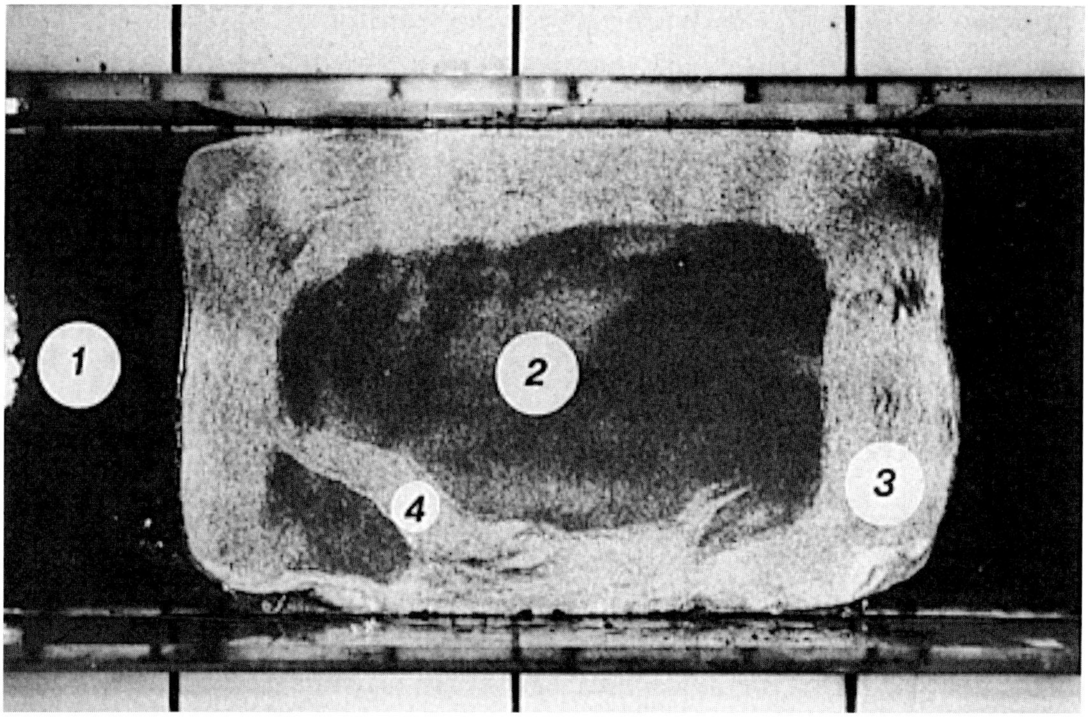

(a)

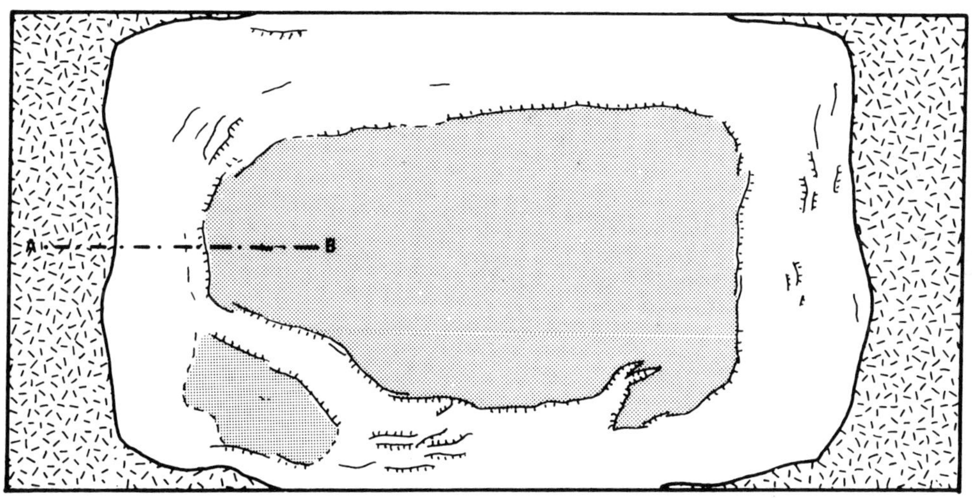

(b)

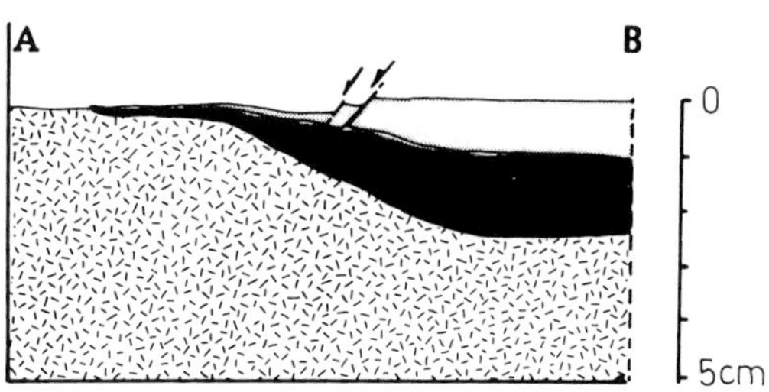

(c)

FIG. 8. Biaxially spreading rectangular model. View of upper surface after 10 hours, equivalent to 10 Ma. (a) shows honey (1), unstretched sand (2), stretched margins (3) and rift (4). Line drawing of same (b), shows main faults and line of section A–B. Balanced cross-section along line A–B, (c), shows honey (ornament), silicone (heavy stipple) and sand (light stipple).

TABLE 1: *Summary of experimental conditions and parameters. All lengths are in cm; velocities in cm hr*$^{-1}$*; stretches, dimensionless*

Experiment	Layering	Basal slope (degrees)	Initial thickness (cm)		Initial length (cm)	Width (cm)	Velocity (cm hr^{-1})	Total stretch (λ)
Uniform horizontal sand model	Sand only	0		5.0	27.4	50		1.50
		0		2.5	29.0	50		1.48
		0		2.5	26.0	50		1.50
		0		1.25	28.0	50		1.50
Inclined uniform sand model	Sand only	14		2.5	21.5	25		1.40
		7		5.0→1.5	22.0	25		≃1.30
		14		2.0→8.0	23.0	25		≃1.30
Sand model with internal silicone layer	Sand Silicone	0	Sand: Silicone:	2.6 0.25	26.0	50	15.0	1.50
	Sand Silicone	0	Sand: Silicone:	1.8 + 2.0 0.3 + 0.3	28.7	50	3.5	1.25
Two-layer models (sand/silicone)	Sand	0	Sand: Silicone:	3.0 3.0	23.0	50	1.8	1.20
	Grey silicone	0	Sand: Silicone:	1.8 2.1	33.0	50	1.0	≃1.20
Three-layer models (stretching)	Lightened sand Pink silicone Honey	0	Sand: Silicone:	1.1 1.5	10.5	20	1.15	1.60
		0	Sand: Silicone:	1.8 2.0	9.0	20	4.60	1.42
	Sand Pink silicone Grey silicone Honey	0	Sand: Pink silicone: Grey silicone:	0.7 1.0 3.6	10.7	20	1.46	1.31
Three-layer models (spreading)	Sand Pink silicone Honey	0	Sand: Silicone:	0.8 1.5	15.0	26	<1.0	1.33 × 1.25

quite simply from buoyancy effects (isostatic readjustment). (vi) Continued extension tends to become localized via minor faults in rift floors, leading eventually to very large extensions. (vii) Unconstrained continental margins spread under their own weight, becoming thinner than they were before continental breakup.

ACKNOWLEDGMENTS: This work was financed jointly by the Compagnie Française des Pétroles (C.F.P.) and the Institut Français du Pétrole (I.F.P.). We thank J.J. Kermarrec, C.N.R.S. technician, for expert technical asisstance in the design, construction and maintenance of laboratory equipment.

References

BRUN, J.P. & CHOUKROUNE, R. 1983. Normal faulting, block tilting and décollement in a stretched crust. *Tectonics*, **2**, 345–56.

CLOOS, E. 1955. Experimental analysis of fracture patterns. *Bull. geol. Soc. Am.* **66**, 241–56.

—— 1968. Experimental analysis of Gulf Coast fracture patterns. *Bull. Am. Assoc. Pet. Geol.* **52**, 420–44.

DIXON, J.M. & SUMMERS, J.M. 1985. Recent developments in centrifuge modelling of tectonic processes: equipment, model construction techniques and rheology of model materials. *J. struct. Geol.* **7**, 83–102.

ELMOHANDES, S.E. 1981. The Central European graben system: rifting imitated by clay modelling. *In*: ILLIES, J.H. (ed.) Mechanism in graben formation. *Tectonophysics*, **73**, 69–78.

FAUGÈRE, E. 1985. La tectonique en extension intracontinentale. Etude de terrain (le Sud Nevada, U.S.A.) et modélisation analogique. *Mémoire des sciences de la Terre, Université Pierre & Marie Curie, Paris*, **85-06**.

FAUGÈRE, E. & BRUN, J.P. 1984. Modélisation expérimentale de la distention continentale. *Compte Rendus de l'Académie des Sciences de Paris*, **299**, **Série II**, 365–70.

FREUND, R. 1974. Kinematics of transform and transcurrent faults. *Tectonophysics*, **21**, 93–134.

HOEPPENER, R., KALTHOFF, E., & SCHRADER, P. 1969. Zur physicalischen tectonik bruchbildung bei verschieden affinen deformationen im experiment. *Geol. Rdsch.* **59**, 179–93.

HORSFIELD, W.J. 1980. Contemporaneous movement along crossing conjugate faults. *J. struct. Geol.* **2**, 305–10.

HUBBERT, M.K. 1937. Theory of scale models as applied to the study of geologic structures. *Bull. geol. Soc. Am.* **48**, 1459–520.

KUSZNIR, N.J. & PARK, R.G. This volume. The extensional strength of the continental lithosphere: its dependence on geothermal gradient, and crustal composition and thickness.

MANDL, G., DE JONG, L.N.J. & MALTHA, A. 1977. Shear zones in granular material. *Rock Mech. Wien.* **9**, 95–144.

MULUGETA, G. 1985. Dynamic models of continental rift valley systems. *Tectonophysics*, **113**, 49–73.

OERTEL, G. 1965. The mechanism of faulting in clay experiments. *Tectonophysics*, **2**, 343–93.

RAMBERG, H. 1967. *Gravity, deformation and the Earth's crust*, 214 pp. Academic Press, London.

SENGOR, A.M.C. & BURKE, K. 1978. Relative timing of rifting and volcanism on earth and its tectonic implications. *Geophys. Res. Lett. Washington*, **5**, **419–21**.

B.V. VENDEVILLE, P.R. COBBOLD, P. DAVY & P. CHOUKROUNE, Centre Armorican d'Etude Structurale des Socles, (Laboratoire C.N.R.S. Conventionné à l'Universite de Rennes), Campus de Beaulieu, 35042 RENNES Cédex, France.

J.P. BRUN, Laboratoire de Tectonique, Université de Paris VII, 4 Place Jussieu, 75007 PARIS, France.

Recent advances in analogue modelling: uses in section interpretation and validation

K. R. McCLAY

Fault Dynamics Project, Geology Department, Royal Holloway University of London, Egham, Surrey, TW20 0EX, UK

Abstract: Scaled analogue modelling is a powerful tool for the understanding of the geometric and kinematic development of extensional, strike-slip and inverted fault systems, as well as for some thrust systems. In this paper the results of 2D and 3D scaled analogue model experiments designed to simulate listric fault systems, rift fault systems, and inverted listric faults are reviewed. Simple 2D rift models are characterized by asymmetric rift graben structures with planar rift border faults and domino-style intra-rift fault systems. 2D listric fault models produce roll-over anticlines with associated crestal collapse grabens whereas the ramp-flat listric fault models produce roll-over anticlines associated with each concave-up segment of the main fault system and also a syncline above the convex-up ramp section of the main detachment. The 2D analogue models are directly comparable to natural examples of extensional fault systems. The results of simple 3D analogue models of an orthogonal and a 60° oblique rift are presented. Orthogonal rift models are characterized by both rift-border and intra-rift faults that are nearly orthogonal to the extension direction. Like-dipping intra-rift faults overlap to form relay ramps that transfer displacement between the faults. In contrast oblique rift models are characterized by en-echelon arrays of segmented extensional faults. Rift border fault systems are parallel to the underlying zone of basement stretching whereas the intra-rift faults are segmented and at a high angle to the extension direction. Oblique accommodation zones formed by interlocking conjugate fault arrays separate individual depocentres within the models. No strike-slip or oblique-slip transfer faults are found in these models. The 3D rift models produce fault patterns that are comparable with natural examples of rift systems. Analogue modelling of inverted listric faults is described and the results compared with natural examples of inverted fault systems. The results of the scaled analogue models described in this paper provide important constraints on section interpretation and restoration in that they permit an understanding of the geometric and kinematic evolution of fault (and fold) structures as well as indicating the sequences of faulting and activity on faults. They provide templates for the interpretation of seismic sections in regions of poor data quality and for projecting surface data into the sub-surface.

Ever since the pioneering model studies of Cadell (1889) who conducted sandbox experiments designed to investigate how the thrust and fold systems of the Scottish Highlands formed, scaled analogue modelling has been commonly (though somewhat intermittently) used to study the geometric, kinematic and dynamic evolution of most geological structures, from extensional fault systems (e.g. Cloos 1928; Cloos 1968; Horsfield 1977, 1980; Faugère & Brun, 1984; Withjack & Jamieson 1986; McClay & Ellis 1987; Vendeville *et al.* 1987; Ellis & McClay 1988; Mandl 1988; Vendeville & Cobbold 1988; Serra & Nelson 1989; McClay 1990*a*, *b*; McClay *et al.* 1991; McClay & Scott 1991; Tron & Brun 1991; Vendeville 1991; Brun & Tron 1993; Lemon & Mahmood 1994; McClay & White 1995; Withlack *et al.* 1995), thrust faulting (e.g. Davis *et al.* 1983; Malavielle 1984; Mulugeta & Koyi 1987; Mulugeta 1988*a*, *b*; Colletta *et al.* 1991; Dixon & Liu 1992; Lallemand *et al.* 1992; Liu 1992; Liu *et al.* 1992; Marshak & Wilkerson 1992; Marshak et al. 1992; Calassou *et al.* 1993; Malavielle *et al.* 1993; Koyi 1995), strike-slip tectonics (e.g. Emmons 1969; Tchalenko 1970; Wilcox *et al.* 1973; Bartlett *et al.* 1981; Odonne & Vialon 1983; Naylor *et al.*, 1986; Reche 1987; Mann *et al.* 1983; Richard & Cobbold, 1990; Richard 1991; Richard *et al.* 1991; Schreurs 1994; Richard *et al.* 1995), inversion (e.g. Koopman *et al.* 1987; McClay 1989; Buchanan & McClay 1991, 1992; McClay & Buchanan 1992, McClay 1995), and salt tectonics (e.g. Talbot 1992; Vendeville & Jackson, 1992*a*, *b*; Weijermars *et al.* 1993; Jackson & Vendeville 1994).

Scaled physical models provide graphic and visual images of how complex geological structures may form in time and space. Such forward models enable the geologist to visualize the geometric and kinematic pathways by which a final complex structural configuration may be reached. In addition and in contrast to most numerical models, scaled analogue models of brittle deformation in

From Holdsworth, R.E. & Turner, J.P. (compilers) *Extensional Tectonics: Faulting and Related Processes*. The Geological Society, Key Issues in Earth Sciences, **2 (2)**, 185–209.
1476-315X/02/$15.00
First published in: "Buchanan, P.G. & Nieuwland, D.A. (eds) Modern Developments in Structural Interpretations, Validation and Modelling, The Geological Society, London, Special Publications **99**, 201–225".

the upper crust allow new faults to form. Analogue models allow the sequences of faulting to be determined and produce templates for the interpretation of complex structures. This understanding is essential if realistic and retrodeformable balanced geological cross-sections are to be constructed through deformed terranes. In addition, detailed analysis of the deformation in analogue models gives an indication of the kinematics and mechanisms of deformation. Such data are essential for the determination of algorithms that can be applied in both forward models of fault systems and particularly in section balancing and restoration software.

The analogue modelling of strike-slip faulting has been recently reviewed in detail by Richard *et al.* (1995) and the reader is referred to that paper for further information. The analogue modelling of inversion structures has been reviewed by McClay (1995) and only a brief summary of the main aspects will be presented here.

Analogue model studies have proved to be extremely valuable in developing an understanding of the geometries and progressive evolution of extensional fault systems (e.g. McClay & Ellis, 1987*a*, *b*; Ellis & McClay 1988; McClay 1990*a*; Tron & Brun 1991; Lemon & Mahmood 1994; McClay & White 1995; Withjack *et al.* 1995; and others). In the present paper 2D analogue models of simple rift systems, listric faults and ramp-flat listric faults and 3D analogue models of rift systems are discussed. The 2D listric and ramp-flat listric fault models typically simulate upper-plate deformation in detached terranes such as those found in progradational deltas or above salt and shale decollements whereas the simple rift models replicate the structural architectures in intracontinental rift systems. 3D analogue modelling of rift systems has focused upon the development of orthogonal and oblique rifts in order to understand the development of fault patterns, the transfer of extensional displacements within the rift systems and in particular the along-strike variations in fault architectures and fault linkages.

Recent advances in scaled analogue models of fault systems have included the use of X-ray tomography (e.g. Richard *et al.* 1989; Colletta *et al.* 1991; Schreurs 1994; Richard *et al.* 1995), 3D reconstruction using advanced computer software (see below), detailed analysis of hangingwall deformation patterns (Lemon & Mahmood, 1994; McClay 1995; Dula 1991; Withjack *et al*, 1995), and improved modelling materials, particularly the use of silicone polymers to simulate plastic deformation in the lower crust (e.g. Weijermars 1986; Davy & Cobbold 1988, 1991; Tron & Brun 1991; White 1993; Schreurs 1994; and others). In particular the modelling of structures in 3D has provided valuable insights into the 3D geometries and evolution of complex fault systems (e.g. Calassou *et al.* 1993; McClay & White 1995; Richard *et al.* 1995) that can be applied to section construction and restoration.

This paper reviews the recent advances in analogue model studies and illustrates how they have been used to determine the geometric and kinematic evolution of complex fault systems in extensional and inverted terranes.

Experimental methods

The 2D analogue modelling experiments were carried out in a glass-sided deformation rig (Fig. 1a) with dimensions of 150 cm long, 20 cm wide and up to 20 cm deep. Initial model dimensions were typically $30 \times 20 \times 10$ cm. The sand models were constructed between the two end walls, one of which was kept fixed and the other was moved by a motor driven worm screw at a constant displacement rate of 4.16×10^{-3} cm sec^{-1}. Models were constructed either on a plastic detachment above a footwall block with fixed geometry (Fig. 1a) or above a rubber sheet that was stretched. For listric fault models two fundamental footwall blocks were used: a simple listric fault and a ramp-flat listric fault, with all models having an upper fault cut-off angle of 60°. The models are constrained by the plastic sheet which simulates a low friction decollement and the hangingwall is translated over the footwall at a constant displacement rate. The models were constructed by carefully sieving alternating layers of coloured sand (190 μm grain size) 2–5 mm thick, or where appropriate, alternating layers of coloured sand 2–5 mm thick with intercalated layers of vermiculite mica flakes (layers approximately 0.5–1 mm thick) into the deformation rig. During extension of the models, sand layers (or sand and mica layers) were added incrementally into the accommodation space created by the extensional deformation in order to maintain a constant, horizontal upper free surface and to simulate syn-extensional sedimentation. The glass side walls of the models were coated with a polymer that reduced frictional drag and minimized smearing of the model. The experiments were recorded by time-lapse 35 mm photography and 16 mm cinematography through the glass side walls of the apparatus. Serial sections of the completed and resin impregnated models were used to determine the detailed final geometry of the deformation.

The 3D analogue modelling experiments were carried out in a glass-sided deformation rig (Fig. 1b) with dimensions of 200 cm long, 60 cm wide and up to 10 cm deep. Initial model dimensions were typically $120 \times 60 \times 10$ cm. As in the 2D experiments a layered sandpack was constructed

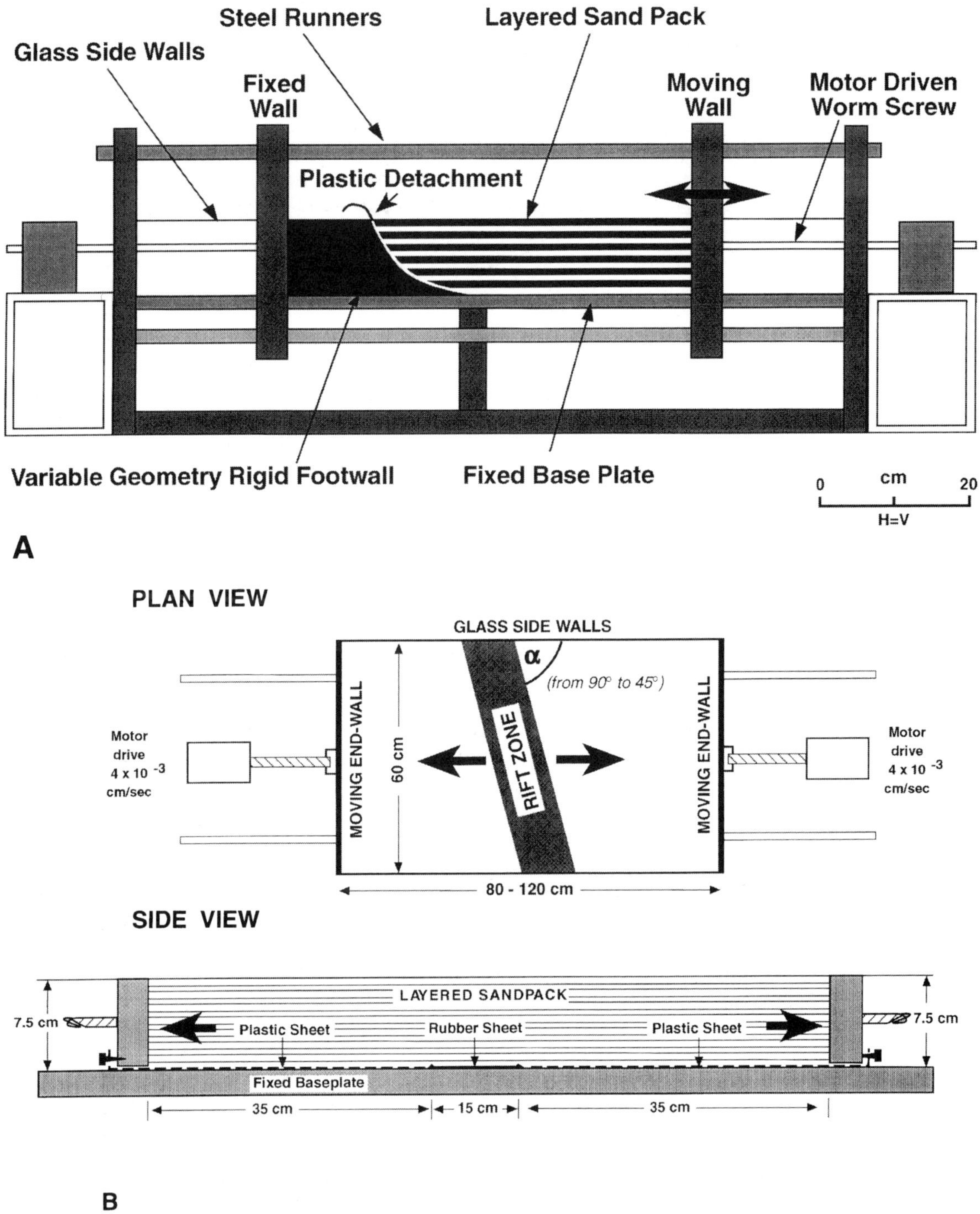

Fig. 1. Experimental apparatus. (**a**) Schematic diagram of the 2D experimental deformation rig. (**b**) Schematic diagram of the 3D experimental apparatus, plan view and side view. Extension can be achieved by moving either one end wall (asymmetric extension) or by moving both endwalls simultaneously (symmetric extension).

between two moveable end walls, one or both of which were moved by a motor driven worm screw at a constant displacement rate of 4.16×10^{-3} cm s^{-1}. The models were constructed of alternating coloured and white sand layers above a basement plastic sheet that contained a central stretching zone formed by a rubber sheet oriented at angles that varied from 90° to 45° with respect to the extension direction (Fig. 1b). During the experiments, syn-extensional sand layers were added incrementally in order to maintain a constant, horizontal upper free surface and to simulate syn-rift sedimentation that kept pace with extension. Serial sections of the completed and resin impregnated models were used to determine the detailed final geometry of the deformation and they were also digitized for 3D reconstruction on a Sun workstation. All of the experiments summarized in this paper have been repeated at least twice with similar results produced each time.

Extensional deformation within sedimentary basins developed within the upper crust is characterized by Navier-Coulomb brittle behaviour. In order to model this, extremely weak, near cohesionless model materials with a similar rheology must be used (Horsfield 1977; Ellis & McClay 1988; McClay 1990*b*). Dry quartz sand is such a material and has been widely used in analogue models studies (Horsfield 1977; Naylor *et al.* 1986; Vendeville *et al.* 1987; McClay & Ellis 1987*a*, *b*; Ellis & McClay 1988; Ellis 1988; McClay 1990*b*; McClay & White 1995). Dry quartz sand is isotropic, and in order to introduce anisotropy and competency contrasts into experiments, dry vermiculite mica and dry china clay may be added respectively (Ellis & McClay 1988). Ellis (1988) and McClay (1990*b*) have demonstrated that these materials have essentially linear Navier-Coulomb rheologies – dry quartz sand (190–275 μm) having an angle of friction of 31°, dry china clay 35° and dry vermiculite mica 30°. In the experiments described in this review homogeneous sand models or sand models with intercalated mica layers were used. If models to prototype ratios of 10^{-5} are used (McClay 1990*a*), 1 cm in the models represents approximately 1 km in nature.

Extensional models: results

The general geometric styles of the analogue model experiments have been described in detail in previous papers (see McClay & Ellis 1987*a*, *b*; Ellis & McClay 1988; McClay 1990*a*; McClay & Scott 1991; McClay & White 1995) and only a brief review is given here.

Simple rift model: 2D

Extension above a horizontal rubber sheet may simulate the development of a simple rift system in 2D. Experiment E-39 (Fig. 2D) shows a typical asymmetric rift model produced by 50% extension

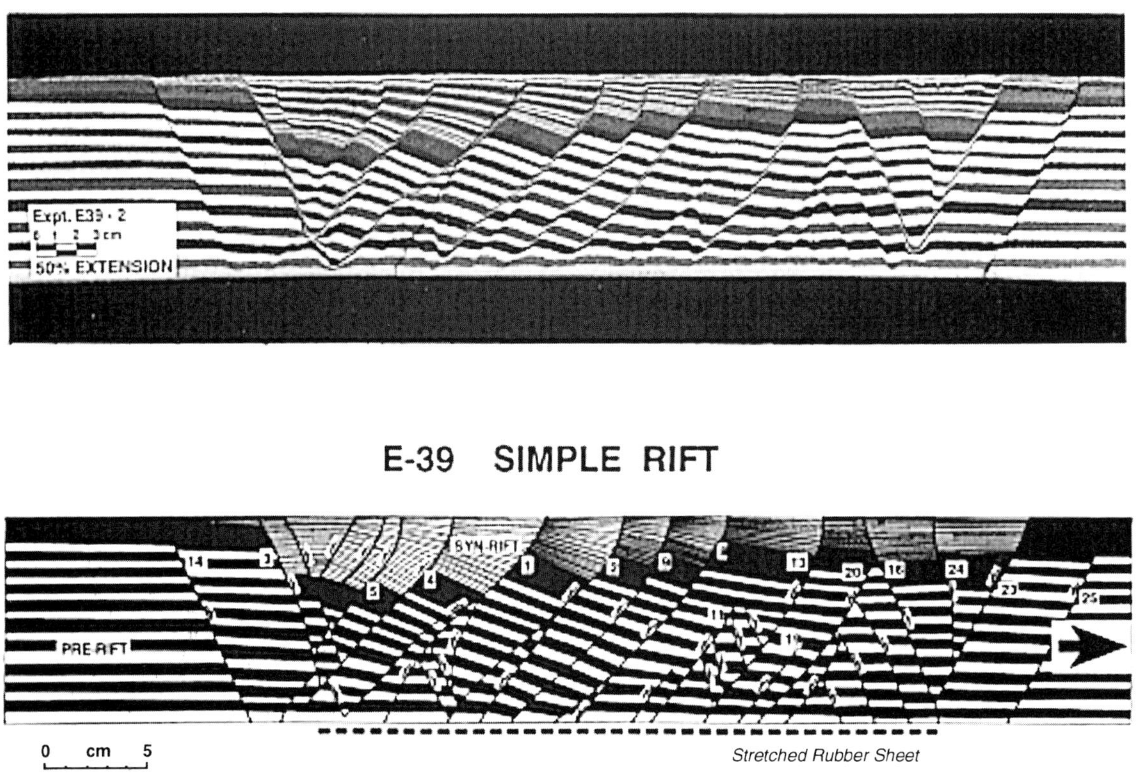

B

Fig. 2. 2D rift model. (**a**) Experiment E39 after 10 cm of extension (50% with respect to the initial length of the stretched rubber sheet at the base of the model). (**b**) Line diagram of Experiment E39 showing the fault patterns and sequence of faulting.

of the basal rubber sheet (right hand end wall moved away to the right). This produces a characteristic half graben bounded by planar rift border faults (Fig. 2a) with domino style intra-rift faults. In the pre-rift strata the intra-rift faults are typically planar but in the syn-rift sequence they are listric growth faults (Fig. 2a). At the left-hand side of the model a small keystone graben (cf. Jackson 1987) has developed (Fig. 2a). The displacement on all of the main extensional faults in this model dies out at the horizontal basal detachment surface that underlies the zone of rifting. The fault sequence diagram (Fig. 2b) shows initial nucleation of faults in the central section of the model with the left-hand rift boundary fault forming early in the deformation history followed by the progressive development of the intra-rift domino fault system and the right-hand rift border fault system (Fig. 2b). Note that there is not a simple sequence of faulting but that the intra-rift domino faults broadly develop from left to right in the direction of extension as increased stretching is progressively taken up upon the basal rubber sheet. This is characteristic of these asymmetric stretching models. The syn-rift growth strata show that many of the intra-rift domino faults were operating synchronously (Fig. 2a). The simple rift model illustrated here is limited in that it does not incorporate footwall uplift associated with individual fault blocks nor as yet can the analogue models incorporate the thermal effects of rift processes.

Simple listric faults: 2D

Extension above a simple listric fault with a horizontal basal detachment is shown for two models in Fig. 3. Hanging-wall deformation above a simple listric fault produces a characteristic rollover anticline with associated crestal collapse graben. In experiment E-80 where the extension is only moderate (10 cm – approximately 50% of the initial length of the model) two simple crestal collapse graben are formed (Fig. 3a) each consisting of a keystone graben bounded by planar antithetic faults (e.g. Nos 2, 6 & 11) and sigmoidal synthetic faults (e.g. Nos 1 & 10). Syn-extensional strata have wedge shaped geometries against the main fault surface (Fig. 3a). The fault sequence diagram (Fig. 3b) shows nucleation of new faults into the crestal collapse graben and the formation of a second crestal collapse graben that is superposed upon the first. Figure 3c shows a model with identical footwall geometry to that shown in Fig. 3a but with 30 cm of extension (approximately 100% of the initial length of the model). In this model the geometry is more complex, with the development of many superposed crestal collapse grabens that migrate towards the main breakaway with increased extension (Fig. 3c). A fan of listric faults synthetic to the main detachment surface form in the syn-rift strata with a panel of highly rotated strata adjacent to the main fault surface (Fig. 3a). The fault sequence diagram for this model (Fig. 3d) shows the progressive superposition of the crestal collapse grabens and their migration towards the main detachment. It should be noted that the simple listric fault models described above have limitations in that the footwall shape is fixed during extension, compaction is not modelled, and that the displacement on the basal detachment surface is constant throughout the deformation history. Despite these limitations many natural examples of listric faults in detached terranes such as progradational delta systems show remarkably similar geometries to these simple analogue models (see below).

Ramp-flat listric faults: 2D

Extension above a ramp-flat fault is shown in Fig. 4. In this style of model thin mica interbeds are placed between the sand layers in order to promote bed-parallel slip and folding as the hangingwall moves down and over the complex ramp-flat surface. A typical model with an intermediate sloping detachment in the middle of the listric footwall block is shown in Fig. 4a. This consists of an upper listric fault segment, an intermediate 20 cm long gently sloping segment which is separated from the lower listric part of the main fault surface by a convex-upwards ramp (Fig. 4a). The upper listric part of the detachment produces a roll-over anticline with crestal collapse graben (Fig. 4a) that superpose in a manner similar to the simple listric faults described above (e.g. Fig. 3c). The lower listric part of the detachment, where it links with the horizontal basal section, has an associated roll-over anticline and crestal collapse graben (Fig. 4a). Where the hanging-wall is translated over the ramp (convex upwards) segment of the detachment a zone of folding and reverse faulting occurs. As a result a marked roll-over anticline with a strong ramp syncline is produced (Fig. 4a). Small reverse faults commonly initiate from the tips of buried crestal collapse graben faults. The fault sequence diagram (Fig. 4b) illustrates the superposition of crestal collapse grabens above the upper listric section of the detachment, and the complexities of fault nucleation and growth over the ramp section of the fault where new reverse faults (e.g. fault numbered 34 & 43) nucleate at the tips of early extensional faults as they pass down the ramp systems. As in the simple listric fault models there are limitations of fixed footwall geometries and constant displacement along the detachment

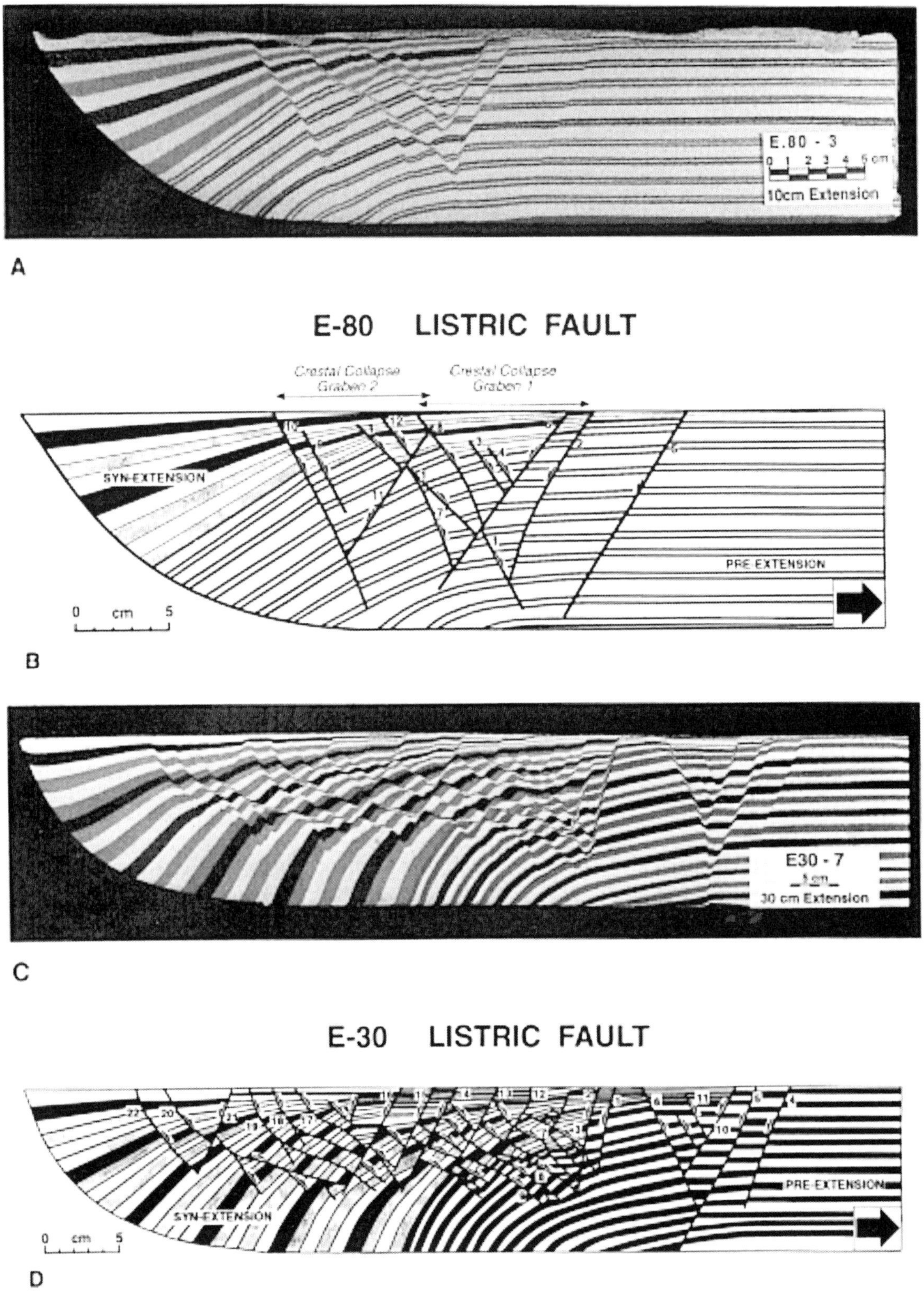

Fig. 3. 2D Listric fault models. (**a**) Experiment E 80 – simple listric fault – 10 cm of extension (approximately 50% with respect to the initial length of model). (**b**) Line diagram of Experiment E 80 showing the fault patterns and sequence of faulting. (**c**) Experiment E 30 – simple listric fault – 30 cm of extension (approximately 100% with respect to the initial length of model). (**d**) Line diagram of Experiment E 30 showing the fault patterns and sequence of faulting.

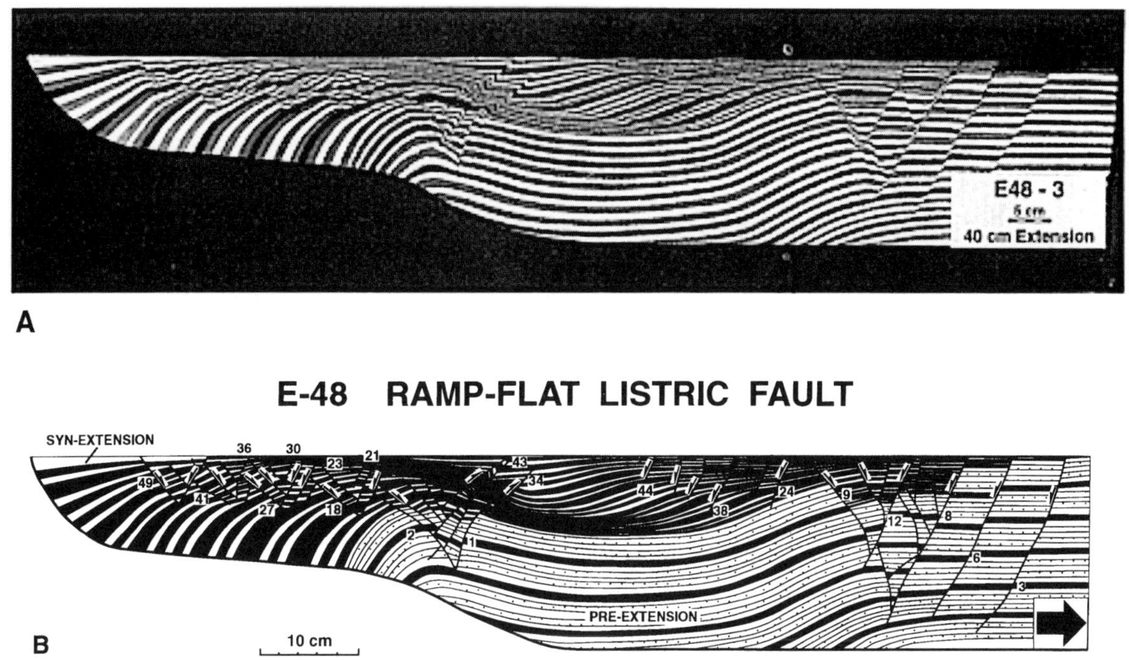

Fig. 4. 2D ramp-flat listric fault models. (**a**) Experiment E 48 after 40 cm of extension (approximately 50% with respect to the initial length of model).(**b**) Line diagram of Experiment E 48 showing the fault patterns and sequence of faulting.

surface for the ramp-flat fault models. These limitations are discussed below.

Particle displacement paths

Hangingwall deformation in 2D extensional fault systems may be analysed by tracking the displacement of individual marker points in the model during deformation (McClay 1995). Triangular marker points formed by coloured sand (with the same material properties as that in the rest of the model) were placed at 1–2 cm intervals in the sand models. Particle paths were monitored during the experiments and analysed. Particle displacement paths for a typical listric extensional fault are shown in Fig. 5. Figure 5a shows the particle paths measured with respect to a reference point in the footwall. These show trajectories parallel to the main detachment surface. However when the particle displacements are calculated with respect to a point in the hanging-wall (Fig. 5b) the relative trajectories of the hangingwall collapse are revealed. For listric fault surfaces these are characteristically curved from moderate dips in the upper part of the model to steep dips near the detachment surface (Fig. 5b). These curved shear trajectories for hanging-wall collapse are markedly different from the vertical or inclined planar shear trajectories commonly assumed for hanging-wall shear during extensional deformation (e.g. Gibbs 1983; 1984; Dula 1991; Withjack & Petersen 1993; Withjack *et al.* 1995)

Synoptic models for 2D extensional fault systems

One of the powerful results from analogue modelling is an understanding of the progressive evolution of extensional fault systems in 2D. This permits the development of forward synoptic models which explain the progressive geometric and kinematic evolution of the faults and fault blocks as well as the sequences of faulting. Such models can then be applied to the interpretation and analysis of deformed terranes such that realistic and valid cross-sections can be constructed. Figure 6a shows the synoptic model for the progressive evolution of a simple rift system where at low extensions the rift border faults and intra-rift faults are dominantly planar. With increased extension new intra rift domino faults develop and deform synchronously. In the syn-rift sequence these intra-rift faults become listric as a result of synchronous rotation of both the hangingwall and the footwall during syn-rift sedimentation (cf. Vendeville & Cobbold 1988). Intra-rift faults may become sigmoidal within the pre-rift sequence as a result of rotation and internal deformation as extension increases.

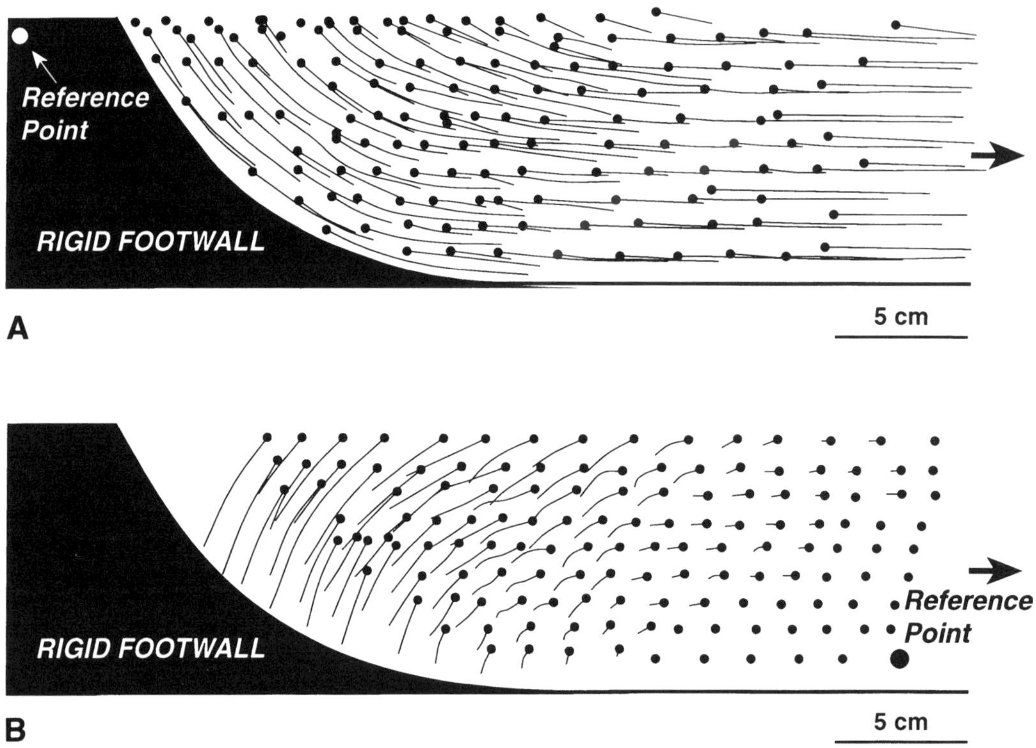

Fig. 5. Hanging-wall deformation in a 2D listric fault model E 80. (**a**) Particle displacement paths measured relative to a fixed point in the footwall block. (**b**) Particle displacement paths measured relative to a fixed point in the hanging-wall block.

Figure 6b shows a synoptic model for a simple listric fault. At low values of extension a simple roll-over and crestal collapse graben develops (Fig. 6b) whereas at high values of extension successive crestal collapse grabens are formed migrating towards the fault breakaway as the hangingwall extends. The superposition of crestal collapse graben produces complicated fault patterns that are difficult to interpret in detail and difficult to restore. The synoptic model for the ramp-flat fault (Fig. 6c) shows the characteristic development of an upper roll-over and crestal collapse graben system together with the ramp syncline and lower roll-over and crestal collapse graben system. With increased extension the ramp syncline may become unfolded (cf. Fig. 4) and reverse faults may nucleate from the tips of the upper crestal collapse graben extensional faults as they pass over the ramp section of the detachment (Fig. 6c). The results of the analogue models and the synoptic diagrams discussed above may be used as templates for the interpretation of seismic sections in areas where data quality is poor.

Examples of extensional fault systems in 2D

Cross-sections through many continental rift systems show many of the features found in the simple 2D analogue models of rift systems (e.g. Fig. 2; see also McClay 1990*a*). In spite of the limitations of sandbox modelling (see discussion below) the 2D sections (e.g. Fig. 2a) show many similarities to cross-sections through the Mozambique rift (McClay 1990*a*), to sections across the East Africa rift lakes (cf. Rosendahl *et al.* 1986; Rosendahl 1987), to sections across the Gulf of Suez (Bosworth 1995) and to sections across the Viking Graben, North Sea (Fig. 7a) (also Scott & Rosendahl 1989; Kusznir *et al.* 1995). In particular in the Northern Viking Graben (as in other intra-continental rifts) the rift is characterised by planar rift border faults and domino style intra-rift fault blocks typically 8–15 km wide (Fig. 7a). The Viking Graben cross-section shows the effects of isostasy and flexure (Kusznir *et al.* 1995) as well as post-rift thermal sag – features which as yet cannot be incorporated into the analogue models (see discussion below).

A typical listric extensional fault with a roll-over and crestal collapse graben is shown in Fig. 7b. This fault system is characterized by wedge geometries of syn-extensional strata in the roll-over together with keystone crestal collapse grabens and well developed planar antithetic faults (Fig. 7b). The architecture is very similar to that developed in the analogue models (Fig. 3) and this style of

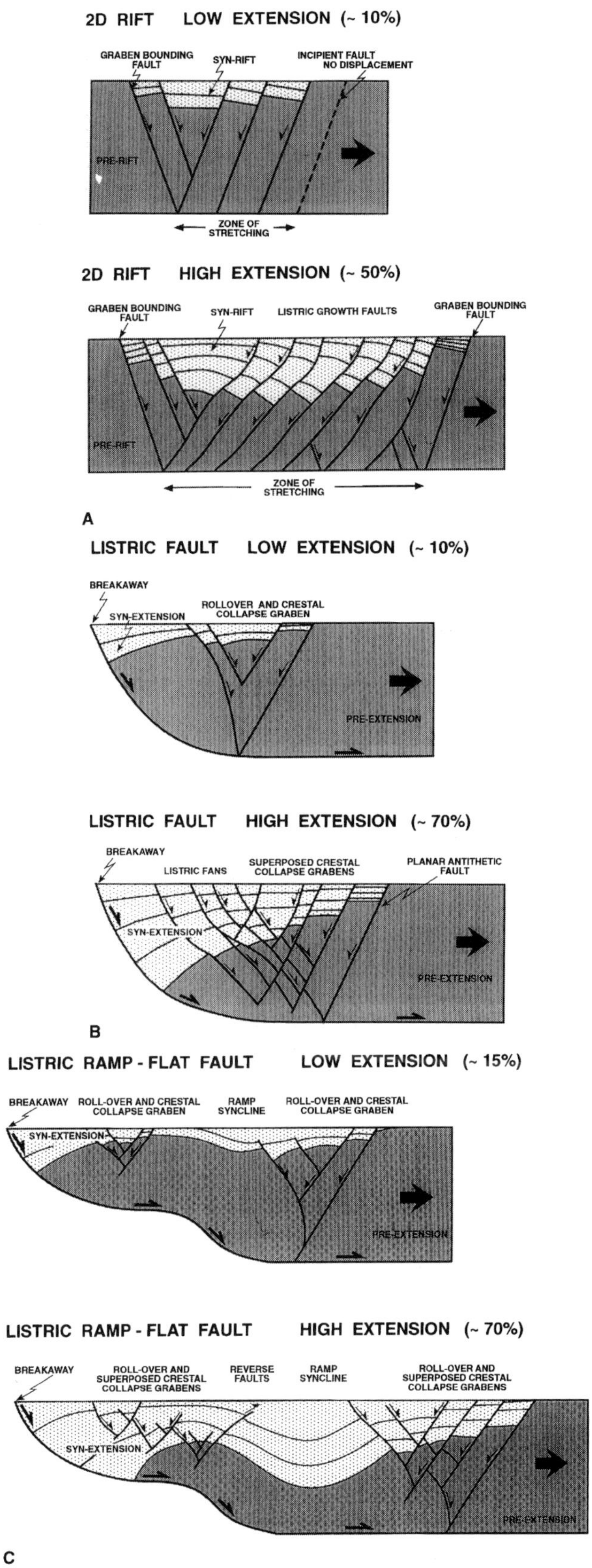

Fig. 6. Synoptic models for 2D extensional fault systems. (**a**) Simple rift model. (**b**) Simple listric fault model. (**c**) Ramp-flat listric fault model.

extensional faulting is characteristic of regional growth faults in shale and salt terranes such as progradational delta systems (e.g. Mississippi Delta: Bruce 1973; Bally 1983; Lopez 1990; Nelson 1991), Nile Delta (Beach & Trayner 1991), Niger Delta (Whiteman 1982) or the Barram Delta, Brunei (Ellenor & James 1984). Excellent seismic examples of this style of listric growth faulting are given in Bally (1983), Ellenor & James (1984), Lopez (1990) and are also discussed in McClay (1990*a*).

Natural examples of ramp-flat extensional faults are not well illustrated in the literature but examples have been interpreted in the Gulf of Mexico (McClay 1990*a*; Withjack *et al.* 1995), and from the Jeanne d'Arc basin, offshore Newfoundland (Tankard *et al.* 1989). Figure 7c shows a section across the ramp-flat Mermaid fault from the North West Shelf of Australia (after Lemon & Mahmood 1994). In this case complex fault systems similar to those in the analogue models (Fig. 4) are developed in the pre-extensional and in the syn-extensional Permian strata (Fig. 7c).

3D rift models: orthogonal rifting

In Experiment M54 (Fig. 8) the zone of rifting was oriented at 90° to the direction of extension and the model was extended to the right achieving 50% extension of the basal rubber sheet (Fig 8a). The surface fault pattern is characterized by long, relatively straight rift border faults (Fig. 8a) and shorter intra-rift faults. Both sets of faults form at, and remain at high angles *c.* 90° to the direction of extension. Along strike the intra-rift faults die out and are overlapped by other faults and form distinct relay ramp structures (Fig. 8a). At either end of the model the tips of the faults are curved due to frictional drag against the sidewalls. Figure 8b shows an interpretation of the surface fault pattern where the rift border faults at each side extend across the whole of the model. The left hand border fault system is the most strongly developed and is terraced (Fig. 8b). Three depocentres separated by indistinct accommodation zones are developed in the model (Fig. 8b). The main sub-basin is bounded by the left hand border fault system and two separate sub-basins are formed towards the end of the right hand border fault system. Within the central part of the rift zone the intra-rift faults remain perpendicular to the extension direction and develop long overlap zones forming relay ramps (Fig. 8b).

3D rift models: oblique rifting

In Experiment M55 the zone of rifting was oriented at 60° to the direction of extension. A distinct left-stepping en-echelon pattern (Fig. 9) of shorter, segmented rift border faults develop (in contrast to the orthogonal rift system in Fig. 8). The rift border faults, developed adjacent to the right hand moving margin of the rift system, are segmented and parallel to the underlying zone of stretching. In contrast the intra-rift faults are arranged in an en-echelon array and the individual faults are at high angles to the extension vector (Fig. 9a). These intra-rift faults are slightly curved, concave towards the extension direction. They nucleate at high angles to the extension direction, rotate slightly during deformation and at the end of the experiment have strikes that are approximately 80° to the extension vector. Both the rift border faults and the intra-rift faults appear to be dominantly dip-slip extensional faults with only small components of oblique slip (it is extremely difficult to determine exact displacement vectors on individual faults in sand). It is probable that the component of lateral motion required to accommodate the oblique rifting is taken up by both the slightly oblique extensional faults and by internal sub-fault scale deformation within the models. The intra-rift faults propagate along strike to form strongly developed overlapping antithetic faults to the main border fault system. Three distinct extensional sub-basins are developed and these are separated by broad accommodation zones oriented at an angle of approximately 45° to the direction of extension (Fig. 9b). In these zones, a fault system with one dip polarity merges into another fault system with opposite dip polarity. As in the orthogonal rift experiments no strike-slip transfer faults were observed. Similar evolutionary patterns of rift border fault systems and intra-rift fault systems were observed in all of the oblique rift experiments (i.e. $\alpha < 90°$ to $\alpha = 45°$). As in the simple 2D rift models described above the 3D rift models are limited in that they do not incorporate footwall uplift nor do they incorporate the thermal effects of rift processes.

Examples of extensional fault systems in 3D

The 3D analogue models of extensional fault systems provide templates for the interpretation of map patterns of extensional faulting. Orthogonal rifting (i.e. the extension direction is perpendicular to the underlying zone of basement stretching produces intra-rift and rift border faults that are 90° to the stretching direction and with large and long zones of overlap. Natural rift systems such as Gregory's Rift, Kenya (Baker 1986; Bosworth *et al.* 1986) where the extension direction is dominantly E–W, show similar fault patterns (Fig. 10a) to the models with long overlapping extensional faults and no strike-slip or oblique slip

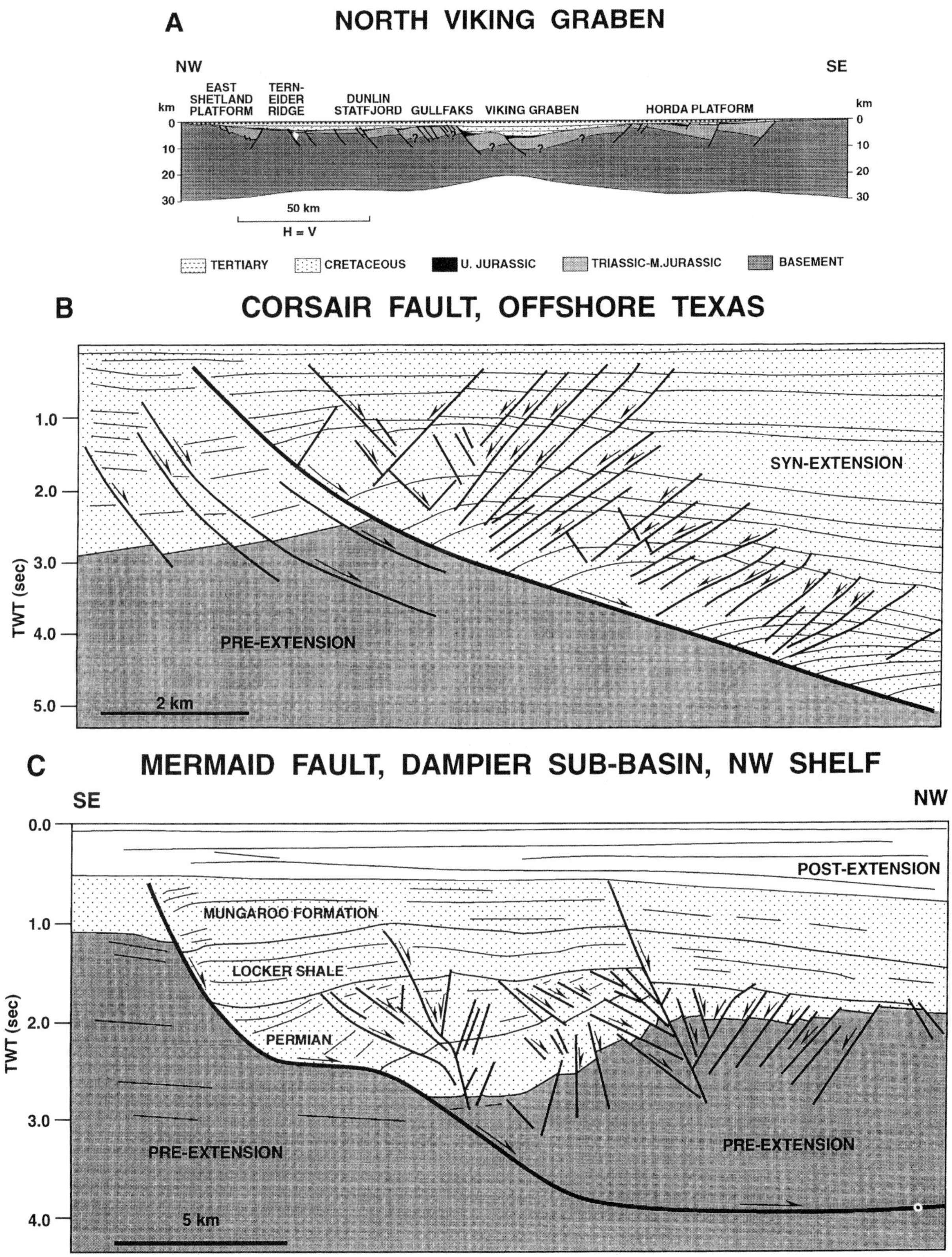

Fig. 7. Examples of extensional fault systems. (**a**) Cross-section through the northern Viking Graben, North Sea (after Kusznir et al. 1995). (**b**) Listric extensional fault, Gulf of Mexico (line drawing of a seismic section in Christiansen 1983). (**c**) Ramp-flat listric extensional fault, Dampier sub-basin, North West Shelf, Australia (line drawing and interpretation of a seismic section in Lemon & Mahmood 1994).

A

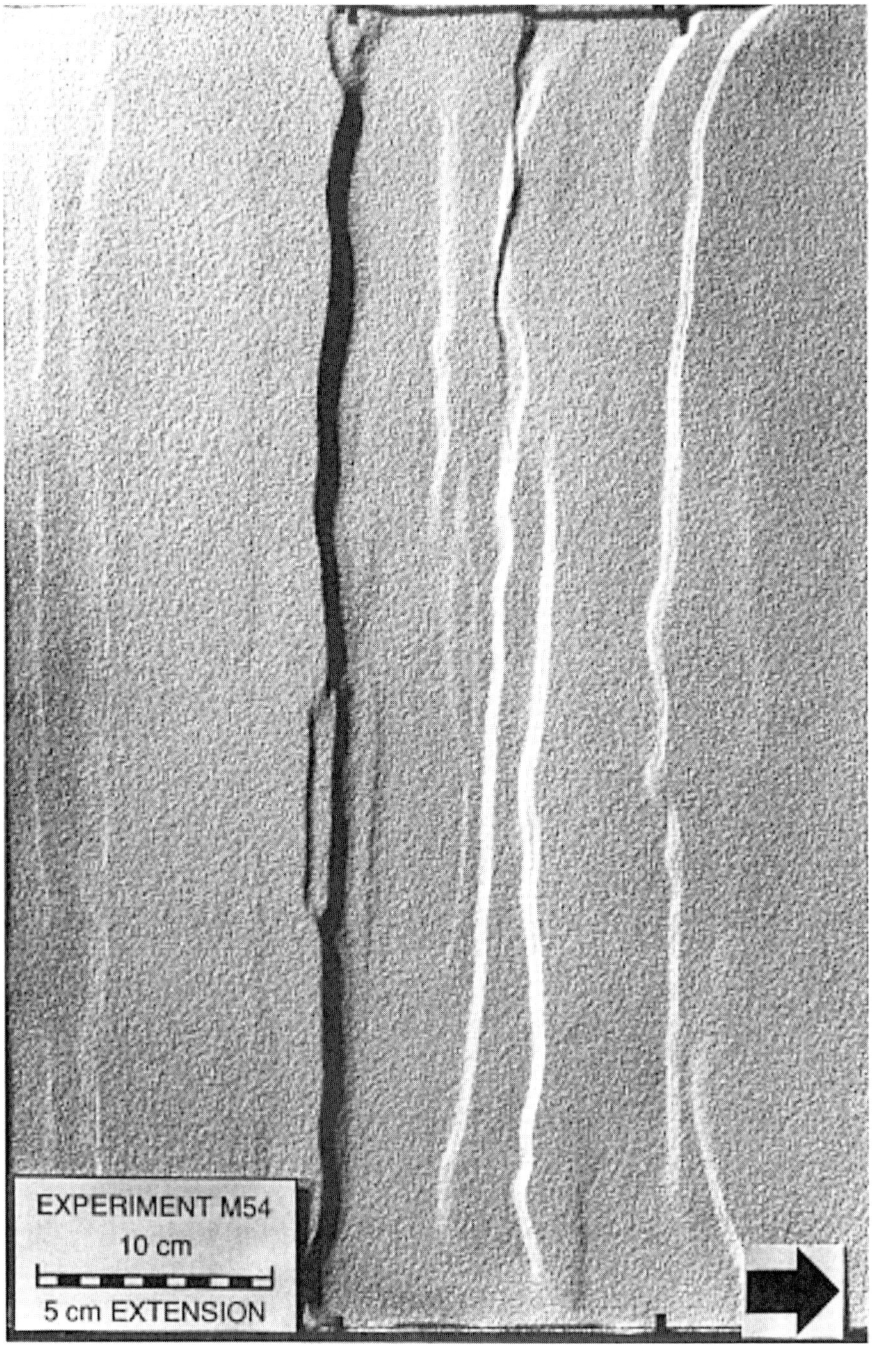

transfer faults. Similarly in regions where the rift axis is oblique to the extension direction such as in parts of Afar (Kronberg 1991) and the East African Rift system (e.g. Morley *et al.* 1992) the rift border faults are segmented and parallel to the rift axis whereas the intra-rift faults are oriented at a high angle to the extension direction (Fig. 10b). In the above cases and in other rift systems (cf. Gulf of Suez) no trans-rift strike-slip or oblique-slip transfer fault systems are found, rather zones of complex faulting generally known as accommodation zones (Bosworth *et al.*, 1986; Morley *et al.* 1990; McClay & White 1995; also see discussion below). It is probable that in oblique rift systems

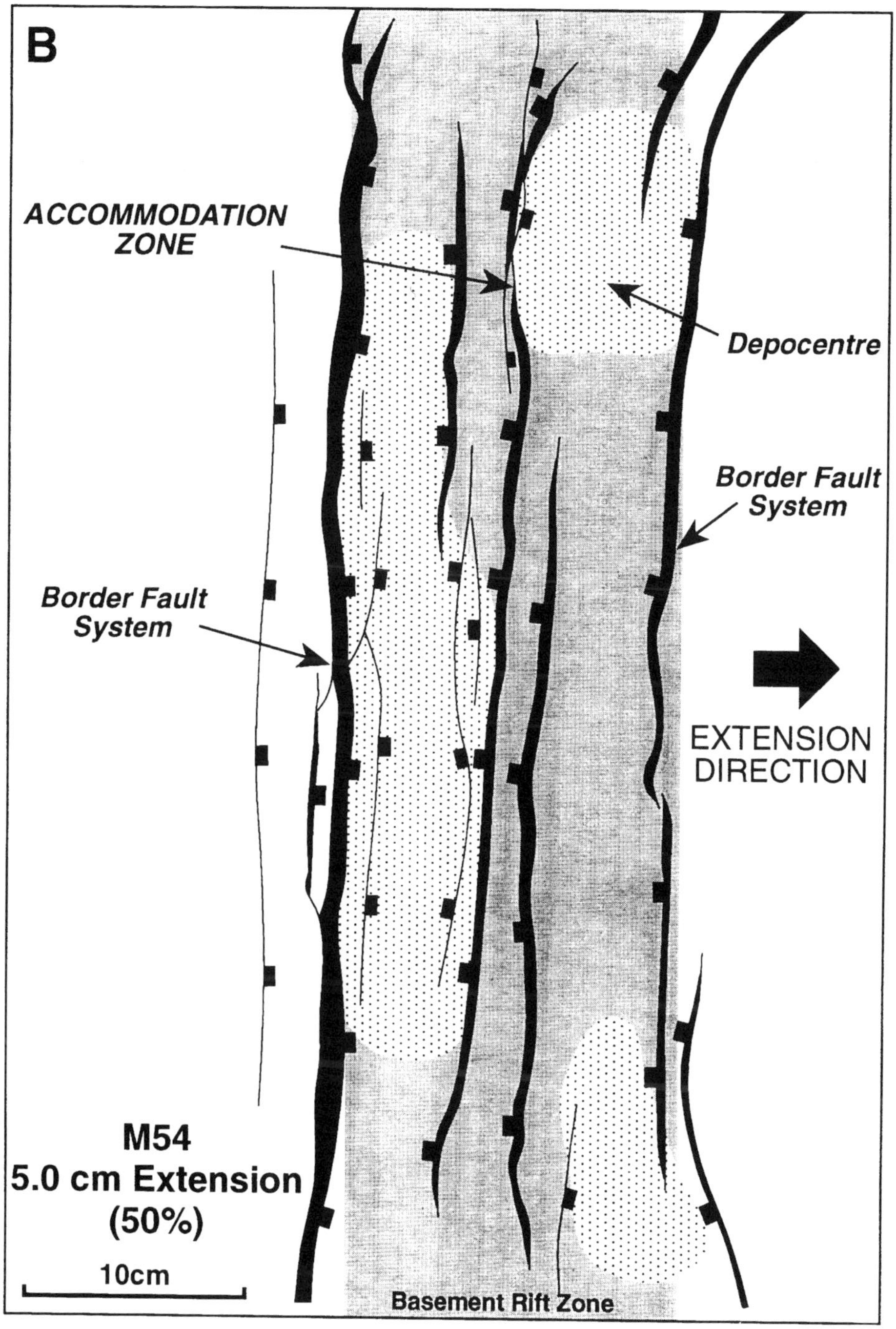

Fig. 8. 90° Orthogonal rift model M54. (**a**) View of the upper surface after 5 cm of extension (50% with respect to the initial length of the stretched rubber sheet at the base of the model). Lighting direction is from the left hand side of the model. (**b**) Map pattern of faults after 5 cm of extension. (from McClay & White 1995, and published with permission of *Marine and Petroleum Geology*)

many of the extensional faults would have components of oblique slip but determination of exact displacement vectors on these fault surfaces is usually extremely difficult. The 3D development of extensional faults and the architectures of accommodation zones are discussed in detail below.

Inversion models – results

Analogue modelling of inverted fault systems involves first extending the models (as described for extensional experiments above) and then horizontally recompressing them such that the extensional faults are reactivated (or partially

A

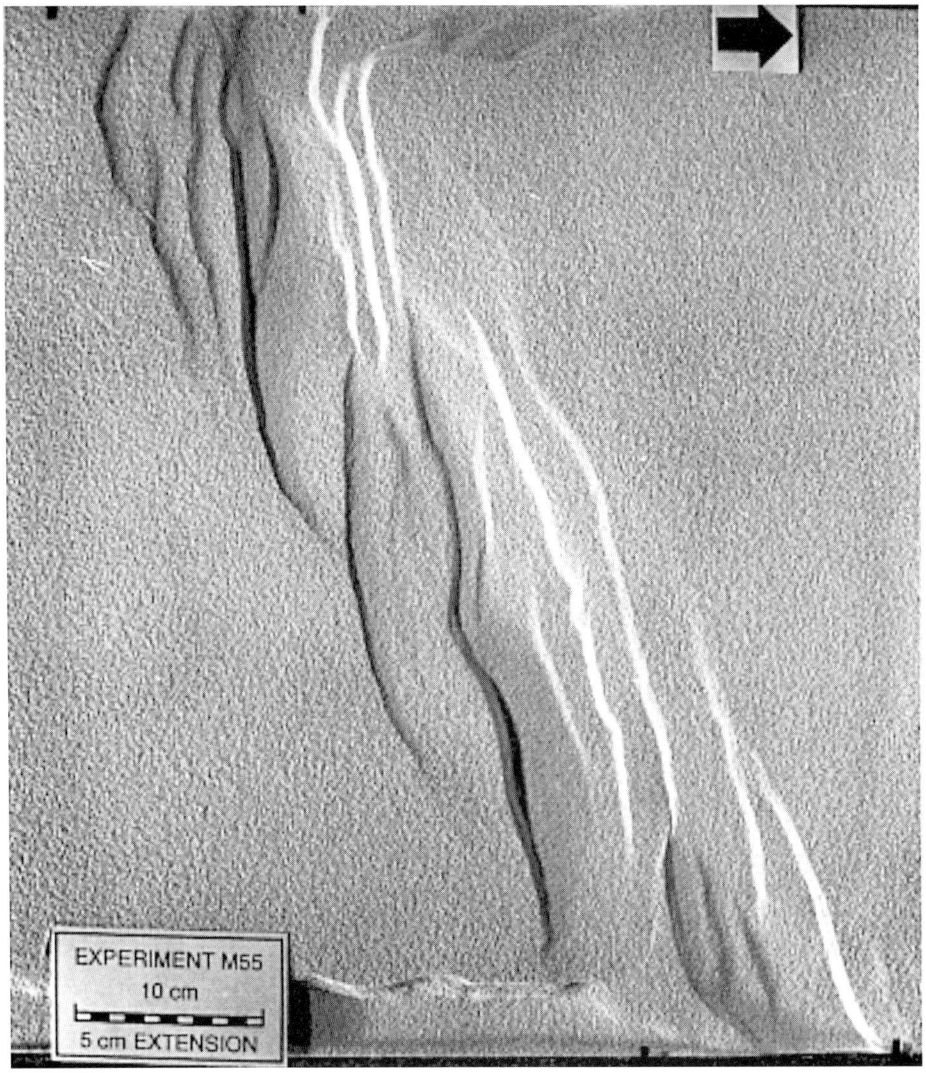

reactivated) as contractional faults (Buchanan & McClay 1991, 1992; McClay & Buchanan 1992; also see McClay 1995 for details). Syn-inversion sedimentation is usually added incrementally during the inversion deformation (cf. McClay, 1995).

Inversion of listric faults

Extension above a listric detachment fault produces a characteristic roll-over structure with an associated crestal collapse graben (cf. descriptions above and Fig. 3a). Inversion of the listric extensional fault model was achieved by horizontal recompression such that the hanging wall is pushed upwards and rotated by reactivation of the main detachment surface to produce a fault-bounded wedge above the main extensional fault breakaway (Fig. 11a). The main detachment fault propagated into the overlying post-rift and syn-inversion strata with the same geometry as the listric extensional segment (Fig. 11a). Small displacement short-cut faults form in the post-rift and syn-rift section in the footwall of the main fault. These short-cut thrust faults flatten upwards in the post-rift and syn-inversion strata. During inversion the crestal collapse graben becomes tighter (i.e. shortened) and small thrust faults nucleate from the tips of the pre-existing crestal-collapse graben extensional faults. This produces a characteristic 'pop-up' structure above the inverted crestal-collapse graben

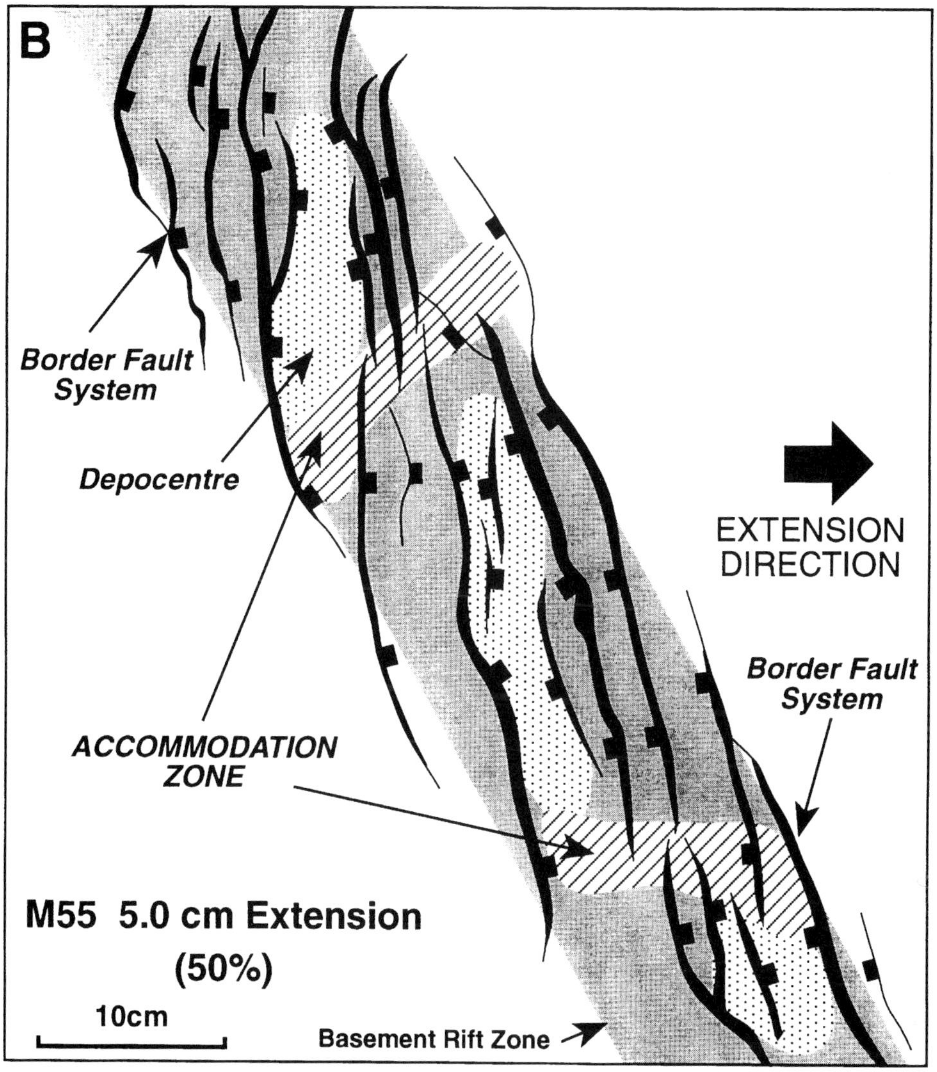

Fig. 9. 60° Oblique rift model M55. (**a**) View of the upper surface after 5 cm of extension (50% with respect to the initial length of the stretched rubber sheet at the base of the model). Lighting direction is from the left hand side of the model. (**b**) Map pattern of faults after 5 cm of extension. (from McClay & White 1995, and published with permission of *Marine and Petroleum Geology*)

(Fig. 11a). Syn-inversion strata thin over the crest of the main fault-bounded wedge and thicken away to either side (Fig. 11a). The fault sequence diagram shows the generation of the extensional fault architecture followed by the inversion architecture. An important aspect to note for section balancing is the changes in displacement history along the faults and in particular the nucleation of thrust faults from the tips of extensional faults (Fig. 11b). An important limitation of this style of inversion model (see also Buchanan & McClay, 1991) is that the rigid footwall block does not permit the detachment to change shape during both extension and inversion and in particular it does not permit footwall short-cut faults to develop in the pre- and syn-extensional sequences during inversion. In natural inverted fault systems footwall shortcut faults are observed (e.g. in southwest Wales, Powell 1989).

Particle displacement paths for inversion

Particle displacement paths for extension above a simple listric fault are shown in Fig. 5. During inversion the particle displacement paths relative to a fixed point in the footwall are parallel the main detachment surface (Fig. 12a) whereas particle displacement paths calculated relative to a fixed point in the hangingwall give trajectories for

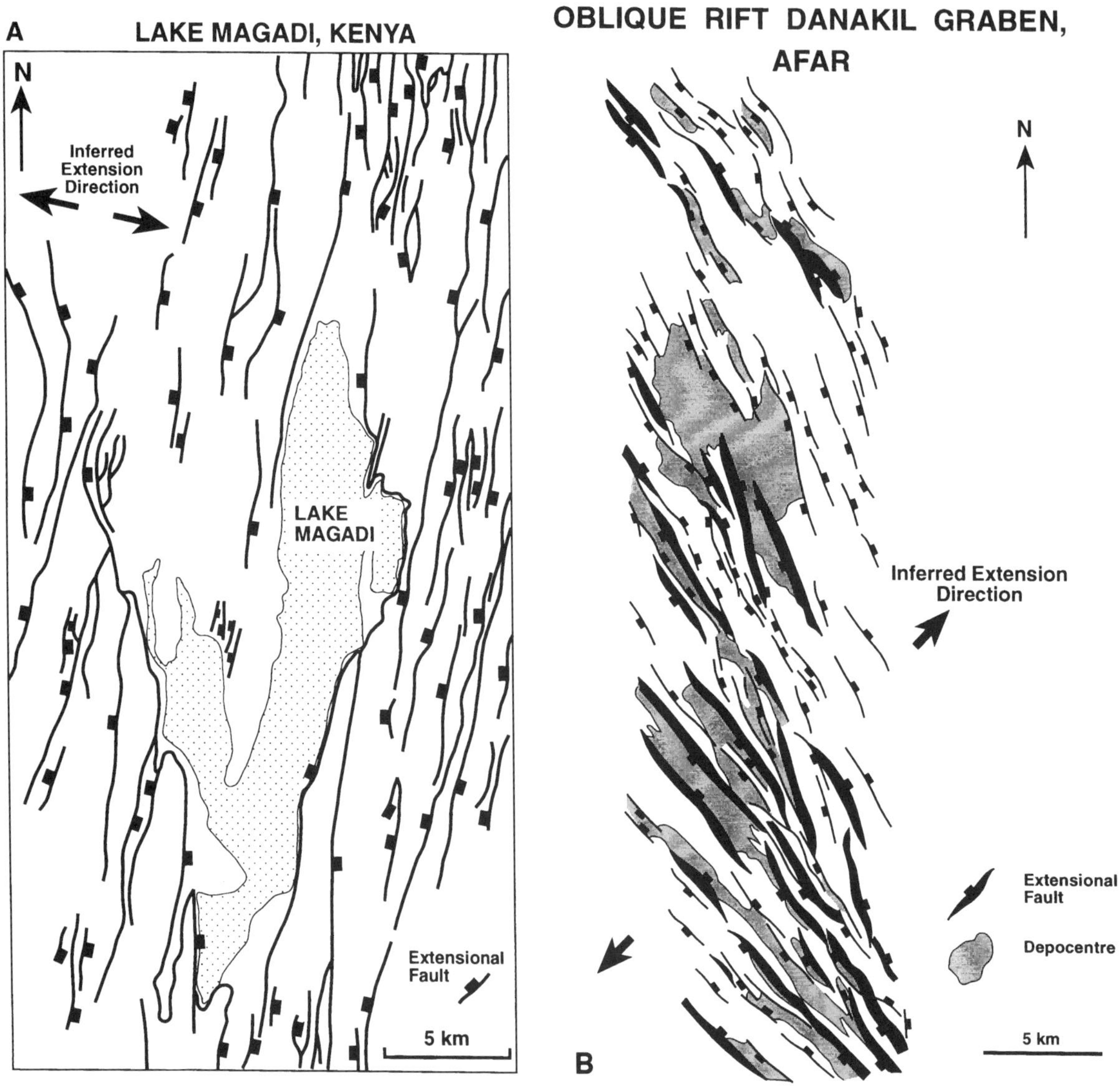

Fig. 10. Examples of rift fault patterns (**a**) Fault pattern in orthogonal rift, Lake Magadi, Gregory Rift Kenya (adapted after Baker 1986). The inferred extension direction is slightly north of E–W. (**b**) Map pattern of faults in an oblique rift, Afar (adapted after Kronberg 1991). The inferred extension direction is NE–SW and the faults are strongly segmented and at a high angle to the extension direction.

inversion that are initially shallow and then steepen upwards with increased contraction (Fig. 12b). The initial flat or shallow apparent shear trajectories reflect an important component of dilation (volume contraction) at the early stages of inversion. It is also important to note that the apparent shear trajectories during inversion are less steep than those during extension thus indicating a significant change to the angle of shear failure between extension and inversion. This has important implications for numerical methods of fault surface predictions and for algorithms used in balancing and restoring inverted terranes (see discussion below).

Synoptic models and examples of inverted listric faults

A synoptic model for an inverted listric fault is presented in Fig. 13a. Note the characteristic arrowhead or harpoon geometry of the inverted syn-extensional sedimentary wedge and the development of a growth anticline above the reactivated main detachment. Natural examples of inverted listric faults are discussed in Buchanan & McClay (1991), McClay & Buchanan (1992) and in McClay (1995). An example of an inverted listric fault from the Natuna Sea is shown in Fig. 13b. This shows many features similar to those developed in the

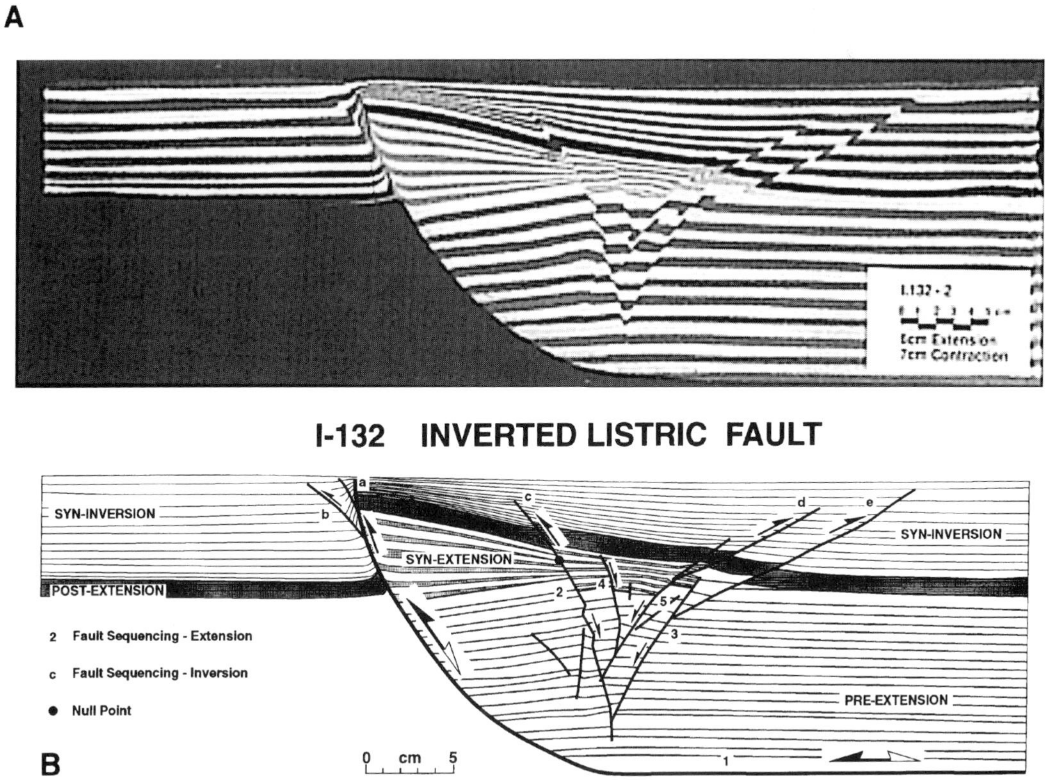

Fig. 11. 2D inversion of a simple listric fault. (**a**) Analogue model I132 after 6 cm of extension (approximately 25% with respect to the initial length of model), followed by 7 cm of inversion. (**b**) Fault sequence diagram for model I132.

analogue models. In particular note the erosion on top of the anticline and growth and onlap of syn-inversion strata onto the anticline (Fig. 13b). The complex structures in the footwall of the inverted listric fault make seismic interpretation difficult and the analogue models are valuable templates for understanding the structural styles in this area.

Discussion

The power of the analogue models that have been briefly reviewed above is shown by their ability to replicate the geometries of many natural examples of extensional and inverted fault systems (e.g. McClay 1990*a*, 1995; Buchanan & McClay 1991, McClay & Buchanan 1992). In particular they show the progressive evolution of structures such that an understanding of both the geometric and kinematic evolution can be attained (McClay 1990*a*, 1995). Similarly the analogue models of strike-slip fault systems (e.g. Naylor *et al.* 1986; Richard *et al.* 1995) and of thrust fault systems/ Coulomb wedges (e.g. Malavielle 1984; Mulugeta & Koyi 1987; Colletta *et al.*1991; Dixon & Liu 1992; Liu *et al.* 1992; Calassou *et al.* 1993; Koyi 1995) have provided great insight as to the geometric and kinematic evolution of these fault systems.

Detailed analysis of the hanging-wall deformation in the analogue models offers the potential for understanding hanging-wall deformation mechanisms in both extensional and inverted fault systems. Most fault reconstruction techniques assume that during extension the hanging-wall deforms either by vertical simple shear (Verrall 1981; Gibbs 1983, 1984) or by inclined simple shear (Dula 1991; Nunns 1991; Kerr & White 1992; de Matos 1993; Withjack & Petersen 1993; Withjack *et al.* 1995). The results of the analysis of hanging-wall deformation in the analogue models, however, shows that above curved extensional faults the shear trajectories are curved. In addition the shear trajectories change from extension to compression becoming less steeply dipping (cf. Fig. 12 and also McClay 1995). These variations in hanging-wall shear trajectories need to be taken into account not only in fault trajectory predictions but also in the algorithms used for section construction and restoration (cf. Gibbs 1983; Rowan & Kligfield 1989).

Advances in modelling techniques, modelling materials, in the use of X-ray tomography, together

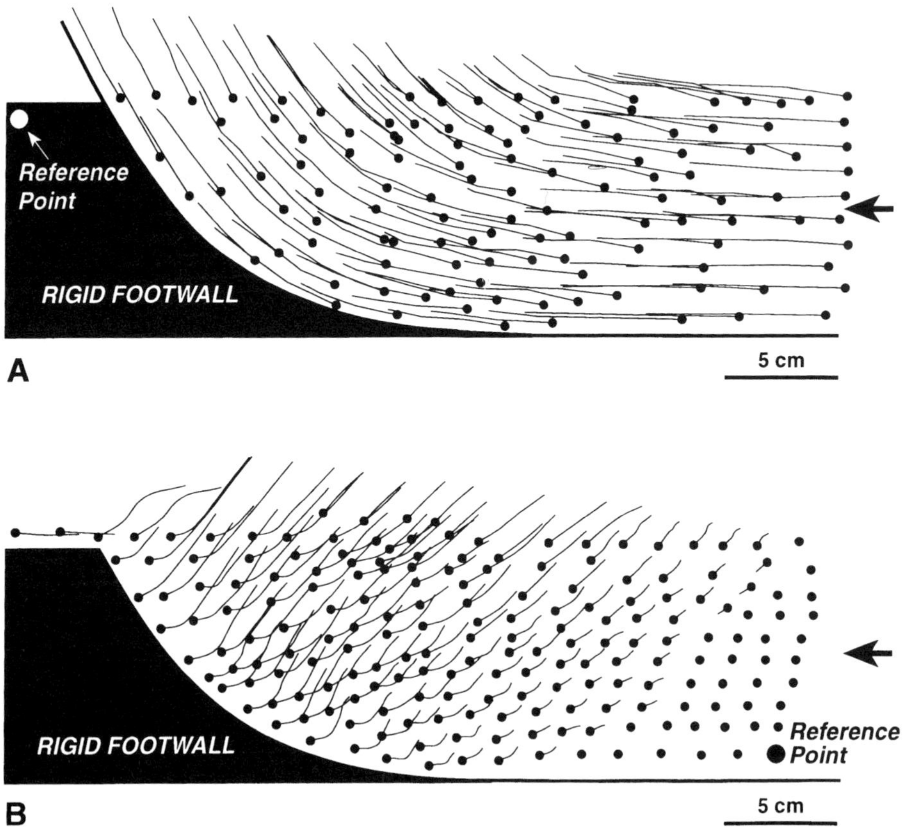

Fig. 12. Hanging-wall deformation in a 2D inverted listric fault. (**a**) Particle displacement paths measured relative to a fixed point in the footwall block. (**b**) Particle displacement paths measured relative to a fixed point in the hangingwall block.

with the analysis of hanging-wall deformation and 3D reconstructions have greatly improved the understanding and applicability of scaled analogue modelling to natural 3D structures. In particular 3D reconstruction using digitized serial sections has greatly enhanced the 3D visualization of complex fault geometries. Figure 14 shows two examples of rift models that have been reconstructed from serial sections cut parallel to the direction of extension. In the reconstruction of the top of the pre-rift surface in 60° oblique rift model M30 (Fig. 14a) the en-echelon arrangement of the rift border faults and the overlap of faults in relay ramp structures is clearly visible. In addition this display shows the elliptical displacement pattern along individual faults and the nature of along-strike fault linkages. These features are characteristic of the analogue models of orthogonal and oblique rift systems (McClay & White 1995). In the orthogonal, oblique and offset rift models no strike-slip fault systems were found – instead displacement transfer occurred by overlapping relay ramp structures (e.g. Fig. 14a and discussed in Larsen 1988; Peacock & Sanderson 1991; McClay & White 1995). The fault geometries in the 3D analogue models of rift systems show close similarities to those found in natural rifts (e.g. Tron & Brun 1991; Brun & Tron 1993; McClay & White 1995). In particular for oblique rifts both the rift border and intra-rift faults are segmented and overlap in relay ramp systems. Further detailed research and analysis required in order to understand the strain partitioning on both the border and intra-rift fault systems within the analogue models and to investigate the effects of ductile substrates on rift fault development (cf. Tron & Brun 1991; Brun & Tron 1993). The analogue models show how the faults grow along strike and how they link and interact and thus place important constraints upon section construction and validation in extensional terranes.

Figure 14b shows the 3D reconstruction of surfaces within an offset orthogonal rift model but with the addition of ray paths for a synthetic seismic line across this 3D structure (cf. Wang & Waltham 1995). 3D analogue modelling and reconstruction of complex fault systems, although still in its infancy (cf. Tron & Brun 1991; McClay

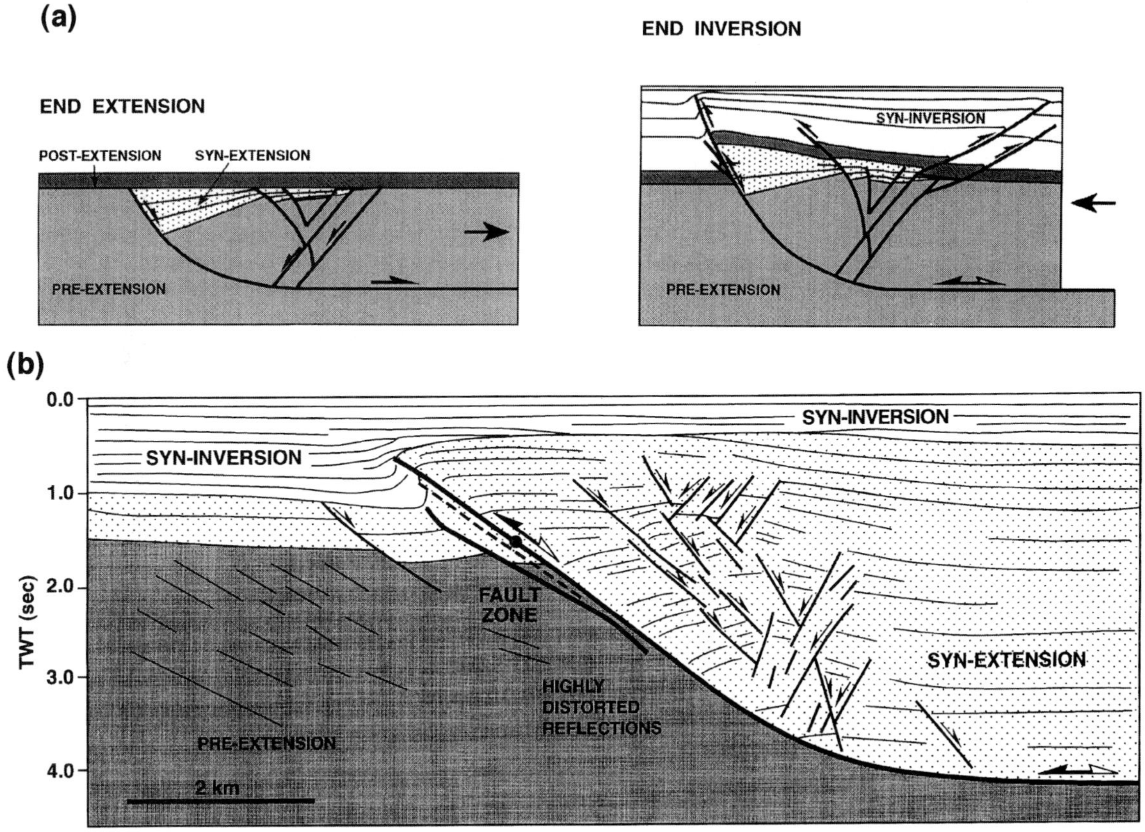

Fig. 13. Listric Fault inversion structures. (**a**) Synoptic model of extension followed by inversion of a simple listric fault. (**b**) Example of an inverted listric fault showing Miocene syn-inversion growth strata in the Natuna Sea, Southeast Asia (interpretation of a seismic section in Bunks *et al.* 1993)

& White 1995), are likely to provide important constraints on fault geometries, kinematics and fault linkages both in plan and in section. In particular the 3D models will provide data on fault displacements that can be tested against natural examples of fault systems and provide important constraints for fault displacement and fault population analyses (e.g. Walsh & Watterson, 1988).

In addition to providing interpretation templates for complex fault systems the analogue models may be used as the starting geometry for seismic modelling of complex structures (e.g. Fig. 14b). An important aspect of the Fault Dynamics research at Royal Holloway, University of London, is seismic modelling of complex structures in order to understand how they image and to determine the limitations of data processing and interpretation in complexly faulted areas (e.g. McClay *et al.* 1991; Waltham *et al.* this volume). Forward analogue modelling combined with synthetic seismic simulations of complex structural geometries will be a powerful tool to aid the exploration geologist in constructing and restoring cross-sections in areas such as inverted and contractional terranes where seismic imaging is commonly difficult.

Strengths and limitations of the analogue models

Scaled physical models are powerful visual tools for the forward modelling of brittle deformation in the upper crust. In particular they can model a wide range of tectonic settings from extension, inversion, strike-slip and contractional terranes as well as mobile salt and shale systems. Analogue models have distinct advantages over numerical models in that they are visual and 3D and are relatively easy to construct. In particular sandbox models allow new faults to form and whole new fault arrays develop and undergo progressive deformation – a feature that is difficult to replicate in most numerical models. Preservation of the models and serial sectioning permit the full 3D geometries of the deformation to be determined. Analysis of 3D structural geometries permits

(a)

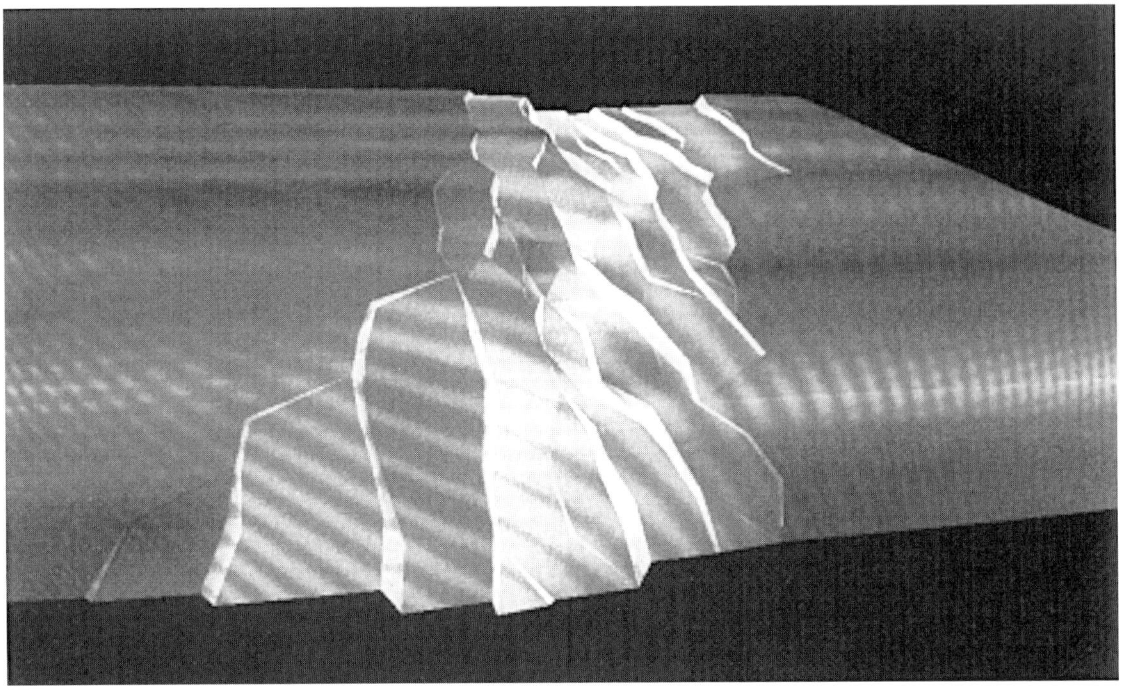

(b)

Fig. 14. 3D reconstructions of analogue models of rift structures. (**a**) Top of the pre-rift surface in analogue model M30 (60° oblique extension). The direction of extension is parallel to the front edge of the surface and the zone of rifting is approximately 25 cm wide. (**b**) Three surfaces in analogue model M26 (90° orthogonal rift with a central offset). The top surface is the pre-rift/syn-rift boundary and the lower two surfaces are within the pre-rift section. The direction of extension is parallel to the front edge of the surface and the zone of rifting is approximately 25 cm wide. The reflected rays from each surface generated by a synthetic 3D seismic survey line (calculated using the 3D Gaussian Beam method (Wang & Waltham 1995)) across this model are also shown.

assessment of fault linkages, fault displacements and complex 3D fault geometries in a manner that the exploration geologist can visualize in 3D in structurally complex areas. X-ray tomography of models whilst they are being deformed allows the internal deformations of the models to be monitored but the small size of the deformation rigs that can be used in the X-ray scanner place serious limitations on the boundary conditions and sizes of models that can be analysed in this way. In addition the resolution of X-ray tomography falls short of that which can be achieved by serially sectioning the models and reconstructing them on a workstation.

There are however important limitations to scaled physical models and these must always be born in mind when interpreting and applying the results of analogue models to natural deformation. Deformation in sandboxes or other experimental apparatus will always be strongly controlled by the boundary conditions at the base and sides of the model. Great care is needed in the construction of physical models to ensure that the boundary conditions in the model replicate those in nature. This feature is commonly a major limitation in the application of scaled physical models to natural deformation. Other important limitations to scaled analogue models include difficulties in correctly scaling the grain sizes and physical properties of natural prototypes, i.e. introducing realistic competency contrasts, elasticities and slip mechanisms. In particular the faults in the analogue models are formed by granular shear processes without grain breakage – a process that does not replicate fault mechanics in nature (cf. Scholtz 1990). Analogue models cannot as yet adequately model flexural and isostatic, and thermal effects of deformation. Compaction of sediments, fluid flow and the development of overpressures are also limitations of the modelling methods that need to be addressed.

Conclusions

From the examples described above it is evident that scaled analogue modelling is a powerful tool for the understanding of the geometric and kinematic development of extensional, and inverted fault systems, as well as for some thrust systems (e.g. Dixon & Liu 1992; Liu *et al.* 1992) and for strike-slip fault systems (e.g. Richard *et al.* 1995). Recent advances in modelling techniques, in modelling materials, in analysis of deformation within the models, and utilization of X-ray tomography and 3D reconstruction have enabled scaled physical models to be more widely applied to complex tectonic regimes and to provide constraints on structural styles, geometries and kinematics.

In particular the results of scaled analogue models can provide significant constraints on the construction, validation and restoration of geological cross sections in extensional and inverted terranes in that they:

(1) Provide an understanding of the geometric and kinematic evolution of fault (and fold) structures during the brittle deformation of the upper crust;
(2) Permit an understanding of the sequences of faulting and activity on faults (i.e. relative slip rates and also synchronous slip on two or more faults);
(3) Provide templates for the interpretation of seismic sections in regions of poor data quality and for projecting surface data into the subsurface;
(4) Can potentially provide an understanding of the mechanisms of hangingwall deformation and the relationships between hanging-wall shear regimes, the shape of the basal detachment surface and the displacement conditions along the fault surface;
(5) Provide templates for seismic modelling of complex structural regimes and templates for numerical modelling of tectonics and sedimentation;
(6) Show that the hangingwall shear orientations change from extension to inversion – an important constraint for section construction in mixed mode tectonic terranes.

The ability of the models to replicate real geological structures is shown by the close geometrical similarities between the models and the natural examples described above and in the literature (e.g. examples in Bally 1983; Lowell 1985; and also discussed in McClay 1990*a*, 1995).

For extensional fault systems the models may simulate structures on a number of scales, from entire rift basins to hanging-wall deformation above a single fault or detachment. The extensional fault models are expected to be particularly powerful for the prediction of structural styles associated with individual detachments, i.e. at the scale of an individual field.

Although still in its infancy, 3D modelling is likely to prove to be extremely valuable in the study of the nucleation, propagation and interaction of extensional fault systems along their strike and at depth and thus provide important constraints on section construction in 3D.

The models reviewed in this paper, nevertheless, have limitations which must be borne in mind when interpreting the results. In the models, the accommodation space in the basins formed by extension is always infilled with syn-rift sediment, a condition which does not always occur in nature. In addition, the listric fault models have constant displacement along the whole length of the detachment whereas in nature displacement rate gradients may be expected to occur on some faults. The models cannot, as yet, incorporate the isostatic, compaction or thermal effects expected for deformation in natural fault systems.

Research summarized in this paper was supported by the Fault Dynamics Research Project sponsored by Arco British Limited, BP Exploration, BRASOIL (UK.) Ltd, Conoco (UK) Ltd, Mobil North Sea Limited and Sun Oil Britain. Kevin D'Souza and Keith Denyer kindly printed the photographs. M. White, N. Deeks, M. Simmons, P. Buchanan and P. Hollings assisted with the analogue modelling and Chris Willacy assisted with the AVS reconstructions. T. K. Wang and X. Wang carried out the synthetic seismic modelling. D. Waltham and P. Richard are thanked for constructive criticism of the manuscript. Fault Dynamics Publication No. 52.

References

BAKER, B. H., 1986. Tectonics and volcanism of the southern Kenya Rift Valley and its influence on rift sedimentation. *In*: FROSTICK, L. E., REBAUT, R. W., REID, I. & TIERCELIN, J.-J. (eds) *Sedimentation in the African Rifts*. Geological Society, London, Special Publications, **25**, 45–58.

BALLY, A. W. 1983. *Seismic expression of structural styles*. American Association of Petroleum Geologists, Studies in Geology, **15**.

BARTLET, W. L., FRIEDMAN, M. & LOGAN, J. M. 1981. Experimental folding and faulting of rocks under confining pressure: Part IX. Wrench faults in limestone layers. *Tectonophysics*, **79**, 255–277.

BEACH, A. & TRAYNER, P. 1991. The geometry of normal faults in a sector of the offshore Nile Delta, Egypt. *In*: ROBERTS, A. M., YIELDING, G. & FREEMAN, B. (eds) *The Geometry of Normal Faults*. Geological Society, London, Special Publications, 56, 173–182.

BOSWORTH, W., 1986. Comment and Reply on Detachment faulting and the evolution of passive continental margins. *Geology*, **14**, 890–892.

——, 1995. The southern Gulf of Suez: a high strain rift. *In*: LAMBIASE, J. J. (ed.) *Hydrocarbon Habitat in Rift Basins*. Geological Society, London, Special Publications, **88**, 91–103.

——, LAMBIASE, J. & KEISLER, R., 1986. A new look at Gregory's Rift: The structural style of continental rifting. *EOS*, **67**, 577–583.

BRUCE, C. H., 1973. Pressured shale and related sediment deformation: Mechanism for development of regional contemporaneous faults. *Bulletin American Association of Petroleum Geologists*, **57**, 878–886.

BRUN, J-P. & TRON, V. 1993. Development of the North Viking Graben: inferences from laboratory modelling. *Sedimentary Geology*, **86**, 31–51.

BUCHANAN, P. G. 1991. *Geometries and kinematic analysis of inversion tectonics from analogue model studies*. PhD thesis, University of London.

—— & MCCLAY, K. R., 1991. Sandbox experiments of inverted listric and planar faults systems. *In*: Cobbold, P. E. (ed.) Experimental and Numerical Modelling of Continental Deformation. *Tectonophysics*, **188**, 97–115.

—— & —— 1992. Experiments on basin inversion above reactivated domino faults. *Marine and Petroleum Geology*, **9**, 486–500.

BUNKS, C., PICHON, P. L., BEN BRAHOM, L. & SAKTI, S. 1993. 2D prestack depth migration: Application to inverted grabens. *Proceedings Indonesian Petroleum Association*, **93**, 1, 22–44.

CADELL, H. M. 1889. Experimental researches in mountain building. *Transactions of the Royal Society of Edinburgh*, **35**, 337–357.

CALASSOU, S., LARROQUE, C. & MALAVIEILLE, J. 1993. Transfer zones of deformation in thrust wedges: an experimental study: *Tectonophysics*, **221**, 325–344.

CHRISTIANSEN, A. F. 1983. An example of a major syndepositional listric fault. *In*: BALLY, A. W. (ed.) *Seismic expression of structural styles*. American Association of Petroleum Geologists, Studies in Geology, **15**.

CLOOS, E. 1968. Experimental analysis of Gulf Coast fracture patterns. *Bulletin American Association of Petroleum Geologists*, **52**, 420–444.

CLOOS, H. 1928. Experimente zur inneren tektonik. *Cetralblatt fur Mineralogie*, Abt. B, 609–621.

COLLETTA, B., LETOUZEY, J., PINEDO, R., BALLARD, J. F. & BALÉ, P. 1991. Computerized X-ray tomography analysis of sandbox models: examples of thin skinned thrust systems. *Geology*, **19**, 1063–1067.

DAVIS, D. M., SUPPE, J. & DAHLEN, F. A., 1983. Mechanics of fold-and-thrust belts and accretionary wedges: *Journal of Geophysical Research*, **88**, 1153–1172.

DAVY, P & COBBOLD, P. R. 1988. Indentation tectonics in nature and experiment, 1. *Bulletin Geological Institution University of Uppsala, N.S*, **14**, 129–147.

—— & —— 1991. Experiments on shortening of a 4-layer model of the continental lithosphere. *Tectonophysics*, **188**, 1–25.

DIXON, J. M. & LIU, S., 1992. Centrifuge modelling of the propagation of thrust faults: *In*: MCCLAY, K. R. (ed.) *Thrust Tectonics*. Chapman & Hall, London, 53–69.

DULA, W. F. 1991. Geometric models of listric normal faults and rollover folds. *American Association of Petroleum Geologists Bulletin*, **75**, 1609–1625.

ELLENOR, D. W. & JAMES, D. M. D. 1984. The oil and gas resources of Brunei. *In*: JAMES, D. M. D. (ed.) *The geology and hydrocarbon resources of Negra Brunei Darussalam*. Muzium Brunei, Kota Batu., 130–139.

ELLIS, P. G. 1988. *Analogue model studies of the kinematics of extensional faulting*. PhD thesis, University of London.

—— & MCCLAY, K. R. 1988. Listric extensional fault systems – results of analogue model experiments. *Basin Research*, **1**, 55–70.

EMMONS, R. C. 1969. Strike-slip rupture patterns in sand models. *Tectonophysics*, **7**, 71–87.

FAUGERE, E. & BRUN, J. P. 1984. Modelisation experimentale de la distention continentale. *Compte Rendu Acadamie des Sciences, Paris*, Series **11**, 299, 365–370.

GIBBS, A. D. 1983. Balanced section constructions from seismic sections in areas of extensional tectonics. *Journal of Structural Geology*, **5**, 2, 153–160.

—— 1984. Structural evolution of extensional basin margins. *Journal of the Geological Society, London*, **141**, 609–620.

HORSFIELD, W. T. 1977. An experimental approach to basement controlled faulting. *Geologie en Mijnbouw*, **56**, 363–370.

—— 1980. Contemporaneous movement along crossing conjugate normal faults. *Journal of Structural Geology*, **2**, 305–310.

JACKSON, J. A. 1987. Active normal faulting and crustal extension. *In*: COWARD, M. P., DEWEY, J. F. & HANCOCK, P. L. (eds) *Continental Extension Tectonics*. Geological Society, London, Special Publications, **28**, 3–17.

JACKSON, M. P. A. & VENDEVILLE, B. C., 1994. Regional extension as a trigger for diapirism. *Geology Society of America Bulletin*, **106**, 57–73.

KERR, H. G. & WHITE, N. 1992. Laboratory testing of an automatic method for determining normal fault geometry at depth. *Journal of Structural Geology*, **14**, 873–885.

KOOPMAN, A., SPEKSNIJDER, A. & HORSFIELD, W. T. 1987. Sandbox model studies of inversion tectonics. *Tectonophysics*, **137**, 379–388.

KOYI, H. 1995. Mode of deformation in sand wedges. *Journal of Structural Geology*, **17**, 293–300.

KRONBERG, P. 1991. Geometries of extensional fault systems, observed and mapped on aerial and satellite photographs of Central Afar (Ethiopia/Djibouti). *Geologie en Mijnbouw*, **70**, 145–161.

KUSZNIR, N. J., ROBERTS, A. M. & MORLEY, C. K. 1995. Forward and reverse modelling of rift basin formation. *In*: LAMBIASE, J. J. (ed.) *Hydrocarbon Habitat in Rift Basins*. Geological Society, London, Special Publications, **88**, 33–56.

LALLEMAND, S. E., MALAVIEILLE, J. & CALASSOU, S., 1992. Effects of oceanic ridge subduction on accretionary wedges: Experimental modeling and marine observations: *Tectonics*. **11**, 1301–1313.

LARSEN, P 1988. Relay structures in a lower Permian basement-involved extension system, East Greenland. *Journal of Structural Geology*, **10**, 3–8.

LEMON, N. M. & MAHMOOD, T. 1994. Two and three dimensional analogue modelling of extensional fault systems with applicayion to the North West Shelf, Western Australia. *Australian Petroleum Exploration Association Journal*, **34**, 555–565.

LIU, H. 1992. *Analogue modelling of thrust wedges*. PhD thesis, University of London.

——., MCCLAY, K. R. & POWELL, D. 1992. Physical models of thrust wedges: *In*: McClay. K. (ed.) *Thrust Tectonics*. Chapman & Hall, London, 71–81.

LOPEZ, J. A. 1990. Structural styles of growth faults in the US Gulf Coast Basin. *In*: BROOKS, J. (ed.) *Classic Petroleum Provinces*. Geological Society, London, Special Publications, **50**, 203–220.

LOWELL, J. D., 1985. *Structural Styles in Petroleum Exploration*. Oil and Gas Consultants International Inc., Tulsa.

MCCLAY, K. R. 1989. Analogue models of inversion tectonics. *In*: COOPER, M. A. & WILLIAMS, G. D., (eds) *Inversion Tectonics*. Geological Society, London, Special Publications, **44**, 41–59.

—— 1990*a*. Extensional fault systems in sedimentary basins. A review of analogue model studies. *Marine and Petroleum Geology*, **7**, 206–233.

—— 1990*b*. Deformation mechanics in analogue models of extensional fault systems. *In*: RUTTER, E. H. & KNIPE, R. J. (eds) *Deformation Mechanisms, Rheology and Tectonics*. Geological Society of London Special Publication, **54**,. 445–454.

——, 1995. The geometries and kinematics of inverted fault systems: a review of analogue model studies. *In*: BUCHANAN, J., G. & BUCHANAN, P. G. (eds) Basin Inversion. Geological Society, London, Special Publications, **88**, 97–118.

—— & BUCHANAN, P. G. 1992. Thrust faults in inverted extensional basins. *In*: MCCLAY, K. R. (ed.) *Thrust Tectonics*. Chapman & Hall, London, 419–434.

—— & ELLIS, P. G. 1987*a*. Analogue models of extensional fault geometries. *In*: COWARD, M. P., DEWEY, J. F. & HANCOCK, P. L. (eds) *Continental extension tectonics*. Geological Society, London, Special Publications, **28**, 109–125.

—— & —— 1987*b*. Geometries of extensional fault systems developed in model experiments. *Geology*, **15**, 341–344.

—— & SCOTT, A. D. 1991. Hangingwall deformation in ramp-flat listric extensional fault systems. *Tectonophysics*, **188**, 85–96.

—— & WHITE, M. 1995. Analogue models of orthogonal and oblique rifting. *Marine and Petroleum Geology*, **12**, 137–151.

——, WALTHAM, D., SCOTT, A. D. & ABOUSETTA, A., 1991. The geometries of listric extensional fault systems. *In*: ROBERTS, A. M., YIELDING, G. & FREEMAN, B. (eds) *The Geometry of Normal Faults*. Geological Society, London, Special Publications, **56**, 231–239.

MALAVIEILLE, J. 1984. Modélisation expérimentale des chevauchements imbriqués: Application aux chaînes de montagnes: *Geological Society of France Bulletin*, **7**, 129–138.

——, LARROQUE, C. & CALASSOU, S. 1993. Modelisation experimentale des relations tectonique/sedimentation entre bassin avant-arc et prisme d'accretion: *Compte Rendu Acadamie Sciences, Paris*, **316**, II, 1131–1137.

MANDL, G. 1988. *Mechanics of tectonic faulting. Models and basic concepts*. Elsevier, Amsterdam.

MANN, P., HEMPTON, P. R., BRADLEY, D. C. & Burke, K., 1983. Development of pull-apart basins. *Journal of Geology*, **91**, 529–554.

MARSHAK, S. & WILKERSON, M. S. 1992. Effect of overburden thickness on thrust belt geometry and development: *Tectonics*, **11**, 560–566.

——, —— & HSUI, A.T. 1992. Generation of curved fold-thrust belts: Insight from simple physical and analytical models: *In*: MCCLAY, K. R. (ed.) *Thrust Tectonics*. Chapman & Hall, London, 83–92.

DE MATOS, R. M. D. 1993. Geometry of the hangingwall above a system of listric normal faults. *Bulletin American Association of Petroleum Geologists*, **77**, 1839–1859.

MORLEY, C. K., CUNNINGHAM, S. M., HARPER, R. M. & WESCOTT, W. A. 1992. Geology and geophysics of the Rukwa Rift, East Africa. *Tectonics*, **11**, 69–81.

——, NELSON, R. A., PATTON, T. L. & MUNN, S. G. 1990. Transfer zones in the East African Rift system and their relevance to hydrocarbon exploration in rifts. *American Association of Petroleum Geologists Bulletin*, **74**, 1234–1253.

MULUGETA, G. 1988*a*. Squeeze box in a centrifuge: *Tectonophysics*, **148**, 323–335.

—— 1988*b*. Modelling the geometry of Coulomb thrust wedges: *Journal of Structural Geology*, **10**, 847–859.

—— & KOYI, H. 1987. Three-dimensional geometry and kinematics of experimental piggyback thrusting: *Geology*, **15**, 1052–1056.

NELSON, T. H. 1991. Salt tectonics and listric normal faulting. *In*: SALVADOR, A. (ed.) *The Gulf of Mexico Basin*. Geological Society of America, The Geology of North America, **J**, 73–89.

NAYLOR, M. A., MANDL, G. & SIJPESTEIN, C. H. K. 1986. Fault geometries in basement induced wrench faulting under different initial stress states. *Journal of Structural Geology*, **8**, 737–752.

NUNNS, A. G. 1991. Structural restoration of seismic and geologic sections in extensional regimes. *Bulletin American Association of Petroleum Geologists*, **75**, 278–297.

ODONNE, F. & VIALON, P. 1983. Analogue models of folds above a wrench fault. *Tectonophysics*, **99**, 31–46.

PEACOCK, D. C. P. & SANDERSON, D. J. 1991. Displacements, segment linkage and relay ramps in normal fault zones. *Journal of Structural Geology*, **13**, 721–733.

POWELL, C. 1989. Structural controls on Plaeozoic basin evolution and inversion in Southwest Wales. *Journal of the Geological Society, London*, **146**, 439–446.

RECHES, Z. 1987. Mechanical aspect of pull-apart basins and push-up swells with applications of the Dead Sea transform. *Tectonophysics*, **141**, 75–88.

RICHARD, P. D., 1991. Experiments on faulting in a two layer cover sequence overlying a reactivated basement fault with oblique (normal-wrench or reverse-wrench) slip. *Journal of Structural Geology*, **13**, 459–469.

—— & COBBOLD, P. R. 1990. Experimental insights into partitioning of fault motions in continental convergent wrench zones. *Annales Tectonicae*, **4**, 35–44.

——, BALLARD, J. F., COLLETTA, B. & COBBOLD, P. R. 1989. Fault initiation and development above a basement strike-slip fault: analogue modelling and tomography. *Compte Rendu Acadamie des Sciences, Paris*, **309** (II), 2111–2118.

——, MOQUET, B. & COBBOLD, P. R. 1991. Experiments on simultaneous faulting and folding above a basement wrench fault. *Tectonophysics*, **188**, 133–141.

——, NAYLOR, M. A. & KOOPMAN, A. 1995. Experimental models of strike-slip tectonics. *Petroleum Geoscience*, **1**, 71–80.

ROSENDAHL, B. R. 1987. Architecture of continental rifts with special reference to east Africa. *Annual Review of Earth and Planetary Sciences*, **15**, 445–503.

——, REYNOLDS, D., LORBER, P., BURGESS, C., MCGILL, J., SCOTT, D., LAMBIASE, J. & DERKSEN, S. 1986. Structural expressions of rifting: lessons from Lake Tanganyika. *In*: FROSTICK, L. E., REBAUT, R. W., REID, I. & TIERCELIN, J.-J. (eds) *Sedimentation in the African Rifts*. Geological Society, London, Special Publications, **25**, 29–43.

ROWAN, M. G. & KLIGFIELD, R. 1989. Cross section restoration and balancing as aid to seismic interpretation in extensional terranes. *Bulletin American Association of Petroleum Geologists*, **73**, 955–966.

SCHOLZ, C. H. 1990. *The mechanics of earthquakes and faulting*. Cambridge University Press, New York.

SCHREURS, G. 1994. Experiments on strike-slip faulting and block rotation. *Geology*, **22**, 567–570.

SCHULTZ-ELA, D. D. 1992. Restoration of cross-sections to constrain deformation processes of extensional terranes. *Marine and Petroleum Geology*, **9**, 372–388.

SCOTT, D. L. & ROSENDAHL, B. R. 1989. North Viking graben: An East African perspective. *Bulletin American Association of Petroleum Geologists*, **73**, 155–165.

SERRA, S. & NELSON, R. A. 1989. Clay modelling of rift asymmetry and associated structures. *Tectonophysics*, **153**, 307–312.

TALBOT, C. J. 1992. Centrifuge models of the Gulf of Mexico profile. *Marine and Petroleum Geology*, **9**, 412–432.

—— 1993. Spreading of salt structures, upward-moving gravity structures. *Tectonophysics*, **228**, 151–165.

TANKARD, A. J., WELSINK, H. J. & JENKINS, W. A. M. 1989. Structural styles and stratigraphy of the Jeanne d'Arc basin, Grand Banks of Newfoundland. *In*: BALKWILL & TANKARD (eds) *Extensional Tectonics and Stratigraphy of the North Atlantic Margins*. American Association of Petroleum Geologists, Memoir, **46**, 95–110.

TCHALENKO, J. S. 1970. Similarities between shear zones of different magnitudes. *Geological Society of America Bulletin*, **81**, 1625–1640.

TRON, V. & BRUN, J-P. 1991. Experiments on oblique rifting in brittle-ductile systems. *Tectonophysics*, **188**, 71–84.

VENDEVILLE, B. 1991. Mechanisms generating normal fault curvature: a review illustrated by physical models. *In*: ROBERTS, A. M., YIELDING, G. & FREEMAN, B. (eds) *The Geometry of Normal Faults*. Geological Society, London, Special Publications, **56**, 241–250.

—— & COBBOLD, P. R. 1988. How normal faulting and sedimentation interact to produce listric fault profiles and stratigraphic wedges. *Journal of Structural Geology*, **10**, 7, 649–659.

—— & JACKSON, M. P. A. 1992*a*. The rise of diapirs during thin-skinned extension. *Marine and Petroleum Geology*, **9**, 331–353.

—— & —— 1992*b*. The fall of diapirs during thin-skinned extension. *Marine and Petroleum Geology*, **9**, 354–371.

——, COBBOLD, P. R., DAVY, P., BRUN, J-P. & CHOUKROUNE, P. 1987. Physical models of extensional tectonics at various scales. *In*: COWARD, M. P., DEWEY, J. F. & HANCOCK, P. L. (eds) *Continental Extensional Tectonics*. Geological Society, London, Special Publications, **28**, 95–107.

VERRALL, P. 1981. *Structural Interpretation with Applications to North Sea Problems*. Course Notes No. 3, Joint Association for Petroleum Exploration Courses (UK).

WALSH, J. J. & WATTERSON, J. 1988. Analysis of the relationship between displacements and dimensions of faults. *Journal of Structural Geology*, **10**, 239–248.

WANG, T. K. & WALTHAM, D. A. 1995. 3D Gaussian beam seismic modelling. *Journal of Geophysical Prospecting*, in press.

WANG, X. & WALTHAM, D. A. 1995. The Stable Beam Ray Tracing Method. *Journal of Geophysical Prospecting*, in press.

WEIJERMARS, R. 1986. Flow behaviour and physical chemistry of bouncing putties and related polymers in view of tectonic laboratory applications. *Tectonophysics*, **124**, 325–358.

——, JACKSON, M. P .A. & VENDEVILLE, B. C. 1993. Rheological and tectonic modelling of salt provinces. *Tectonophysics*, **217**, 143–174.

WHITE, M. J. 1993. *Physical modelling and analysis of the geometries and kinematics of orthogonal and oblique rift systems*. PhD thesis, University of London.

WHITEMAN, A. 1982. *Nigeria: Its petroleum geology, resources and potential* (Vol. 2). Graham & Trotman, London.

WILCOX, R. E., HARDING, T. P. & SEELY, D. R., 1973. Basic wrench tectonics. *American Association of Petroleum Geologists Bulletin*, **57**, 74–96.

WITHJACK, M. O. & JAMISON, W. R., 1986. Deformation produced by oblique rifting. *Tectonophysics*, **126**, 99–124.

—— & PETERSEN, E. T. 1993. Prediction of normal fault geometries – A sensitivity analysis. *Bulletin American Association of Petroleum Geologists*, **77**, 1860–1873.

——, ISLAM, Q. T. & LA POINTE, P. R. 1995. Normal faults and their hangingwall deformation: An experimental study. *Bulletin American Association of Petroleum Geologists*, **79**, 1–18.

Estimating inversion: results from clay models

GLORIA EISENSTADT & MARTHA OLIVER WITHJACK
Mobil Research and Development Corporation, PO Box 650232, Dallas, TX 75265, USA

Abstract: Physical models using wet clay show that, although both fold and fault geometries change with increasing amounts of inversion, it is difficult to estimate the magnitude of inversion. During extension, a listric main normal fault and a rollover fold cut by secondary normal faults developed above two diverging, basal plates. With 50% inversion (4 cm extension, 2 cm shortening), the main normal fault underwent reverse displacement, and an asymmetric syncline formed in the pre-growth layers in the hangingwall. Secondary antithetic and synthetic normal faults passively rotated. Faults on the synclinal limb near the main normal fault rotated away from the fault, whereas those on the opposing limb rotated towards the fault. A broad anticline formed at shallow levels above the original half graben. Compressional reactivation of the main normal fault ceased between 50% and 100% inversion (4 cm extension, 4 cm shortening). Instead, low-angle thrust faults with small displacements accommodated most shortening. The shallow anticline above the original half graben expanded laterally and vertically. With 200% inversion (4 cm extension, 8 cm shortening), low-angle thrust faults with large displacements deformed the hangingwall and footwall of the main normal fault. Thrust faults cut the main normal fault and many of the secondary normal faults. The geometry of the hangingwall fold varied with depth. A broad anticline deformed the upper growth layers, changing to a syncline in the lower growth layers and an anticline in the pre-growth layers.

Estimates of inversion from null point analysis, amount of uplift above regional datum, or line length balancing, grossly underestimate the amount of shortening in the models. In addition, both the 50% and 100% inversion models lack the large-scale deformation features typically associated with compression. These results suggest that both quantitative and qualitative techniques used to calculate inversion magnitude can significantly underestimate the amount of shortening. Furthermore, if the inversion uplift is eroded or if many small-scale deformation features are not imaged seismically, then inversion may go unrecognized.

Quantifying the magnitude of shortening during inversion is a fundamental problem in basin reconstruction and modelling. The most widely cited technique for estimating inversion magnitude involves identifying a null point, which is a level of no stratigraphic offset between normal and reverse motion along a fault (De Paor & Eisenstadt 1987; Williams *et al.* 1989). This method assumes that all shortening occurs along a single normal fault. Therefore, the null point should reflect the relative importance of extension versus contraction (Mitra 1993). Other cited methods of estimating inversion magnitude involve measuring the amount of uplift above a regional datum (Stoneley 1982; Mitra 1993) or estimating shortening by line-length restoration techniques (Badley *et al.* 1989; Chadwick 1993), effectively treating inversion structures as compressional structures. Chadwick (1993) has tried to quantify the former technique by suggesting that shortening can be calculated as a function of the uplift amplitude above the regional datum and the dip of the planar, controlling fault.

In practice, most geologists make qualitative judgments to estimate inversion magnitude. These judgments are based primarily on how much the inversion geometry differs from typical extensional structures, or resembles typical compressional structures. Figure 1 shows two examples of compressional inversion structures. Cartwright (1989) suggested that the structure in the Danish Central Graben (Fig. 1a) represents mild inversion because of its similarity to typical extensional structures. The structure in Fig. 1b, however, appears to represent greater inversion because of its resemblance to a compressional structure.

In this paper we address the questions of how fault and fold geometries vary with increasing inversion and how well quantitative and qualitative methods determine the magnitude of inversion. We approached the problem experimentally with a series of inversion models of listric normal faults using wet, layered clay as the modelling material. For comparison we also performed pure extensional and contractional

From Holdsworth, R.E. & Turner, J.P. (compilers) *Extensional Tectonics: Faulting and Related Processes*. The Geological Society, Key Issues in Earth Sciences, **2 (2)**, 211–228.

First published in: "Buchanan, J.G. & Buchanan, P.G. (eds) *Basin Inversion*, The Geological Society, London, Special Publications **88**, 119–136".

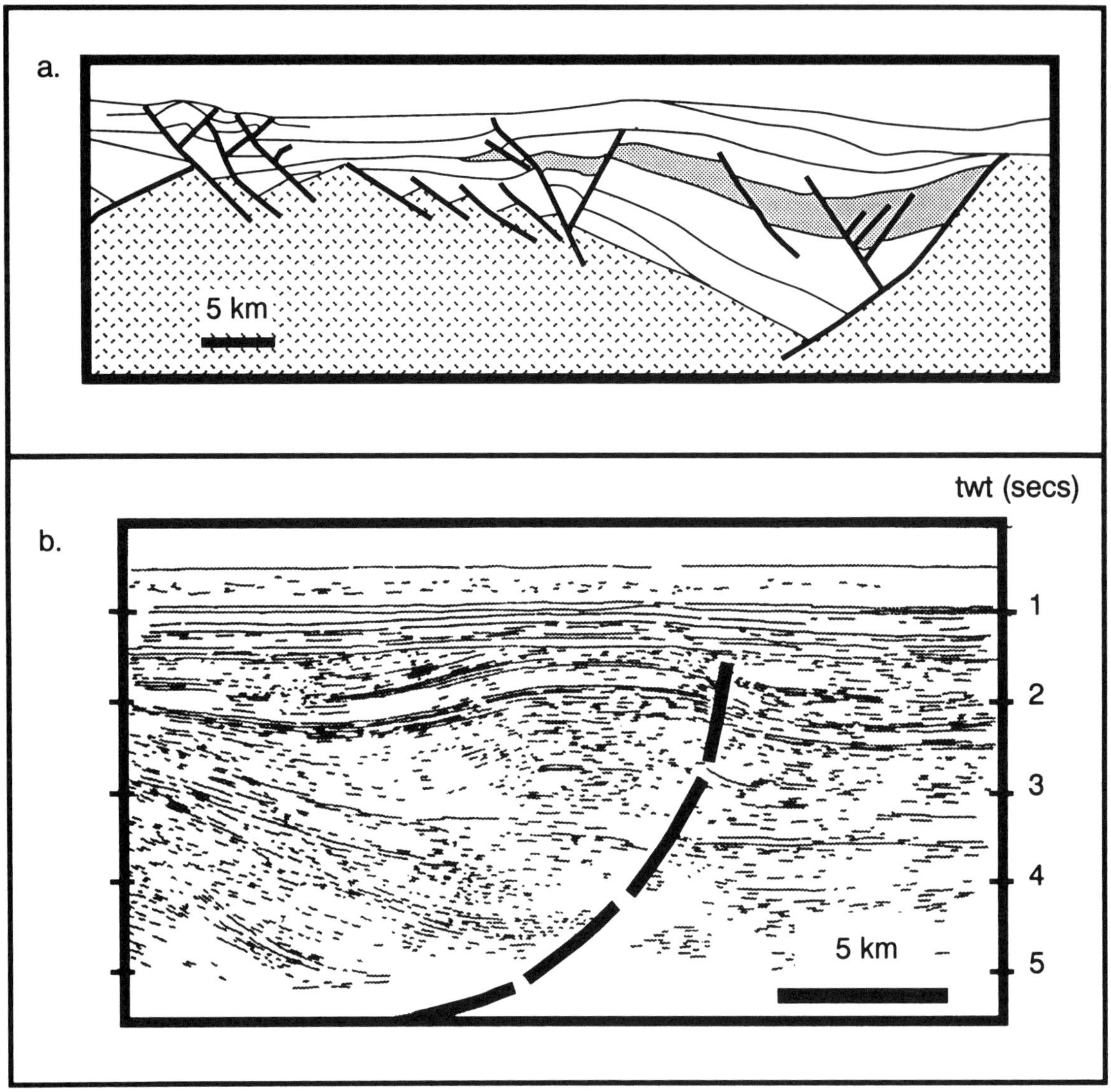

Fig. 1. Examples of inversion structures. (a) Inversion structure from Danish Central Graben, considered to be mild (after Cartwright 1989). Vertical exaggeration is approximately 1 : 1. (b) Line drawing of proprietary time-migrated seismic line showing inversion structure believed to have undergone moderate amount of inversion. Vertical exaggeration is approximately 1 : 1 at 2 seconds TWT.

experiments. Although other physical models of inversion structures above listric normal faults have been reported, these models focused on hangingwall deformation using a rigid footwall. Some of those models used sand (McClay 1989; Buchanan 1991), and other models used unlayered clay that extruded sideways during inversion, causing loss of area in cross sections (Mitra 1993). In our clay models both the hangingwall and footwall could deform, the model's side surfaces were constrained to prevent extrusion and, because the clay was layered, we were able to examine cross sections free of edge effects.

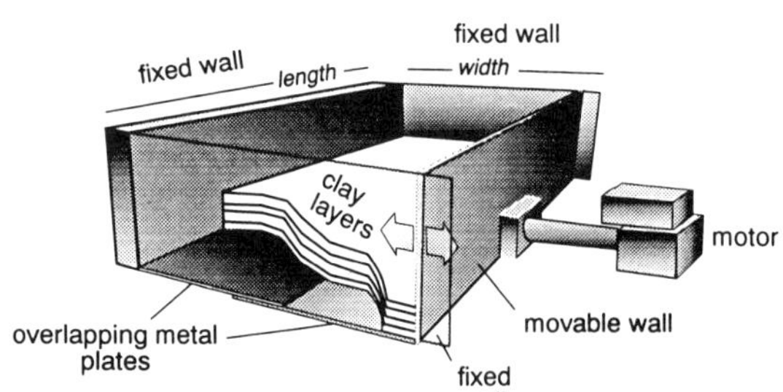

Fig. 2. Diagram of experimental apparatus showing cut-away view of clay layers above thin, overlapping metal plates. The upper metal plate attaches to a fixed wall. The bottom plate is attached to a movable wall.

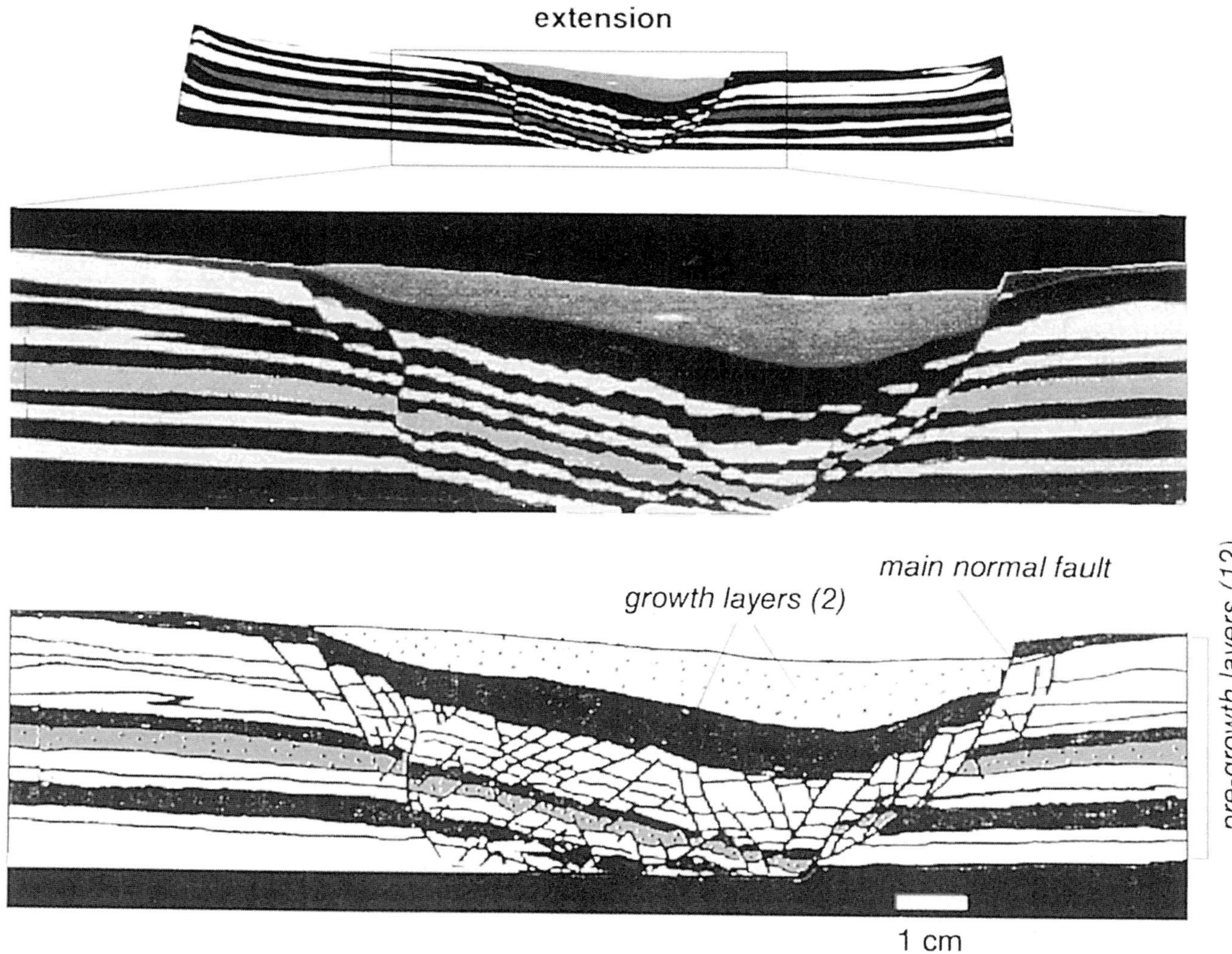

Fig. 3. Extensional model after 4 cm of displacement of the movable wall. (a) Photograph of section through model. Only central part of model is shown. (b) Interpreted line drawing of same cross-section showing main normal fault, rollover and secondary antithetic and synthetic normal faults. The sixth layer from the bottom is highlighted (purple) for comparison with the other models.

Experimental procedure

The experimental apparatus had three fixed walls, a movable wall, and two thin, overlapping basal metal plates (Fig. 2). The lower basal plate was attached to the movable wall and the upper basal plate was attached to the opposing, fixed wall. Clay, consisting of water (50% by weight), kaolin, powdered nepheline-syenite and powdered flint, covered the metal plates. Its cohesive strength was about 40 Pa (Sims 1993). The clay, initially 4 cm thick, 60 cm long and 24.5 to 32 cm wide, was composed of twelve thin, coloured layers of identical composition.

During modelling, displacement of the movable wall and the attached metal plate produced extension and/or contraction in the overlying clay. The model length remained constant while the width changed (Fig. 2). The edge of the upper plate determined the location of the main normal fault during extension and localized contractional deformation during shortening. In each model, the displacement rate of the movable wall was 4 cm/hr (0.001 cm/s). Additional coloured clay layers were added during extension to simulate deposition and record the timing of normal faulting. Each growth layer was added such that its upper surface was flat and horizontal. After each model run, the clay was left to dry for approximately one week. It was then serially sectioned, lacquered and photographed. Although the extent of edge effects (as evidenced by curved faults along the fixed walls) appear to only extend 2 to 4 cm from the walls, measurements of strain markers on the surface showed that the edge effects actually extended 8 to 10 cm from the fixed walls. Consequently only sections through the central part of the models were used for analysis.

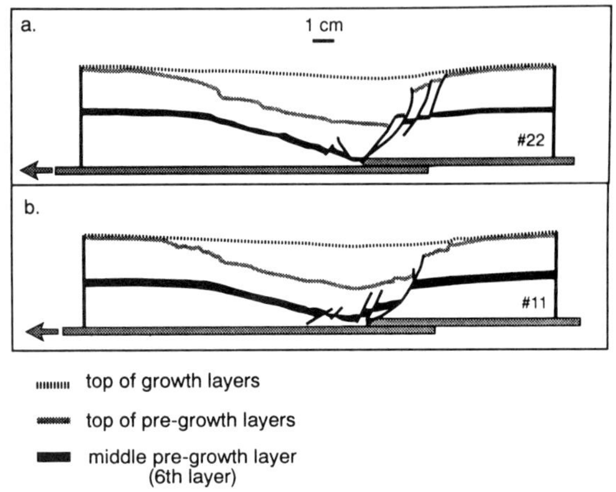

Fig. 4. Simplified line drawings of cross-sections from extensional model showing variation in fold geometry of pre-growth layers along stike. (a) Cross-section where all displacement occurred along the main normal fault, creating a rollover into the fault. (b) Cross-section where displacement occurred along several fault splays, creating a syncline in the hangingwall next to the main normal fault.

Extensional experiments

The initial width of the extensional models was 24.5 cm, and they were extended to 28.5 cm. In these models a main normal fault formed above the edge of the upper metal plate and propagated upward through the clay (Fig. 3). The main normal fault initially formed as discontinuous fault segments along strike. With greater extension, some normal faults coalesced along strike to form a through-going main normal fault. Other normal faults, bypassed by the main normal fault, became inactive. As the experiment continued, secondary synthetic and antithetic normal faults and a rollover fold formed in the hangingwall of the main normal fault. The main normal fault and most secondary normal faults dipped 60° to 70°. The geometry of the hangingwall fold varied along strike. A hangingwall rollover formed in the lower pre-growth layers where most of the displacement occurred on the main normal fault (Fig. 4a). Where displacement was accommodated amongst the main normal fault and several fault splays, a hangingwall syncline formed instead (Fig. 4b).

The extensional models resemble those published by H. Cloos (1928, 1930) and E. Cloos (1968). Unlike these published experiments, however, our models were internally layered, permitting us to analyse changes in fold and fault geometries laterally through the models. Additionally, the layers added during extension allowed us to compare structural geometries in the pre-growth and growth sections. Surface and side observation showed no apparent difference between the behaviour of the clay in layered and unlayered models, suggesting that the layering exerted no influence on the deformation.

Withjack *et al.* (1995) demonstrated that some boundary conditions have little effect on fold and fault geometries, whereas other boundary conditions profoundly affect resulting geometries. For example, they showed that extensional deformation patterns that form over overlapping basal plates are similar whether the footwall is rigid or deformable. In contrast, extensional physical models with a mylar sheet along the main fault surface have very different secondary fault patterns. One of the major differences is the absence of a crestal collapse graben in models without a mylar sheet.

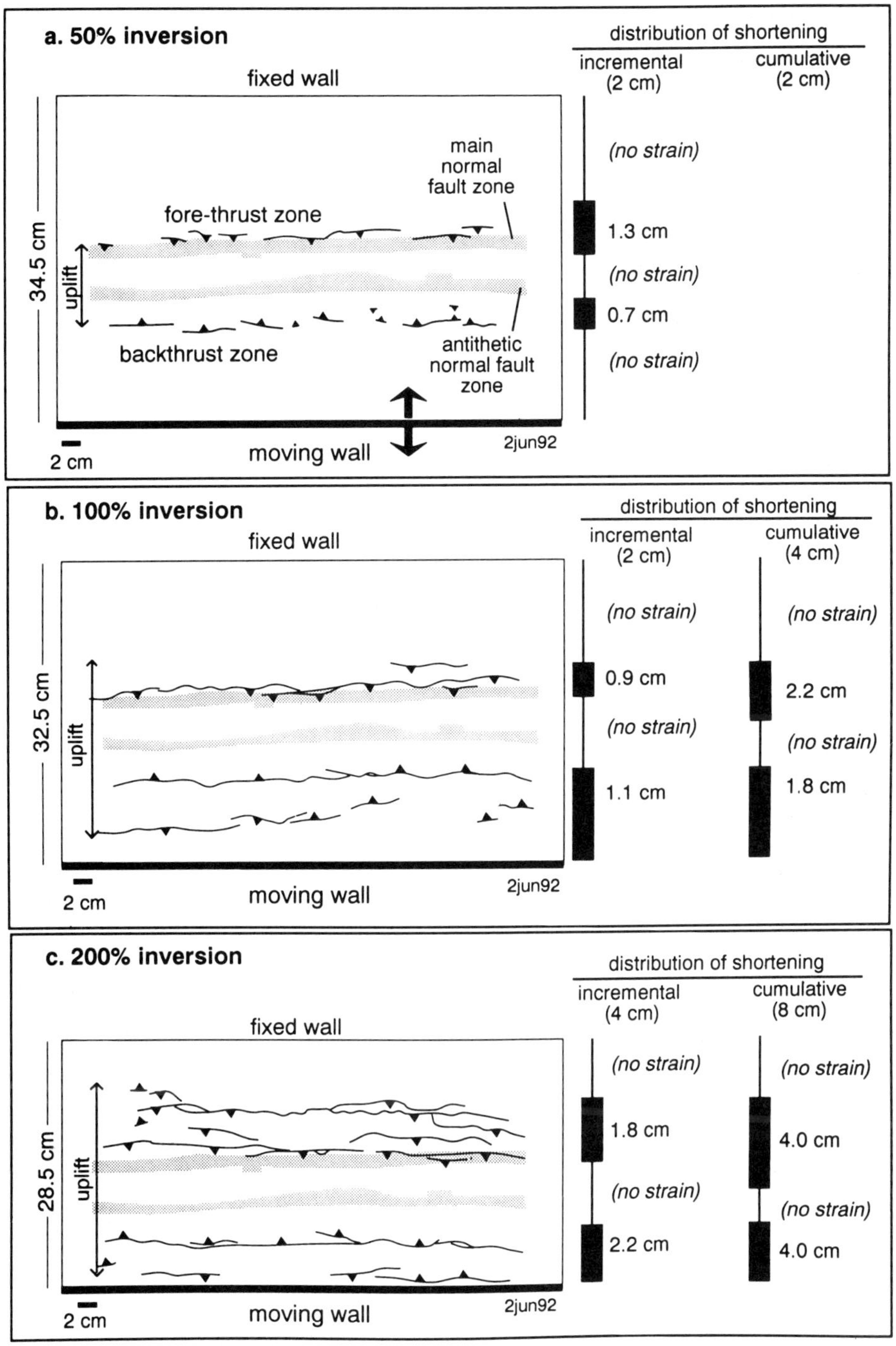

Fig. 5. Map views of progressive deformation in 200% inversion model. Original main normal fault and antithetic normal fault zones shown in gray. (a) At 50% inversion, numerous small fore- and backthrusts are visible. The fore-thrust zone overlies the old main normal fault zone and accommodates twice as much shortening as the backthrust zone. An uplift develops between the fore- and backthrust zones. (b) At 100% inversion, thrusts link along strike, new thrusts develop further away from the half graben, and the uplift grows laterally and vertically. The distribution of shortening along the fore- and backthrust zones has almost equalized. (c) At 200% inversion, new thrusts form. The proximity of the moving wall hinders the development of backthrusts.

Cloos (1968) showed that the overall geometry of extensional sand and clay models with overlapping basal plates is very similar; the different materials cause only minor variations. Our preliminary tests indicate that this conclusion is not true for inversion models; clay and

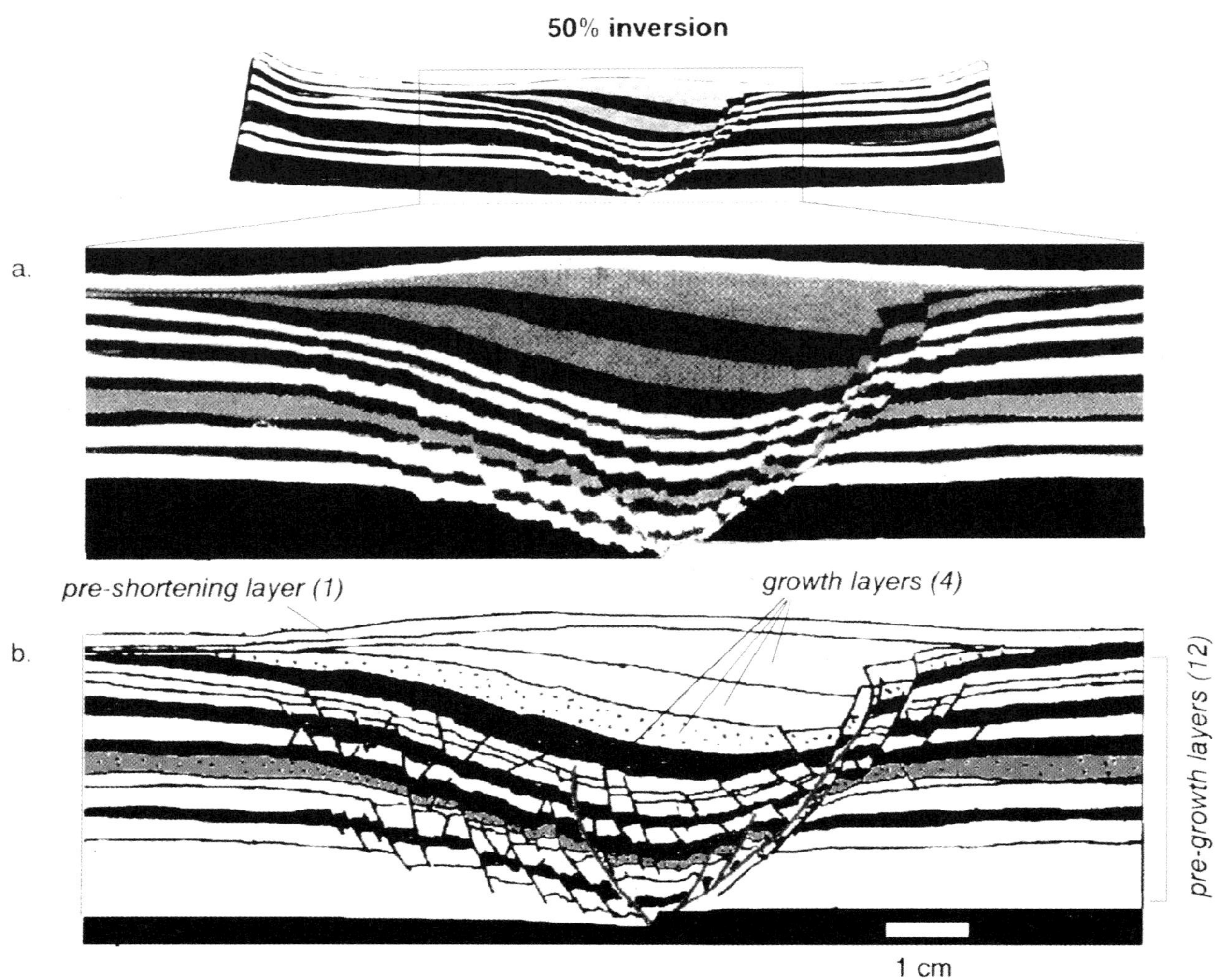

Fig. 6. (a) Photograph of cross-section through 50% inversion model. Only central part of model is shown. (b) Interpreted line drawing: black lines are old normal faults; red lines are normal faults reactivated as reverse faults. The sixth layer from the bottom is highlighted (purple) for comparison with the other models.

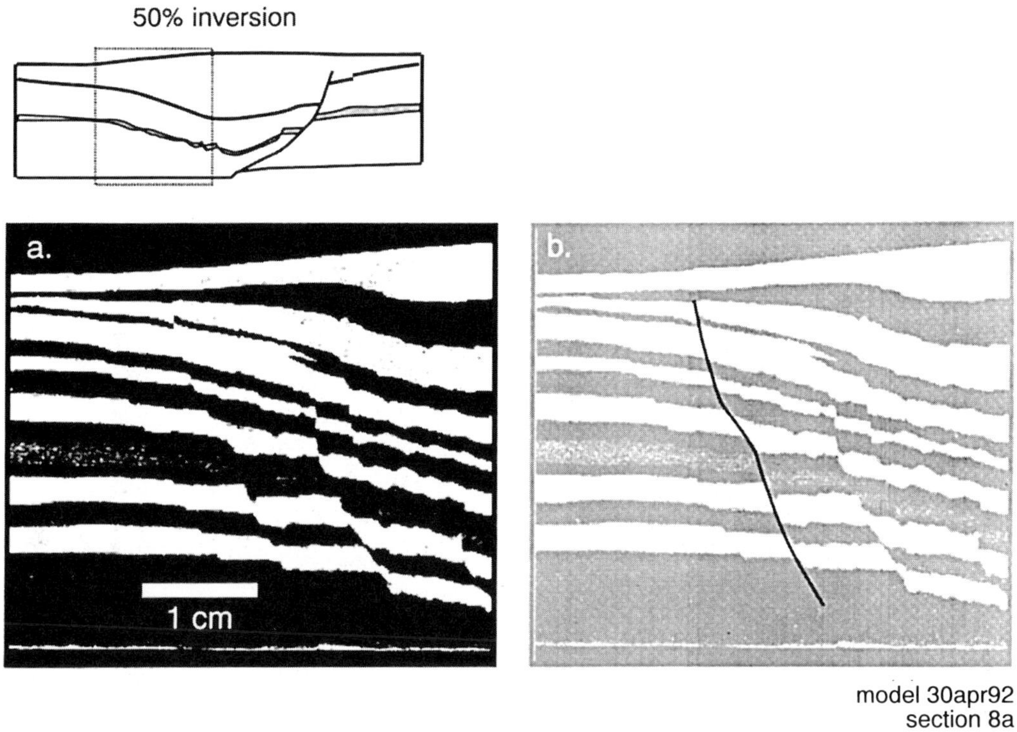

Fig. 7. (a) Close-up photograph of 50% inversion model. Sketch indicates location. (b) Same photograph highlighting old normal fault reactivated as reverse fault. Note normal displacement of lower layers and reverse displacement of upper layers.

sand inversion models formed over overlapping basal plates have very different fault geometries. Sand inversion models show only a minor amount of reactivation of older normal faults and develop thrust faults much sooner than clay models. Faults in clay seem to both reactivate more readily and to accommodate much more displacement than those in sand.

Inversion experiments

Each inversion experiment started with an extensional growth model 28.5 cm wide before shortening (except for the 200% inversion model that was 36.5 cm wide). A thin layer of clay (0.25 cm thick) covered the model after extension and before shortening. Each model was initially extended 4 cm, and the amount of shortening varied: the 50% inversion model was shortened 2 cm; the 100% inversion model was shortened 4 cm; and 200% inversion model was shortened 8 cm. The surface of the inversion models was photographed after every 0.25 cm of displacement to record fold, fault and strain patterns.

Map view

Figure 5 shows the progression of deformation in map view of the 200% inversion model. The first thrust faults were visible after 1.75 cm of displacement of the movable wall (44% inversion). At 2 cm of displacement (50% inversion), fore-thrusts and backthrusts appeared on either side of an inversion uplift (Fig. 5a). The fore-thrusts developed above the main normal fault zone of the extensional models. The backthrusts, however, were not spatially associated with the zone of antithetic normal faults. Shortening was not equally partitioned between the fore-thrust and backthrust zones, demonstrating that preferential movement occurred along the original main normal fault zone.

By 4 cm of displacement (100% inversion), the early-formed thrust faults had coalesced along strike and new fore-thrusts and backthrusts began to form away from the inversion uplift, causing the uplift to broaden (Fig. 5b). Shortening was confined to the fore- and backthrust zones and was almost equally partitioned between the two zones. By 8 cm of displacement (200% inversion), additional fore- and backthrusts developed away from the inversion uplift, causing the uplift to further broaden. At 200% inversion, shortening was equally partitioned between the fore- and backthrust zones (Fig. 5c).

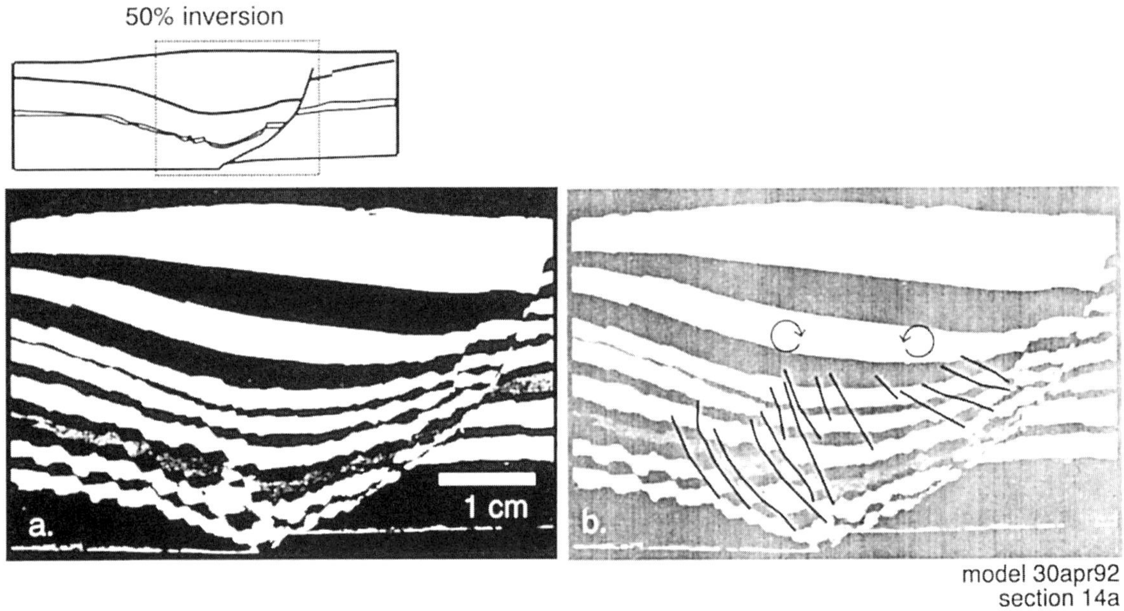

Fig. 8. (a) Close-up photograph of 50% inversion model. Sketch indicates location. (b) Same photograph highlighting rotated antithetic normal faults. Antithetic faults on the synclinal limb near the original main normal fault rotate counter-clockwise (away from the fault) and become more gently dipping, whereas antithetic faults on the opposing limb rotate clockwise (towards the fault) and become more steeply dipping as the normal fault reverses movement.

50% inversion – cross-sectional view

In cross-sectional view, the 50% inversion model lacked deformation features typically associated with compression (Fig. 6). Shortening reactivated the main normal fault and other secondary normal faults as reverse faults (compare position of highlighted layer in the extensional model (Fig. 3) and 50% inversion model (Fig. 6)). Most faults maintained a normal sense of offset, although some faults showed reverse offset in the uppermost layers (Fig. 7). Inversion produced an uplift in the upper growth and post-growth layers above the original half graben. At deeper levels, in the pre-growth layers, an asymmetric syncline formed in the hangingwall of the main normal fault. This hangingwall syncline developed in all cross sections, regardless of whether the original extensional geometry was a rollover monocline into the fault or a gentle hangingwall syncline. The development of this hangingwall syncline during inversion caused secondary normal faults on the limb near the main normal fault to rotate counter-clockwise (away from the fault). As a result the dip of synthetic faults increased and the dip of antithetic faults decreased (Fig. 8). Similarly, secondary normal faults on the opposing limb of the syncline rotated clockwise (towards the fault), causing the dip of synthetic faults to decrease and the dip of antithetic faults to increase.

Only a few thrust faults can be seen in cross-sections of the 50% inversion model, all with minute displacements. Although the model surface showed two discrete zones of thrusts (Fig. 5a), these zones are not evident in cross-sectional view. The lack of correlation between surface and cross-sectional deformation may be due to few pre-existing normal faults in the uppermost growth layers. The upper growth layers accommodated the shortening along new thrusts while the interior of the original half graben accommodated the shortening by reverse motion along old normal faults.

100% inversion – cross-sectional view

Reverse motion along the original main normal fault stopped between 50% and 100% inversion (Fig. 9), leaving most normal faults with a normal sense of offset. The hangingwall syncline in the pre-growth layers became very gentle and those layers almost resumed their pre-extensional geometry. The syncline in the growth layers became more pronounced with increased shortening. The inversion uplift expanded laterally and vertically, and pervasive low-angle thrust faults cut the hangingwall and footwall of the original main normal fault (Figs 10 & 11). These

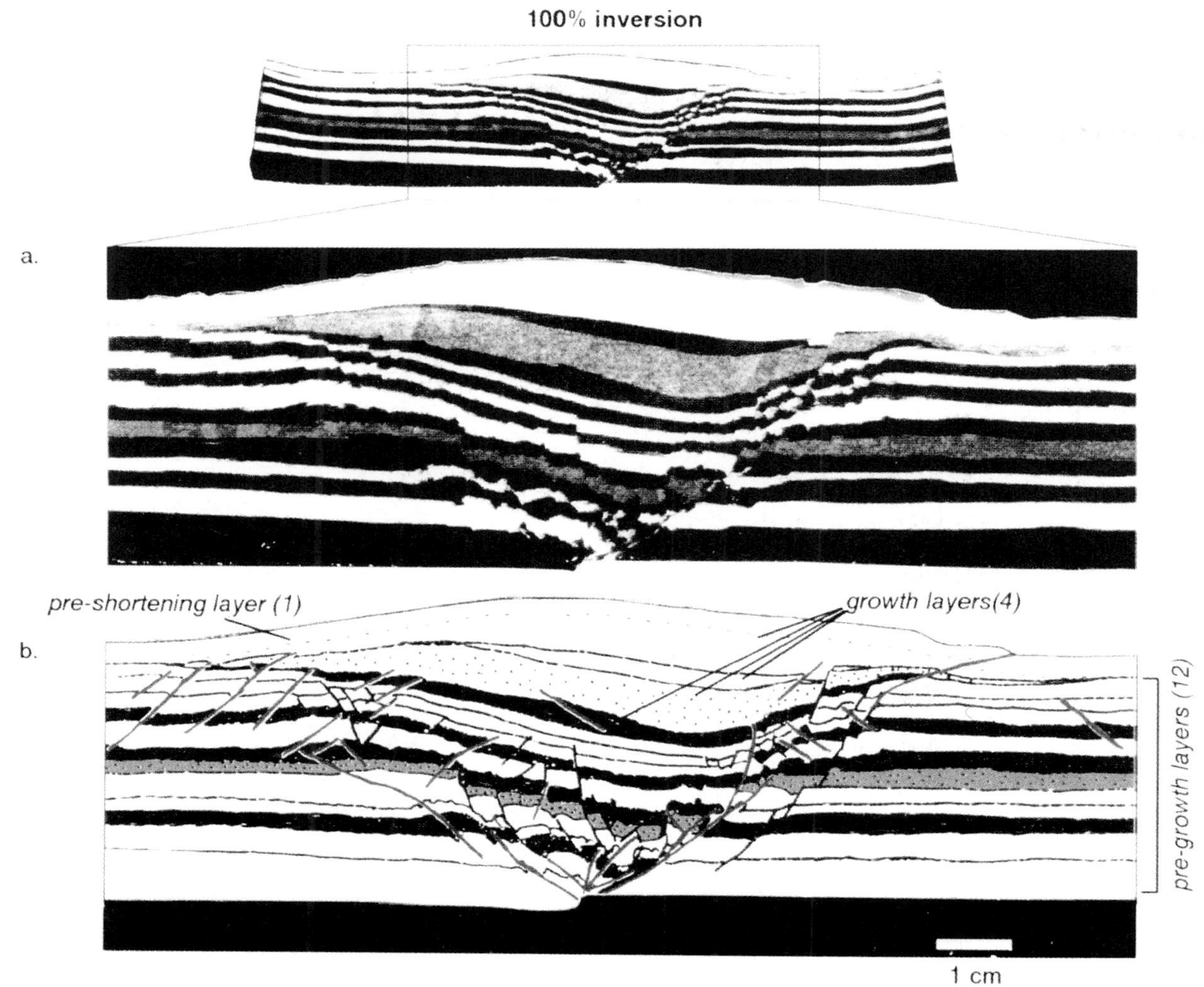

Fig. 9. (a) Photograph of cross-section through 100% inversion model. Only central part of model is shown. (b) Interpreted line drawing: black lines are old normal faults; red lines are normal faults reactivated as reverse faults and newly formed thrust faults. The sixth layer from the bottom is highlighted (purple) for comparison with the other models.

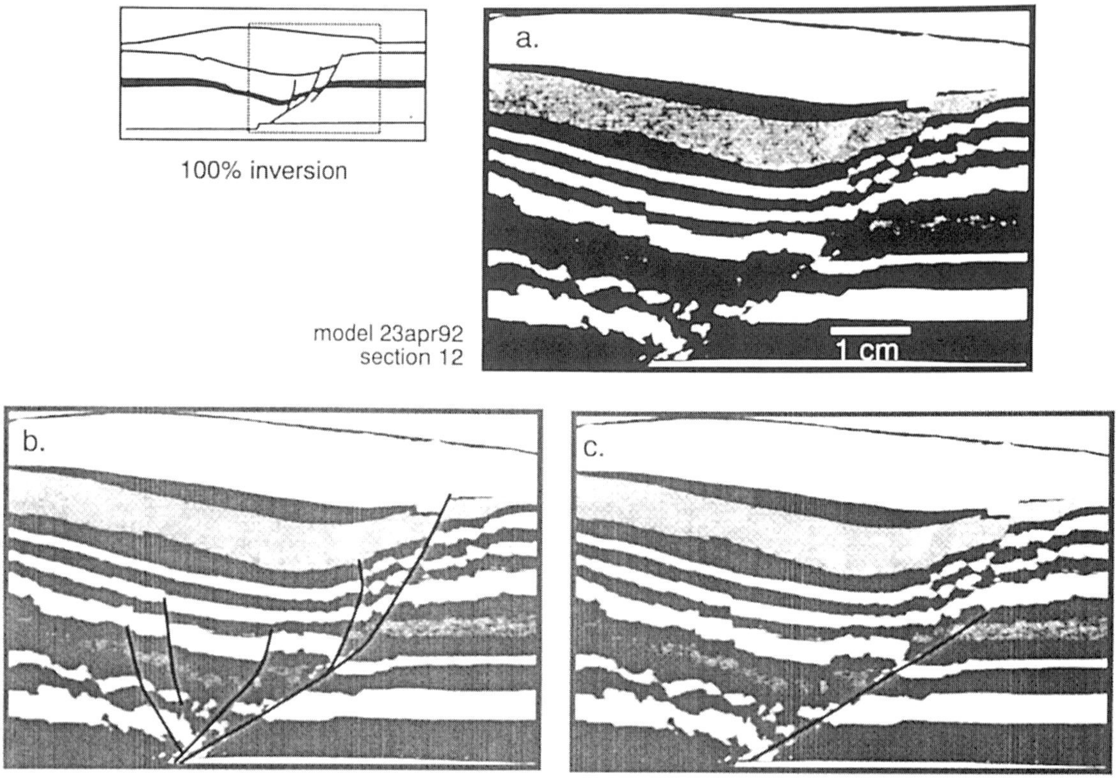

Fig. 10. (a) Close-up photograph of 100% inversion model. Sketch shows location of photograph. (b) Same photograph highlighting old normal faults, and (c) same photograph showing one of several newly formed thrust faults cutting the old main normal fault zone.

thrusts verge toward the centre of the half graben. The thrust faults mapped at the surface of the 100% inversion model cannot be traced at depth in cross-sectional view. These surface thrust faults appear to acommodate the shortening in the upper growth layers (Fig. 5b).

200% inversion – cross-sectional view

By 200% inversion, low-angle thrust faults accommodated the shortening throughout the model (Fig. 12). Several thrust faults cut the main normal fault, giving it the appearance of a folded fault with a lower overall dip of about 45° (Fig. 13). With increasing shortening, the inversion uplift expanded laterally and vertically. The fold geometry changed dramatically with depth: a broad anticline developed in the uppermost layers, the syncline in the lower growth layers became more asymmetric, and an asymmetric anticline formed in the lower pre-growth layers. The central part of the half graben was uplifted and rotated towards the main normal fault. Many unusual structures exist in the 200% inversion model, although they do not all appear on every cross section. For example, on some sections small inverted grabens in the footwall of the original main normal fault have structural characteristics of flower structures (Fig. 14), and on other sections fault-propagation folds at depth appear to link upward into normal faults (Fig. 15).

Influence of pre-existing structures

To investigate the influence of pre-existing zones of weakness on inversion geometries, we compared two models shortened the same amount, a 100% inversion model and a contractional model. The inversion model was initially 24.5 cm wide. It was then extended 4 cm, and then shortened 4 cm. The contractional model was initially 28.5 cm wide, and then shortened 4 cm. Comparisons of the 100% inversion model with the contractional model show that the pre-existing zones of weakness associated with extension profoundly affected the fault patterns during shortenings (Figs 9 & 16). Cross-sections through the contractional model show that faulting occurred in two discrete zones on either side of a symmetrical pop-up structure (Fig. 16). In the 100% inversion experiment, however, deformation was more evenly distributed throughout the model as reactivated normal faults and newly formed thrust faults (Fig. 9). The strain distribution in map view also differed

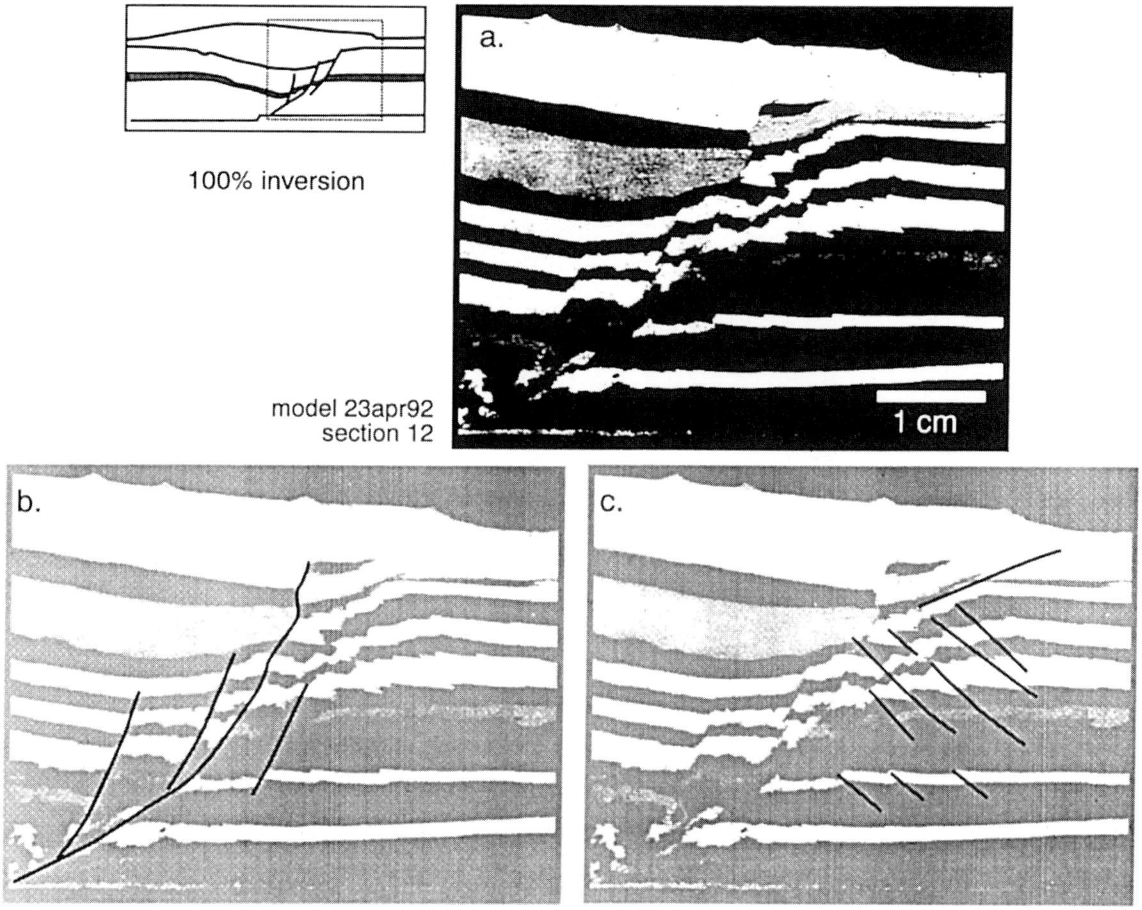

Fig. 11. (a) Close-up photograph of 100% inversion model. Sketch shows location of photograph. (b) Same photograph highlighting old normal faults, and (c) same photograph showing a few of the newly formed thrust faults deforming the footwall.

markedly in the inversion and contractional models. In the 100% inversion model, shortening occurred unequally in two discrete zones (Fig. 5b), with most initial shortening occurring over the original main normal fault. In the contractional model, strain was initially evenly distributed throughout the model (Fig. 17a). With increasing shortening the strain then became partitioned equally into two discrete zones (Fig. 17b). It is clear from a comparison of these maps that the original extensional faults control the early distribution of shortening in the inversion models.

Discussion of modelling results

The layered clay models show that fold geometries change with increasing amounts of inversion (Fig. 18). In the extensional models, either a rollover or a syncline forms in the lower pre-growth layers in the hangingwall of the main normal fault (Fig. 18a). With up to 50% inversion, the main normal fault experiences reverse movement, and an asymmetric syncline develops in the pre-growth layers in the hangingwall, regardless of the original extensional geometry (Fig. 18b). A broad anticline forms at shallow levels above the original half graben. With 50% inversion, fold geometries generally resemble those associated with extension rather than compression (compare Figs 4b & 18b). Compressional reactivation of the main normal fault ceases between 50% and 100% inversion. The anticline above the original half graben becomes higher and broader, whereas the hangingwall syncline in the pre-growth layers becomes very gentle (Fig. 18c). At 200% inversion, the geometry of the hangingwall fold varies with depth (Fig. 18d). An asymmetric anticline exists in the lower pre-growth layers, a syncline affects the lower growth layers, and a broad anticline deforms the uppermost growth and post-growth layers.

The models also show that fault geometries

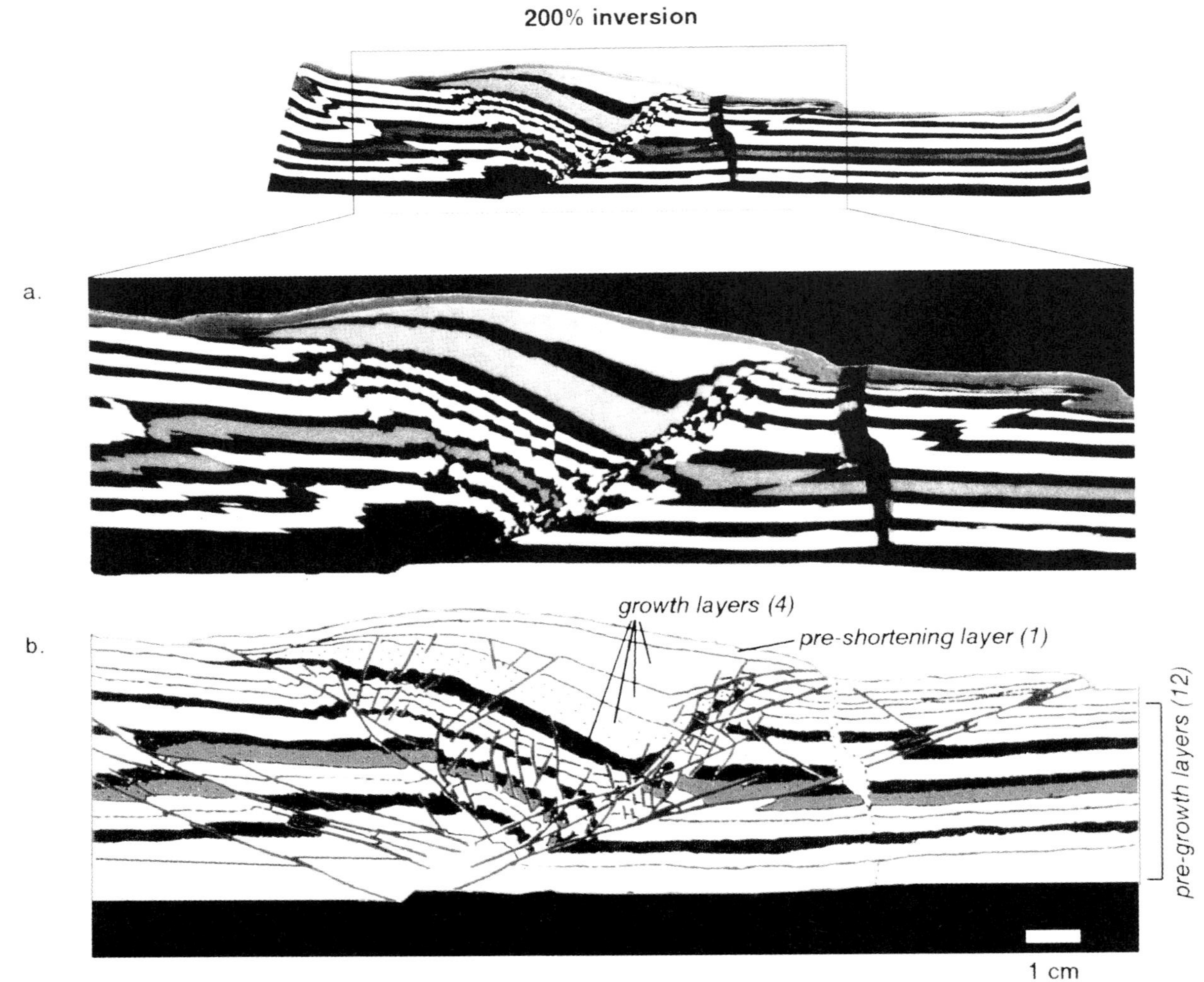

Fig. 12. (a) Photograph of cross-section through 200% inversion model. Only central part of model is shown. (b) Interpreted line drawing: black lines are old normal faults; red lines are normal faults reactivated as reverse faults and newly formed thrust faults. The sixth layer from the bottom is highlighted (purple) for comparison with the other models.

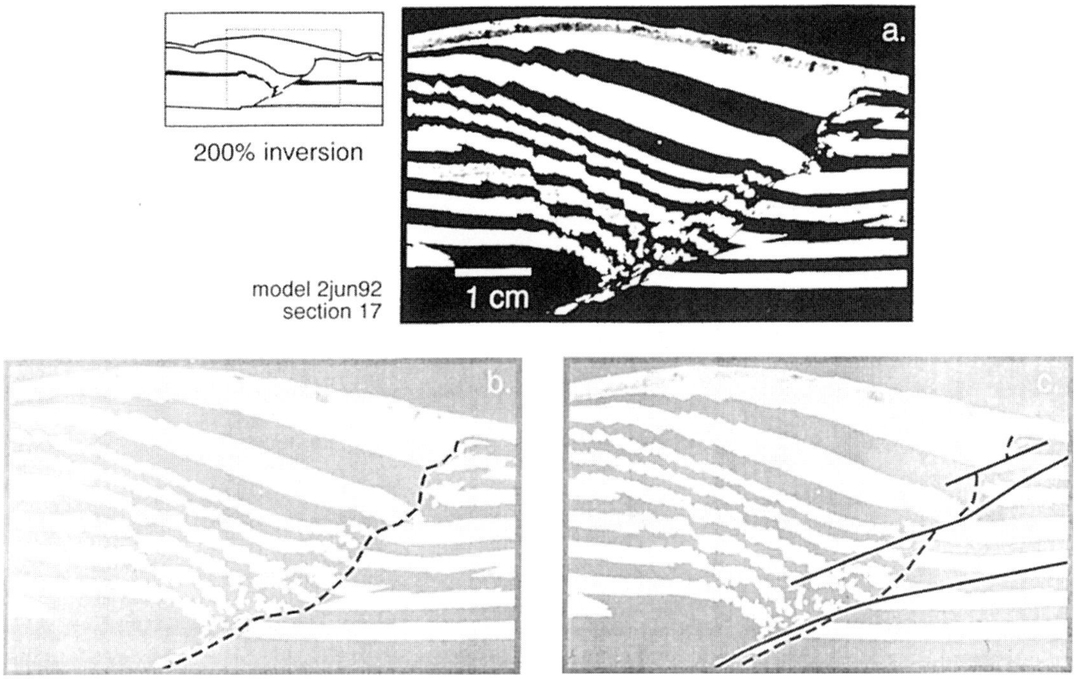

Fig. 13. (a) Close-up photograph of 200% inversion model. Sketch shows location of photograph. (b) Same photograph showing erroneous interpretation of main normal fault being folded or having ramp/flat geometry. (c) Same photograph showing correct interpretation with old main normal fault being cut by several low-angle thrusts.

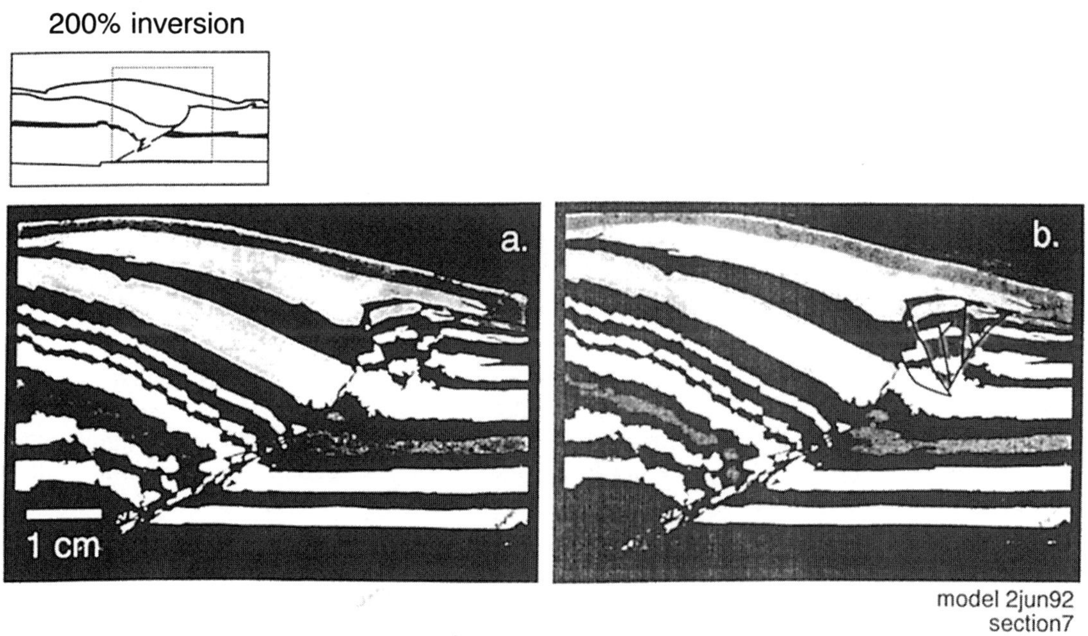

Fig. 14. (a) Close-up photograph of 200% inversion model. Sketch shows location of photograph. (b) Same photograph highlighting inverted graben in the footwall resembling a flower structure.

change with increasing amounts of inversion (Fig. 19). A main normal fault and secondary normal faults, having dips of 60° to 70°, develop in the extensional models. With 50% inversion, secondary antithetic and synthetic normal faults passively rotate as the hangingwall reverses displacement along the main normal fault and forms a syncline in the pre-growth layers (Fig. 19b). Faults on the synclinal limb near the main normal fault rotate counter-clockwise (away

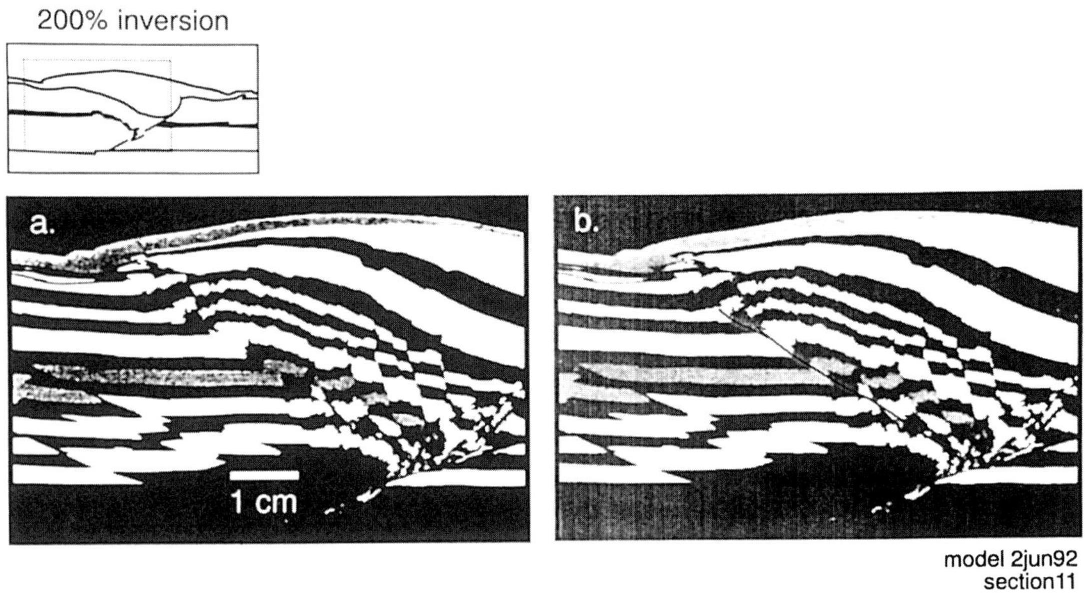

Fig. 15. (a) Close-up photograph of 200% inversion model. Sketch shows location. (b) Same photograph showing newly formed fault-propagation fold (solid line) linking upsection to older normal fault (dashed line).

from the fault), whereas those on the opposing limb rotate clockwise (towards the fault). This rotation produces anomalous high and low fault dips. By 100% inversion, reverse motion has ceased along the main normal fault. Instead, low-angle thrust faults with small displacements accommodate most of the shortening and deform both the hangingwall and footwall (Fig. 19c). These thrusts faults form on both sides of the original half graben, far from the end walls, and have a vergence towards the half graben. With 200% inversion, the hangingwall and footwall of the main normal fault are cut by low-angle thrust faults with large displacements, these faults have a vergence away from the half graben. These later thrust faults have the same dip and vergence as the thrusts formed in the pure shortening model. Some thrust faults cut the main normal fault, causing it to appear folded and more gently dipping (Fig. 19d).

Techniques for determining inversion magnitude (i.e. null point analysis, line-length restoration and uplift calculations) underestimate the amount of shortening in these inversion models. Null-point analysis along the main normal fault suggested that the amount of extension was greater than the amount of shortening in the three inversion models. This prediction is correct for the 50% inversion model, but erroneous for the 100% and 200% inversion models. Null-point analysis fails because shortening during inversion is accommodated by newly formed thrust faults and reactivated secondary normal faults that are distributed throughout the model, not just by displacement along the reactivated main normal fault. In cases where the main normal fault has been cut by low angle thrust faults (200% inversion), a null point analysis becomes meaningless.

Line-length restoration assumes that layer thickness remains constant during deformation and all the shortening occurs along visible thrust faults or folds. Neither assumption is valid for these inversion models and therefore the technique grossly underestimates the amount of shortening. The calculated amount of shortening for the upper layers in the 50%, 100% and 200% inversion models is less than 0.1 cm, 0.2 cm, and 1.6 cm, respectively. The actual amounts are 2 cm, 4 cm and 8 cm respectively. The difference between the measured and actual amount of shortening suggests that much of the deformation is occurring at a scale too small for observation.

Measuring the area of uplift provided the best estimate, although still too low, of the amount of inversion. The technique assumes that the area of uplift above the regional datum divided by the detachment depth is equivalent to the amount of shortening associated with inversion. This method suggests that the amount of shortening in the 50%, 100% and 200% inversion models was about 0.6 cm, 1.6 cm and 4.8 cm respectively. Although area is preserved in the cross-sections during inversion, uplift calculations fail to accurately predict the amounts of shortening in the physical models because both the footwall and hangingwall are elevated during inversion.

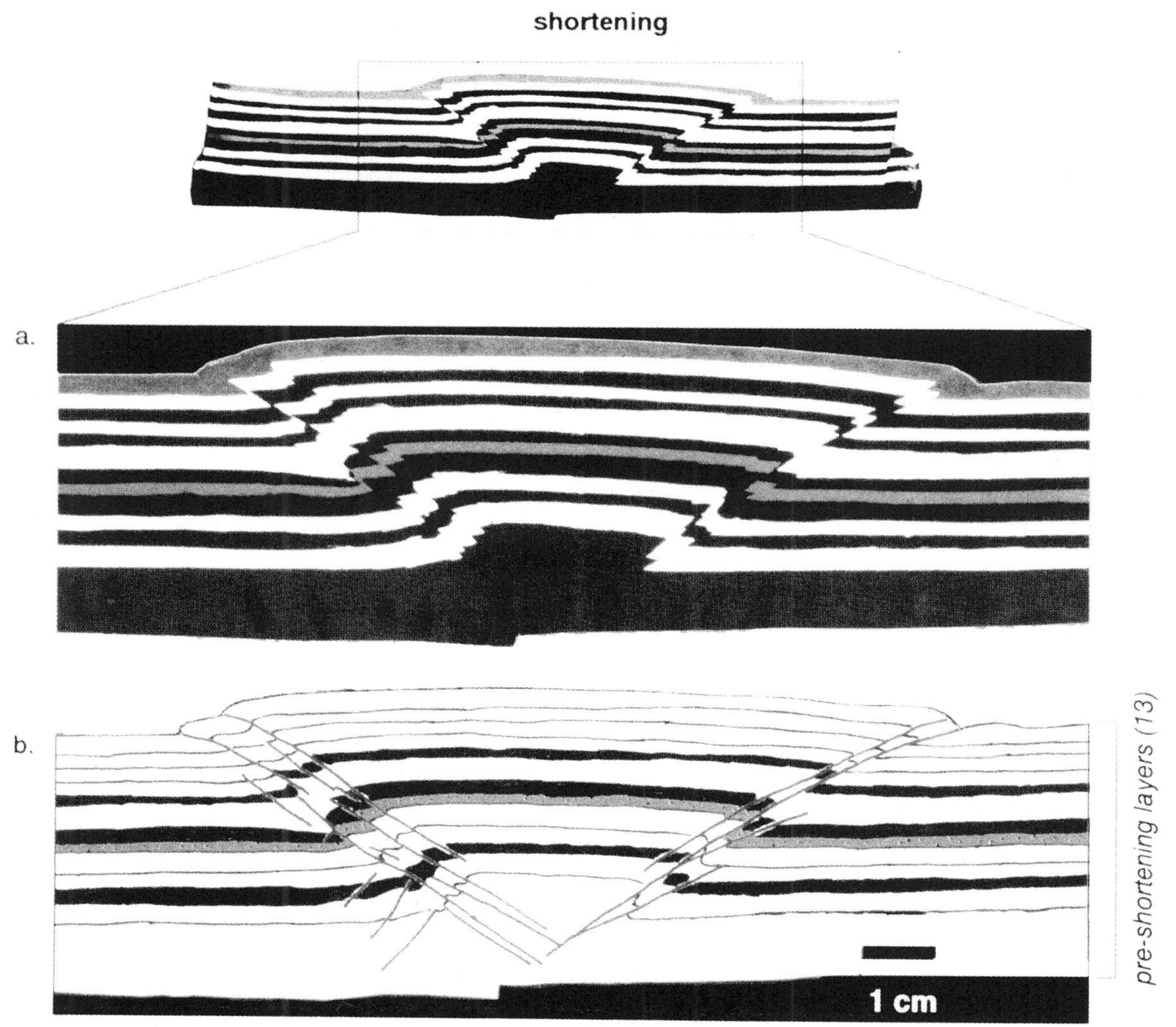

Fig. 16. Contractional model after 4 cm of displacement of the movable wall. (a) Photograph of centre of cross-section through model. (b) Interpreted line drawing of same cross-section showing thrust faults (red). The sixth layer from the bottom is highlighted (purple) for comparison with the other models.

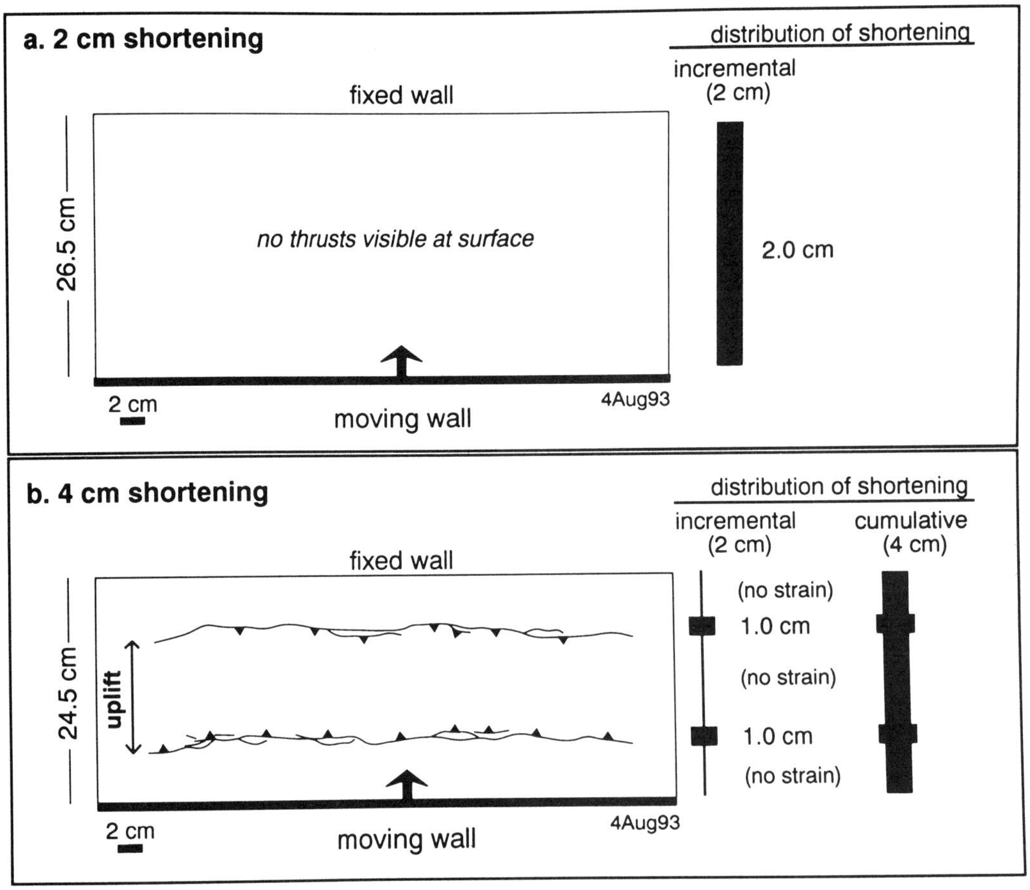

Fig. 17. Map view of shortening model showing thrusts, uplift and distribution of shortening. (a) After 2 cm of displacement of the movable wall, no thrusts are visible. Shortening is distributed evenly throughout the model. (b) After 4 cm of displacement, fore- and backthrusts bound a well-developed uplift. Shortening is distributed equally along the two fault zones. Continued shortening causes new thrusts to form further away from the centre of the model, increasing both the height and width of the uplift.

Thus, the interpreted regional datum is too high and the measured amount of uplift above the regional datum is too low. In real examples there is also the danger that a seemingly undeformed regional datum has been uplifted during inversion.

Qualitative judgements based on how much the geometry of an inversion structure resembles either an extensional or compressional structure would also underestimate the magnitude of inversion. For example, the structural features in the 50% and 100% inversion models more closely resemble those in the extensional model than those in the contractional model. Consequently subtle inversion structures, like the feature in the Danish Central Graben (Fig. 1a), might actually have undergone significant amounts of shortening. Erosion of the uplift associated with mild inversion would leave only structures with extensional geometries.

Conclusions

The clay models show that although quantitative techniques fail to estimate correctly amounts of inversion, there are systematic changes in fold and fault patterns with increasing magnitudes of inversion. These changes in structural geometries, along with the recognition of small-scale deformation, may help define the magnitude of inversion in real examples. Initially shortening during inversion is accommodated along pre-existing normal faults and small-scale thrust faults. For less than 100% inversion, the main normal fault experiences reverse displacement, and an asymmetric hangingwall syncline forms in the pre-growth layers. Secondary antithetic and synthetic normal faults passively rotate. Faults on the synclinal limb near the main normal fault rotate away from the main normal fault, whereas those on the opposing limb rotate

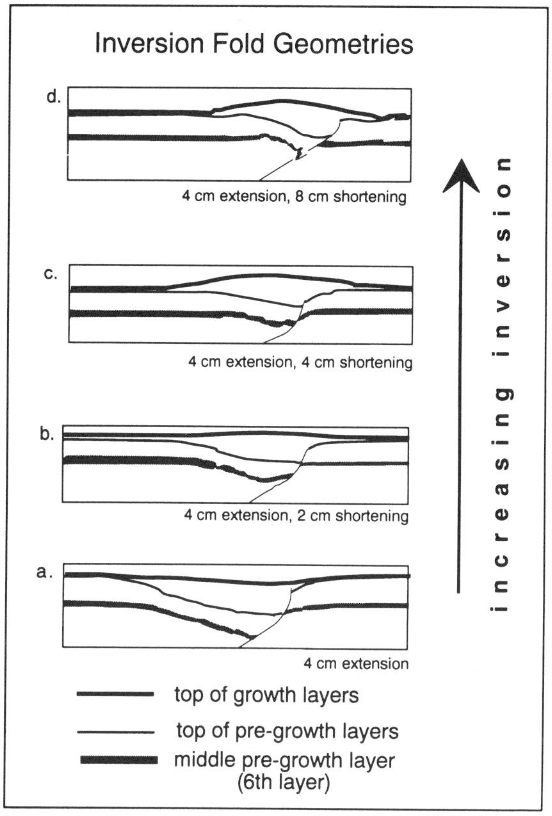

Fig. 18. Diagram summarizing changes in fold geometries with increasing inversion. (a) A rollover into the main normal fault is developed in the pre-growth layers during extension.
(b) Characteristics of the 50% inversion model include reactivation of the main normal fault as a reverse fault creating a hangingwall syncline in pre-growth layers, and the development of a broad surface uplift. (c) Characteristics of the 100% inversion model include continued reactivation of the main normal fault as a reverse fault, continued development of the surface uplift and flattening of pre-growth layers. (d) Characteristics of the 200% inversion model include continued development of the surface uplift, development of a hangingwall anticline in the pre-growth layers, and development of a hangingwall syncline in the growth layers.

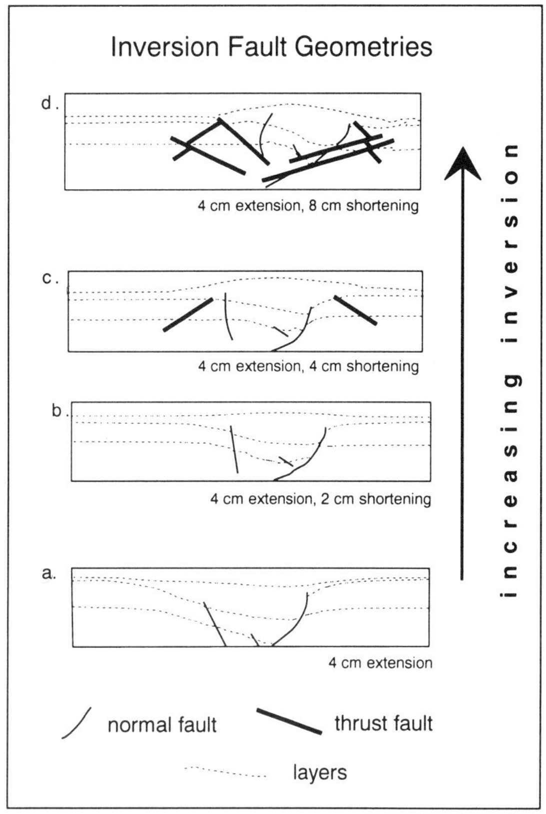

Fig. 19. Diagram summarizing changes in fault geometries with increasing inversion. (a) Extensional antithetic and synthetic faults have dips of 60° to 70°.
(b) Characteristics of 50% inversion include reactivation of main normal fault as a reverse fault, causing rotation of secondary normal faults with anomalous high and low fault dips. (c) Characteristics of 100% inversion include continued rotation of secondary normal faults and pervasive development of fore- and backthrusts (verging towards centre of model) in both hangingwall and footwall.
(d) Characteristics of 200% inversion include offset of main normal fault by low-angle thrusts (causing it to appear more gently dipping), continued uplift and rotation of secondary normal faults, and new fore- and backthrusts verging away from the centre of the model.

towards the fault. A broad anticline forms at shallow levels above the original half graben. Compressional reactivation of the main normal fault ceases between 50% and 100% inversion, and low-angle thrust faults with small displacements accommodate most shortening. The broad anticline above the original half graben expands laterally and grows vertically. With 200% inversion, the hangingwall and footwall of the main normal fault are cut by low-angle thrust faults with large displacements. This causes the main normal fault to appear to be more gently dipping with a ramp/flat or folded trajectory. Thrust faults cut the main normal fault and many of the secondary normal faults. The geometry of the hangingwall fold varies with depth. A tight anticline in the pre-growth layers changes to a syncline in the lower growth layers and a broad anticline in the upper growth and post-growth layers.

The clay models suggest that both quantitative and qualitative methods would consistently underestimate inversion magnitude. If the footwall and hangingwall are elevated during inversion, then techniques using the amount of uplift above the regional datum would underestimate inversion magnitude. Similarly, if pervasive small-scale thrust faults and reactivated normal faults accommodate much of the shortening, then techniques using line-length restoration or null-point analysis would also underestimate the amount of shortening associated with inversion. A broad, gentle anticline and small-scale thrust faults are the only evidence of shortening in both the 50% and 100% inversion models. Amounts of inversion for either model would be underestimated because the large-scale faults and folds in those models are more typically associated with extension rather than compression.

We thank Kris Meisling, Bruno Vendeville and Chris Banks for their thoughtful and critical reviews. Peter Hennings, Eric Peterson and Joana Vizgirda provided helpful discussions, and Charlie Wall assisted with the physical modelling. We also thank Mobil Research and Development Corporation and their management for support of this research and permission to publish.

References

BADLEY, M. E., PRICE, J. D. & BACKSHALL, L. C. 1989. Inversion, reactivated faults and related structures: seismic examples from the southern North Sea. *In*: COOPER, M. A. & WILLIAMS, G. D. (eds) *Inversion Tectonics*. Geological Society, London, Special Publication, **44**, 201–219.

BUCHANAN, P. 1991. *Geometries and Kinematic Analysis of Inversion Tectonics from Analogue Model Studies*. PhD thesis, Royal Holloway and Bedford College, University of London.

CARTWRIGHT, J. A. 1989. The kinematics of inversion in the Danish Central Graben. *In*: COOPER, M. A. & WILLIAMS, G. D. (eds) *Inversion Tectonics*, Geological Society, London, Special Publication, **44**, 153–175.

CHADWICK, R. A. 1993. Aspects of basin inversion in southern Britain. *Journal of the Geological Society,* **150**, 311–322.

CLOOS, E. 1968. Experimental analysis of Gulf Coast fracture patterns. *American Association of Petroleum Geologists Bulletin,* **52**, 420–444.

CLOOS, H. 1928. Experimente zur inneren Tektonik. *Centralblatt für Mineralogies,* **Abt B**, 609–621.

—— 1930. Kunstliche Gebirge, II. *Natur und Museum,* **60**, 258–269.

DE PAOR, D. G. & EISENSTADT, G. 1987. Stratigraphic and structural consequences of fault reversal: An example from the Franklinian Basin, Ellesmere Island. *Geology,* **15**, 948–949.

MCCLAY, K. R. 1989. Analogue models of inversion tectonics. *In*: COOPER, M. A. & WILLIAMS, G. D. (eds) *Inversion Tectonics*. Geological Society, London, Special Publication, **44**, 41–59.

MITRA, S. 1993. Geometry and kinematic evolution of inversion structures. *American Association of Petroleum Geologists Bulletin,* **77**, 1159–1191.

SIMS, D. 1993. The rheology of clay: a modeling material for geologic structures. *Eos,* Transactions, American Geophysical Union, 569.

STONELEY, R. 1982. The structural development of the Wessex Basin. *Journal of the Geological Society,* **139**, 543–554.

WILLIAMS, G. D., POWELL, C. M. & COOPER, M. A. 1989. Geometry and kinematics of inversion tectonics. *In*: COOPER, M. A. & WILLIAMS, G. D. (eds) *Inversion Tectonics,* Geological Society, London, Special Publication, **44**, 3–15.

WITHJACK, M. O., ISLAM, Q. & LA POINTE, P. 1995. Normal faults and their hangingwall deformation: an experimental study. *American Association of Petroleum Geologists Bulletin*, **79**, 1–18.

Intra-reservoir faulting and fluid flow

Fault sealing processes in siliciclastic sediments

Q. J. FISHER & R. J. KNIPE

Rock Deformation Research Group, Department of Earth Sciences, University of Leeds, Leeds LS2 9JT, UK

Abstract: The microstructure and petrophysical properties of fault rocks from siliciclastic hydrocarbon reservoirs of the North Sea are closely related to the effective stress, temperature and sediment composition at the time of deformation, as well as their post-deformation stress and temperature history. Low permeability fault rocks may develop due to a combination of processes including: the deformation induced mixing of heterogeneously distributed fine-grained material (principally clays) with framework grains, pressure solution, cataclasis, clay smear, and cementation. Fault rocks can be classified into various types (disaggregation zones, phyllosilicate-framework fault rocks, cataclasites, clay smears, and cemented faults/fractures) based upon their clay and cement content as well as the amount of cataclasis experienced. In the absence of extensive cementation, the distribution of fault rock types along a fault plane can often be predicted from a detailed knowledge of the reservoir sedimentology. The permeability of fault rocks can vary by over six orders of magnitude, depending on the extent to which the porosity reduction processes have operated. Utilizing the strong link between the petrophysical properties of fault rocks and their geohistory allows the risks associated with fault seal evaluation to be reduced.

Faults are a major heterogeneity in sedimentary sequences and play an important role in the entrapment and movement of hydrocarbons on both geological and production time-scales (Smith 1966, 1980; Berg 1975; Watts 1987). Barriers or restrictions to fluid flow may be formed by the juxtaposition of reservoir rocks against impermeable horizons (juxtaposition seal) or by the development of a fault rock with a high sealing capacity (fault seal *sensu stricto*). Juxtaposition seals are one of the most widely recognized hydrocarbon traps and are frequently targeted during exploration. The importance of intra-reservoir faults (which maintain reservoir/reservoir juxtapositions) is less well recognized and often ignored. An understanding of their sealing capacity is, however, important to both assess the likely distribution of hydrocarbons in a particular area and to plan an optimum production strategy. It is therefore surprising how little published work is available on the key factors which control the types and petrophysical properties (porosity, permeability, capillary pressure) of fault rocks within siliciclastic hydrocarbon reservoirs. Notable exceptions have discussed clay smears (Knipe 1992; Gibson 1994; Berg & Avery 1995; Yielding *et al.* 1997), clay gouges (Morrow *et al.* 1984) and cataclasites (Aydin 1978; Pittman 1981; Underhill & Woodcock 1987; Knipe 1992; Antonellini & Aydin 1994). Deformation features present within impure sandstones have only occasionally been studied (e.g. Sverdrup & Bjørlykke 1992).

The scarcity of published data in the petrophysical properties of fault rocks has led to the situation where fault sealing is often treated in an empirical manner. Beginning with the technical vocabulary, faults are usually either classified as sealing or non-sealing. The term 'sealing' is used to cover a range of faults from those that act as minor production baffles to those which have supported large hydrocarbon columns over geological time. This classification of fault sealing is over-simplistic, and neglects previous work which has shown that faults may act as membrane seals with finite sealing capacities (Watts 1987).

The lack of published data on the types and petrophysical properties of fault rocks limits the extent to which fluid distribution patterns and production data within established hydrocarbon fields may be interpreted. For example, a knowledge of the petrophysical properties of the various fault rock types is needed if production data are to be interpreted to assess which fault rock types control sealing capacity. Finally, an understanding of the fundamental processes which control the petrophysical properties of faults is necessary in order to extrapolate results gained from studying the sealing behaviour of a single fault, to other faults within a field and then to other fields.

The purpose of this paper is to describe the microstructural characteristics of fault rocks within siliciclastic sedimentary rocks and to outline the fundamental factors controlling their

From Holdsworth, R.E. & Turner, J.P. (compilers) *Extensional Tectonics: Faulting and Related Processes*. The Geological Society, Key Issues in Earth Sciences, **2 (2)**, 231–248.
1476-315X/02/$15.00
First published in: "Jones, G., Fisher, Q.J. & Knipe, R.J. (eds) *Faulting, Fault Sealing and Fluid Flow in Hydrocarbon Reservoirs*, The Geological Society, London, Special Publications **147**, 117–134".

petrophysical properties. We show that a wide range of processes may reduce the permeability and increase the capillary pressure of fault rocks. The extent to which these operate is shown to vary significantly depending upon the sediment composition, as well as the stress and temperature histories experienced pre-, syn- and post-deformation. Understanding these processes and relationships will provide a valuable tool for the prediction of fault seal capacity.

The paper uses microstructural and physical property analyses obtained, over the last five years, from faults and fractures present within cores from 38 hydrocarbon reservoirs from the North Sea. To supplement this database, some results obtained during the study of an oil field in Australasia have also been included. The fields studied are diverse both in terms of their reservoir rocks and burial and deformation histories experienced. This paper integrates these observations with published data on sandstone diagenesis and rock deformation experiments.

Methods

Laboratory based analysis was aimed at characterizing the microstructural and petrophysical properties inside and outside of faults and fractures sampled, identifying the deformation mechanisms and sealing processes experienced, and constraining the timing of deformation relative to the diagenetic history.

Microstructural examination

The microstructure of samples were examined by secondary electron (SE), back-scattered electron (BSE) and cathode luminescence (CL) imagery using the same instrumentation as outlined in Peacock *et al.* (this volume). The modal composition of deformed and undeformed sediments was determined by image analysis of BSE and Cl micrographs as well as X-ray dot maps, as outlined in Fisher *et al.* (in press). Point counting was used to quantify mineral phases which could not be differentiated using image analysis techniques.

Pore aperture size measurement and threshold pressure determinations

The pore aperture size distributions of samples was measured using mercury porosimetry. The Micromeritics Autopore II 9220 porosimeter used injects mercury into evacuated samples in pressure increments up to 60 000 psi. After reaching equilibrium at each pressure increment, the volume of mercury intruded was recorded. The pressure was plotted against the mercury saturation to produce an injection curve from which the threshold pressure, and hence capillary pressure, was determined (Katz & Thompson 1986, 1987).

Ideally, separate plugs from inside and outside the fault zone should be used. Most faults examined were, however, too narrow to cut a separate sample. In such cases, a plug was cored with the fault zone cutting horizontally across its centre. The sample was sealed with epoxy resin on all sides except its base. The mercury injection curves from these sealed samples have two threshold pressure indications. The first threshold is usually characteristically low and represents the host sediment. The second threshold represents the pressure at which the fault is breached.

Water permeability

Traditional methods of gas permeametry, such as the Hassler sleeve, are unable to measure permeabilities below about 0.01 mD, and the results below 1 mD are probably not very accurate because of lateral sealing problems and uncertainties in the Klinkenberg correction. Many fault rocks have far lower permeabilities; hydraulic conductivity tests were therefore performed using a custom-designed flow-pump permeameter (Olson & Daniel 1981; Olsen *et al.* 1991). The computerized permeameter, which uses an accurate constant-rate-of-flow pump, and a highly linear differential pressure transducer, is capable of measuring permeabilities in the range of a few darcies to a few hundredths of a microdarcy.

Permeability measurements were made on 27 mm diameter plugs or $\sim$ 25 mm $\times$ 50 mm rectilinear blocks. The samples were cut so that flow was measured across the fault. The cylindrical cores were placed into tightly fitting silicone rubber sleeves. The rectilinear samples were confined in polymer sleeves individually moulded to fit the specimens. The sleeved samples were placed in a secondary rubber or neoprene membrane and confined in a cell at a pressure of around 800 kPa to prevent any leakage of permeant occurring down the sides of the sample.

The fault rocks examined tended to be thinner than the sample size used for testing. The permeability of the faults was therefore calculated by assuming that the permeability of the sample containing the fault represented the harmonic mean of the fault rock and the host sediment; the permeability of the latter was measured separately.

The water flow permeameter is accurate to <5% down to permeabilities of <1 μD. Inaccuracies in permeability analysis may, however, occur due to the measurement of fault rock thickness, temperature variation in the testing laboratory, the presence of microcracks formed during sample preparation or as a result of stress relaxation following coring. It seems likely, however, that the accuracy of these permeability measurements is far higher than the intrinsic permeability variation that exists within individual specimens. This is consistent with the excellent correlation obtained between permeability and mercury–air threshold pressures.

Results

The microstructure and petrophysical properties of most faults examined, as well as the sealing processes experienced, were found to be highly dependent on the composition of the sediment at the time of deformation. The main situation where this proved not to be the case was where cements precipitated from fault- and fracture-transported fluids. The fault rocks described later are therefore divided in terms of the composition of the host sediment at the time of deformation. Faults rocks cemented by possibly fault-transported fluids are then considered.

Faults within clean sandstones

Faults within clean sandstones (<5% clay at the time of deformation) occur either in isolation or in dense clusters. The latter are similar to the banded deformation bands identified in the field by Jamison and Sterns (1982). The microstructure of the faults varies depending upon the extent of grain-fracturing incurred during deformation and the extent of post-deformation quartz cementation. In addition, variations exist in the manner in which the grain-size distribution changes across individual faults. Two end-members were identified. The first, have a relatively constant cross-fault grain-size distribution (Fig. 1a). These occur either in isolation or in dense clusters and account for over 98% of the fault rocks examined from clean sandstones in the North Sea. The second, are composed of a gouge which itself has a banded sub-structure, containing discrete slip planes (such as described by Logan *et al.* 1992) along which strain was concentrated (Fig. 1b); these were only observed within the densely clustered faults.

Reservoir sandstones within the North Sea often have not experienced significant diagenetic alteration between the depths of 0.5 km and

Fig. 1. BSE micrographs showing faults gouges in clean sandstones with (**a**) an homogeneous microstructure and a broad grain-size distribution, (**b**) discrete slip bands (arrows), (**c**) no grain-size reduction, (**d**) a large grain size reduction but a narrow grain size distribution.

~3 km. Using petrographic data alone, it is therefore difficult to accurately constrain the burial depth at which deformation occurred. The information available does, however, provide important clues as to the depth at which they formed. Namely, only homogeneous fault gouges were identified within sandstones that have been buried to <2.8 km. Fault gouges containing discrete slip bands were only identified in reservoirs that had experienced deeper burial.

The fault rocks identified ranged from those that experienced no grain fracturing (Fig. 1c) to those in which every grain was extensively fractured (Fig. 1d); a full range of intermediates also exist. Faults that experienced no grain-size reduction have an homogenous structure, similar to that of the host sediment. These features are termed disaggregation zones and deformation probably occurred by particulate flow under low effective stress conditions (Borradaile 1981). Fault rocks that experienced grain fracturing are termed clay-free cataclasites. Fault rocks that experienced a partial grain-size reduction also tend to have a homogenous macrostructure with a broad grain-size distribution due to the presence of large, mostly unfractured, grains and small fragments produced by cataclasis. The grain fragments within these fault rocks tend to form the matrix between the undeformed grains. Small elongate particles are also very common which probably formed as a result of spalling (see Knipe & Lloyd 1994).

Extensive grain-size reductions occur mainly within the banded faults. In the fault gouges that lack discrete slip surfaces, the fragments produced by grain fracturing are often not widely separated from their parent grain (Fig. 2). The fragments are, however, separated sufficiently from each other to produce a texture that appears dilational. It is important to realize that in porous sandstones the grain-fragments within these faults have collapsed into (and filled) macroporosity resulting in net compaction. The overall displacement along individual faults is often too small to be identified in hand specimen (i.e. ≪0.5 mm).

The most extensive grain-size reductions were confined to the discrete slip planes within the banded fault gouges, such as shown in Fig. 1b. The grains within some discrete slip planes are better sorted than those within the host sandstone. The fault gouge outside these discrete slip zones often has a broad grain-size distribution similar to the homogeneous faults.

In the majority of samples (>99%) examined, no evidence of distributed microcracking was observed within the host sediment adjacent to the fault. The only exception to this was adjacent

(a) (b)

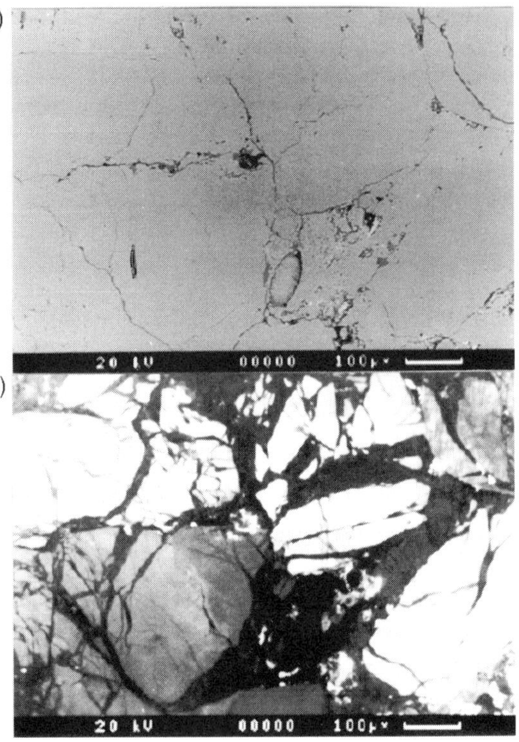

Fig. 2. BSE (**a**) and CL (**b**) images of the same area of a quartz-cemented cataclastic fault. Note the dark quartz cement in the CL image. Also, note that the fabric appears dilational because the grain fragments have separated from each other. In fact, the fragments have compacted into what was macroporosity.

to a small number of thrust faults which formed following the initiation of quartz overgrowth precipitation (see Knipe & Lloyd 1994).

Mesocrystalline quartz cement is concentrated within cataclastic faults in clean sandstones that had been buried to >3 km. These fault rocks often have very low porosities (≪3%) and, when viewed using BSE imagery, have the appearance of quartz-cemented veins (Fig. 2a). CL images reveal, however, that the faults are composed of fragments of detrital grains cemented by mesocrystalline quartz (Fig. 2b). Detrital fragments are not supported by the cement. Instead, the quartz appears to have preferentially precipitated between fragments in mutual contact and along intragranular fracture surfaces. All of the quartz-cemented cataclastic faults studied have lower intragranular volumes (see Houseknect 1987 for a definition) than their host sandstone, suggesting that they experienced a deformation-induced porosity collapse.

The amount of quartz cement within the cataclastic faults often appears to be related to the amount of grain-size reduction experienced. On the whole, faults that have experienced only very small amounts of grain fracturing appear

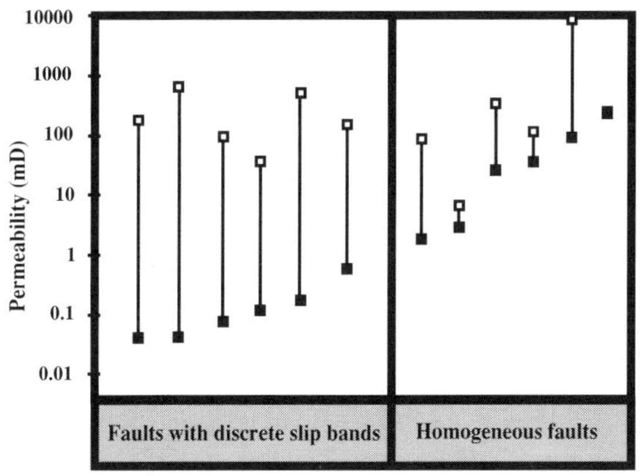

Fig. 3. Diagram showing the range of permeabilities obtained from cataclastic faults (solid symbols) from an Australasian oil field. These fault rocks have not experienced post-faulting quartz cementation or pressure solution. The results have been divided according to whether the fault gouges have an homogenous sub-structure or contain discrete slip planes. The permeability of their host sediment (open symbols) has been included for comparison.

less cemented than those that experienced pervasive cataclasis.

Cataclastic faults with a grain-size of $>5\,\mu$m did not show evidence of having experienced enhanced post-faulting pressure solution compared to their host sediment. The microstructure of cataclasites with a grain-size of $<5\,\mu$m could not, however, be resolved using CL. The relative amount of pressure solution and quartz cementation which contributes to the final microstructure of these very fine-grained cataclasites could therefore not be established.

Not surprisingly, the petrophysical properties of fault rocks within clean sandstones show enormous variation. Those that did not experience cataclasis have very similar petrophysical properties to their host sediment. Figure 3 shows a range of permeability values obtained from faults rocks within clean sandstones from an Australasian oil field in which cataclasis was the principle porosity reduction mechanism (a North Sea example was not used because most of the faults that have extensive grain-size reductions are also quartz cemented). The permeabilities obtained range from >100 mD to <0.05 mD. The principle factor which affects these permeability values is the final grain-size and grain-sorting of the gouge (see Knipe *et al.* 1997). The high permeability values are from faults that have experienced little grain-size reduction, whereas the low permeability values are from cataclasites that have experienced extensive grain-size reductions. The faults containing discrete slip bands have the lowest permeability values.

The permeabilities of cataclastic fault rocks within clean sandstones that have experienced post-faulting quartz cementation are on the whole far lower than the uncemented faults (Fig. 4). The extent to which the permeability of fault rocks is further reduced by quartz cementation depends both on the extent of grain-size reduction and the quantity of cement present. Cataclastic faults in clay-free sandstones from the North Sea have mercury–air threshold pressures of 15 psi to 2000 psi.

Fault rocks within sandstones containing low concentrations of clays

Fault rocks within sandstones containing low concentrations of phyllosilicates (5 to 15%) have a similar structure and range of grain-size distributions to those within clean sandstones. These fault rocks sometimes also contain discontinuous domains in which phyllosilicates partly fill the intragranular volume. These domains form by the deformation induced mixing of phyllosilicate material derived from the host sediment with the framework grains. In reservoirs that have been buried to >3 km, these domains usually show signs of enhanced pressure solution compared to their host sediment.

The volume of phyllosilicates within the host sandstone is insufficient for phyllosilicate-rich domains to be continuous along the entire fault plane. The faults therefore contain domains which have the same microstructure as observed within the phyllosilicate-free equivalents. The petrophysical properties of these fault rocks are therefore similar to those within the clean sandstones.

Fault rocks within impure sandstones

Faults within impure sandstones (those with 15–40% clay) also experienced a wide range of grain-size reductions, although the very large grain-size reductions experienced by the faults within the less clay-rich sandstones were not observed in the specimens studied. The microstructure and petrophysical properties of these fault rocks often differ markedly from their host sediment, regardless the extent of grain fracturing.

The permeability of impure sandstones is usually as not as high as clay-free sandstones from the same reservoir. Nevertheless, they can have good to moderate reservoir properties, because the phyllosilicates present often have a heterogeneous distribution leaving interconnected clay-free channels (e.g. Fig. 5a). Deformation of

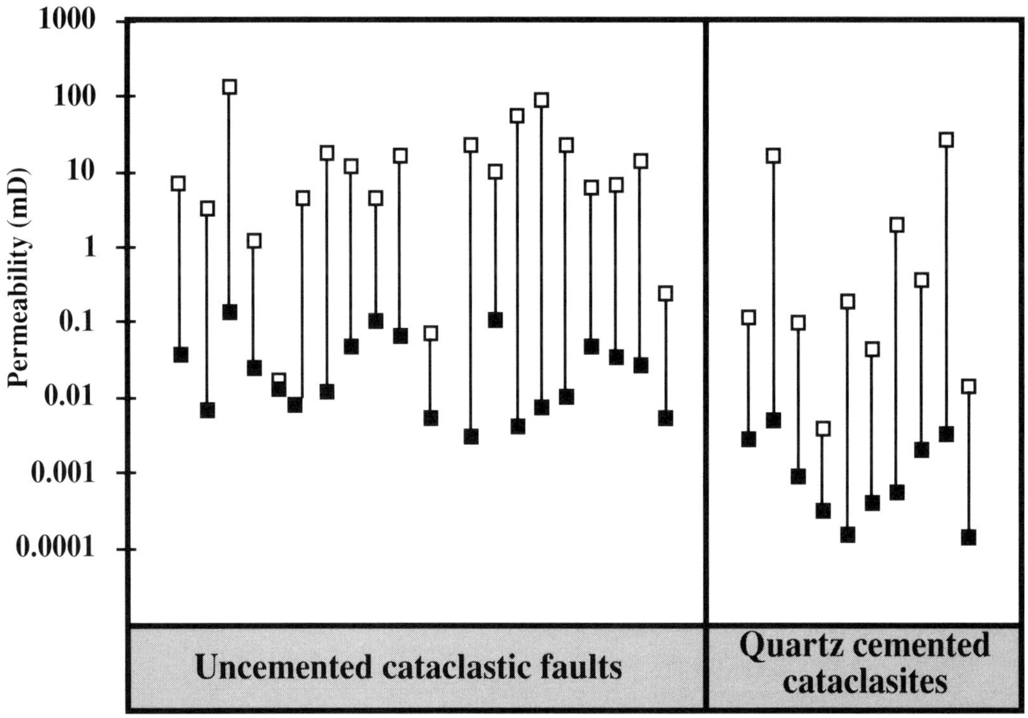

Fig. 4. Diagram showing the range of permeabilities obtained from cataclastic faults (solid symbols) from the North Sea. The fault rocks have been subdivided according to whether or not they have experienced post-deformation quartz cementation. The permeability of their host sediment (open symbols) has been included for comparison.

these impure sandstones results in the mixing of clays with framework grains and produces a fault rock that has a more homogenous distribution of clays than the host sediment (Fig. 5b). This process we have termed deformation-induced mixing and produces a distinct group of deformation features which have been termed phyllosilicate-framework fault rocks (Knipe *et al.* 1997)

In reservoirs that have been buried to >3 km, these phyllosilicate-rich fault rocks have often experienced enhanced pressure solution compared to the host sandstone (Fig. 5c & d). The extent of enhanced pressure solution often varies significantly both along and between individual faults within reservoirs. The faults which experienced most pressure solution are within sandstones with moderate concentrations (~15 to 25%) of heterogeneously distributed phyllosilicates. Faults within sediments which contain >25% intergranular clays, or quartz grains which have extensively developed clay coats, often do not experience significant enhanced pressure solution.

Sutured grain contacts, formed by enhanced pressure solution within fault rocks, frequently have preferred orientations. The grain contacts in most fault rocks examined are predominantly sub-horizontal (Fig. 5c). Fault rocks which have dominantly sub-vertical grain contacts have also been identified (Fig. 5d). Grain elongation as a result of pressure solution can be very significant. For example, quartz grains with aspect ratios of >5 have formed by enhanced pressure solution of what were originally sub-rounded grains.

Faults within impure sandstones that experience grain fracturing sometimes contain slightly more quartz cement than their host sediment. This tends to be concentrated along quartz fracture surfaces that are not coated by clay minerals. It should be emphasized that the extent of post-deformation quartz cementation of these fault rocks is far less than observed within the less clay-rich sediments.

The permeability of 53 fault rocks produced by the deformation of impure sandstones examined from the North Sea has been measured. The values obtained range from 0.9 mD to 0.2 μD. Their mercury–air threshold pressure varies between 75 and 1600 psi. The extent to which post-faulting pressure solution has contributed to the reduction in permeability of the fault rocks is difficult to assess. Nevertheless, some indication of this is gained by the observation that most of the framework-phyllosilicate fault rocks that have been buried to <2.5 km have Hg–air threshold pressures of <300 psi. However, most framework-phyllosilicate fault rocks buried to >3.5 km have threshold pressures of >350 psi.

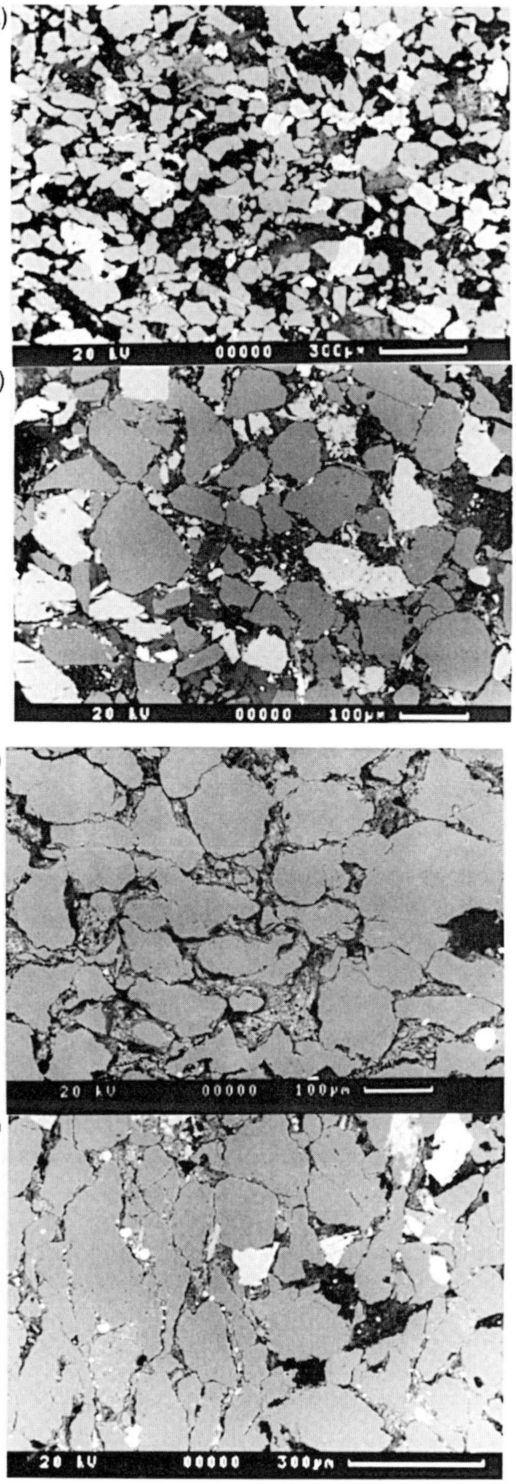

Fig. 5. BSE micrographs showing: (**a**) the undeformed phyllosilicate-rich sandstones close to (**b**) a fault rock that has experienced porosity collapse by disaggregation and deformation induced mixing; (**c**) fault rock that has experienced enhanced grain-to-grain dissolution along sub-horizontal grain-to-grain contacts; (**d**) fault rock that has experienced enhanced grain-to-grain dissolution along sub-vertical grain-to-grain contacts.

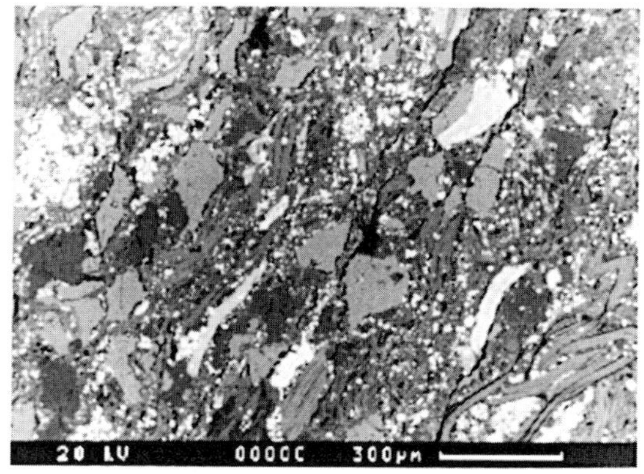

Fig. 6. BSE image showing the alignment of phyllosilicates within a clay smear.

Deformation features in clay-rich sediments

Sediments containing >40% phyllosilicates deform to produce fault rocks containing domains of aligned phyllosilicates (Fig. 6). These fault rocks frequently become smeared between the hangingwall and footwall cut-offs of phyllosilicate-rich horizons. In addition, faulting of some very phyllosilicate-rich horizons has resulted in the injection of a clay gouge into less clay-rich sediment. These fault rocks are clay smears and have been described elsewhere in the literature (e.g. Smith 1980; Bouvier *et al.* 1989; Sassi *et al.* 1992; Jev *et al.* 1993; Lindsay *et al.* 1993; Gibson 1994).

The microstructure and continuity of clay smears examined from hydrocarbon fields within the North Sea varies enormously. For example, discontinuous clay smears were identified along faults with displacements of less than twice the thickness of the clay-rich layer present. However, coherent clay smears were identified along faults with displacement of >15 times the thickness of the clay-rich layer present. The main difference between these cases was the grain-size of the phyllosilicates present. The continuous clay smears contained coarse-grained phyllosilicates (>20 μm), whereas the clay within the discontinuous smears was very fine-grained (<0.5 μm).

The permeability and threshold pressure of seventeen clay smears have been analysed during this study. Twelve yielded permeability values of <0.0005 mD. Clay smears with higher permeability values later proved to be discontinuous on the hand-specimen scale. The permeability value of 0.0005 mD must, however, be taken as a maximum, as it is close to the resolution of the water flow permeameter used. All phyllosilicate smears analysed had mercury-air threshold pressures in excess of 2000 psi, with fourteen of the measurements being over 10000 psi.

Cemented deformation features present within North Sea hydrocarbon reservoirs

A variety of cement types have precipitated along faults and fractures within the North Sea reservoirs, including: anhydrite, ankerite, barite, calcite, dolomite, kaolin, microcrystalline quartz, pyrite, siderite, sphalerite, and mesocrystalline quartz. These cements can be found in a range of structures including: cataclastic faults, veins, and fluid escape features. Mesocrystalline quartz is by far the most common cement concentrated within cataclastic faults in clean sandstones that have not experienced reactivation.

Anhydrite, ankerite, barite, calcite, dolomite, pyrite, siderite, and sphalerite occur as vein-filling cements within lithified sandstones and reactivated cataclastic faults. Calcite veins were found to be particularly common within calcite cemented sandstones (Fig. 7a). In the Rotliegendes of the southern North Sea, anhydrite, ankerite, siderite and barite are particularly common within fractures formed (Fig. 7b), and cataclastic faults reactivated (Fig. 7c), during inversion.

Microcrystalline quartz (Fig. 7d) and pyrite cement are concentrated within faults formed during very early burial associated with fluid escape. These deformation features often have very high minus cement porosities (>40%) suggesting that they experienced dilation during cement precipitation. Microcrystalline quartz-cemented deformation features generally occur within sandstones containing large concentrations of biogenic silica, such as sponge spicules, suggesting that the cement is a product of its dissolution.

Veins usually have extremely low permeabilities (<1 μD) and high mercury-air threshold pressures (>1500 psi). Higher permeability, and lower threshold pressure values were only obtained from veins which microstructural analysis revealed to be discontinuous. In the case of reactivated faults, cements are usually concentrated within dilational jogs. Portions of such faults that have not experienced extensive dilation remain uncemented. The ultimate sealing capacity of the fault is therefore controlled by the capillary pressure of the fault rock between cemented sections. For example, Fig. 8 shows a cataclastic fault which has been reactivated and cemented by barite. Barite is only present along the parts of the fault that experienced dilation, and the sealing capacity of the fault rock is controlled by the pore structure of the cataclasite between.

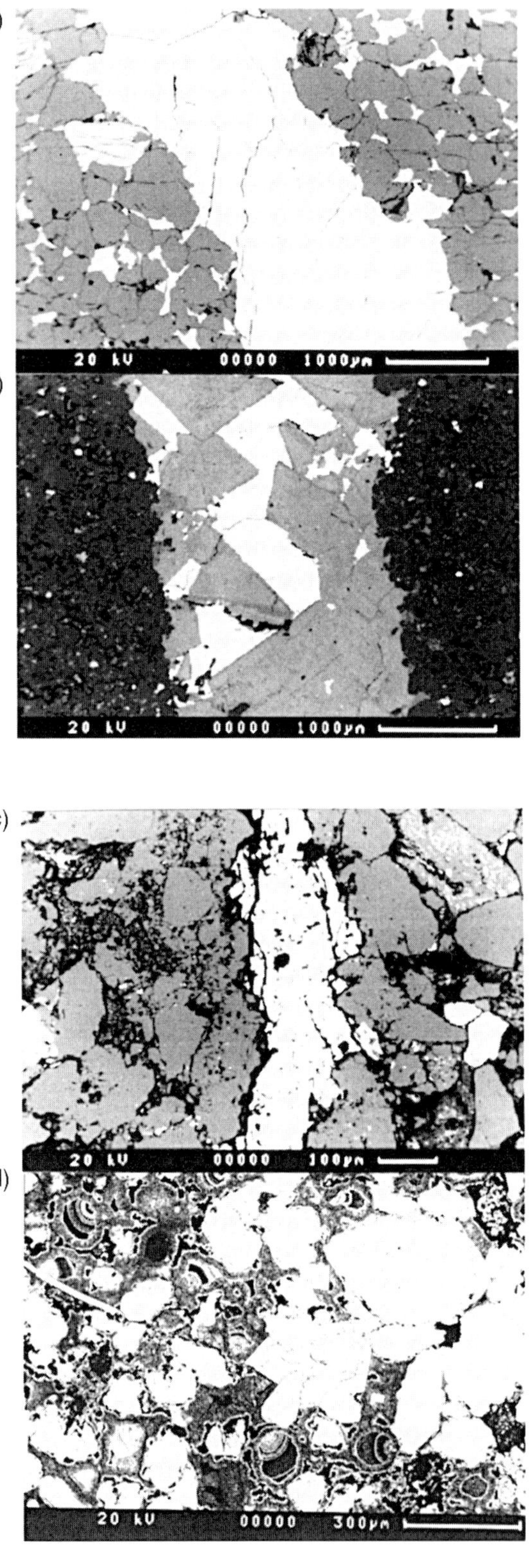

Fig. 7. BSE micrographs showing: (**a**) a calcite vein in a calcite cemented sandstone; (**b**) a dolomite and anhydrite vein in a lithified sandstones; (**c**) an anhydrite cemented reactivated cataclastic fault; (**d**) a microcrystalline quartz-cemented water-escape structure;

Fig. 8. BSE photomontage of a cataclastic fault that has been reactivated. The dilated areas produced during reactivation have been cemented by barite. Note that the barite cement is discontinuous therefore the sealing potential of the fault is still controlled by the pore structure of the cataclastic fault.

The permeability of microcrystalline quartz cemented water-escape structures varies dramatically (between 2 mD and <0.1 μD), depending upon cement morphology. Quartz cements within these deformation features range from highly porous networks of microcrystalline quartz to non-porous chert veins. These water escape structures are usually discontinuous; however, they often occur in high abundance and may greatly increase the overall tortuosity of fluid flow paths within sponge spicule-rich hydrocarbon reservoirs.

Discussion

Fault seal processes and their controls

The results presented earlier show that the porosity and permeability of fault rocks may be decreased and their threshold pressure increased by a number of processes, including:

(i) deformation induced porosity reduction by disaggregation and mixing;
(ii) pressure solution;
(iii) cataclasis;
(iv) cementation, and
(v) clay smearing.

This section discusses the results and identifies the main factors which control the extent to which these processes operate.

Porosity collapse by deformation induced mixing. The deformation of clean sandstones by particulate flow without grain-fracturing does not result in a long term change in their petrophysical properties, because the grain-size and grain-sorting of the fault is not significantly changed. Temporary dilation may occur due to changes in grain packing, however, if the fault rock is not quickly cemented, mechanical compaction usually results in a return of the grains to their former packing arrangement. Deformation of impure sandstones results in the mixing of fine-grained phyllosilicates with framework grains and the production of a more homogenous microfabric in which macroporosity is partly or wholly replaced by clay minerals and microporosity. This may occur during early burial by particulate flow without the need for grain-fracturing. Macroporosity destruction results in a reduction in permeability and average pore-aperture size. The most important factor controlling the ability of this process to create a continuous barrier to flow is the amount of fine-grained phyllosilicates present. Only sediments containing >15% phyllosilicates deform by particulate flow without grain-fracturing to produce a low permeability fault gouge in which all intragranular porosity is partly filled by fine-grained phyllosilicates. If grain fracturing occurs, less phyllosilicates are required to fill intergranular porosity.

Pressure solution and quartz cementation. The extent to which the fault rocks have experienced post-faulting quartz cementation or pressure solution depends on both their clay content and distribution at the time of burial, as well as on the depth/temperature history experienced. Faults within sandstones containing <5% phyllosilicates experienced enhanced quartz cementation but evidence of enhanced pressure solution was not observed. Faults within sandstones containing ~5 to 15% phyllosilicates experienced both enhanced quartz cementation and pressure solution. Faults within sandstones containing ~15 to 25% experienced enhanced pressure solution but not, in most cases, extensive quartz cementation. Faults in sandstones containing very large quantities of phyllosilicates (>25%) were usually not significantly affected by either of these processes.

The effect that clays have on the rate of quartz cementation and pressure solution in sandstones is well established (Heald 1955; Dewers & Ortoleva 1991; Bjørkum 1996; Oelkers *et al.* 1996). In particular, small concentrations of phyllosilicates at grain-contacts increase the rate of pressure solution (e.g. Heald 1955, 1959; Thompson 1959; de Boer *et al.* 1977; Chang & Yortsos 1994; Bjørkum 1996; Oelkers *et al.* 1996). In contrast, grain-coating clay minerals suppress quartz cementation (Cecil & Heald 1971; Tada & Siever 1989) by reducing the reactive quartz surface area and hence reducing the overall rate of quartz precipitation (Walderhaug 1996).

Theoretical considerations suggest that silica solubility and the rate of fluid flow in sandstones are not sufficient to account for the mobility of large volumes of silica by advection. For example, Bjørlykke (1994) showed that 3×10^9 pore volumes of fluid would be required to completely cement a single pore volume with quartz. Local processes are therefore required to explain the magnitude of quartz cementation and pressure solution observed in sandstones (Bjørlykke 1994). It seems likely that silica generated by pressure solution precipitates locally (i.e. at a distance of <1 m) as mesocrystalline quartz cement (Oelkers *et al.* 1996).

The link between quartz cementation and pressure solution is emphasized by the observation that faults rocks within the North Sea that have experienced these processes have been buried to >3 km. This finding reflects the fact that quartz cementation in sandstones, which is necessary for pressure solution, occurs at negligible rates below temperatures of 90 °C (Rimstidt & Barnes 1980; Giles 1997).

The susceptibility of a fault rock to pressure solution can also be related to variations in the textural distribution of phyllosilicates. Clays within many sandstones are either pore-filling or grain-coating, both of which inhibit quartz cementation; as they are not present between grain contacts, they do not increase the rate of pressure solution. During deformation, some of these clays become emplaced at quartz grain-to-grain contacts, rendering the quartz more susceptible to pressure solution. Faults in very clay-rich sandstones ($>25\%$ clay) do not experience enhanced pressure solution, because quartz cementation in the surrounding sediment is inhibited by the presence of the clays, which lowers the surface area available for quartz cementation.

It may be tempting to suggest that enhanced quartz cementation of cataclasites occurs due to the precipitation of silica from fluids focused along the fault. The quartz-cemented cataclastic faults identified during this study have, however, a far lower finite intergranular volume than their host sediment. There is therefore no reason why fluids would be focused along these faults as opposed to the more permeable host sandstone. This suggests that the faults sampled did not act as long term conduits for fluid flow and therefore cementation is unlikely to have occurred from non-local fluids focused along these features (Fisher *et al.* in press). The reason why cataclastic faults within clean sandstones often become preferentially quartz cemented requires an understanding of the kinetics of quartz cementation.

The rate of quartz cementation in most sandstones is inhibited either due to the presence of small quantities of clays (Cecil & Heald 1971) or other surface pollutants such as Al^{3+} (Iller 1979). For example, Mullis (1991) showed that the precipitation constant for silica was up to three orders of magnitude lower in the Jurassic Dogger Beta Formation from northern Germany than the value derived in laboratory experiments using clean sands. Cataclastic deformation of a clean sandstone produces a fault gouge that not only has a larger quartz surface area than the host sediment but also the newly formed fracture surfaces will contain less contaminants (clay etc.) than are present on the surface of the undeformed detrital quartz grains. Hence, quartz can precipitate faster in the cataclastic gouge than in the surrounding host sediment.

As advective processes cannot supply large volumes of silica required for quartz cementation in sandstones (Bjørlykke 1994; Giles 1997), it seems likely that the silica in the cataclasites was produced by processes such as pressure solution near the fault. Integrating the microstructural observations made during this study, it appears that faults in clean sandstones often act as sinks for silica generated by diagenetic processes in the host sediment. However, faults in impure sandstones act as sources of silica for the precipitation of quartz cement in the adjacent sediment.

The hypothesis that cataclastic fault gouges offer kinetically favourable sites for the precipitation of silica generated by pressure solution in the host sediment implies that the timing of faulting may be an important control on their sealing capacity. In particular, if deformation occurs below the temperature that quartz can rapidly precipitate ($\sim$90°C), the fracture surfaces generated during cataclasis may become polluted by the growth of clay minerals, etc. Indeed, the cataclasites examined during this study that experienced most quartz cementation are those which also experienced the highest grain-size reductions and probably deformed at the deepest depths.

Pressure solution and quartz precipitation may be suppressed by the emplacement of hydrocarbons (Griggs 1940) which lowers the rate of silica diffusion from grain contacts to precipitation sites (Worden *et al.* 1998). This implies that early hydrocarbon migration may reduce the capacity of faults above the hydrocarbon–water contact to seal by post-faulting pressure solution and quartz cementation.

Cataclasis. Cataclasis is the dominant process responsible for reducing the porosity and permeability and increasing the threshold pressure of porous, clay-free sandstones during faulting. These changes occur because cataclasis causes macroporosity to collapse, average grain-size to be reduced and grain-sorting to become poorer. It is difficult to determine the key factors which control the grain-size distribution and hence the petrophysical properties of cataclasites from microstructural observations of natural fault rocks alone. Integration of microstructural data with published accounts of triaxial deformation experiments (Engelder 1974; Zoback & Byerlee 1976; Logan 1992; Gu & Wong 1994; Wong *et al.* 1997; Zhu & Wong 1997) has, however, the potential to provide such information.

Sandstone deformation experiments show that low confining pressures favour brittle faulting in which failure occurs along single slip planes (Handin *et al.* 1963; Scott & Nielson 1991). High confining pressures favour ductile deformation by distributed cataclastic flow without the formation of discrete slip planes (Handin *et al.* 1963; Scott & Nielson 1991). A transitional regime exists by which deformation occurs along multiple slip planes (Scott & Nielsen 1991). The effective pressure at which this transition occurs is related to the porosity and grain-size of the sandstone (Wong *et al.* 1997). In particular, it has been shown that fine grain sizes (Zhang *et al.* 1990), low porosities (Rutter & Hadizaddeh 1991), and high cement contents (Ménendez *et al.* 1996) favour brittle faulting as the dominant deformation process. The relative manner in which these factors vary during burial can be complex. For example, confining pressures tend to increase during burial but porosity tends to decrease (e.g. Jamison & Stearns 1982).

The microstructure of fault gouges within clean sandstones is also dependant upon the amount of shear strain. For example, Logan (1992) showed that under high effective pressures (equivalent to >2000 m of overburden), deformation becomes concentrated along discrete slip planes within the gouge zones even at low shear strains. Strains of >100 are, however, required to form discrete shear planes within gouge zones under lower confining stress conditions.

The faults within the clean sandstones examined during this study can be subdivided according to whether they occur in isolation or in dense clusters, and whether the fault gouges themselves have a homogeneous grain size distribution or contain discrete slip planes. The micromechanical work presented by Ménendez *et al.* (1996) suggests that isolated faults are produced by single brittle faulting events at low confining pressures. In contrast, dense clusters of cataclastic faults probably deformed at higher confining pressures and are associated with more distributed brittle deformation. It seems likely that the structure of natural cataclastic faults is also controlled by the effective pressure conditions during deformation.

Discrete slip planes within fault gouges were observed mainly within the densely packed cataclastic faults. It is suggested that these also deformed under high effective pressures, although Logan (1992) suggests that these fault rocks may also have accommodated more strain than those with a more homogeneous sub-structure.

The permeability of sands and porous sandstones decreases as strain is increased during high pressure triaxial deformation experiments (Zoback & Byerlee 1976; Zhu & Wong 1997). The extent to which the permeability decreases varies depending upon whether the sandstone deforms in a localized, transitional or distributed manner. For example, triaxial experiments on the Berea sandstone (porosity = 21%), show that under low mean effective pressures (<10 MPa), porosity and permeability decrease slowly as strain is increased until failure occurs (Zhu & Wong 1997). At failure, the samples lose cohesive strength, porosity increases slightly but permeability continues to fall by up to an order of magnitude (Zhu & Wong 1997). Under high effective pressures (160 MPa), the porosity and permeability decrease slowly with increasing strain until a critical stress is reached, after which a dramatic fall in permeability of around two orders of magnitude occurs but the samples strain harden (Zhu & Wong 1997). Under intermediate effective pressures (~40 MPa) the permeability behaves in a similar manner to the high pressure experiments but at a critical pressure the samples strain softened (Zhu & Wong 1997). Integration of the microstructural observations of Ménendez *et al.* (1996) with the permeability results of Zhu & Wong (1997) suggest that the sudden fall in permeability in the samples deformed in a distributed and transitional manner occurs due to pervasive porosity collapse and grain fracturing.

Microstructural studies of fault gouges have shown that the particle size of cataclastic faults decreases with increasing shear strain and confining pressure (Engelder 1974; Sammis *et al.* 1987; Marone & Scholtz 1989). The non-quartz-cemented faults within the clay-free sandstones with the lowest permeabilities, were also the ones which experienced the largest grain size reductions. These fault rocks were usually from dense clusters or contained discrete slip planes that probably deformed under high effective pressures. It therefore seems likely that high effective stresses at the time of deformation favour the production of cataclastic faults with low permeabilities and high capillary pressures. It should, however, be emphasized that experimental work shows that low porosity sandstones are likely to deform by localized brittle deformation (Rutter & Hadizaddeh 1991). The porosity and permeability of sandstones decreases due to chemical compaction beyond around 90 °C. It is therefore possible that sandstones that have experienced severe porosity reduction due to burial diagenesis would undergo localized faulting even when deformed under very high effective stresses. The results presented by Zhu & Wong (1997) suggest that these would not produce the high permeability reductions as observed within the sandstones that have deformed in a transitional manner.

Experimental studies have not yet fully explored the effect that mineralogy has on sediment susceptibility to grain-size reduction and porosity collapse by cataclastic processes. Deformation of feldspar-rich rocks produces cataclasites with a lower average grain size than less feldspathic lithologies under similar conditions (Blenkinsop 1991; Antonellini *et al.* 1994). Microstructural work undertaken during this study also suggests that sediments containing concentrations of phyllosilicates tend to be less prone to cataclasis than clean sandstones. This is because clay-rich sediments can deform by grain-sliding, grain rotation, and the plastic deformation of phyllosilicates without the need for significant dilation and grain crushing.

Cementation. The microstructure of the cemented faults and fractures provides clues for identifying the origin of the cements, the role of exotic fluids in cementation, as well as conditions under which fault- and fracture-related fluid flow occurs. Except for the quartz-cemented cataclastic faults, cements are only concentrated within faults and fractures that had a higher intergranular volume than their host sediment following deformation. This suggests that prior to cementation they had a higher porosity and permeability than their host sediment. Indeed, if the faults were not more permeable that the host sediment there would be no reason why focused fluid flow and cement precipitation would occur.

The cemented faults and fractures that were identified during this study fall into two categories. Those that formed in lithified sediment and those which formed in poorly lithified sediment. In these two cases, fundamental differences may exist in the mechanism by which dilation was maintained during cementation.

In the case of the veins, the wall rock was sufficiently strong or the fluid pressure was sufficiently high, either due to early cementation or later lithification, to allow fractures to remain open following initial deformation (Bjørlykke & Høeg 1997). For example, all of the cemented fractures within the Rotliegendes sandstones of the southern North Sea studied appear to have formed during structural inversion after the sandstone had become well lithified. These veins had a higher permeability than their host sediment and were capable of acting as conduits for fluid flow. There are no obvious local sources of solutes for the precipitation of some vein-filling cements, suggesting that they precipitated from exotic fluids. For example, the anhydrite, ankerite, barite, and sphalerite cements in the Rotliegendes of the southern North Sea may have required external sources of solutes such as the overlying Zechstein evaporites or the underlying Carboniferous sediments (Sullivan *et al.* 1990; Gaupp *et al.* 1993). In other cases, such as the calcite veins in calcareous sandstones, local cement sources are available and the open veins may have merely provided porosity for cementation to occur.

The cemented water-escape structures have very high intergranular volumes, suggesting that they experienced significant fluid pressure-induced dilation, because the host sandstone was poorly lithified at the time of deformation. As with the veins, cementation in fluid escape structures may result from decreases in pore pressure and temperature, or due to the mixing of fluids.

Other than mesocrystalline quartz, cements are not concentrated within faults within porous sandstones that were normally pressured at the time of deformation. This observation suggests that under these conditions, faulting does not result in the creation of fluid flow conduits for sufficient lengths of time for cementation to occur. This is entirely consistent with the results of sandstone deformation experiments (e.g. Zhu & Wong 1997).

Continuous cemented faults and fractures have been shown to have very low permeabilities and high threshold pressures. Microstructural evidence shows however, that they are rarely continuous – even on the hand specimen scale. In the case of reactivated faults, cements are usually concentrated within dilational jogs. Regions along such faults that have not experienced dilation may be uncemented, in which case the ultimate sealing capacity of the fault may still be controlled by the petrophysical properties of the fault between cement zones.

The continuity of fractures may also be very limited. For example, Peacock *et al.* (this volume) show an example where an array of calcite veins are linked by pressure solution seams. In this example, the sealing capacity of the vein array is probably controlled by the pressure solution seams and not the veins.

Hydrocarbon field examples are also inconsistent with the idea that faults and fractures cemented by externally derived fluids are sufficiently continuous to control the sealing capacity of faults on a large scale. The capillary pressures measured during this study suggest that continuous, cemented faults would be capable of supporting hydrocarbon column heights of hundreds of metres. Such large column height differences are seldom seen across faults in hydrocarbon fields in areas of known cementation, suggesting that their sealing capacity is usually controlled by more permeable fault rock types.

Clay smear. Continuous clay smears have very low permeabilities and high threshold pressures, emphasizing that they provide extremely effective barriers to fluid flow. The continuity of clay smears has, however, been found to be highly variable. The more continuous clay smears identified during this study contain coarse-grained phyllosilicates (>20 μm), whereas the clay within the discontinuous smears are very fine-grained (<0.5 μm). It seems likely that the continuity of clay smears will be dependent upon a large number of factors, including: sediment lithification, effective stress, confining pressure, strain rate, mineralogy etc.

The prediction of the presence and continuity of clay smears in the sub-surface is usually based on parameters such as clay smear potential – CSP (Bouvier *et al.* 1989), shale smear factor – SSF (Lindsay *et al.* 1993), and clay gouge ratio – CGR (Gibson 1994). The large variations in the proportion of clay required to form an effective clay smear emphasizes that it is important to calibrate these techniques based on known hydrocarbon distributions, fluid pressures and production data from within the area of interest.

Fault rock types, fault sealing processes and their relationship to sediment composition

The results and discussion presented above illustrate that there is often a distinct link between sediment composition and the sealing processes that occur within faults both during and following faulting. In particular, faults within clean sandstones tend to seal by cataclasis and post-deformation quartz cementation. Faults within impure sandstones tend to seal by deformation induced mixing, cataclasis and post-faulting pressure solution. Faults within very clay-rich sediments tend to deform by disaggregation and grain-boundary sliding to produce clay smears. We have therefore divided fault rocks developed in siliciclastic sediments into three series (clean sand series, impure sand series and clay-rich series) according to the composition of the host sediment at the time of deformation (Fig. 9).

Each of these fault rock series can be sub-divided, based partly on the classification of Sibson (1977) into a number of fault rock types according to the extent of cataclasis experienced. For example, Fig. 9, shows that the clean sand series can be divided into disaggregation zones, protocataclasites, cataclasites and ultracataclasites depending upon the proportion of fractured grains.

Each of the fault rocks can also be sub-divided according to the extent of post-deformation lithification using the scheme adopted by Knipe (1986). Again, it is important to note that the extent to which post-faulting processes may operate is a function of the fault rock composition. Cataclastic faults in clean sandstones are susceptible to quartz cementation. Faults in impure sandstones are susceptible to post-faulting pressure solution. Faults in ultra clay-rich faults become lithified by mechanical compaction and clay mineral diagenesis.

The classification scheme described is sometimes difficult to apply without the use of electron microscopes. For example, in ultra clay-rich sandstones it is impossible to establish the extent of grain fracturing without the use of transmission electron microscopes. We have therefore outlined a series of fault rock types (clay-free disaggregation zones, clay-free cataclasites, phyllosilicate-framework fault rocks, clay smears and cemented faults) which can, in the majority of circumstances, be used to classify fault rocks on both a hand-specimen and micro-structural scale. These are described below:

(1) Clay-free disaggregation zones are formed during shallow burial; strain is accommodated by particulate flow without extensive grain fracturing. In clay-free sandstones, disaggregation zones do not contain sufficient phyllosilicates to render them susceptible to extensive porosity reduction, either by deformation induced mixing or by later pressure solution.
(2) Clay-free cataclastic faults have porosities and permeabilities reduced mainly by cataclasis and post-deformation quartz cementation.

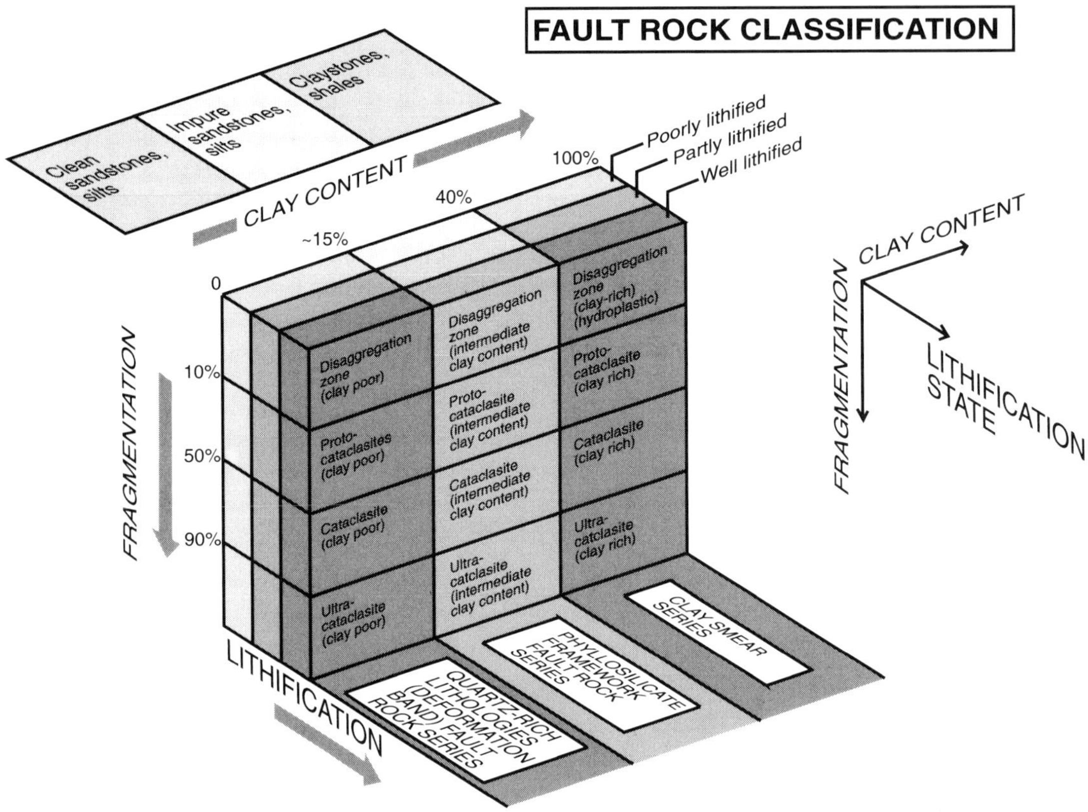

Fig. 9. Diagram showing the types of fault rocks developed in the North Sea and their relationship to the composition of the host sediment and the extent of grain-size reduction and post-deformation lithification experienced.

Cataclastic faults in clay-free sandstones identified within North Sea reservoir sandstones have permeability values of >100 mD to <1 μD and mercury–air threshold pressures of 15 psi and 2000 psi.

(3) Phyllosilicate-framework fault rocks are a new class of deformation feature (Knipe *et al.* 1997) introduced to describe fault rocks that are developed within impure sandstones (>15% phyllosilicates). The porosity and permeability of phyllosilicate-framework fault rocks are reduced mainly by deformation-induced mixing of framework grains with fine-grained phyllosilicates and often later pressure solution. Cataclasis may also reduce porosity and permeability of these fault rocks. Phyllosilicate-framework fault rocks are therefore used here to describe both clay-rich disaggregation zones and clay-rich cataclasites. Phyllosilicate-framework faults can be regarded as transitional to clay smears as the length and continuity of domains of aligned phyllosilicates increases. The phyllosilicate-framework fault rocks examined during this study have permeabilities of 0.9 mD to 0.2 μD and mercury-air threshold pressures of 75 to 1600 psi.

(4) Clay smears contain coherent domains of aligned phyllosilicates that form by the deformation of extremely phyllosilicate-rich sediments (i.e. >40% phyllosilicates).

(5) Cemented faults/fractures, are those in which the main porosity reduction mechanism is cementation. The features classified in this group are those which experienced significant dilation during deformation, resulting in the presence of open fractures along which cement could precipitate. Quartz cementation of cataclastic faults is an integral part of the evolution of a cataclasite, and therefore this type of fault rock is not grouped with the other cemented deformation features. The continuous cemented faults and fractures analysed during this study have permeabilities of <0.1 μD and threshold pressures of >1500 psi.

Prediction of fault sealing capacity

The relationship between fault rock type, fault sealing mechanisms and sediment composition suggests that, with a knowledge of the fault displacement distributions and sediment distribution within a field, it should be possible to predict the distribution of fault rock types (Knipe 1997; Knipe *et al.* 1997). The extent to which each of the fault sealing mechanisms operates for a given sediment composition depends upon the interaction between the stress, strain and temperature history of the reservoir. Only comprehensive knowledge obtained by microstructural and physical property analysis of faults within cored sections of reservoirs allows the sealing capacity of faults to be accurately predicted.

In the majority of individual fields examined, the relative importance and extent to which each of the fault sealing mechanisms operate (and hence the petrophysical properties of each fault rock type) is quite restricted. For example, faults present within clean sandstone in the Sleipner Field have permeabilities of ~0.4 to 43 mD (Ellevset *et al.* this volume). Only in five out of the thirty eight fields we have studied have the petrophysical properties of each fault rock type been highly varied, with permeabilities varying over four orders of magnitude. This is probably because faulting occurred throughout their burial history. Even in such cases, a knowledge of the range of petrophysical properties of each fault rock type may have an important impact on both exploration targets and production strategies.

Future work

The present paper has concentrated on the processes which reduce permeability and increase the threshold pressure of fault rocks. Microstructural and petrophysical analysis of core material can provide extremely important data on the likely properties of faults in the various lithologies. In addition to petrophysical properties of individual fault rocks types, the sealing capacity of field-scale faults will depend on other factors such as the distribution of open fractures in the subsurface and the continuity of fault rock types (particularly clay smears); these are important areas of future research.

Conclusions

(1) The principle fault sealing processes which operate within siliciclastic hydrocarbon reservoirs are: the deformation induced mixing of heterogeneously distributed fine-grained material with framework grains, pressure solution, cataclasis, clay smear, and cementation.

(2) The potential for fault sealing by each of these processes is dependent upon the sediment composition at the time of deformation as well as their pre- syn-, and post-deformation stress, strain and temperature histories. Clay-free sandstones tend to seal by cataclasis and post-deformation quartz cementation. Impure sandstones tend to seal by deformation induced mixing and post-faulting pressure solution.

(3) Fault rocks in siliciclastic sediments do not experience significant quartz cementation or pressure solution unless they are buried to >90 °C.

(4) The extent of cataclasis is particularly dependent upon the effective stress and the porosity at the time of deformation. Deformation of high porosity sandstones under high effective stress conditions leads to extensive porosity collapse and grain-size reductions.

(5) Excluding sediment composition, many of the principle factors which control the final petrophysical properties of a fault rock are common to an individual field as a whole (i.e. maximum temperature, stress state at the time of deformation, etc.). Microstructural and physical property analysis of deformation features within cores can therefore be used to reduce risks associated with fault seal analysis and can provide realistic fault rock properties for robust reservoir modelling.

The authors would first of all like to thank our industrial sponsors (Agip, Amoco, Arco, British Gas, British Petroleum, Conoco, Chevron, Elf, JNOC, Mobil, Phillips, Texaco and Statoil) who provided financial support and core material for this study. We would also like to thank M. Antonellini, E. Sverdrup and D. Peacock for providing thoughtful reviews which greatly improved this manuscript. B. Clennell, B. Kidd, A. Harrison, D. Condliffe and R. Jones are thanked for conducting the physical property analyses.

References

ANTONELLINI, M. & AYDIN, A. 1994. Effect of faulting on fluid flow in porous sandstones: petrophysical properties. *American Association of Petroleum Geologists Bulletin*, **78**, 335–377.

——, —— & POLLARD, D. D. 1994. Microstructure of deformation bands in porous sandstones at Arches National Park, Utah. *Journal of Structural Geology*, **16**, 941–959.

AYDIN, A. 1978. Small faults formed as deformation bands in sandstone. *Pure and Applied Geophysics*, **116**, 913–942.

BERG, R. R. 1975. Capillary pressure in stratigraphic traps. *American Association of Petroleum Geologists Bulletin*, **59**, 939–956.

—— & AVERY, A. H. 1995. Sealing properties of Tertiary growth faults, Texas Gulf coast. *American Association of Petroleum Geologists Bulletin*, **79**, 375–393.

BJØRKUM, P. A. 1996. How important is pressure in causing dissolution of quartz in sandstones? *Journal of Sedimentary Research*, **66**, 147–154.

BJØRLYKKE, K. 1994. Pore-water flow and mass transfer of solids in solution in sedimentary basins. *In*: PARNELL, A. (ed.) *Geofluids: Origin, Migration and Evolution of Fluids in Sedimentary Basins*. Geological Society, London, Special Publications, **78,** 127–140.

—— & HØEG, K. 1997. Effects of burial diagenesis on stresses, compaction and fluid flow in sedimentary basins. *Marine and Petroleum Geology*, **14**, 267–276.

BLENKINSOP, T. G. 1991. Cataclasis and processes of particle size reduction. *Pure and Applied Geophysics*, **136**, 60–86.

BORRADAILE, G. J. 1981. Particulate flow and the generation of cleavage. *Tectonophysics*, **72**, 306–321.

BOUVIER, J. D., SIJPESTEIJN, K., KLEUSNER, D. F., ONYEJEKWE, C. C. & VAN DER PAL, R. C. 1989. Three-dimensional seismic interpretation and fault sealing investigations, Nun River field, Nigeria. *American Association of Petroleum Geologists Bulletin*, **73**, 1397–1414.

CECIL, C. B. & HEALD, M. T. 1971. Experimental investigation of the effects of grain-coatings on quartz over-growth. *Journal of Sedimentary Petrology*, **41**, 582–584.

CHANG, J. & YORTSOS, Y. C. 1994. Lamination during silica diagenesis – effects of clay content and Oswald rippening. *American Journal of Science*, **294,** 137–172.

DE BOER, R. D., NAGTEGAAL, P. J. C. & DUYVUS, E. M. 1977. Pressure solution experiments on quartz sand. *Geochimica et Cosmochimica Acta*, **41**, 249–256.

DEWERS, T. & ORTOLEVA, P. J. 1991. Influences of clay minerals on sandstone cementation and pressure solution. *Geology*, **19**, 1045–1048.

ELLEVSET, S. O., KNIPE, R. J., OLSEN, T. S., FISHER, Q. J. & JONES, G. 1998. Fault controlled communication in the Sleipner Vest Field, Norwegian Continental Shelf; detailed, quantitative input for reservoir simulation and well planning. *This volume*.

ENGELDER, J. T. 1974. Cataclasis and the generation of fault gouge. *Geological Society of America Bulletin*, **85**, 1515–1522.

FISHER, Q. J., KNIPE, R. J. & WORDEN, R. In press. The microstructure of deformed and undeformed sandstones from the North Sea: its implications for the origin of quartz cement. *In*: MORAD, S. & WORDEN, R. (eds) *Quartz Cement: Origin and Effects on Hydrocarbon Reservoirs*. International Association of Sedimentology, Special Publication

GAUPP, R., MATTER, A., PLATT, J., RAMSEYER, K. & WALZEBUCK, J. 1993. Diagenesis and fluid evolution of deeply buried Permian (Rotliegende) gas reservoirs, northwest Germany. *American Association of Petroleum Geologists Bulletin*, **77**, 1111–1128.

GIBSON, R. G. 1994. Fault-zone seals in siliclastic strata of the Columbus Basin, offshore Trinidad. *American Association of Petroleum Geologists Bulletin*, **78**, 1372–1385.

GILES, M., 1997. *Diagenesis: a Quantitative Perspective, Implications for Basin Modelling and Rock Property Prediction*. Kluwer, Dordrecht.

GRIGGS, D. T. 1940. Experimental flow of rocks under conditions favouring recrystallisation. *Geological Society of America Bulletin*, **51**, 1001–1022.

GU, Y. & WONG, T-F. 1994. Development of shear localisation in simulated quartz gouge: effect of cumulative slip and gouge particle size. *Pure & Applied Geophysics*, **143**, 387–423.

HANDIN, J., HAGER, R.V., FRIEDMAN, M. & FEATHER, J. N. 1963. Experimental deformation of sedimentary rocks under confining pressure: pore pressure effects. *American Association of Petroleum Geologists Bulletin*, **47**, 717–755.

HEALD, M. T. 1955. Stylolites in sandstones. *Journal of Geology*, **63**, 101–114.

—— 1959. Significance of stylolites in permeable sandstones. *Journal of Sedimentary Petrology*, **29**, 251–253.

HOUSEKNECT, D. W. 1987. Assessing the relative importance of compaction processes and cementation to reduction of porosity in sandstones. *American Association of Petroleum Geologists Bulletin*, **71**, 633–642.

ILLER, R. K. 1979 *The Chemistry of Silica*. John Wiley and Sons, UK.

JAMISON, W. R. & STEARNS, D. W. 1982. Tectonic Deformation of Wingate sandstone, Colorado National Monument. *American Association of Petroleum Geologists Bulletin*, **66**, 2584–2608.

JEV, B. I., KAARS-SIJPESTEIJN, C. H., PETERS, M. P. A. M., WATTS, N. L. & WILKIE, J. T. 1993. Akaso field, Nigeria: use of integrated 3-D seismic, fault slicing, clay smearing, and RFT pressure data on fault trapping and dynamic leakage. *American Association of Petroleum Geologists Bulletin*, **77**, 1389–1404.

KATZ, A. J. & THOMPSON, A. H. 1986. Quantitative prediction of permeability in porous rock. *Physical Reviews*, **B34**, 8179–8181.

—— & —— 1987. Prediction of rock electrical conductivity from mercury injection measurements. *Journal of Geophysical Research*, **92**, 599–607.

KNIPE, R. J. 1986. Deformation mechanism path diagrams for sediments undergoing lithification. *In*: MOORE, J. C. (ed.) *Structural Fabrics in Deep Sea Drilling Project Cores From Forearcs*. Geological Society of America, Memoirs, **166**, 151–160.

—— 1992. Faulting processes and fault seal. *In*: LARSEN, R. M., BREKKE, H., LARSEN, B. T. & TALLERAAS, E. (eds) *Structural and Tectonic Modelling and its Application to Petroleum Geology*. Norwegian Petroleum Society, Special Publications, **1**, 325–342.

—— 1997. Juxtaposition and seal diagrams to help analyse fault seals in hydrocarbon reservoirs. *American Association of Petroleum Geologists Bulletin*, **81**, 187–195.

—— & LLOYD, G. E. 1994. Microstructural analysis of faulting in quartzite, Assynt, NW Scotland: implications for fault zone evolution. *Pure & Applied Geophysics*, **143**, 229–254.

—— FISHER, Q. J., JONES, G., CLENNELL, M. R., FARMER, A. B., KIDD, B., MCALLISTER, E., PORTER, J. R. & WHITE, E. A. 1997. Fault seal analysis, successful methodologies, application and future directions. *In*: MØLLER-PEDERSON, P. & KOESTLER, A. G. (eds) *Hydrocarbon Seals — Importance for Exploration and Production*. Norwegian Petroleum Society (NPF), Special Publications, **7**, 15–37.

LINDSAY, N. G., MURPHY, F. C., WALSH, J. J. & WATTERSON, J. 1993. Outcrop studies of shale smears on fault surfaces. International Association of Sedimentologists, Special Publication, **15**, 113–123.

LOGAN, J. M. 1992. The influence of fluid flow on the mechanical behaviour of faults. *In*: TILLERSON, J. R. & WAWERSIK, W. R. (eds) *Rock Mechanics*. Balkema, Rotterdam, 141–149.

MARONE, C. & SHULTZ, C. H. 1989. Particle-size distribution and microstructures within simulated fault gouge. *Journal of Structural Geology*, **11**, 799–814.

MENÉNDEZ, B., ZHU, W. & WONG T.-F. 1996. Micromechanics of brittle faulting and cataclastic flow in Berea sandstone. *Journal of Structural Geology*, **18**, 1–16.

MORROW, C. A., SHI, L. Q. & BYERLEE, J. D. 1984. Permeability of fault gouge under confining pressure and shear stress. *Journal of Geophysical Research*, **89**, 3193–3200.

MULLIS, A. M. 1991. The role of silica precipitation kinetics in determining the rate of quartz pressure solution. *Journal of Geophysical Research*, **96**, 10 007–10 013.

OELKERS, E. H., BJØRKUM, P. A. & MURPHY, W. M., 1996, A petrographic and computational investigation of quartz cementation and porosity reduction in North Sea sandstones. *American Journal of Science*, **296**, 420–452.

OLSON R. E. & DANIEL, D. E. 1981. Measurement of the hydraulic conductivity of fine-grained soils. *In*: ZIMMIE, T. F. & RIGGS, C. O. (eds) *Permeability and Groundwater Contaminant Transport*, American Society for Testing of Materials, STP 746, 18–64.

——, GILL, J. D., WILLDEN, A. T. & NELSON, K. R. 1991. Innovations in hydraulic conductivity measurements. In: Proceedings of the Transportation Research Board 70th Annual Meeting., Washington D.C. Paper 910367.

PEACOCK, D. C. P., FISHER, Q. J., WILLEMSE, E. J. M. & AYDIN, A. 1998. The relationship between faults and pressure solution seams in carbonate rocks, and the implications for fluid flow. *This volume*.

PITTMAN, E. D. 1981. Effect of fault-related granulation on porosity and permeability of quartz sandstones, Simpson Group (Ordovician), Oklahoma. *American Association of Petroleum Geologists Bulletin*, **65**, 2381–2387.

RIMSTIDT, J. D. & BARNES H. L. 1980. The kinetics of silica–water reactions. *Geochimica et Cosmochimica Acta*, **44**, 1683–1699.

RUTTER, E. H. & HADIZADEH, J. 1991. On the influence of porosity on the low temperature brittle–ductile transition in siliciclastic rocks. *Journal of Structural Geology*, **13**, 609–614.

SAMMIS, C. G., KING, G. & BIEGEL, R. 1987. The Kinematics of Gouge Deformation. *Pure and Applied Geophysics*, **125**, 777–812.

SASSI, W., LIVERA, S. E. & CALINE, B. R. P. 1992. Reservoir compartmentation by faults in Cormorant Block IV, U.K. northern North Sea. *In*: LARSEN, R. M., BREKKE, H., LARSEN, B. T. & TALLERAAS, E. (eds) *Structural and Tectonic Modelling and its Applications to Petroleum Geology*. Norwegian Petroleum Society, Special Publications, **1**, 355–364.

SCOTT, T. E. & NIELSON, K. C. 1991. The effects of porosity on the brittle–ductile transition in sandstones. *Journal of Geophysical Research*, **96**, 405–414.

SIBSON, R. H. 1977. Fault rocks and fault mechanisms. *Journal of the Geological Society, London*, **133**, 199–213.

SMITH, D. A. 1966. Theoretical consideration of sealing and non-sealing faults. *American Association of Petroleum Geologists Bulletin*, **50**, 363–374.

—— 1980. Sealing and non-sealing faults in Louisiana Gulf Coast salt basin. *American Association of Petroleum Geologists Bulletin*, **64**, 145–172.

SULLIVAN, M. D., HASZELDINE, R. S. & FALLICK, A. E. 1990. Linear coupling of carbon and strontium isotopes in Rotliegend Sandstone, North Sea: Evidence for cross-formational fluid flow. *Geology*, **18**, 1215–1218.

SVERDRUP, E. & BJØRLYKKE, K. 1992. Small faults in sandstones from Spitsbergen and Haltenbanken. A study of diagenetic and deformational structures and their relation to fluid flow. *In*: LARSEN, R. M., BREKKE, H., LARSEN, B. T. & TALLERAAS, E. (eds) *Structural and Tectonic Modelling and its Application to Petroleum Geology*. Norwegian Petroleum Society, Special Publications, **1**, 507–518.

TADA, R. & SIEVER, R. 1989. Pressure solution during diagenesis. *Annual Reviews of Earth and Planetary Sciences*, **17**, 89–118.

THOMPSON, A. 1959. Pressure solution and porosity. *In*: IRELAND, H. A. (ed.) *Silica in Sediments*. Society of Economic Paleontologists and Mineralogists Special Publications, **7**, 92–110.

UNDERHILL, J. R. & WOODCOCK, N. H. 1987. Faulting mechanisms in high porosity sandstones; New Red Sandstone, Arran, Scotland. *In*: JONES, M. E. & PRESTON, R. M. F. (eds) *Deformation of sediments and sedimentary rocks*. Geological Society, London, Special Publications, **29**, 91–105.

WALDERHAUG, O. 1996. Kinetic modeling of quartz cementation and porosity loss in deeply buried sandstone reservoirs. *American Association of Petroleum Geologists Bulletin*, **80**, 731–745.

WATTS, N. L. 1987. Theoretical aspects of cap-rock and fault seals for single and two phase hydrocarbon columns. *Marine and Petroleum Geology*, **4**, 274–307.

WONG, T.-F. DAVID, C. & ZHU, W. 1997. The transition from brittle faulting to cataclastic flow in porous sandstones: Mechanical deformation. *Journal of Geophysical Research*, **102**, 3009–3025.

WORDEN, R. H., OXTOBY, N. H. & SMALLEY, P. C. 1998. Can oil emplacement stop quartz cementation in sandstones? *Petroleum Geoscience*, **4**, 129–138.

YIELDING, G., FREEMAN, B. & NEEDHAM, D. T. 1997. Quantitative fault seal prediction. *American Association of Petroleum Geologists Bulletin*, **81**, 897–917.

ZHANG, J., WONG, T.-F. & DAVIS, D. M. 1990. Micromechanics of pressure-induced grain crushing in porous rocks. *Journal of Geophysical Research*, **95**, 341–352.

ZHU, W. & WONG T.-F. 1997. The transition from brittle faulting to cataclastic flow: Permeability evolution. *Journal of Geophysical Research*, **102,** 3027–3041.

ZOBACK, M. D. & BYERLEE, J. D. 1976 Effect of high-pressure deformation on permeability of Ottawa Sand. *American Association of Petroleum Geologists Bulletin*, **60**, 1531–1542.

Physical character and fluid-flow properties of sandstone-derived fault zones

RICHARD G. GIBSON

Amoco Exploration and Production Co., 501 Westlake Park Blvd, Houston, TX 77079, USA

Abstract: Evaluation of hydrocarbon entrapment and production patterns in faulted sandstone reservoirs requires understanding of the nature and fluid-flow properties of sandstone-derived fault zones. This study documents the interrelationships between sandstone composition, deformation mechanisms, fault-zone character, and fluid-flow properties (permeability and capillary properties) using a global sandstone dataset.

Quartz-rich sandstones deform by cataclasis (most commonly), diffusive mass transfer, or a combination of these processes to form deformation bands. The fluid-flow properties of these zones depend on deformation mechanism(s). Faulting of mineralogically immature sandstones results in the formation of clay-matrix gouge zones by a combination of processes, including cataclasis, intergranular sliding in clay-rich materials, and diffusive mass transfer. Clay-matrix gouge zones generally have lower permeabilities and higher capillary displacement pressures than deformation bands.

Most deformation bands have capillary properties sufficient to maintain hydrocarbon column-height differences of less than 75 m across them, whereas clay-matrix gouge zones can potentially seal hydrocarbon columns with heights of several hundred metres. Both low-permeability deformation bands and clay-matrix gouge zones are likely to influence production patterns, although the magnitude of these effects will depend on the spatial distribution and abundance of faults and the permeabilities of the fault zones and undeformed sandstone.

Information about fault zone fluid-flow properties (e.g. permeability, porosity, capillary properties) in siliciclastic strata is necessary for accurate evaluation of fault-seal risk, model simulation of faulted petroleum reservoirs, and basin-scale fluid-flow modelling. This paper contains the results of a petrographic and laboratory study aimed at determining the fluid-flow properties of natural fault zones and evaluating the factors that influence these properties. Because significant difficulties are inherent in trying to collect coherent, representative samples of macroscopic fault zones for fluid-flow tests, the approach utilized in this study is to document the character and properties of small-displacement faults derived from sandstones of various compositions. Such fault zones are primarily mixtures of quartz, feldspar and phyllosilicates (micas + clay minerals) that vary significantly in character and show many features in common with macroscopic fault zones in siliciclastic stratigraphic sequences. Fault zones similar to some of those examined in this study have been studied by a number of previous workers (e.g. Aydin 1978; Pittman 1981; Jamison & Stearns 1982; Nelson 1985; Chester & Logan 1986; Underhill & Woodcock 1987; Trevena 1989; Knipe 1993; Antonellini & Aydin 1994). However, no previous study has addressed the spectrum of variability observed among the samples considered in this paper.

The data presented here demonstrate that faults formed from porous sandstones range in character from quartz-rich cataclastic zones to clay-matrix gouge zones, generally reflecting the composition of the protolith. Capillary pressure and permeability data collected on a suite of samples spanning this spectrum show systematic variations as a function of both sandstone and fault-zone composition, as well as the deformational processes active during faulting. Clay-matrix gouge zones consistently have low permeabilities and the highest capillary pressures (smallest effective pore-throat radii). All of the fault zones studied have permeabilities potentially low enough to influence hydrocarbon production from a sandstone reservoir. However, clay-matrix gouge zones also have small enough pore-throats to form capillary seals for significant hydrocarbon columns.

Sample description

The specimens used in this study include both outcrop and core samples collected from a number of different geographic areas (Table 1). Arrays of small-displacement faults like those

From Holdsworth, R.E. & Turner, J.P. (compilers) *Extensional Tectonics: Faulting and Related Processes*. The Geological Society, Key Issues in Earth Sciences, **2 (2)**, 249–263.
1476-315X/02/$15.00
First published in: "Coward, M.P., Daltaban, T.S. & Johnson, H. (eds) *Structural Geology in Reservoir Characterization*, The Geological Society, London, Special Publications **127**, 83–97".

Table 1. *List of samples, localities, descriptive information, permeability (k), and pore-throat radius (R) data*

Sample number	Stratigraphic Unit and Location (well names and depths in italics)	Fault type*	Fault displ.†	k (mD) (host ss.)	k (mD) (fault zone)	R (μm) (CH_4 inj.)	R (μm) (Hg inj.)
Deformation bands							
Cataclastic							
DAK-1	Dakota Fm, Wet Mountains, W of Canon City, CO, USA	R	meso	–	0.369	1.09	–
DAK-5	Dakota Fm, Wet Mountains., W of Canon City, CO, USA	R	micro	–	0.241	0.18	–
ENT-1	Entrada Fm., Colorado Plateau, near Moab, UT, USA	N	meso	–	–	–	2.1
FOUN-1	Fountain Fm., Front Range, NE of Canon City, CO, USA	N + SS	meso	–	0.123	0.42	–
NAV-1	Navajo Fm, Colorado Plateau, near Moab, UT, USA	N	meso	–	–	–	1.14
NUB-1	Nubia Fm, Gulf of Suez, Egypt (*July 62–86, 11387′*)	N	micro	–	8.002	10.9	–
OCR-1	Oil Creek Ss., Arbuckle Mountains, Mill Creek, OK, USA	SS?	meso	–	0.715	0.66	–
OCR-2	Oil Creek Ss., Arbuckle Mountains, Mill Creek, OK, USA	SS?	meso	3175	2.63	–	2.6
RUD-2	Rudeis Fm, Gulf of Suez, Egypt (July 8, *8812′*)	N	micro	–	1.48	2.17	–
TRIN-1	Gros Morne Fm, offshore SE Trinidad (*OPR-13, 12400′*)	N	meso	17.8	1.16	0.72	2.7
TRIN-5	Gros Morne Fm, offshore SE Trinidad (*Samaan B-1, 6812′*)	N	micro	–	–	–	2.7
TRIN-7	Gros Morne Fm, offshore SE Trinidad (*Poui A-4, 10514′*)	N	micro	210	7.21	–	–
Solution							
FUL-1	Fulmar Fm, North Sea, offshore UK (*30/12b-4, 12968′*)	N	meso	–	–	–	0.15
NUB-3	Nubia Fm, Gulf of Suez, Egypt (*Sidki-7, 10924′*)	N	micro	165	0.084	–	2.7
RUD-3	Nubia Fm, Gulf of Suez, Egypt (*July 8, 9067′*)	N	micro	–	0.081	0.53	–
WIL-1	Wilcox Fm, south TX, USA (*Amoco Peters Estate-1, 13344′*)	N	micro	1.20	0.206	0.80	–
WIL-1	Wilcox Fm, south TX, USA (*Amoco Peters Estate-1, 13352′*)	N	micro	0.314	0.032	0.31	–
Complex							
LYONS-3	Lyons Fm, Front Range, Colorado Spgs., CO, USA	R	micro	–	0.0053	<0.008	–
NUG-1	Nugget Fm, Sevier belt, WY, USA (*Champlin 404-C-1, 14154′*)	R	meso	96.2	0.0155	–	0.23
NUG-2	Nugget Fm, Sevier belt, WY, USA (*Champlin 404-C-1, 14186′*)	R	micro	5.79	0.0106	–	0.23
NUG-3	Nugget Fm, Sevier belt, WY, USA (*Champlin 404-C-1, 14214′*)	R	meso	–	0.0208	0.053	–
Clay-matrix gouge zones							
CA-1	Painted Cyn. fault zone, Mecca Hills, CA, USA	SS	macro	–	–	–	0.04
CA-2	San Andreas fault zone, Mecca Hills, CA, USA	SS	macro	–	0.055	<0.01	0.03
GOR-1	Pliocene ss. near San Andreas fault, S. Of Gorham, CA, USA	SS	micro	–	0.0002	<0.01	0.07
HEM-2	Hemlock Fm, Cook Inlet, AK, USA (*Mid. Ground Shoal-25, 8216′*)	R	micro	1-20	0.002	<0.025	–
HEM-3	Hemlock Fm, Cook Inlet, AK, USA (*Mid. Ground Shoal-25, 8049′*)	R	meso	–	–	–	1.08
MOAB-1	Culter Fm in Moab fault zone, N of Moab, UT, USA	N	macro	–	0.048	<0.01	0.12
MOAB-4	Culter Fm in Moab fault zone, N of Moab, UT, USA	N	macro	–	–	–	0.12
MYAN-1	Tertiary ss., Chindwin basin, Myanmar	N?	micro	5.5-45	0.024	0.034	–
MYAN-2	Tertiary ss., Chindwin basin, Myanmar	N?	micro	5.5-45	0.028	0.053	–
ROM-85	Mageresti Fm Carpathian belt, central Romania	R	micro	–	–	–	0.04
TRIN-9	Gros Morne(?) Fm, offshore SE Trinidad (*N. Poui 2, 11878′*)	N	micro	4.80	0.034	0.015	0.28

Dashes indicate that parameter was not measured.

* Fault type: N = normal; R = reverse; SS = strike-slip

† Fault displacement magnitude: micro ≤ *c.* 10 cm; meso = 10 cm to 10 m; macro ≥ *c.* 10 m

studied here are typical of deformed sandstones in regions of localized high strain, such as in areas of high bedding curvature and within deformation halos adjacent to macroscopic faults (Aydin 1978; Jamison & Stearns 1981; Chester & Logan 1986; Antonellini & Aydin 1994; Knipe *et al.* 1996). Fault types (normal, reverse, strike-slip) and ranges of displacement magnitude are listed in Table 1 for each of the specimens included within the study. In the case of most samples, the fault displacement is small enough that it is clear that the fault-zone material was derived directly from the adjacent sandstone. Two samples (CA-1, CA-2) were collected within major displacement zones within the San Andreas fault system and their protoliths are not known.

The samples were compositionally characterized by (1) point-counting of impregnated, stained thin sections of the protolith sandstone and (2) collecting semi-quantitative X-ray diffraction (XRD) data on the fault-zone material and, where possible, its sandstone protolith. These data are summarized in Table 2 and shown graphically in Fig. 1. XRD analyses have analytical uncertainties on the order of $\pm 10\%$. Significant XRD mineralogical differences between some of the fault-zone material and the corresponding protoliths are shown in Table 2 and are indicated by arrows in Fig. 1b. Faulted sandstones exhibiting obvious diagenetic modification after faulting were avoided during the sample selection process. Therefore, no distinction is made in either the point-count or XRD data between detrital and authigenic phases, since both contribute to the character of subsequent faults cross-cutting these sandstones.

For descriptive purposes, the suite of fault zones is divided into two broad textural categories, deformation bands and clay-matrix gouge zones (Table 1). The distinction between these two groups is based primarily on fault-zone appearance in outcrop or drill-core and secondarily on characteristics observed in thin section. The criteria by which they are distinguished are outlined below.

Deformation bands

Deformation bands (Aydin 1978) are narrow (< 2 mm thick), planar displacement zones that are a common feature in porous sandstones and have been referred to in the literature by a variety of names (see Pittman 1981; Jamison & Stearns 1981; Nelson 1985; Underhill & Woodcock 1987; Fowles & Burley 1994). In drill-core or outcrop, they are typically lighter coloured and better indurated than the surrounding sandstone and occur as either complex, intersecting arrays or zones comprising multiple, sub-parallel bands (Fig. 2a). They occur preferentially in mineralogically mature, quartz-rich ($>65\%$ of solids) sandstones with less than 30% clay + mica + lithic fragments (Fig. 1a, Table 2). XRD data show that the deformation bands have total phyllosilicate contents below 25% (most below 10%) and do not differ significantly in mineralogy from their sandstone protoliths (Fig. 1b, Table 2). On the basis of the deformation mechanism(s) responsible for their formation, the deformation bands can be sub-divided into three groups.

Cataclastic deformation bands consist of fractured and fragmented detrital grains. They are finer grained, more poorly sorted, and contain smaller pores than the protolith sandstone. Deformation associated with these zones is extremely localized, with sharp boundaries between the zone of grain fragmentation and adjacent undeformed rock (Fig. 2b). Grain fragments within the deformation bands show no apparent preferred orientation and are typically sub-equant, with grain boundaries that do not interpenetrate with neighbouring fragments (Fig. 2c and d). The dominant process responsible for the formation of these deformation bands is interpreted to be cataclasis (see also Aydin 1978; Pittman 1981; Jamison & Stearns 1981; Underhill & Woodcock 1987; Antonellini & Aydin 1994; Fowles & Burley 1994). Of the three types of deformation bands, this is the most common variety within the sample set.

Solution deformation bands consist of tightly packed grains (mostly quartz) with serrated, interpenetrating grain boundaries and irregular concentrations of opaque material and/or phyllosilicates along grain boundaries (Fig. 3a). These zones show little or no indication of fracturing or fragmentation of detrital quartz grains. Instead, grains within them have a weak to strong dimensional alignment parallel to the stylolitic seams and fault-zone margins. The long axes of these aligned grains are comparable in length to the diameter of the detrital grains within the undeformed sandstone (Fig. 3a). The welded grain contacts and grain-shape alignment are interpreted to reflect quartz dissolution along grain boundaries sub-parallel to the fault-zone margin. This diffusive mass transfer may have been superimposed onto pre-existing zones formed by grain-boundary sliding, such as those described by Sverdrup & Prestholm (1990).

Table 2. *Compositional data for samples included in this study*

Sample Number	X-Ray diffraction data				Thin-section point-count data					
	Quartz	Fsp	Phyl	Other	Quartz	Fsp	Lithics	Phyl	Other	Porosity
Deformation bands										
Cataclastic										
DAK-1	98	0	2	0	76.1	0.0	0.0	0.0	0.0	23.9
DAK-5	95	0	5	0	61.9	0.0	0.0	24.0	0.3	13.8
ENT-1	–	–	–	–	76.8	3.2	0.0	3.7	0.4	15.8
FOUN-1	65	13	23	0	50.9	30.3	0.0	10.8	0.6	7.4
NAV-1	–	–	–	–	65.4	8.1	0.0	7.8	0.8	17.8
OCR-1	98	0	2	0	67.1	0.0	0.0	10.3	0.0	22.6
OCR-2	95	0	4	1	–	–	–	–	–	–
RUD-2	90	0	10	0	61.6	1.1	1.2	1.9	19.0	15.2
TRIN-1	95	0	5	0	66.6	3.4	2.3	5.7	0.4	21.6
TRIN-5	–	–	–	–	60.2	2.6	5.0	2.0	1.4	28.8
TRIN-7	–	–	–	–	55.7	4.0	6.8	9.8	0.7	23.1
Solution										
FUL-1	–	–	–	–	60.0	28.0	tr	1.0	2.0	9.0
NUB-3	–	–	–	–	68.5	0.0	0.0	24.3	0.0	7.3
RUD-3	90	0	10	0	65.8	1.7	1.2	4.9	12.5	13.9
WIL-1	85	0	15	0	66.9	0.0	13.5	6.7	2.1	10.8
WIL-2	85	0	15	0	58.4	0.0	16.8	9.3	3.4	12.1
Complex										
LYONS-3	90	3	7	0	59.5	10.2	6.4	10.4	0.0	13.5
NUG-1	80	5	15	0	79.7	8.1	0.4	7.3	0.2	4.3
NUG-2	80	5	15	0	–	–	–	–	–	–
NUG-3	95	3	3	0	82.4	5.5	1.1	2.1	0.0	8.9
Clay-matrix gouge zones										
CA-1	30	5	40	25	–	–	–	–	–	–
CA-2	25	0	55	20	–	–	–	–	–	–
GOR-1	25 (35)	10 (25)	65 (40)	tr	28.7	52.0	0.0	17.7	0.2	1.4
HEM-2	40	15	45	0	34.3	22.3	32.1	6.3	5.0	tr
HEM-3	60 (50)	10	30 (40)	tr	49.5	9.0	35.5	2.0	tr	4.1
MOAB-1	60	5	35	0	32.1	15.9	0.0	48.0	4.1	tr
MOAB-4	25 (35)	0 (15)	75 (40)	0 (10)	29.0	34.1	0.4	30.4	6.1	tr
MYAN-1	40	30	30	0	30.6	16.2	18.2	7.2	14.7	13.1
MYAN-2	50	30	20	0	37.6	25.0	11.8	20.3	3.4	1.8
ROM-85	–	–	–	–	36.4	5.8	19.2	13.6	24.0	1.0
TRIN-9	70	3	28	0	48.0	13.4	26.8	3.1	0.2	8.5

X-ray diffraction data are for fault-zone material and protolith (protolith value in parentheses only where >10% (weight) different from that of fault zone). Point-count data are for undeformed sandstone adjacent to fault zones.

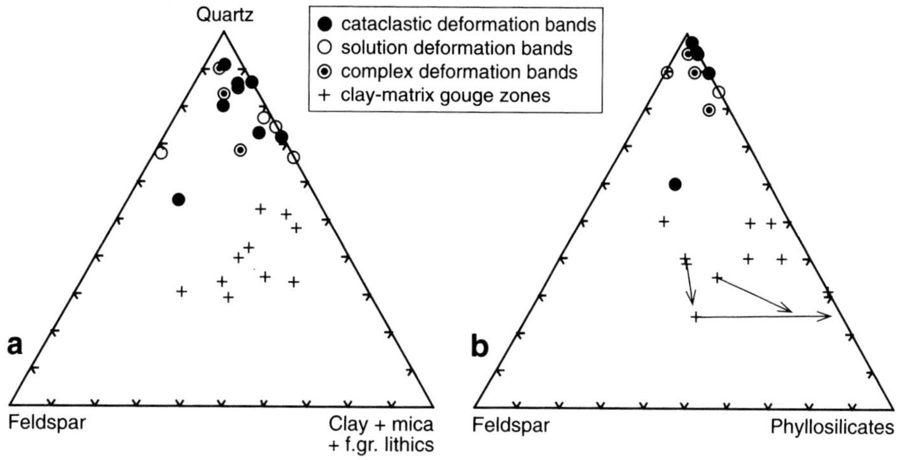

Fig. 1. Porosity-free compositional plots of sandstone and fault-zone samples used in this study. (**a**) Point-counts of thin sections of unfaulted sandstone. (**b**) X-ray diffraction data of sandstone and fault gouge: dashed lines connect sandstone composition (at symbol) with that of fault gouge (arrowhead) for cases where the two are not coincident within analytical uncertainty.

Complex deformation bands are zones of extremely fine grain size with thin, irregular laminae of opaque material parallel or at a low angle to their margins (Fig. 3b). Unlike the cataclastic deformation bands, boundaries of grains within the fine-grained zones tend to be sutured and interlocking (Fig. 3c) and there is locally a grain-shape alignment. Discontinuous mica

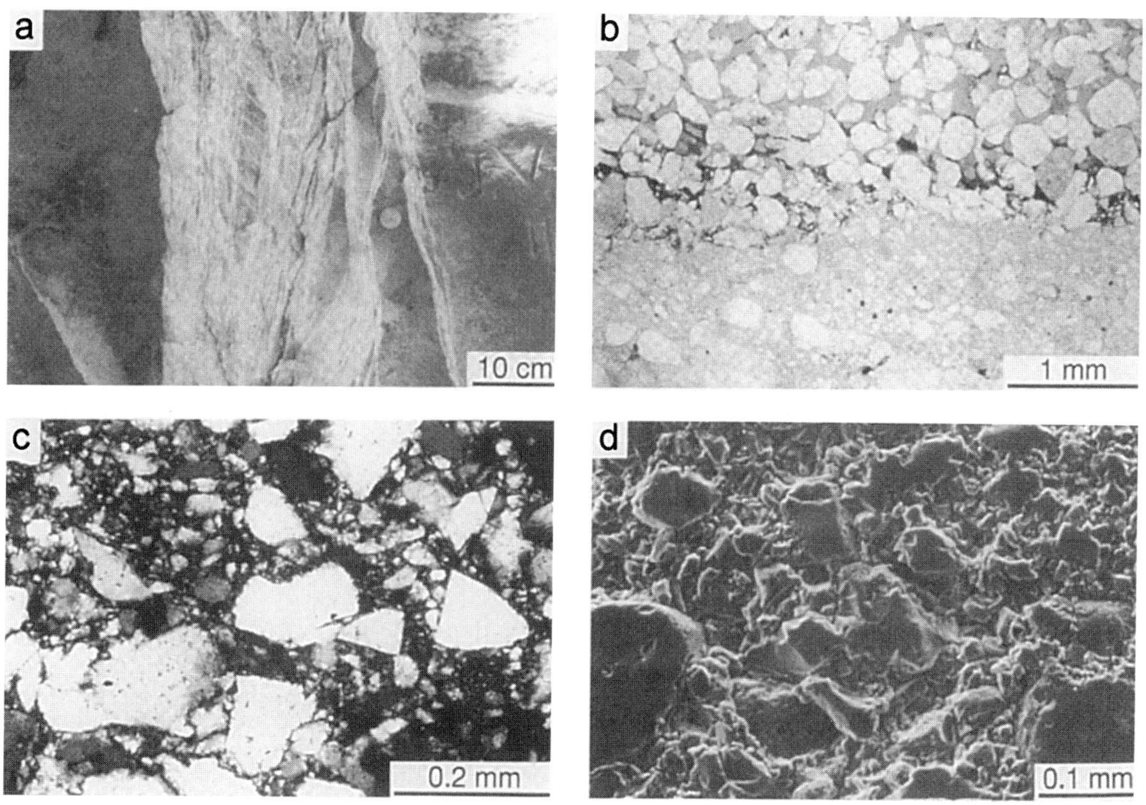

Fig. 2. Cataclastic deformation bands: (**a**) zone of deformation bands in outcrop of Dakota Sandstone, Canon City, Colorado, USA; (**b**) thin section of (a) showing cataclastic deformation band (bottom) and adjacent sandstone (plane light photomicrograph); (**c**) enlargement of (b) showing fractured, angular quartz grains in cataclastic deformation band (crossed polars photomicrograph); (**d**) scanning electron micrograph of (b) showing conchoidal fracture faces on quartz fragments and discrete nature of grain boundaries between fragments.

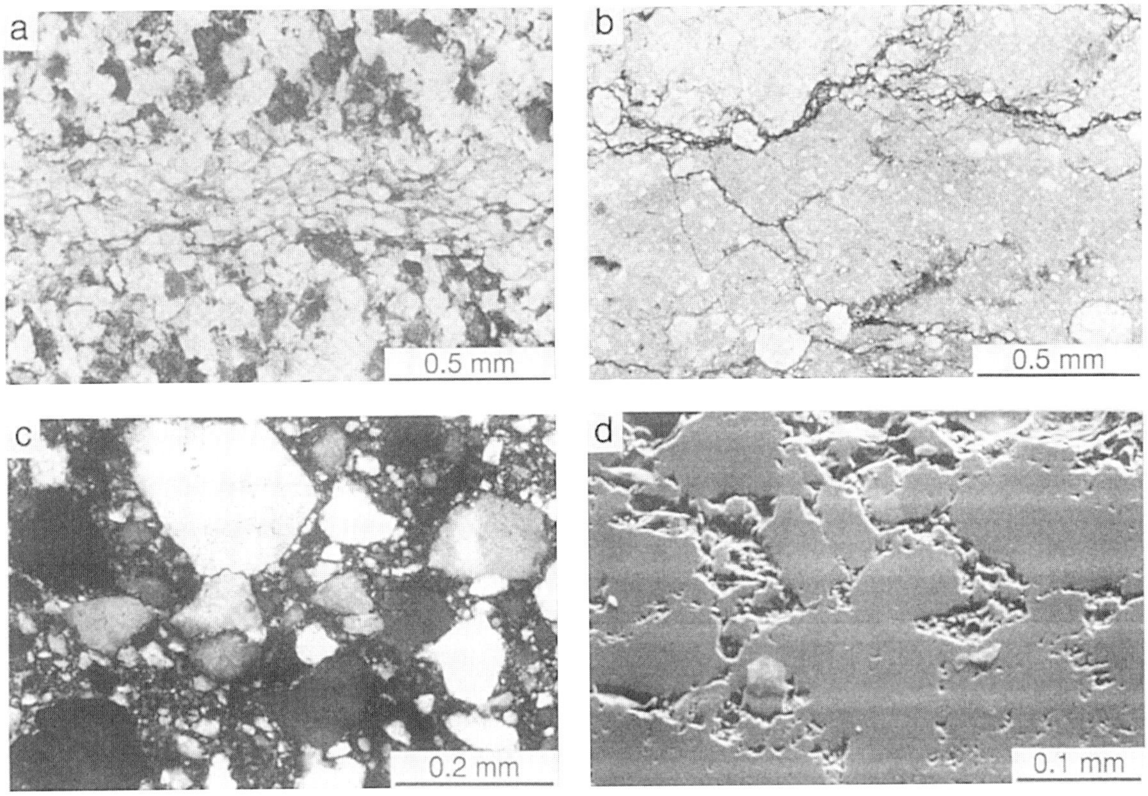

Fig. 3. Solution and complex deformation bands: (**a**) solution deformation band (across centre) showing aligned, unfractured grains bounded by dark, phyllosilicate-rich selvedges, Wilcox Formation (Fm), south Texas, USA (plane light photomicrograph); (**b**) complex deformation band with dark stylolites parallel to deformation band margin (vertical), Nugget Fm, western Wyoming, USA (plane light photomicrograph); (**c**) enlargement of (b) showing sutured quartz grain boundaries within complex deformation band (crossed polars photomicrograph); (**d**) scanning electron micrograph of (b) showing indistinct nature of quartz fragment boundaries resulting from pressure solution and local cementation.

films occur along quartz grain boundaries that are sub-parallel to the fault-zone margin. The textural contrast between these zones and cataclastic deformation bands is especially evident in scanning electron micrographs, where the boundaries between adjacent fragments in the deformed zone appear to be partially to entirely welded (Fig 3d). The microscopic character of these zones is that of cataclastic deformation bands that have been overprinted and modified by pressure solution. Therefore, these zones are interpreted to have formed by a combination of cataclasis and diffusive mass transfer.

Clay-matrix gouge zones

Clay-matrix gouge zones appear in outcrop, drill-core and hand sample as shiny, lineated, dark-coloured surfaces or zones (Fig. 4a) that commonly serve as planes of preferential parting. In many cases, the fault-zone material is poorly indurated and readily disaggregates into platy, lenticular fragments upon handling. Thin clay-matrix gouge zones in various orientations sometimes form intersecting arrays that separate less-deformed sandstone into polygonal blocks. They are preferentially developed from mineralogically immature sandstones, specifically those with greater than 30% clay + mica + lithics and less than 60% quartz (Fig. 1a, Table 2). Mineralogically, the gouge zones range in composition from 20 to 80% phyllosilicates and, in some cases, show significant differences from their protoliths (Fig. 1b, Table 2).

In thin section, the clay-matrix gouge zones vary in character both within individual samples and between samples from different locations. The principal feature common to all samples is the presence of a fine-grained, clay-bearing matrix within which the phyllosilicate grains are strongly aligned (Fig. 4b). The foliation defined by these phyllosilicates is oriented at low angles or parallel to the fault zone margin (Fig. 4c). In some cases, the phyllosilicate fabric is accompanied by irregular laminae of opaque

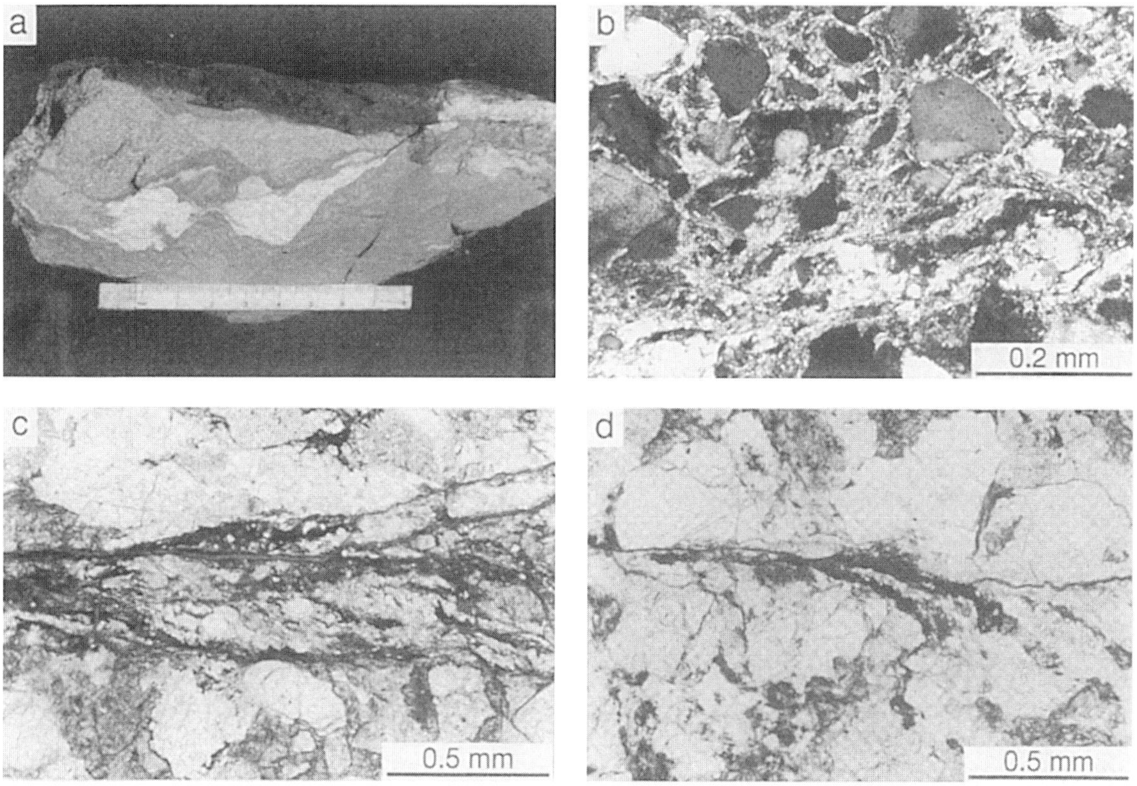

Fig. 4. Clay-matrix gouge zones: (**a**) cut slab of dark-coloured clay-matrix gouge in micaceous Cutler Sandstone from Moab fault zone, central Utah, USA (scale in centimetres); (**b**) thin section of (a) showing aligned, fine-grained phyllosilicates in clay-matrix gouge (crossed polars photomicrograph); (**c**) prominent oblique foliation defined by phyllosilicate aggregates, Hemlock Fm, Cook Inlet, Alaska, USA (plane light photomicrograph); (**d**) deflection of phyllosilicate-rich lithic clast into incipient clay-matrix gouge zone, Hemlock Fm, Cook Inlet, Alaska (plane light photomicrograph).

material. Remnants of the protolith, such as shattered detrital grains and pieces of intact sandstone, occur as porphyroclasts within this foliated matrix. Margins of the gouge zones vary from sharp to diffuse, locally displaying a gradual decrease of foliation intensity into the adjacent rock. Within the sandstone outside of the gouge zones, micaceous detrital grains (e.g. lithic fragments) commonly show evidence of ductile grain distortion and dimensional alignment into parallelism with the foliation (Fig. 4d). In some cases, feldspar grains are partially or completely altered to fine-grained phyllosilicate aggregates in close proximity to the gouge zones.

Sandstone-derived fault rocks similar to the clay-matrix gouge zones discussed above have been previously described by Chester & Logan (1986), Trevena (1989), and Knipe (1993). The complex character of these zones is interpreted to reflect their formation by the simultaneous activity of several deformation mechanisms, including cataclasis, diffusive mass transfer, and intragranular sliding within phyllosilicate-rich domains (detrital fragments or gouge matrix).

Analysis of fluid-flow properties

Fluid-flow tests on the fault-zone samples included analysis of specific permeability and capillary properties. Since permeability is the fundamental factor controlling the rate of fluid flow in response to an applied differential head (see Hubbert 1956), it is the parameter likely to be most important in dictating fluid-flow patterns within a reservoir during production. In contrast, capillary properties dictate the magnitude of the differential hydrocarbon buoyancy pressure (typically manifest as differential hydrocarbon column height) that can be maintained across a lithologic boundary. Capillary properties control the ability or inability of a lithologic boundary to act as a barrier to migrating hydrocarbons (see overview by Schowalter (1979)) on a long-term time frame. Therefore, measurement of permeability and capillary properties provides information on the potential influence of fault zones on fluid flow over two different time scales.

Permeability

Where possible, 1.9 to 2.5 cm diameter plugs were cut approximately perpendicular to the fault zones, and the plugs were trimmed to maximize the portion consisting of fault-zone material. These plugs were used for Klinkenberg-corrected permeability measurement at 20.7 MPa (3000 psi) hydrostatic confining pressure. Measurement error associated with the permeability tests is estimated to be <15% of the measured value for samples with permeabilities greater than 1 mD and up to 35% for lower permeability samples (G. Potter, Amoco, pers. comm.). Despite these large uncertainties in absolute permeability values, the data are considered sufficient to detect order-of-magnitude differences between samples within the dataset.

Many of the plugs used for permeability analysis consisted of both protolith sandstone and fault-zone material. In these cases, the permeabilities measured in the laboratory represent a harmonic average of the fault-zone and sandstone permeabilities (see Amyx & Bass (1962) for discussion of permeability averaging in heterogeneous media). True fault-zone permeabilities for these samples were calculated from a knowledge of the host sandstone permeability (estimated from previous core analysis results or measured on companion plugs of unfaulted material) and the proportion of the plug (measured parallel to the axial flow direction) consisting of fault-zone material.

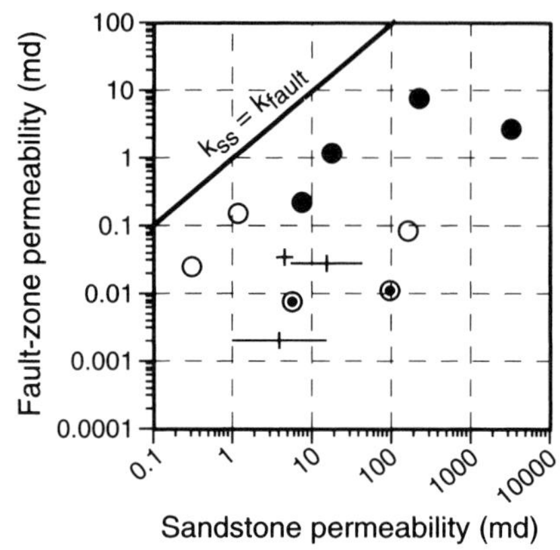

Fig. 5. Comparison of sandstones and fault-zone permeabilities. Symbols as in Fig. 1.

Fault-zone permeabilities span five orders of magnitude (Table 1). For all of the samples in which the permeability of both the sandstone and fault zone were measured, the fault zone has lower permeability (Fig. 5). These results are consistent with the conclusions of other authors (e.g. Aydin 1978; Pittman 1981; Nelson 1985; Trevena 1989; Antonellini & Aydin 1994; Fowles & Burley 1994) who have demonstrated significant permeability reduction associated with faulting of porous sandstones. In the sample set considered here, the permeability reduction relative to the protolith sandstone is one to three orders of magnitude for cataclastic and solution deformation bands, and two to four orders of magnitude for clay-matrix gouge zones and complex deformation bands.

The fault-zone permeabilities generally decrease with decreasing protolith quartz content (Fig. 6a) and increasing fault-zone phyllosilicates (Fig. 6b). Deformation-band permeabilities span

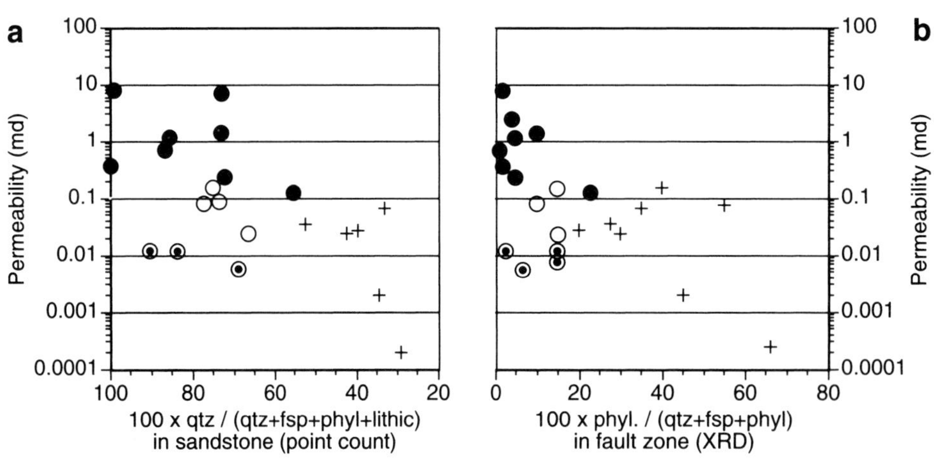

Fig. 6. Fault-zone permeability at 20.7 MPa (3000 psi) hydrostatic confining pressure as a function of (**a**) sandstone protolith composition and (**b**) fault-zone composition. Symbols as in Fig. 1.

a wide range, from 0.008 to 9 mD, showing a strong dependence on the deformation mechanisms active in band formation. Cataclastic deformation bands have the largest permeabilities, whereas the complex deformation bands (cataclasis plus diffusive mass transfer) have the smallest values. Solution deformation bands have permeabilities intermediate between the other two types. Clay-matrix gouge zone permeabilities are below 0.07 mD and, with the exception of a couple of data points, show a crude tendency to decrease with increasing phyllosilicate content (Fig. 6b). The upper part of their permeability range overlaps with that of solution and complex deformation bands.

Measured permeabilities of the cataclastic deformation bands (0.1–9 mD) are similar to those previously determined for cataclastic deformation bands in the Oil Creek Sandstone (Pittman 1981), the Nubia Sandstone (Harper & Moftah 1985), quartzose sandstones in the Colorado Plateau (Antonellini & Aydin 1994), and Permian sandstones of northwest England (Fowles & Burley 1994). Permeabilities of complex deformation bands are similar to other values reported for faults in the Nugget Sandstone (Nelson 1985). The only clay-rich fault-zone permeabilities reported in the literature are those of Morrow *et al.* (1981) on San Andreas fault gouge. They reported a brine permeability of approximately 0.00001 mD, more than three orders of magnitude lower than the San Andreas gouge permeability measured here (CA-2, Table 1). This difference may be related, in part, to extreme water sensitivity of clays in San Andreas gouge samples.

Capillary properties

Capillary pressure data were collected using two different techniques: (1) methane injection into brine-saturated plugs, and (2) mercury injection into evacuated samples. Because these two test methods utilize different fluid systems, comparison of the resulting pressure measurements cannot be done directly (see Schowalter 1979). In order to allow comparison, the pressure results were converted to effective pore-throat radii (R) using (Washburn 1921):

$$R = 2\gamma \cos\theta / P_d$$

where P_d is the measured pressure at any point during a test, γ is the interfacial tension between the two fluids (480 dyn cm^{-1} for mercury–air; values from Hough *et al.* (1951) for methane–brine), and θ is the hydrocarbon–brine–solid contact angle (140° for mercury–air; 0° for methane–brine). Capillary pressure results throughout the remainder of this paper are expressed as effective pore-throat radii.

Methane injection was used on plug samples that could be adequately sealed in a Hassler sleeve. In these tests, a brine-saturated (50 000 ppm NaCl, 5000 ppm CaCl) core plug was put under 20.7 MPa (3000 psi) hydrostatic confining pressure and methane was applied to one end. The methane pressure was increased in a stepwise fashion to simulate the build-up of hydrocarbon buoyancy pressure adjacent to a seal. The pressure at the first appearance of methane on the downstream end of the plug is considered the capillary breakthrough pressure of the fault zone. These tests were terminated either at methane breakthrough or when the applied methane pressure was equal to the confining pressure, in which case only a minimum breakthrough pressure could be determined.

Samples that could not be satisfactorily plugged were tested using high-pressure mercury injection. In these tests, the samples were unconfined and mercury could generally enter the sample from all directions. Therefore, these tests measure the pore-throat population of both the fault zone and any undeformed host rock adjacent to it. As a result, interpretation of the fault-zone capillary properties is less straightforward than in the case of methane injection. Figure 7a illustrates the basis for interpretation of composite plugs (sandstone + fault zone). Three samples from the same depth interval within one drill-core of Nugget Sandstone were tested, one consisting of unfaulted sandstone, a second cut entirely from a zone of complex deformation bands, and a third composed of sandstone transected by a complex deformation band. Changes in curve slope indicate significant increases in mercury saturation at 0.4 and 3.4 MPa (60 and 500 psi) for the sandstone and fault-zone samples, respectively. These pressures are interpreted to be the capillary entry pressures of the sandstone and fault zone, and correspond to pore-throat radii of approximately 2 and 0.2 μm. The mercury injection curve for the compound sample shows multiple bends, one at a pressure corresponding to mercury entry into the host sandstone and the other corresponding to entry into the fault zone. Dual inflections of this sort are observed on the mercury injection curves for other compound (sandstone + fault) samples. In these cases (e.g. Fig. 7b), the inflection at the smaller pore-throat radius is interpreted to correspond to the entry of mercury into the finer-grained fault-zone material.

Several samples were tested using both mercury and methane injection in order to understand the

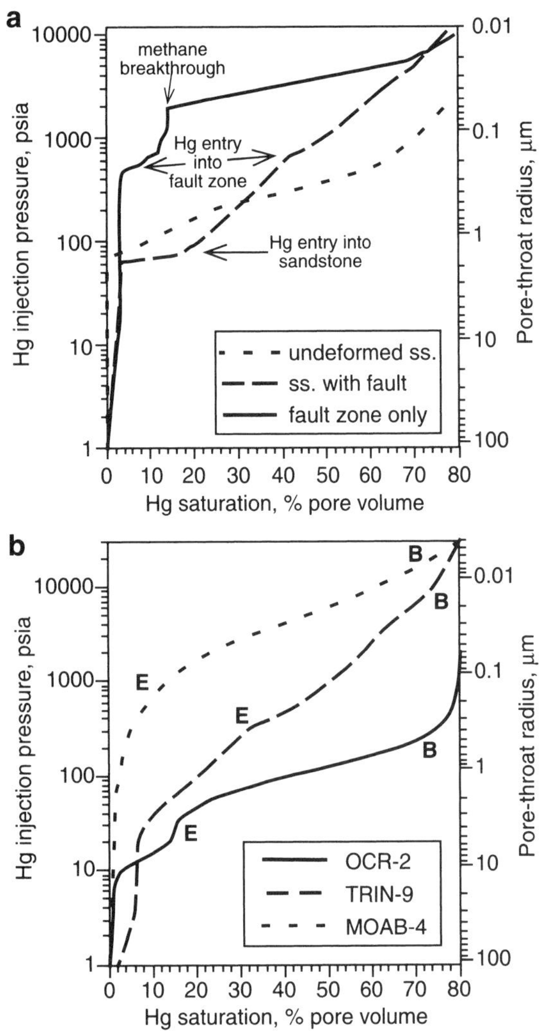

Fig. 7. Examples of capillary pressure data: (**a**) mercury injection data for three samples of Nugget Sandstone discussed in text; (**b**) data for three samples showing comparison between entry pressures (E) determined by mercury injection and breakthrough pressures (B) determined by methane injection.

relationship between the information gained using both techniques. Figure 7a and b illustrate these comparisons. Pore-throat radii corresponding to non-wetting phase entry (mercury injection) into the fault zones are three to 20 times larger than those corresponding to non-wetting phase breakthrough (methane injection). Such disparity between the two techniques is not surprising for several reasons: (1) a greater uncertainty is associated with identifying the fault-zone entry pressures from the mercury-injection data (see above); (2) significant non-wetting phase saturation is typically required to form a connected fluid filament through the sample before breakthrough occurs (Schowalter 1979; Katz & Thompson 1987); (3) the higher confining pressure (20.7 MPa) used in the methane injection tests could have caused pore-throat collapse relative to the samples used for mercury injection; and (4) the samples (especially the clay-matrix gouge zones) are anisotropic and it is likely that the highest capillary pressures would be required for non-wetting phase breakthrough across the fault zones. Despite these data quality limitations, the results are believed to be sufficient for identifying order-of-magnitude differences between similarly analysed samples within the dataset.

Measured fault-zone pore-throat radii corresponding to non-wetting phase entry (small symbols) and breakthrough (large symbols) are plotted as a function of sandstone and fault-zone composition in Fig. 8. Measured values span more than three orders of magnitude, with an overall trend of decreasing pore-throat radii as a function of decreasing sandstone quartz content and increasing fault-zone phyllosilicates. The contrast between capillary entry and breakthrough pressures for materials of equivalent composition clearly shows up on these plots. The general distribution of the data in

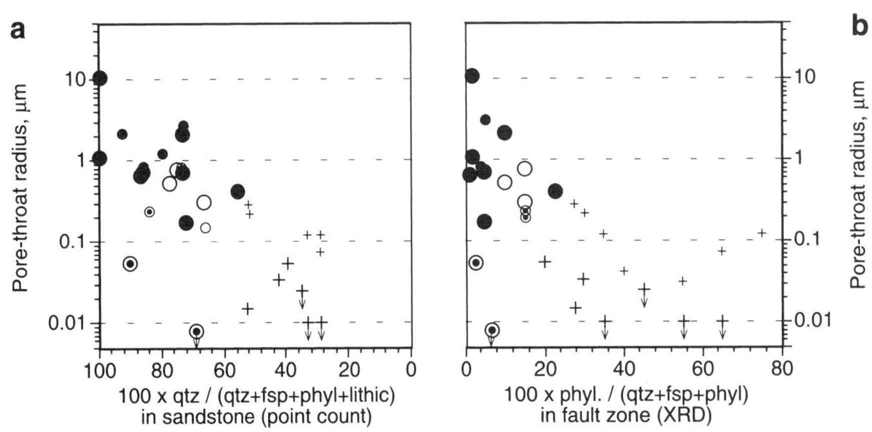

Fig. 8. Effective pore-throat radii of fault zones as a function of (**a**) sandstone composition and (**b**) fault-zone composition. Symbols as in Fig. 1. Arrows connected to some symbols indicate that these are maximum values.

Fig. 8 is similar to that of the permeability data: cataclastic deformation bands have the largest pore-throats, complex deformation bands and clay-matrix gouge zones have the smallest pore-throats, and solution deformation bands have intermediate values.

Discussion

Ranges of permeability (k) and effective pore-throat radius (R) for clay-matrix gouge zones and the various types of deformation bands are summarized in Fig. 9. These results show that the fluid-flow properties of quartz-phyllosilicate fault gouge depend on the gouge composition and deformation mechanisms involved in its formation. Clay-matrix gouge zones, derived from phyllosilicate-bearing sandstones and consisting of more than *c.* 20% phyllosilicates, have the smallest pore throats and lowest permeabilities ($k < 0.08$ mD, $R \ll 0.3\,\mu$m). The fluid-flow properties of deformation bands derived from quartzose sandstones are more variable ($k = 0.008$–9 mD, $R = 0.02$–$10\,\mu$m) and show a strong dependence on deformation mechanism(s). Within this compositional range, only those fault zones exhibiting evidence of both cataclastic and diffusive mass transfer processes, have fluid-flow properties comparable to those of the clay-matrix gouge zones. Specimens deformed by either individual deformation mechanism (cataclasis or diffusive mass transfer) have greater permeabilities and larger effective pore-throat radii.

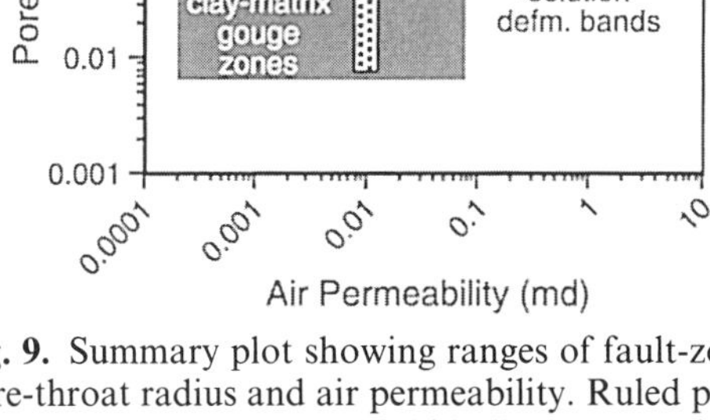

Fig. 9. Summary plot showing ranges of fault-zone pore-throat radius and air permeability. Ruled portion of clay-matrix gouge zone field indicates pore-throat radii based on mercury injection; solid portion is based on methane injection tests.

Seals for hydrocarbon entrapment

In order to understand the implications of Fig. 9 for long-term hydrocarbon seals, it is important to consider how much hydrocarbon column could be sealed by fault zones with various pore-throat radii. Capillary pressure equations presented by Schowalter (1979) can be manipulated to the form:

$$H = 2\gamma(\cos\theta)/R(\Delta\rho)$$

which expresses the maximum sealable hydrocarbon column height (H) as a function of seal pore-throat radius (R), hydrocarbon–brine interfacial tension ($g\gamma$), hydrocarbon–brine–solid contact angle (θ), and *in situ* density difference between the hydrocarbons and brine ($\Delta\rho$). This form of the equation assumes that the capillary properties of the reservoir rock can be neglected. Figure 10 is a plot of the equation shown above using input values appropriate for a gas and brine system at the temperature and pressure specified. Superimposed onto this plot are general domains corresponding to the pore-throat radius values determined in this study for cataclastic deformation bands and clay-matrix gouge zones. The clay-matrix gouge zones have pore-throats that are sufficiently small to seal gas columns of 270 m or more, whereas the deformation bands are unlikely to seal columns larger than 140 m. *In situ* hydrocarbon–brine interfacial tension values are likely to be somewhat lower than those predicted

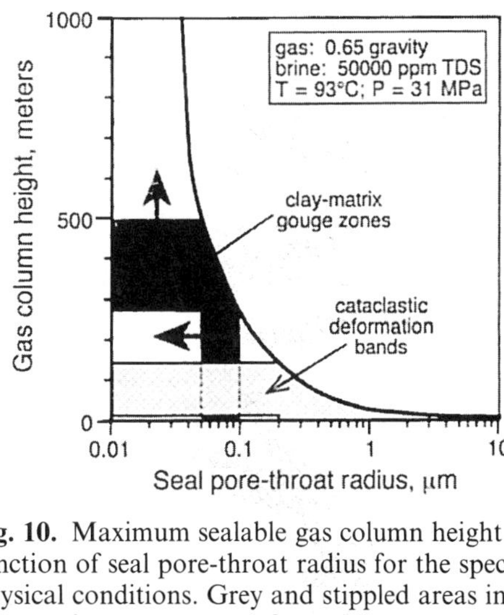

Fig. 10. Maximum sealable gas column height as a function of seal pore-throat radius for the specified physical conditions. Grey and stippled areas indicate ranges of each parameter for clay-matrix gouge zones and cataclastic deformation bands based on experimental data. Only the upper edge of the pore-throat radius range is shown for clay-matrix gouge zones; arrows indicate possible continuation of this field to smaller pore-throats and larger column heights

based on laboratory interfacial tension data (see Firoozabadi & Ramey,1988), which were used in the generation of Fig. 10. Therefore, under natural sub-surface conditions, sealable column heights for the fault zones considered are likely to be closer to the low end of the ranges indicated. For different hydrocarbon–brine systems and different pressure–temperature conditions, the absolute column height values will change, although clay-matrix gouge zones should be able to seal significantly longer columns than cataclastic deformation bands under all conditions.

Solution deformation bands and complex deformation bands are not considered in Fig. 10. Complex deformation bands have very small pore-throat radii and could seal hydrocarbon columns comparable to those maintained by clay-matrix gouge zones. However, their occurrence cannot be predicted solely from rock compositional data. Pore-throat radii of solution deformation bands are intermediate between those of cataclastic bands and clay-matrix gouge zones (Fig. 9). Thus, they should have a correspondingly intermediate seal capacity.

Previous studies of petroleum fields in siliciclastic sedimentary sections (Weber *et al.* 1987; Smith 1980; Bouvier *et al.* 1989; Gibson 1994; Fristad *et al.* 1996) have shown that the best fault seals exist where a phyllosilicate-rich fault zone is present across the faulted reservoir termination. In these field examples, the fault zones are typically formed by a 'shale-smear' mechanism involving ductile deformation of shale beds intercalated with the reservoir sands (Fig. 11a). The clay-matrix gouge zones examined in this study can be viewed as small-scale, natural models of these macroscopic fault zones, with the relatively ductile, phyllosilicate-bearing components of the sandstones playing the same role as the shale layers at a larger scale (Figs 4d and 11b). At the macroscopic scale, shale-smear zones appear to form along faults cutting stratigraphic sections containing in excess of 15–30% shale (see Weber 1987; Lindsay *et al.* 1993; Gibson 1994; Fristad *et al.* 1996). Similarly, the small-scale clay-matrix gouge zones examined here formed from sands containing more than 20–30% phyllosilicate-bearing components. The correspondence between the amount of phyllosilicate-rich material necessary for the generation of fault gouge with good sealing properties is striking, and supports the similarity of processes at two markedly different scales.

One implication of the data presented here is that fault zones cutting through mineralogically immature sandstones can form high-quality seals, even where shale interbeds are not present. Knipe (1993) reached a similar conclusion and illustrated the microstructural change associated with development of a fault rock similar to the clay-matrix gouge zones discussed in this paper. Figure 12 is a schematic cross-section that illustrates one case where the presence of sandstone-derived, clay-matix gouge zones could be important. The cross-section shows two

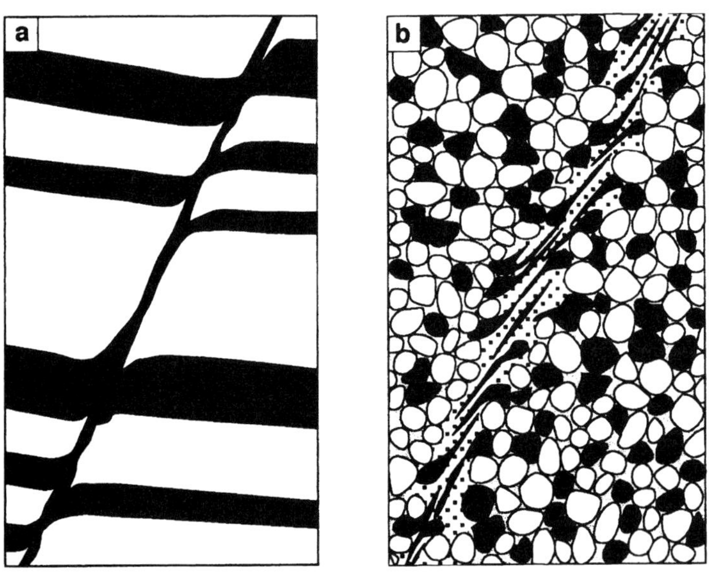

Fig. 11. Comparison of (a) 'shale-smear' mechanism of phyllosilicate-rich fault zone formation from interbedded sands and shales (black) with (b) formation of clay-matrix gouge from sandstone containing phyllosilicate-bearing fragments (black). Figure (a) modified from Weber *et al.* (1978).

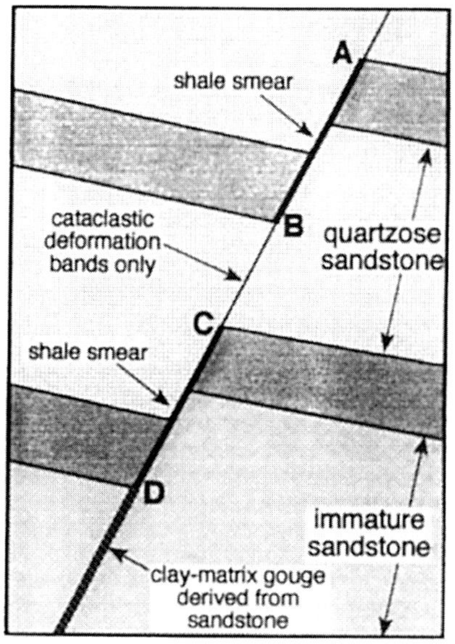

Fig. 12. Schematic cross-section showing phyllosilicate-rich fault zone distribution in siliciclastic sequences containing mineralogically mature sands (top) and immature sands (bottom). Shale beds are

normal-fault traps, the shallower one involving a quartzose sandstone and the deeper one involving a mineralogically immature sandstone. For the shallower trap, the spatial extent of a sealing phyllosilicate-rich fault zone includes only that portion of the fault along which a shale smear is developed (A–B in Fig. 12). Between B and C (Fig. 12), the fault zone is likely to consist solely of deformation bands derived from the quartzose reservoir sandstone. Because of the relatively large pore-throats of most deformation bands, the potential hydrocarbon column that can be sealed in this trap will probably be limited by a spill point near the base of the shale smear (point B). For the deeper reservoir, however, a phyllosilicate-rich fault zone will exist along the entire fault termination of the sandstone, even where no shale has been displaced along the fault (below D, Fig. 12). Thus, no cross-fault spill point exists for this trap, and there is potential to seal a longer hydrocarbon column than in the shallower sandstone. Thus, for a specified fault displacement, fault seals are likely to have greater spatial continuity in siliciclastic sections composed of mineralogically immature sandstones than in sections containing quartzose sandstones.

Seals for hydrocarbon production

From the perspective of hydrocarbon production from a sandstone reservoir, the most important facet of the fault zones discussed in this paper is that they are of lower permeability than the sands from which they are derived (see also Aydin 1978; Pittman 1981; Nelson 1985, Antonellini & Aydin 1994; Fowles & Burley 1994). The presence of planar zones of low permeability in an otherwise permeable reservoir will influence both the overall (bulk) reservoir permeability and the permeability anisotropy. Permeability parallel to a dominant fault set will be an arithmetic (thickness-weighted) average permeability (Amyx & Bass 1962). Since the fault zones are generally thin, this value will be dominated by sandstone permeability and will probably be only slightly lower than that of the undeformed sandstone. In contrast, flow across the faults will be controlled by a harmonic average permeability (Amyx & Bass 1962) calculated from the permeabilities of the sandstone and fault zones.

Figure 13 illustrates how a harmonic average permeability depends on the proportion of the flow path composed of fault-zone material and the permeabilities of the two components. One of the things evident in this diagram is that as little as 10% low-permeability fault-zone material along a flow path can result in dramatic bulk permeability reduction. The degree of permeability reduction (i.e. number of times reduction) varies depending on the permeability values of the host sandstone and fault zone. Therefore, the presence of fault zones of an arbitrary permeability (e.g. 1 mD) will have a much more damaging effect on the flow characteristics of a high-permeability reservoir than of a low-permeability one (compare solid and dotted curves, Fig. 13). In addition, the lower the permeability of the fault zone cutting a given sandstone, the greater the degree of permeability

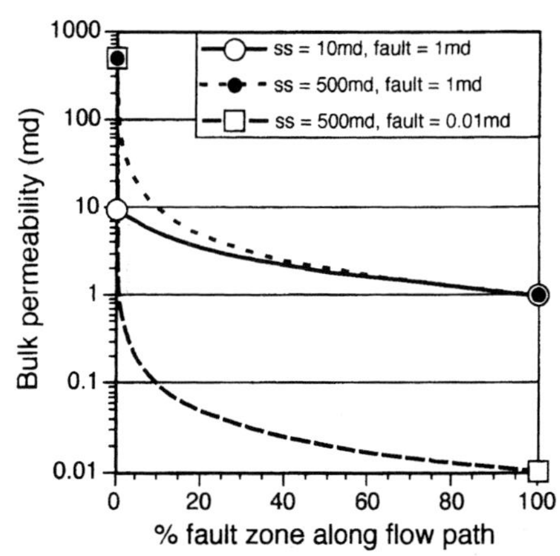

Fig. 13. Effect of low permeability fault zones on bulk permeability of a reservoir along a linear flow path. See text for discussion.

reduction (compare dotted and dashed curves, Fig. 13).

Because of relative permeability effects, the amount of fault-related permeability reduction implied by the test results presented in this paper may underestimate those in a sub-surface hydrocarbon reservoir. At a given elevation in a hydrocarbon column, the contrasting capillary properties of the sandstone and fault zone (e.g. Fig. 6a) will cause the fault zones to either have no hydrocarbon saturation (if the buoyancy pressure in the hydrocarbon column is less than the fault-zone entry pressure) or lower saturation than the undeformed sandstone (if the hydrocarbon buoyancy pressure exceeds the fault-zone entry pressure). In the case of zero fault-zone saturation, which is likely for clay-matrix gouge zones and complex deformation bands, the fault-zone permeability will be effectively zero, producing a no-flow barrier within the reservoir. In cases where the fault zone is partially hydrocarbon saturated, the reduced saturation relative to the undeformed sandstone will result in greater effective permeability contrast between the sandstone and fault zone than indicated by the specific permeabilities presented here. This effect should increase the impact of the faults on the bulk reservoir permeability, although partially hydrocarbon-saturated fault zones may remain somewhat permeable and serve as only partial flow barriers.

In addition to the potential effects outlined above, the overall influence of faults on production behavior will also depend on the spatial fault distribution within the reservoir, including fault interconnectivity and clustering (e.g. Watterson *et al.* 1995; Manzocchi *et al.* 1998; Walsh *et al.* 1998). If the faults are widely distributed throughout the reservoir, the effect may be a general permeability reduction, possibly in conjunction with a significant permeability anisotropy. On the other hand, fault-related deformation localized into narrow zones of sufficient thickness could produce discrete permeability barriers that prevent effective hydrocarbon drainage from portions of the reservoir.

Conclusions

An understanding of the nature and fluid-flow properties of sandstone-derived fault zones is necessary for evaluation of hydrocarbon entrapment and production patterns in faulted sandstone reservoirs. This study of naturally deformed sandstones from various areas shows that deformation of quartzose sandstones results in deformation bands, whereas faulting of mineralogically immature sandstones produces clay-matrix gouge zones. Clay-matrix gouge zones generally have lower permeabilities and higher capillary displacement pressures than deformation bands. These fluid-flow properties are largely dependent on rock composition, but show a superimposed influence of deformation mechanism, especially in the case of deformation bands.

Most deformation bands have capillary properties sufficient to maintain only small hydrocarbon column-height differences across them, whereas clay-matrix gouge zones can potentially seal hydrocarbon columns with heights of several hundred metres. However, both the low-permeability deformation bands and clay-matrix gouge zones can potentially influence fluid-flow patterns within a producing reservoir, although the magnitude of these effects will depend on the spatial distribution and abundance of faults and the permeabilities of the fault zones and undeformed sandstone.

X-ray diffraction analyses were conducted by G. Powers of the Amoco Exploration and Production Technology Group. Permeability and capillary pressure tests were done at Core Laboratories (Tulsa, OK, and Carrollton, TX), K&A Laboratories (Tulsa, OK), and Petrotech Associates (Houston, TX). Outcrop samples from Myanmar were provided by S. Serra. The Amoco Exploration and Production Technology Group is thanked for permission to publish this paper. Helpful reviews of the manuscript were provided by R. J. Knipe and R. A. Nieuw.

References

Amyx, J. W. & Bass, D. M. Jr 1962. Properties of Reservoir Rocks, *In*: Frick, T. C. (ed.) *Petroleum Production Handbook, Volume II*. McGraw-Hill, Chapter 23, 140.

Antonellini, M. & Aydin, A. 1994. Effect of faulting on fluid flow in porous sandstones: petrophysical properties: *AAPG Bulletin*, **78**, 355–377.

Aydin, A. 1978. Small faults formed as deformation bands in sandstone. *Pure and Applied Geophysics*, **116**, 913–930.

Bouvier, W. F., Kaars-Sijpesteijn, C. H., Kluesner, D. F., Onyejekwe, C. C. & Van Der Pal, R. C. 1989. Three-dimensional seismic interpretation and fault-sealing investigations, Nun River field, Nigeria. *AAPG Bulletin*, **73**, 1397–1414.

Chester, F. M. & Logan, J. M. 1986. Implications for mechanical properties of brittle faults from observations of the Punchbowl fault zone, California. *Pure and Applied Geophysics*, **124**, 79–106.

Firoozabadi, A. & Ramey, H. J. Jr 1988. Surface tension of water–hydrocarbon systems at reservoir conditions. *Journal of Canadian Petroleum Technology*, **27**(3), 41–48.

FOWLES, J. & BURLEY, S. 1994. Textural and permeability characteristics of faulted, high porosity sandstones. *Marine and Petroleum Geology*, **11**, 608–623.

FRISTAD, T; GROTH, A.; YIELDING, G. & FREEMAN, B. 1996. Quantitative fault seal prediction – a case study from Oseberg Syd. *In*: NORWEGIAN PETROLEUM SOCIETY (eds) *Hydrocarbon Seals – Importance for Exploration and Production* (abstracts). Norwegian Petroleum Society, Oslo.

GIBSON, R. G. 1994. Fault-zone seals in siliciclastic strata of the Columbus basin, offshore Trinidad. *AAPG Bulletin*, **78**, 1372–1385.

HARPER, T. R. & MOFTAH, I. 1985. Skin effect and completions options in the Ras Budran reservoir. *Society of Petroleum Engineers, Middle East Oil; Technical Conference and Exhibition*, Bahrain, 211–219.

HOUGH, E. W., RZASA, M. J. & WOOD, B. B. 1951. Interfacial tensions at reservoir pressures and temperatures, apparatus and the water–methane system. *American Institute of Minine Engineers, Petroleum Transactions*, **192**, 57–60.

HUBBERT, M. K. 1956. Darcy's law and the field equations of the flow of underground fluids. *American Institute of Mining Engineers, Petroleum Transactions*, **207**, 222–239.

JAMISON, W. R. & STEARNS, D. W. 1982. Tectonic deformation of Wingate sandstone, Colorado National Monument. *AAPG Bulletin*, **66**, 2584–2608.

KATZ, A. J. & THOMPSON, A. H. 1987. Prediction of rock electrical conductivity from mercury injection measurements, *Journal of Geophysical Research, Series B*, **92**, 599–607.

KNIPE, R. J. 1993. The influence of fault zone processes and diagenesis on fluid flow. *In*: HORBURY, A. D. & ROBINSON, A. G. (eds) *Diagenesis and Basin Development*. AAPG, Studies in Geology, **36**, 135–151.

KNIPE, R. J., FISHER, Q. J., JONES, G., *et al.* 1996. Fault seal prediction: methodologies, applications and successes. *In*: NORWEGIAN PETROLEUM SOCIETY (eds) *Hydrocarbon Seals – Importance for Exploration and Production* (abstracts). Norwegian Petroleum Society, Oslo.

LINDSAY, N. G., MURPHY, F. C., WALSH, J. J. & WATTERSON, J. 1993. *Outcrop Studies of Shale Smears on Fault Surfaces*. International Association of Sedimentologists, Special Publications, **15**, 113–123.

MANZOCCHI, T., RINGROSE, P. S. & UNDERHILL, J. R. 1998. Flow through fault systems in high-porosity sandstones. *This volume*.

MORROW, C., SHI, L. Q. & BYERLEE, J. 1981. Permeability and strength of San Andreas fault gouge under high pressure. *Geophysical Research Letters*, **8**, 325–328.

NELSON, R. A. 1985. *Geologic Analysis of Naturally Fractured Reservoirs*. Gulf, Houston, TX, 320 pp.

PITTMAN, E. D. 1981. Effect of fault-related granulation on porosity and permeability of quartz sandstones, Simpson Group (Ordovician), Oklahoma. *AAPG Bulletin*, **65**, 2381–2387.

SCHOWALTER, T. T. 1979. Mechanics of secondary hydrocarbon migration and entrapment. *AAPG Bulletin*, **63**, 723–760.

SMITH, D. A. 1980. Sealing and non-sealing faults in Louisiana Gulf Coast salt basin. *AAPG Bulletin*, **64**, 145–172.

SVERDRUP, E. & PRESTHOLM, E. 1990. Synsedimentary deformation structures and their implications for stylolitization during deeper burial. *Sedimentary Geology*, **68**, 201–210.

TREVENA, A. S. 1989. Small faults as barriers to permeability in sandstone gas reservoirs, Pattani Basin, Thailand. *AAPG Bulletin*, **73**, 553.

UNDERHILL, J. R. & WOODCOCK, N. H. 1987 Faulting mechanisms in high-porosity sandstones; New Red sandstone, Arran, Scotland. *In*: JONES, M. E. & PRESTON, R. M. F. (eds) *Deformation of Sediments and Sedimentary Rocks*. Geological Society, London, Special Publications, **29**, 91–105.

WALSH, J. J., WATTERSON, J., HEATH, A., GILLESPIE, P. A. & CHILDS, C. 1998. Assessment of the effects of sub-seismic faults on bulk permeabilities of reservoir sequences. *This volume*.

WASHBURN, E. W. 1921. Note on a method of determining the pore sizes in a porous material. *Proceedings of the National Academy of Science*, **7**, 115–116.

WATTERSON, J., HEATH, A., KNIGHT, S., FLINT, S., WALSH, J., CHILDS, C. & GILLESPIE, P. A. 1995. Estimation of the effects of sub-seismic faults on effective permeabilites of reservoir sequences. *Abstracts of Structural Geology in Reservoir Characterization and Field Development Conference*, Imperial College, London, 34.

WEBER, K. J. 1987. Hydrocarbon distribution patterns in Nigerian growth fault structures controlled by structural style and stratigraphy. *Journal of Petroleum Science and Engineering*, **1**, 91–104.

——, MANDL, G., PILAAR, W. F. & PRECIOUS, R. G. 1978. The role of faults in hydrocarbon migration and trapping in Nigerian growth fault structures. *Offshore Technology Conference Proceedings*, 2643–2653.

Fault controlled communication in the Sleipner Vest Field, Norwegian Continental Shelf; detailed, quantitative input for reservoir simulation and well planning

S. OTTESEN ELLEVSET[1], R. J. KNIPE[2], T. SVAVA OLSEN[1], Q. J. FISHER[2] & G. JONES[2]

[1] *Statoil, PO Box 300, 4001 Stavanger, Norway*
[2] *Rock Deformation Research Group, Department of Earth Sciences, University of Leeds, Leeds LS2 9JT, UK*

Abstract: The Sleipner Vest field, located in blocks 15/6 and 15/9 on the Norwegian Continental Shelf, contains hydrocarbons (mainly gas condensates) within the marginal marine units of the Middle Jurassic Hugin Formation. The field is segmented into fault-bounded compartments, which exhibit differences in gas–water contacts of between 10 and 100 m. Microstructural analysis of core samples has identified three principal fault types; cataclasites, developed from clean sandstones, framework phyllosilicate fault rocks created from impure sandstones and clay smears developed from phyllosilicate-rich units. The distribution of these has been linked to the phyllosilicate content of the undeformed Hugin reservoir at the time of deformation. Petrophysical property analysis has been used to quantify the representative permeabilities and threshold capillary pressures of the fault rocks and their undeformed equivalents. Juxtaposition/seal diagrams and detailed fault plane maps were also constructed and provide a basis for mapping sand–sand juxtapositions and the distributions of the different fault rock types with assigned permeabilities and capillary threshold pressures. The results provide an explanation of the mapped variations in gas–water contacts. An important correlation exists between the type of fault rock predicted to dominate the fault plane near the hydrocarbon–water interface and the difference in hydrocarbon depths across the faults. Compartments separated by faults with windows of juxtaposed clean sandstone cataclasites have small hydrocarbon–water level differences (7–11 m), suggesting that the low capillary threshold pressure (more permeable) cataclastic fault rocks control communication. Compartments separated by faults predicted to have extensive phyllosilicate-rich fault rocks (developed from impure sandstones and clay rich units) with high capillary threshold pressures correlate to larger hydrocarbon–water differences (>39 m), reflecting the reduced communication.

In recent years, fault seal analysis has become an increasingly important element in hydrocarbon exploration and production (Bouvier *et al.* 1989; Knipe 1992; Knott 1993; Knipe *et al.* 1997). Understanding fault behaviour is now recognized as critical to successfully locating hydrocarbon accumulations and for their efficient and cost-effective development. In this paper, we discuss the results of an integrated fault seal study of the Sleipner Vest (west) field in the Norwegian North Sea, which was initiated to assess the transmissibility of a number of mapped sealing faults as the field entered a programme of production and development.

Sleipner Vest is a large, structurally complex gas-condensate field which is located in the eastern South Viking Graben (blocks 15/6 and 15/9) in the central North Sea Basin (Fig. 1). Mapping of recently acquired 3D seismic data (1994) has shown that the field is extensively segmented by faults (Fig. 2). Differences in hydrocarbon contacts in the order of metres to tens of metres have been observed within the Middle Jurassic (Callovian) marginal marine reservoir sandstones of the Hugin Formation. Larger contact differences occur across faults which juxtapose reservoir against non-reservoir (sealing) lithologies. On the basis of these differences and the mapped fault pattern, a model for compartmentalization of the Sleipner Vest field has been proposed (Fig. 2). The successful appraisal and development of the field, which was brought onstream in 1995, was therefore considered to be strongly dependent on understanding the nature and degree of fluid communication across these compartmentalizing faults under production drawdown conditions.

The analysis of fault zone behaviour in Sleipner Vest reported here was based on a detailed evaluation of 3D seismic data (including extensive fault plane mapping), the characterization of deformation features present in cores and

From Holdsworth, R.E. & Turner, J.P. (compilers) *Extensional Tectonics: Faulting and Related Processes*. The Geological Society, Key Issues in Earth Sciences, **2 (2)**, 265–279.

First published in: "Jones, G., Fisher, Q.J. & Knipe, R.J. (eds) *Faulting, Fault Sealing and Fluid Flow in Hydrocarbon Reservoirs*, The Geological Society, London, Special Publications **147**, 283–297".

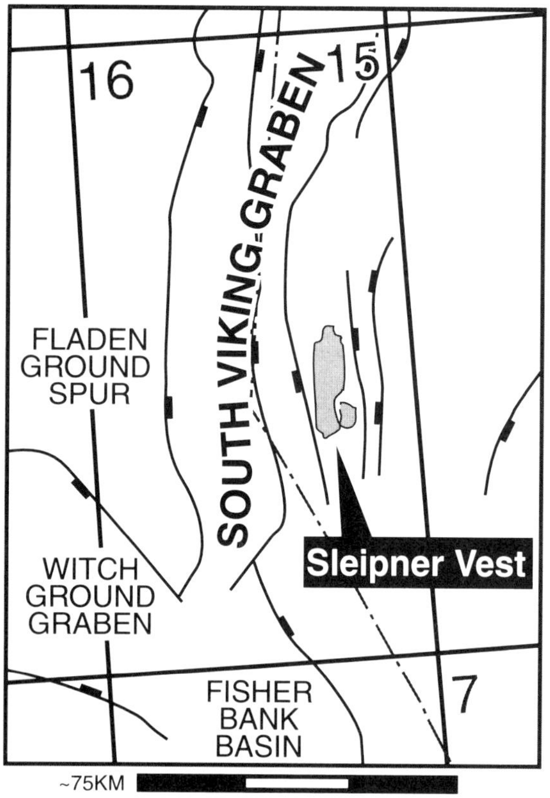

Fig. 1. Structural depth map of the Top Hugin Fm, contoured inside the GWC.

also the microstructural and physical property analysis of selected fault rocks. We also introduce the applied use of a new graphical technique for analysis of cross-fault lithological juxtapositions in fault zones (Knipe 1997). The overall objective was to combine data from these different scales and provide detailed input to well planning and reservoir simulation studies.

Fault seal assessment methodology

The sealing behaviour of faults has often been assessed by relating the probability of a fault plane sealing to a number of input parametres, which are then transferred to a map of the fault plane. Common parametres used include the displacement along the fault surface, the fault zone thickness (Knott 1993), the lithology either side of the fault, (often expressed as a net/gross ratio) and also some factor which estimates the effect of deforming and smearing clay units (e.g. the Smear Gouge Ratio, SGR; Bouvier *et al.* 1989; G. Skerlec unpublished report; Yielding *et al.* 1997). Such analysis is usually completed by mapping the connectivity across the fault using (vertical) fault plane diagrams (Allan 1989), in order to assess the transmissivity. Where known, hydrocarbon–water contacts and pressure data are used to calibrate the depths where transmissivity occur (Yielding *et al.* 1997).

Whilst these methods do help to reduce risk during exploration, they do not provide enough detail for reservoir evaluation and modelling. As pointed out in Knipe (1992), fault seal mechanisms, and thus petrophysical properties and sealing behaviour, will vary over a given fault surface and depend not only on the factors described above, but also as a result of the internal structure of fault zones and their geohistory. The aim of any fault seal study should be to map this variation, and ultimately, to quantify variations in transmissibility as a basis for understanding reservoir communication across the faults.

In this study, intra-reservoir communication across the faults separating the main compartments of the Sleipner Vest Field has been systematically analysed. Communication across a total of 61 selected faults or fault segments has been assessed. The analysis presented has involved the integration of structural geological data obtained from core to seismic scales, with detailed microstructural and petrophysical property analysis of fault rocks obtained from core. Initially, structural logging of cored wells highlighted the nature of the fault zones and the main fault rock and fault seal types present in Sleipner Vest. The diagenetic and microstructural evolution of selected fault rocks and their petrophysical properties were characterized using the techniques described in Fisher & Knipe (this volume).

The detailed fault rock characterization exercise described has been integrated with the mapping of cross-fault reservoir juxtapositions. This exercise was undertaken using two different methodologies. Firstly, graphical juxtaposition analysis (triangular Juxtaposition Diagrams of Knipe 1997) were used to assess the effects of deforming typical Hugin Formation reservoir stratigraphies. This technique has the advantage of providing rapid visualization of the distribution of unit juxtapositions and can also highlight the critical throws at which particular juxtaposition types (e.g. sand–shale) are developed. Also, the technique is not dependent on other factors such as the quality and resolution of the seismic data, the availability of seismic reflectors for mapping and interpreter preference.

The second methodology involved creating vertical fault plane diagrams (Allan 1989), based on the 3D seismic fault and horizon mapping, using the FAPS software (Badley Earth

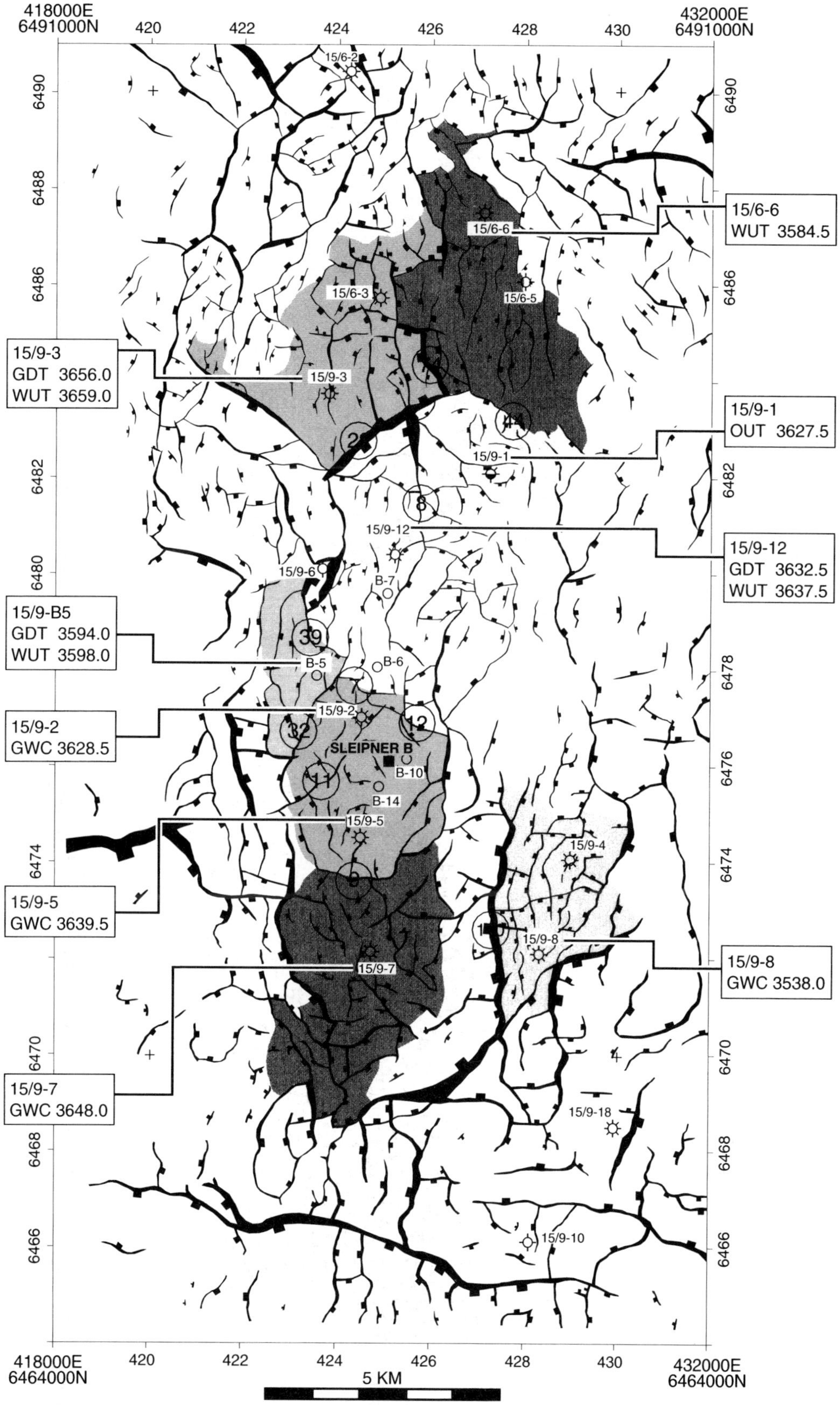

Fig. 2. Suggested compartmentalization of the field, based on the fault pattern and observed differences in GWC.

Sciences). This allows throw variations along the fault zone to be assessed and defines the approximate depths at which the critical juxtapositions identified occur in the subsurface. The top Hugin reservoir is only a moderately reliable reflector, and reservoir units have been isopached up from a strongly reflective coal unit which lies close to the base of the reservoir sequence. Vertical seismic resolution over Sleipner Vest is in the order of 30 m, which could generate some important possible variations in reservoir topography.

Tectonostratigraphic evolution

The Sleipner Vest field lies in a tectonically complex part of the central North Sea Basin that has experienced a long history of deformation. The fundamental structural elements in this area are probably of Early Palaeozoic age. These include deep-seated fault systems aligned along a WNW trend, developed as part of the Tornquist structural zone in northern Europe (Pegrum 1984). Early Mesozoic (Lower Triassic) extension was probably an important element in the initial development of the N–S trending South Viking Graben, possibly re-activating Devonian basins (Coward 1994). Subsequently, during the early Middle Jurassic (Toarcian–Aalenian), the structural evolution of the Sleipner area was markedly influenced by uplift and erosion resulting from thermal doming of the central North Sea area (Underhill & Partington 1993). This resulted in the removal of Lower Jurassic sediments.

The Sleipner area was characterized by an essentially W–E oriented coastline system during the deposition of the mainly Callovian age Hugin Formation reservoir, in which marginal marine to offshore sediments were deposited from south to north, respectively. Comparatively little tectonic activity is recorded at the seismic scale during reservoir deposition, and a progressively deepening marine environment is recorded throughout the Upper Jurassic. Widespread extension occurred on the Sleipner terrace area during the latest Jurassic and earliest Cretaceous, which generated both N–S faults and WNW cross-faults in the Hugin reservoir. The Sleipner terrace presently forms a down-thrown area, (*c.* 1000 m offset), along a NNE–SSW fault zone, relative to the Gamma High in the east. Extension was probably assisted by the presence of evaporites at depth. Although presently unproved by drilling, seismic data suggests fault-controlled salt pillows may be present. Most faults cut through the regionally diachronous 'base Cretaceous' reflector (Rattey & Hayward 1993) and appear to terminate in the Early Cretaceous shale units.

The Sleipner area shows some evidence of compressional tectonic activity during the Early Cretaceous. This is possibly related to re-activation of the Tornquist zone, which may have switched to dextral transpression from earlier regional transtension (Pegrum 1984; Pegrum & Ljones 1984). Additional, poorly understood compressional reactivation of normal faults also took place in the Late Cretaceous and Early Tertiary, as a result of intra-plate stresses resulting from both Alpine collision and North Atlantic opening.

Reservoir diagenesis

The diagenetic history of the Hugin Formation has proved remarkably consistent across the Sleipner Field (Knipe *et al.* unpublished report). Four diagenetic processes have affected the samples examined:

(i) quartz overgrowth development,
(ii) kaolin neoformation;
(iii) K-feldspar dissolution; and
(iv) illite precipitation.

In addition to these field extensive diagenetic processes, the Hugin sandstones have also experienced more localized precipitation of pyrite, calcite, siderite, dolomite and barite. This diagenetic history is summarized in Fig. 3. The timing of deformation, in relation to the diagenetic history, was established for each sample. This provides a framework for the assessment of fault rock development in the field. The deformation features preserved in the cores can be assigned to different combinations of a four phase history involving:

(i) early deformation of poorly consolidated sediments;

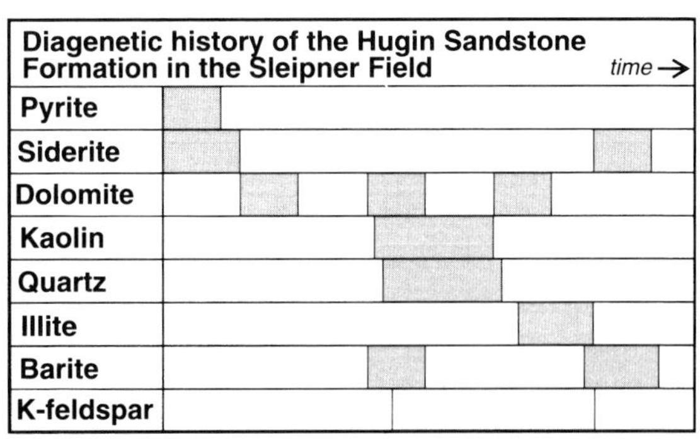

Fig. 3. The diagenetic history of the Hugin Formation in the Sleipner Vest Field. The dark coloured boxes refer to precipitation events. The light coloured box is a dissolution event.

(ii) intermediate stage faulting which pre-dates the main phase of kaolin and quartz precipitation;
(iii) late stage faulting which post-dates the main period of meso-diagenesis and
(iv) inversion.

Fault rocks and fault seal

Most faults are best considered as membranes (Watts 1987) or flow retarders with variable transmissibility properties. This view is in contrast to the sealing/non-sealing assessment of some recent studies (Knott 1993; Skerlec unpublished report). In the analysis reported here, the distribution of capillary threshold entry pressures is used to evaluate cross-fault communication, as the sealing capacity of the fault rocks can be related to the capillary threshold pressure (Watts 1987) and a permeability to threshold entry pressure relationship can also be established. No hydrocarbon flow through the fault zone will take place unless the threshold pressure is reached; flow is then controlled by the permeability. Microstructural and petrophysical analysis has allowed the fault rocks to be classified according to the deformation mechanisms involved in their formation as well as their present-day properties. The classification used is based on that presented in Knipe *et al.* (1997) and reviewed in Fig. 4. Seals present in the Sleipner Vest field cores include:

(i) phyllosilicate smears developed from clay-rich units;
(ii) phyllosilicate framework fault rocks developed from impure sandstones;
(iii) cataclastic collapse faults developed from clean sandstones;
(iv) cemented deformation features.

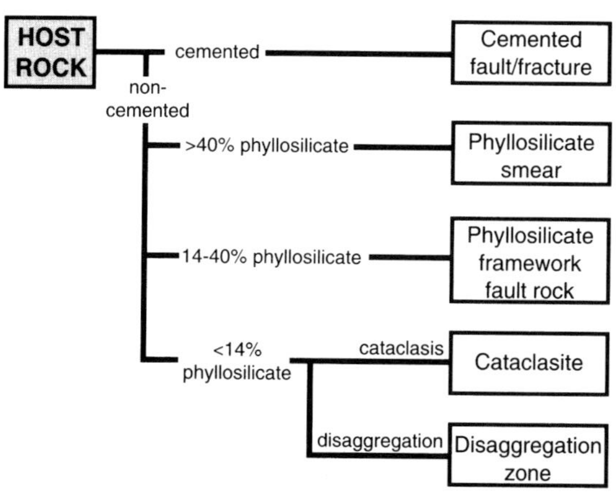

Fig. 4. Host rock lithology and fault rock type. The relation between phyllosilicate and microcrystalline quartz content of the sandstone and the fault rock development.

One of the most important aspects of this classification is the recognition of the relation between phyllosilicate content of the host sandstone and the seals which develop. A more detailed discussion on the classification and properties of fault rocks is given in Fisher & Knipe (this volume). In addition, regional compression (inversion faulting) events have in some cases induced a further decrease in the petrophysical properties of fault zones by localized pressure solution processes. There is also evidence for a late rupturing of early formed cataclastic fault rocks, which may be ascribed to this late inversion. Halite has been found in faults within the Sleipner Formation and the Skaggerak Formations but not in the overlying Hugin Formation reservoir. The salt sampled appears most likely to be a secondary feature associated with core recovery and drying. Each of the main fault rock classes and their role in the Sleipner Vest field is outlined below.

Cemented deformation features

Extensive development of cemented fault zones will invalidate the analysis of fault seals based on the simple prediction of fault rocks from the host undeformed lithologies (Knipe 1993; Knipe *et al.* 1997; Sverderup & Bjørlykke 1997). However, in many cases the continuity of cements may not be extensive enough to impact on communication, but may be restricted to parts of a fault plane. Cemented sandstone horizons, which contain cemented (calcite) fractures, are present in the Sleipner Vest field. This material may be redistributed and reprecipitated during deformation. However, the core data suggest that during faulting, this redistribution is likely to be restricted to an area close to the hanging wall and footwall cut-off regions of the cemented units. Given the patchy nature of the undeformed cement, such redistribution will probably be laterally discontinuous. In Sleipner Vest, cemented deformation features present have a range of permeabilities from 1 mD to >0.0001 mD, which is controlled by the micro-architecture of the cement. For our analysis of fault zone permeability in Sleipner Vest, we have assumed that the extent of any cementation on the fault plane does not exceed three times the thickness of the cemented sedimentary bed which acts as a source.

Phyllosilicate smear

In Sleipner, sedimentary units containing more than approximately 40% phyllosilicates (clay, micas) develop phyllosilicate smears when faulted. Physical property analysis of these rocks was not obtained directly from Sleipner Vest, but similar material sampled at equivalent depths usually has permeabilities well below 0.001 mD and capillary threshold pressures of $\gg$1500 psi. ($\sim$100 bars, or $\sim$10 MPa) for the Hg–air system (Fisher & Knipe this volume). For our analysis of the distribution of clay smears in Sleipner Vest, we have used the criterion that the smear is discontinuous when the effective fault throw exceeds three times the thickness of the clay-rich unit (see Knipe *et al.* 1997).

Phyllosilicate framework fault rocks (PFFR)

Faults in sedimentary rocks with 14–40% phyllosilicate contents develop phyllosilicate framework fault rocks (PFFRs) from impure sandstone lithologies and are composed of microsheared phyllosilicate domains (see Knipe 1992; Knipe *et al.* 1997; Fisher & Knipe this volume). The permeability of the measured phyllosilicate framework fault rocks developed in impure sandstones in the Sleipner Vest Field ranges from 0.3 to 0.004 mD and Hg threshold capillary pressures are above 120 psi ($\sim$1 MPa, $\sim$10 bars). There are two areas of a fault plane where these PFFR are likely to impact on sealing behaviour. The first of these is where impure sandstones are directly juxtaposed against the fault zone, and the second is along faults between the hanging wall and footwall cut-offs of the impure sandstone units. This latter situation is analogous to the formation of clay smears. The effectiveness of the PFFRs to form barriers to flow depends on their continuity, which is likely to decrease with increasing throw in the same manner as clay smears. There will be a zone near to the hanging wall and footwall cut-offs to the impure sandstones where there is an increase in the probability of these fault rocks forming seals. In Sleipner Vest, these zones of high potential for development of phyllosilicate framework fault rocks are taken to be effective for a distance (up or down fault planes) equal to the thickness of the impure sandstone units which create the fault rocks.

Cataclasites developed from clean sandstones

Sedimentary rocks containing less than 14% phyllosilicates develop either cataclasites or disaggregation zones during deformation. There is a very large variation in the permeability of this type of fault rock, depending upon the geohistory of the fault (Knipe *et al.* 1997; Fisher & Knipe this volume). In Sleipner Vest, these fault rocks have high permeabilities, with a range from 43–0.4 mD and they form the least effective flow barriers. The high permeability cataclasites are associated with early faulting at shallow depths of burial, and thus the lower permeability range (<1–5 mD) are more representative of the larger offset, seismically visible faults which were active at deeper levels. The fault plane windows dominated by cataclastic fault rocks developed from clean sandstones are likely to occupy the parts of fault planes rocks where sand–sand juxtapositions occur away from clay smears and phyllosilicate framework fault rocks.

Fault seal evaluation at sampled wells

Based on the detailed core, microstructural and petrophysical analysis, as well as the lithological description of the reservoir, a fault seal evaluation has been possible. The lithological description from the nearest well(s) can be used to produce a diagram showing the varying juxtapositions of the different reservoir stratigraphies with different fault throws. Using the results of core analysis, it is possible to predict which fault rocks will develop for a specific reservoir stratigraphy, and for a range of throw values. It is then possible to identify the throw windows that will develop local areas dominated by the three important fault rocks in Sleipner Vest; i.e. cataclasites, phyllosilicate framework fault rocks, and phyllosilicate smears (see Knipe 1997). The RDR Group has developed a software programme called ‘Juxtaposition’ that allows rapid visualization and assessment of juxtaposition and fault seal patterns generated by faulting of varying reservoir stratigraphies. An example of a juxtaposition and fault seal triangular plot from well 15/9-8 of the Sleipner Vest Field is given in Fig. 5.

Upscaling of physical properties

The hydrocarbon filling history of the Sleipner Vest Field is complex, with the possibility of several source areas contributing to the fluid pool. The timing of the main filling was from Tertiary until present and very little burial has occurred after filling of the structures. The measured capillary threshold pressures for the

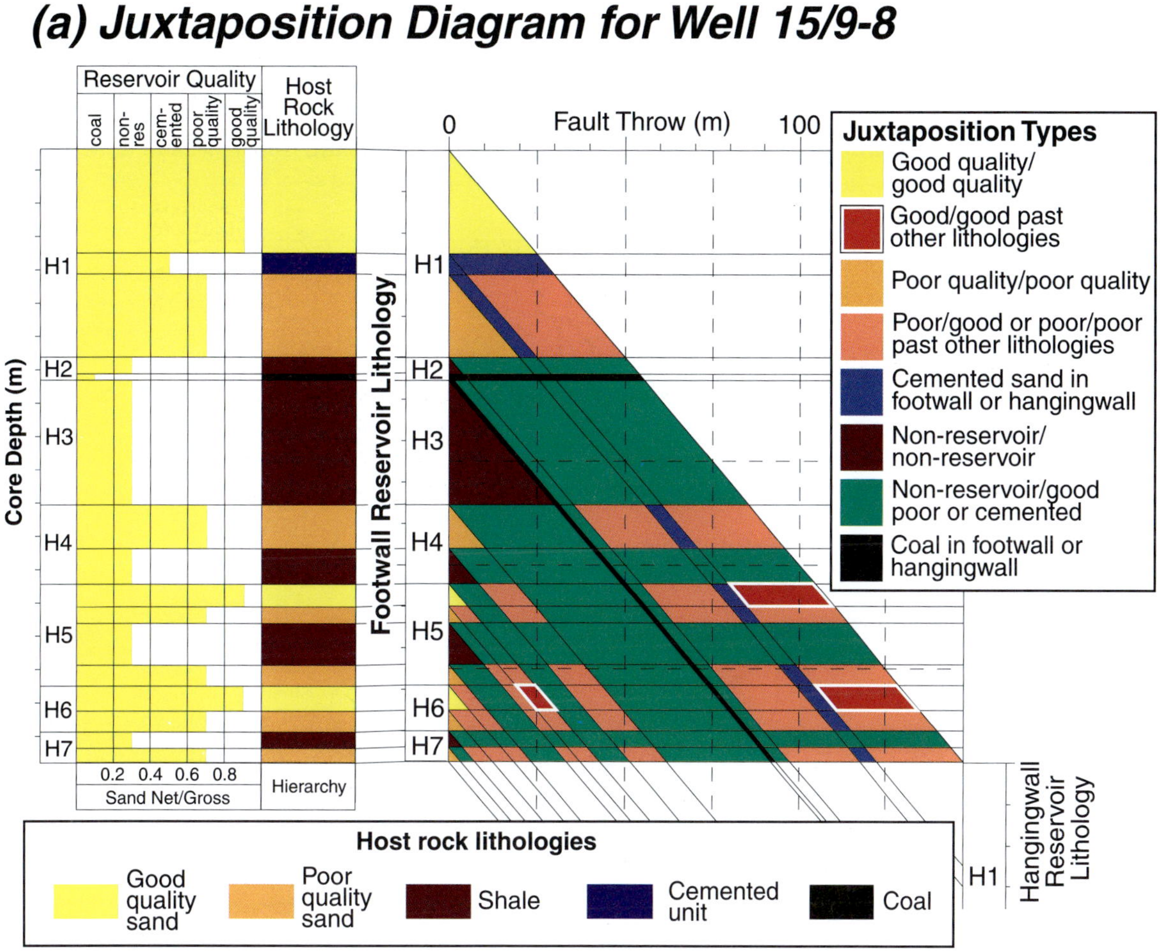
(a) Juxtaposition Diagram for Well 15/9-8
Reservoir Quality
coal
non-res
cem-ented
poor quality
good quality
Host Rock Lithology
Core Depth (m)
H1
H2
H3
H4
H5
H6
H7
0.2 0.4 0.6 0.8
Sand Net/Gross
Hierarchy
Footwall Reservoir Lithology
0
Fault Throw (m)
100
Hangingwall Reservoir Lithology
Juxtaposition Types
Good quality/ good quality
Good/good past other lithologies
Poor quality/poor quality
Poor/good or poor/poor past other lithologies
Cemented sand in footwall or hangingwall
Non-reservoir/ non-reservoir
Non-reservoir/good poor or cemented
Coal in footwall or hangingwall
Host rock lithologies
Good quality sand
Poor quality sand
Shale
Cemented unit
Coal

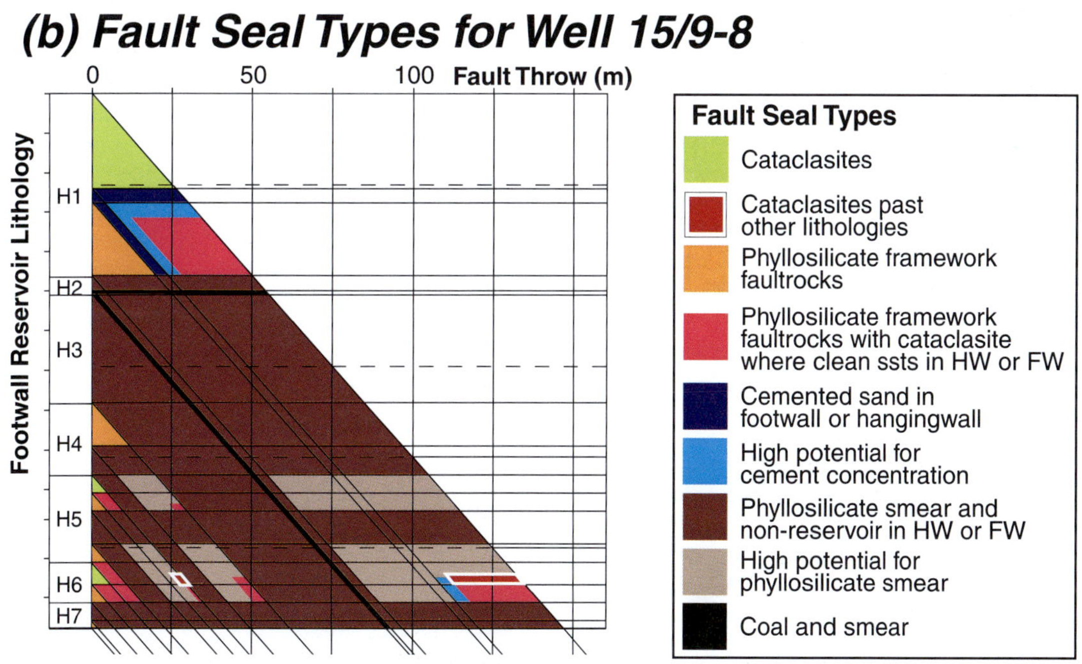
(b) Fault Seal Types for Well 15/9-8
0
50
100
Fault Throw (m)
Footwall Reservoir Lithology
H1
H2
H3
H4
H5
H6
H7
Fault Seal Types
Cataclasites
Cataclasites past other lithologies
Phyllosilicate framework faultrocks
Phyllosilicate framework faultrocks with cataclasite where clean ssts in HW or FW
Cemented sand in footwall or hangingwall
High potential for cement concentration
Phyllosilicate smear and non-reservoir in HW or FW
High potential for phyllosilicate smear
Coal and smear

faults are therefore expected to be representative of the fault rocks at the time of filling, apart from in the case of reactivated faults. The differences in hydrocarbon column heights can thus be used for a calibration and validation for the up-scaling of fault rock properties.

The distribution of fault rocks near the GWCs across the field was investigated using the results from the Juxtaposition software, together with simple Allan-type diagrams. The predicted fault rock distribution map is shown in Fig. 6. The cataclasites are the most leaky fault rocks,

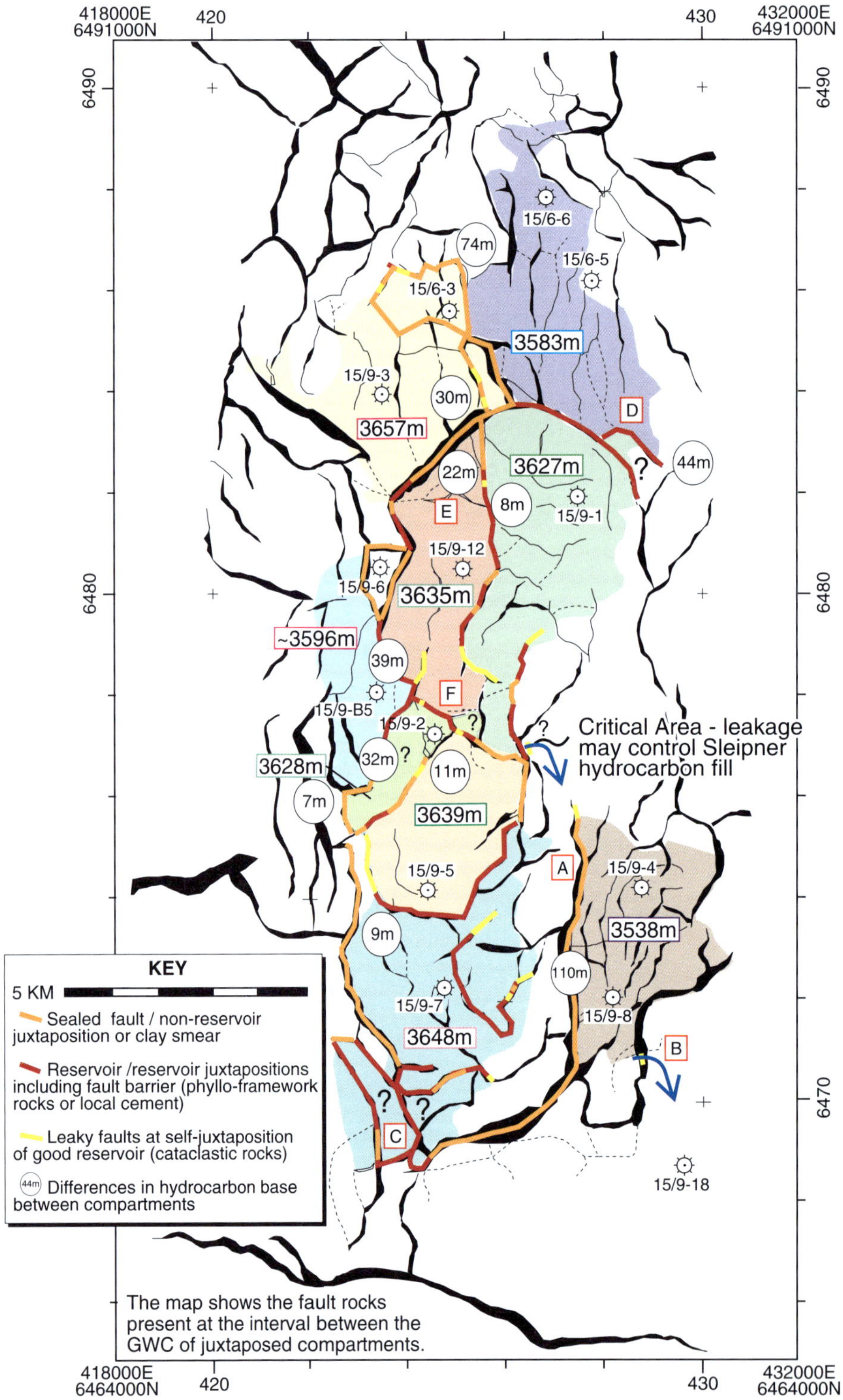

Fig. 6. Distribution of the most 'leaky' fault rocks at and near the GWC.

and are believed to control the differences in the hydrocarbon column heights present in the southern part of the Sleipner Vest Field. The map in Fig. 6 shows the distribution of the different types of fault rocks at the depth interval near the gas–water contacts. The figure illustrates that in areas where the fault zones are dominated by the cataclasites, they are associated with cross-fault differences of 7–11 m in the hydrocarbon column. A prediction of the sealing capacities (column heights of hydrocarbons which can be held below the seal) of these fault rocks can be made from the measured threshold capillary pressures and their equivalent permeabilities. The predicted sealing capacities correspond well to the observed GWC differences. For example, a 7 m difference in GWC is equivalent to a fault rock permeability ~1 mD.

A good correlation between the predicted sealing capacity of the phyllosilicate framework fault rocks and the values indicated from the field GWC data is also apparent. The measured permeabilities on small displacement phyllosilicate framework fault rocks vary from 0.3 to 0.004 mD. According to fault rock distribution mapping (Fig. 6), the sections of faults where phyllosilicate framework fault rocks dominate near the GWC depth intervals, create differences in the hydrocarbon base of 22–44 m. This corresponds to a range of laboratory permeabilities of ~0.05–~0.01 mD, which is nearer the upper, high permeability values for the measured range. This emphasizes, as expected, that the higher permeabilities in the measured range dominate the large scale sealing (membrane) behaviour of the fault zones. Smaller areas of the fault plane will be characterized by the lower permeabilities (higher threshold capillary pressures), as measured in selected specimens, but the localized distribution of these fault rocks will not allow them to control the large-scale effective fault permeability. To assess the wide range of physical properties associated with this class of fault rocks, permeabilities in the range of 0.3 mD–0.005 mD have been selected for input for the reservoir simulation.

Phyllosilicate smears

Phyllosilicate smears from Sleipner Vest are expected to have permeabilities of ≪0.001 mD. The phyllosilicate smear fault rocks will have such very low permeabilities, even for small faults, and up-scaling seems unnecessary as these faults probably act as seals during production. This is supported by the correspondence between the GWC differences (>44 m) in the areas of the field where smears dominate. The maximum sealing capacity possible associated with the low permeability end of these phyllosilicate smears is >300 m.

Calibration against production data and pressure history

The production history of the Sleipner Vest Field is too short for calibrating the detailed permeabilities of the distributed fault rocks with the pressure history. However, two wells from a neighbouring field, that is also producing from the Hugin Formation, provide an initial means of calibration. Faults are present between the two wells and are considered to represent membrane seals consisting of phyllosilicate framework fault rocks, as they are believed to fault an impure sandstone within the Hugin Formation. The pressure history of the two wells (A and B) is shown in Fig. 7. The data illustrate that there is some communication between the two wells, because the pressure dropped in well B before production started in this well in October 1995, inferring a pressure drawdown from another well. There was a pressure difference of about 10 bars when well B (Fig. 7) started production. A pressure of 10 bars for the Sleipner hydrocarbon–water system is equivalent to a 61 bar (~900 psi) capillary pressure for the Hg–air system. This falls within the range measured for the phyllosilicate framework fault rocks in Sleipner. A good history match was made using 3 bars for the hydrocarbon–water capillary pressure on a fault separating the two wells (H. Hansen, pers. comm).

Distribution of fault rocks on fault planes

The recognition that different phyllosilicate contents in sandstones create different fault rock types by faulting is important to seal evaluation (Knipe 1992; Gibson 1994; Yielding *et al.* 1997). We have used FAPS software (Freeman *et al.* 1989; Needham *et al.* 1996) to provide an estimate of the sealing potential distributions on faults, as the percentage of shale within the slipped interval changes. That is, the GOUGE RATIO (hereafter named GR) has been estimated for different parts of the same fault. The overall procedure was as follows:

(i) the depth model obtained from the IRAP^{TM} model of the Sleipner Vest Field, with the subdivision of the Hugin Fm reservoir into 8 reservoir zones, was imported into the FAPS database;

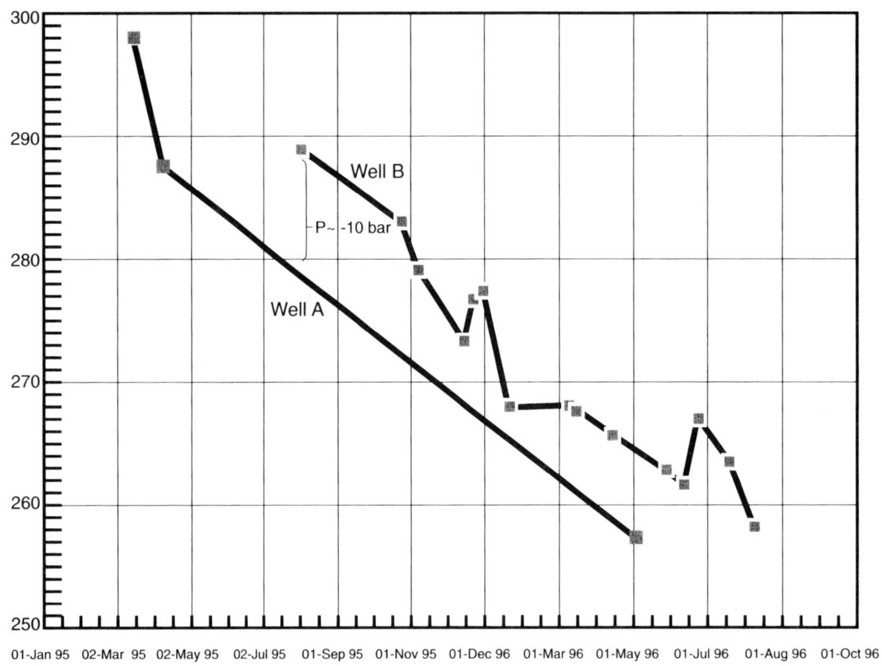

Fig. 7. Pressure history from two wells producing from the Hugin Formation on a neighbouring field.

(ii) A detailed stratigraphic subdivision of the reservoir was separated into three different classes (good, poor and non-reservoir), based on the clay-content of the rock in the nearest well;

(iii) the sealing potential (or GR) was calculated and contoured for a number of points on the mapped fault surface, using equation 1 below.

$$GR = \{(\Sigma V_d \cdot Dz)/t \times 100\% \quad (1)$$

V_d is the clay volume fraction in each interval of thickness, t is the fault throw at that point.

The assumption of a correlation between the volume of shale in the undeformed section and the volume of phyllosilicates in the fault is made. The GR values were divided into three groups (0–14, 14–40 and >40), and the expected distribution of the different fault rock types contoured on the whole fault plane, or only at the reservoir–reservoir juxtapositions as shown in Figs 8 a & b. Due to a large spread of permeabilities, the phyllosilicate framework fault rock class has been subdivided into three groups (14–20%, 20–30% and 30–40%) and decreasing permeabilities assigned to increasing phyllosilicate contents and increasing GR values. The contouring of the fault plane makes it possible to estimate the differences in communication for different juxtaposed sandstones. Figure 8b illustrates that lower permeabilities (i.e. higher entry pressures) are required for the thin sands juxtaposed in the lower part of the fault plane compared to the cleaner sands juxtaposed in the top of the fault plane.

Tectonic reactivation

Reactivation of faults is observed both in the microstructural data and on seismic data from the Sleipner Vest. A systematic pattern for the distribution of the reactivation has been difficult to map from seismic data. Microstructural analysis on Sleipner Vest has detected both an increase and decrease of the petrophysical properties related to late inversion. The fault orientation relative to the inversion stress axes are believed to control the permeability reduction or increase. The late inversion has a maximum horizontal compressional stress orientated approximately N–S (Pegrum & Ljones 1984). This will give compression on the E–W faults and extension on the N–S faults. Better transmissibilities may therefore be expected for the same fault rock types on N–S striking faults, compared to E–W striking faults.

Communication across faults in Sleipner Vest

The map in Fig. 9 shows the distribution of the most leaky fault rocks on the main faults of the Sleipner Vest Field. It is based on the type

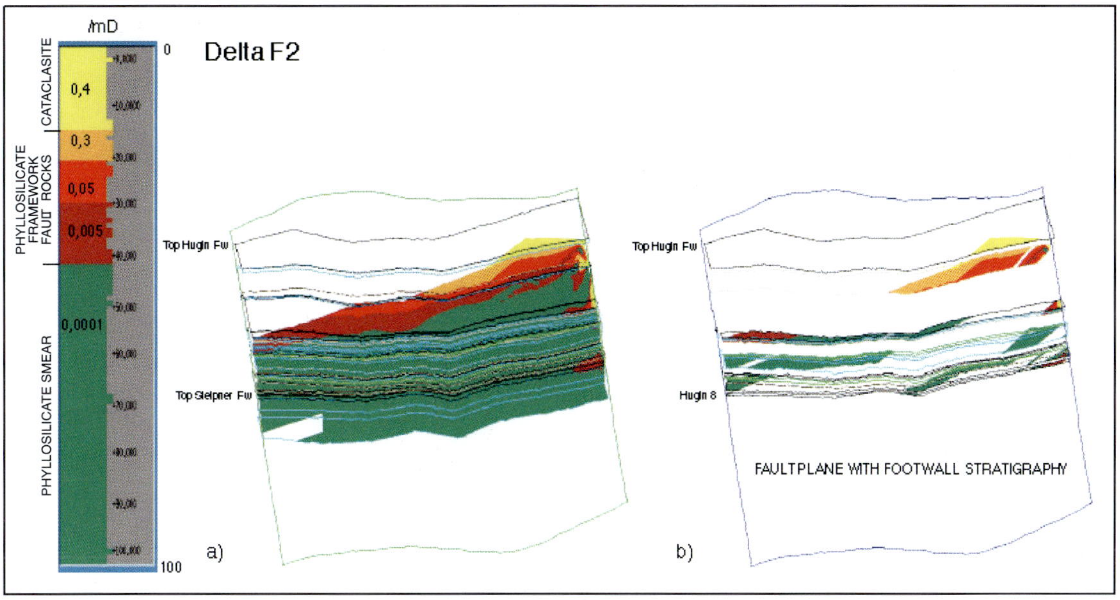

Fig. 8. Fault rock type distributions contoured on a fault plane.

of fault plane diagrams shown in Figs 8a & b. In the case of Sleipner Vest (primarily a gas-condensate field) a cut-off permeability value of 0.1 mD can be used as a general seal / non seal value in the reservoir model. For oil, a permeability cut off of 1 mD is used. A brief review of the more detailed controls on the communication in different parts of the field is presented.

Southern Sleipner Vest area

In the southern part of the field, the faults with reservoir–reservoir juxtaposition are dominated by cataclasites and relatively high permeability phyllosilicate framework fault rocks, with expected permeabilities and threshold capillary pressures of ~0.3 mD and ~2 bars, respectively, for the hydrocarbon–water system in Sleipner Vest. Good communication across intra-reservoir faults is expected, especially in the upper parts of the reservoir, where the cleanest sands are present. There are no pressure data indicating lithological controlled horizontal barriers in the field, but this cannot be excluded.

Central Sleipner Vest area

The faults limiting the fault block of well 15/9-B5 (Fig. 9) are considered to be dominated by phyllosilicate framework fault rocks with permeabilities of 0.05–0.005 mD and threshold capillary pressures of ~4 to ~10 bars for the hydrocarbon–water system in Sleipner. The difference in the hydrocarbon contacts of 39 m confirms the poorer communication across the fault blocks (e.g. in the central western part of the field).

For the southern/central part of the field, high permeability phyllosilicate framework fault rocks dominate and communication is expected to be good. Faults with throws exceeding 50 m are, however, likely to produce large areas with low permeability phyllosilicate framework fault rocks. Permeabilities and entry pressures of 0.05–0.005 mD and ~4 to ~10 bars for the hydrocarbon–water system in Sleipner, respectively, are predicted, which will retard communication.

Northern Sleipner Vest area

In the north of the Sleipner field, communication appears to be more difficult as impure sandstones and shale layers of the 'distal' facies are more common, which will create lower permeability phyllosilicate framework fault rocks and phyllosilicate smears. The expected permeabilities and capillary entry pressures are 0.3–0.005 mD and ~2 to ~10 bars and ≪0.001 mD and >56 bars (for the hydrocarbon–water system in Sleipner), respectively. Communication appears difficult between the east and west areas in the north. Communication is, however, good on the flanks of the northern part of the field and this indicates that water support will be effective.

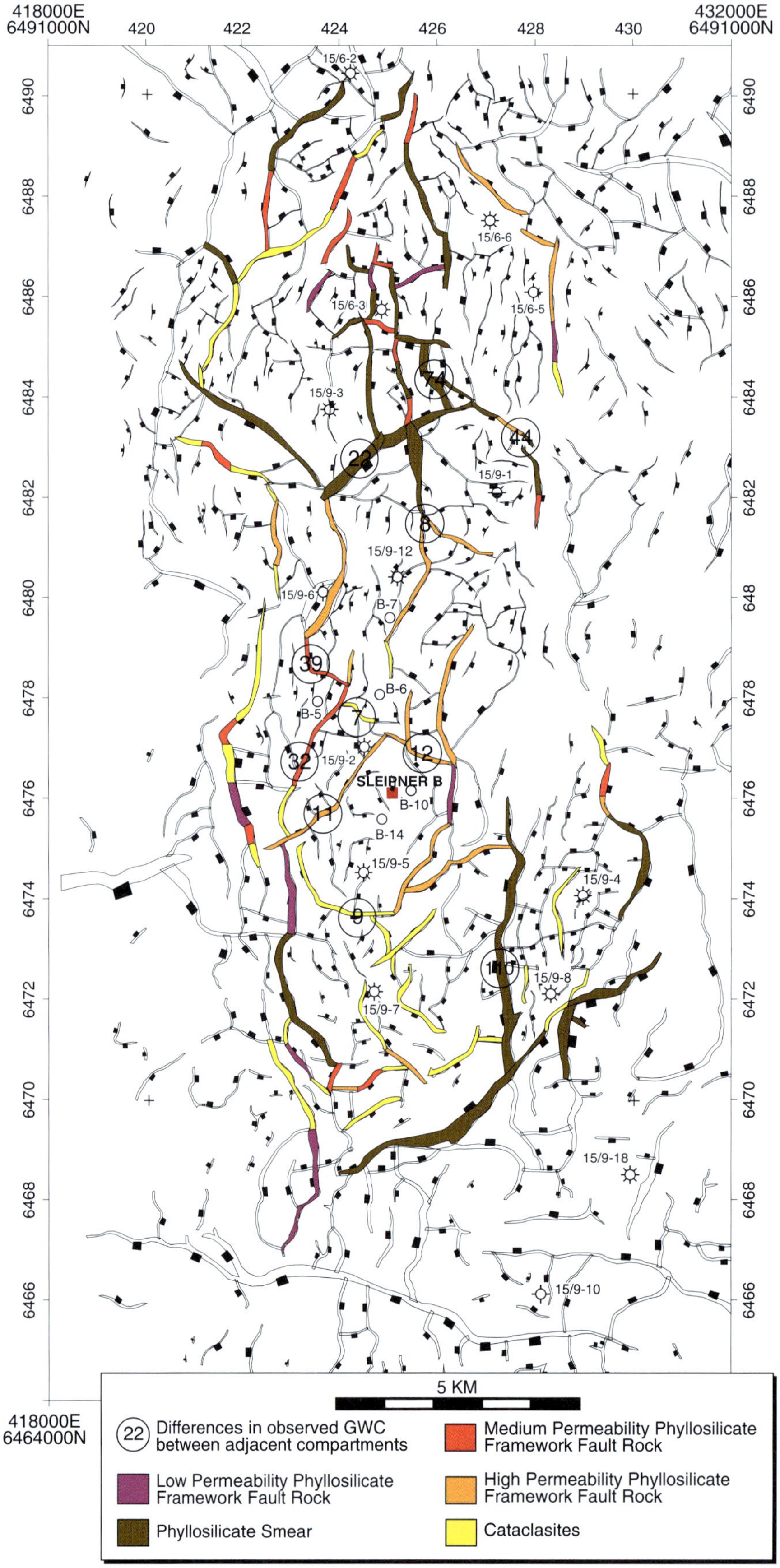

418000E 6491000N
420
422
424
426
428
430
432000E 6491000N
6490
6488
6486
6484
6482
6480
6478
6476
6474
6472
6470
6468
6466
418000E 6464000N
15/6-2
15/6-6
15/6-5
15/6-3
15/9-3
15/9-1
15/9-12
15/9-6
B-7
B-6
B-5
15/9-2
SLEIPNER B
B-10
B-14
15/9-5
15/9-4
15/9-8
15/9-7
15/9-18
15/9-10
74
22
44
8
39
7
32
12
11
9
110
5 KM
22 Differences in observed GWC between adjacent compartments
Medium Permeability Phyllosilicate Framework Fault Rock
Low Permeability Phyllosilicate Framework Fault Rock
High Permeability Phyllosilicate Framework Fault Rock
Phyllosilicate Smear
Cataclasites

Although salt invasion along the fault zones has been recognized as a possibility for the northern part of the Sleipner Vest Field, salt invasion in the reservoir fault rocks is not required to explain the differences in GWC in this region. During the production of the field, a pressure drop in the range of 200 bars is to be expected. Given the conclusions from this study, this appears adequate to allow some communication across all the intra-reservoir faults with the possible exception of the phyllosilicate smear fault rocks. The initial modelling suggests that many months of production is required to induce the pressures needed for production through low permeability phyllosilicate framework fault rocks.

Discussion

The GR is similar to the shale/gouge ratio (SGR) of Skerlec and with the clay smear potential (CSP) of Bouvier *et al.* (1989) and Jev *et al.* (1993). Skerlec (unpublished report) studied known sealing and non-sealing faults affecting the Upper Jurassic in the Block 35/11 area in the North Viking Graben, and concluded that non-sealing faults had SGR <29 whereas SGR >40 implies sealing faults. These conclusions are very similar to those conclusions of Bouvier *et al.* (1989) who also calibrated the CPS against known sealing and non-sealing faults. CPS <28 have low sealing potential while CPS >36 have high sealing potential. Jev *et al.* (1993) used the same technique on the Akaso Field and quoted CPS <15 as non-sealing while CPS >30 as leaking. Fristad *et al.* (1997) have studied the Oseberg Syd area using similar methods to this study, and have found that SGR values of 15–18 are consistent with adjacent fault blocks having small pressure differentials (typically less than ~1–2 bars for the hydrocarbon–water system). Values of 18 or greater correspond to significant fault seal barriers, 8 bars of pressure differential or 100 m difference in oil–water contacts.

In the Sleipner Vest study reported here, a GR >40 is associated with clay smearing, and sealing in general. The data also imply that the low permeability PFFR (GR 30–40) can hold considerable hydrocarbon columns but may be transmissible for gas late in the proposed production history. The cataclastic fault rocks (GR 0–14) and the high permeability PFFR (GR 14–20) can in general be considered as being non-sealing for hydrocarbons. The medium permeability PFFR (GR 20–30) can also be considered as non-sealing for gas, but will probably be sealing for oil.

The results from the Sleipner Vest Field are in good agreement with these previous studies but contain more information and detail on the communication across faults. The GWC data allow validation of the core based petrophysical analysis of faults and sealing. In addition, the study has emphasized the need to consider the complexities of communication and the likely impact of the variation in transmissibility on different parts of the same fault plane. In the Sleipner Vest field, the overall hydrocarbon level is probably primarily controlled by a topographic spill point. In this situation, the differences in the GWCs recorded are likely to be related to the sealing properties of the faults near to the GWCs.

Despite the success of the integrated study reported here, and the demonstration that small-scale core studies can make a significant contribution to both the prediction and the modelling of fault sealing, there are still uncertainties. The most important ones are listed below:

(1) Seismic interpretation of fault throw. In Sleipner Vest, a large proportion of the intra-reservoir faults have interpreted throws in the range of seismic resolution (~30 m). This is a critical limitation in mapping fault geometries and estimating fault throws. The detailed analysis reported involved the evaluation of a number of possible/viable geometries to both the stratigraphy and the faults (including damage zone structures) in critical areas identified by the initial (triangular) Juxtaposition analysis.

(2) The VSH-log used to differentiate the phyllosilicate content of the host rock. The correlation with the true phyllosilicate content depends upon the resolution and ability to detect different phyllosilicate types. Given the critical link between the clay content and sealing properties in this field, careful calibration of this V-Shale data is essential.

(3) The GR calculation in the FAPS module assumes that the fault gouge is accumulating equal amounts of sand and shale and that equal amounts are accumulated from hanging-wall and footwall. This may be an over-simplication of the complex processes which take place in fault zones, but the pressure

Fig. 9. Distribution of the fault rocks on the main faults of the Sleipner Vest field. The map is based on fault plane diagrams as in Fig. 8 and shows the projected location of high permeability windows.

or inter-well permeability data from across a fault zone will incorporate this complexity.

(4) The strengths of the phyllosilicate smears are uncertain. Quantification of the strength of the different seals is important for the estimation of gas communication and water support during production. The first well planned on the adjacent Delta structure will provide an excellent opportunity to test the communication across these fault rocks, and will provide early information needed for back-calculating fault behaviour. With limited vertical communication in Sleipner Vest, and no water communication expected across the faults with phyllosilicate smears, the thin and poor quality sands in the deeper sections will be very difficult to produce, and possibly 1/3 of the volumes could be unproducible. Production testing from the first well on the Delta structure should be used to assess how much pressure build-up the low permeability phyllosilicate framework fault rocks can hold before leaking, either because the threshold capillary pressure has been reached, or because the seal experiences a mechanical failure.

Conclusions

The study has illustrated that prediction of the variations in hydrocarbon–water contact levels across Sleipner derived from careful evaluation of fault rock properties is possible. The validation of predictions of GWCs over the field support the integrated approach to fault seal analysis, which combines detailed microstructural and petrophysical characterization with seismic data on fault geometry. In addition, the amalgamation of the results with field information of hydrocarbon distribution creates an important platform for future modelling of the reservoir behaviour. The study has also identified the possible future directions which will continue this integration as the field development proceeds.

In summary:

(i) faults in the Sleipner Vest field do not have the ability to completely seal gas over much of the crest of the field. However, the fault rocks developed on intra-reservoir faults provide flow retarders or barriers which will restrict flow by variable degrees;

(ii) faults probably act as seals to gas in the central northern parts of the field, due to more extensive distribution of phyllosilicate smears on the faults in this lower reservoir net/gross region;

(iii) the study provides a consistent explanation for the observed differences in GWC in different fault blocks and highlights the importance of clay content in creating seals and flow retarders in this field.

The support of Statoil and the Sleipner Vest field partners is gratefully acknowledged by the Rock Deformation Research Group. Discussions with T. Harper and E. Lundin, as well as with other members of the RDR group are acknowledged. Constructive comments and reviewing by G. Yielding and S. Digert are also appreciated.

References

ALLAN, U. S. 1989. Model for hydrocarbon migration and entrapment within faulted structures. *American Association of Petroleum Geologists Bulletin*, **73**, 803–811.

BOUVIER, J. D., SIJPESTEIJN, K., KLEUSNER, D. F., ONYEJEKWE, C. C. & VAN DER PAL, R. C. 1989. Three-dimensional seismic interpretation and fault sealing investigations, Nun River field, Nigeria. *American Association of Petroleum Geologists Bulletin*, **73**, 1397–1414.

COWARD, M. P. 1994. Inversion Tectonics. *In*: HANCOCK, P. L. (ed.) *Continental Deformation*. Pergammon Press, 289–304.

FISHER, Q. J. & KNIPE, R. J. 1998. Fault sealing processes in sliliciclastic sediments. This volume.

FREEMAN, B., YIELDING, G. & BADLEY, M. 1989. Fault correlation during seismic interpretation. First Break, 8 (3), 87–95.

FRISTAD, T., GORTH., A., YEILDING, G. & FREEMAN, B. 1997. Quantitative fault seal prediction – a case study from Oseberg Syd. *In*: MOLLER-PEDERSON, P. & KOESTLER, A. G. *Hydrocarbon Seals – Importance for Exploration and Production*. NPF Special Publication **7**, Elsevier, Singapore, 107–124.

GIBSON, R. G. 1994. Fault zone seals in siliclastic strata of the Columbus Basin, Offshore Trinidad. *American Association of Petroleum Geologists Bulletin*, **78**, 1372–1385.

JEV, B. I., KARS-SIJPESTEIJN, C. H., PETERS, M. P. A. M., WATTS, N. L. & WHITE, J. T. 1993. Akaso field, Nigeria: Use of integrated 3D seismic, fault slicing, clay smearing, and RFT pressure data on fault trapping and dynamic leakage. *American Association of Petroleum Geologists Bulletin*, **77**, 1389–1404.

KNIPE, R. J., 1992. Faulting processes and fault seal. *In*: LARSEN, R. M., BREKKE, H., LARSEN, B. T. & TALLERAS, E. (eds) *Structural and Tectonic Modelling and its application to Petroleum Geology*. NPF Special Publications **1**, Elsevier, Amsterdam, 325–342.

—— 1993. The influence of fault zone processes and diagenesis on fluid flow. *In*: HORBURY, A. D. & ROBINSON, A. G. (eds) Diagenesis and Basin Development. *American Association of Petroleum Geologists Bulletin*, **36**, 135–154.

—— 1997. Juxtaposition and seal diagrams to help analyze fault seals in hydrocarbon reservoirs. *American Association of Petroleum Geologists Bulletin*, **81** (2) 187–195.

——, Fisher, Q. J. Jones, G., Clennell, M. B., Farmer, A. B., Harrison, A. Kidd, B., McAllister, E., Porter, J. R. & White, E. A. 1997. Fault seal analysis: successful methodologies, application and future directions. *In*: Moller-Pederson P. & Koestler, A. G. *Hydrocarbon Seals – Importance for Exploration and Production*. NPF Special Publications, **7**, Elsevier, Singapore, 15–38.

Knott, S. D. 1993. Fault seal analysis in the North Sea. *American Association of Petroleum Geologists Bulletin*, **77**, 778–792.

Needham, D. T, Yielding, G. & Freeman, B. 1996. Analysis of fault geometry and displacement patterns. *In*: Buchanan, P. G. & Nieuwland, D. A. (eds) *Modern Developments in Structural Interpretation Validation and Modelling*. Geological Society, London, Special Publications, **99**, 189–200.

Pegrum, R. M. 1984. The extent of the Tornquist Zone in the Norwegian North Sea. *Norsk Geologisk Tidsskrift*, **64**, 39–68.

—— & Ljones, T. E. 1984. 15/9 Gamma gas field offshore Norway, new trap type for the North Sea basin with regional structural implications. *American Association of Petroleum Geologists Bulletin*, **68**, 874–902.

Rattey, R. P. & Hayward, A. B. 1993. Sequence stratigraphy of a failed rift system; the Middle Jurassic to Early Cretaceous basin evolution of the Central and Northern North Sea. *In*: Parker J. R. (ed.) *Petroleum Geology of Northwest Europe: Proceedings of the 4th Conference*, Geological Society, London, 215–250.

Sverdrup, E. & Bjørlykke, K. A.1997. Fault properties and the development of cemented fault zones in sedimentary basins. Field examples and predictive models. *In*: Moller-Pederson, P. & Koestler, A. G. (eds) *Hydrocarbon Seals - Importance for Exploration and Production*. NPF Special Publications, **7**, Elsevier, Singapore, 91–106.

Underhill, J. R. & Partington, M. A. 1993. Jurassic thermal doming and deflation in the North Sea: implications of the sequence stratigraphic evidence. *In*: Parker, J. R. (ed.) *Petroleum Geology of Northwest Europe: Proceedings of the 4th Conference*. Geological Society, London, 337–326.

Watts, N. L. 1987. Theoretical aspects of cap-rock and fault seals for single and two-phase hydrocarbon columns. *Marine and Petroleum Geology*, **4**, 274–307.

Yielding, G., Freeman, B. & Needham, D. T. 1997. Quantitative fault seal prediction. *American Association of Petroleum Geologists Bulletin*, **81**, 897–917.

Flow through fault systems in high-porosity sandstones

T. MANZOCCHI[1,2], P. S. RINGROSE[1,3] & J. R. UNDERHILL[4]

[1] *Department of Petroleum Engineering, Heriot-Watt University, Edinburgh EH14 4AS, UK*
[2] *Present address: Fault Analysis Group, Department of Earth Sciences, University of Liverpool, Liverpool L69 3BX, UK*
[3] *Present address: Statoil Research Centre, Postuttak, Trondheim 7005, Norway*
[4] *Department of Geology and Geophysics, The University of Edinburgh, Edinburgh EH9 3JW, UK*

Abstract: Small-scale faults in high-porosity sandstones form highly connected systems over a range of length scales. The effective permeability and oil recovery in such fault systems are strongly controlled by their geometrical architecture. This paper describes partially sealing faults in terms of: (a) characterization of their spatial distribution; (b) their effects on reservoir compartmentalization; and (c) their significance for fluid flow. Four suites of numerical flow simulations in highly compartmentalized fault systems are used to assess the influence of geometrical variability on single- and two-phase flow. The single-phase suites demonstrate that effective permeability is approximately linearly related to the number of fault-enclosed compartments present in a fault system. Use of a fault heterogeneity measure allows effective permeability for a geometrical case to be estimated from fault density and fault and matrix permeability. The two-phase simulations model water-floods, and show the relationships between length scale, fault system geometry and oil recovery. In a self-similar fault system, oil recovery is preferentially inhibited at the smaller length scales. Oil recovery is influenced by compartment volume distribution and is therefore sensitive to fault clustering. The fault density necessary to severely affect either single- or two-phase flow is likely to occur only close to structures which are large relative to the scale under consideration. This improved understanding of the relative influences of fault system geometry, density and petrophysics should lead to significantly improved hydrocarbon recovery from faulted high-porosity sandstones.

Fault system architecture

High-porosity quartz arenitic sandstones often show high fault densities both in outcrop and core. Although individual fault displacements are often less than a centimetre, these faults are characterized by significant reduction in porosity, grain sorting, grain sphericity and pore throat size over a 1–2 mm wide zone. The mechanics of deformation in quartz-rich siliciclastics has been extensively reviewed (e.g. Aydin & Johnson 1983; Underhill & Woodcock 1987; Antonellini *et al.* 1994). Because the strength of a sandstone is proportional to porosity (Dunn *et al.* 1973) and high porosity favours fault strain-hardening, small-scale faulting will tend to be most prevalent in the best quality sandstone reservoirs, and faulting there will be predominantly small-scale.

Figure 1 shows an example of faulting in the Permian Penrith Sandstone in the north of England at George Gill near Espland (Knott 1994). Single-event deformation bands (granulation seams) often form multiple-event zones of bands (Fig. 1a). The faults form in characteristic orientations, and interaction of bands and zones of different orientations produces polyhedral fault-bounded sandstone compartments (Fig. 1b). Although the petrophysical properties of the faults vary with provenance, porosity can be reduced by up to 80%, permeability is frequently around three orders of magnitude less than that of the unfaulted sandstone matrix, and capillary pressures across faults are much higher than in matrix samples (e.g. Pittman 1981; Jamison & Stearns 1982; Underhill & Woodcock 1987; Antonellini & Aydin 1994; Fowles & Burley 1994).

Figure 2 illustrates the nature of faulting in porous sandstones at a range of length scales. The smallest-scale image is a computerized tomograph (CT) scan of a 1 cm thick zone of deformation bands from a sample recovered from core from the North Sea (Fig. 2a). The image has a pixel resolution of 0.2 mm × 0.2 mm and has been processed to display only CT values above an arbitrary threshold. The CT number correlates directly with density, and therefore a high CT number indicates low porosity in monomineralic rock. At this scale the anastomosing nature of the zones of bands is evident, with frequent fault splays and patches of higher porosity sandstone between zones of high density fault gouge.

The second fault map is one order of magnitude larger, and comes from the Penrith Sandstone at

From Holdsworth, R.E. & Turner, J.P. (compilers) *Extensional Tectonics: Faulting and Related Processes*. The Geological Society, Key Issues in Earth Sciences, **2 (2)**, 281–298.
1476-315X/02/$15.00
First published in: "Coward, M.P., Daltaban, T.S. & Johnson, H. (eds) *Structural Geology in Reservoir Characterization*, The Geological Society, London, Special Publications **127**, 65–82".

(a)

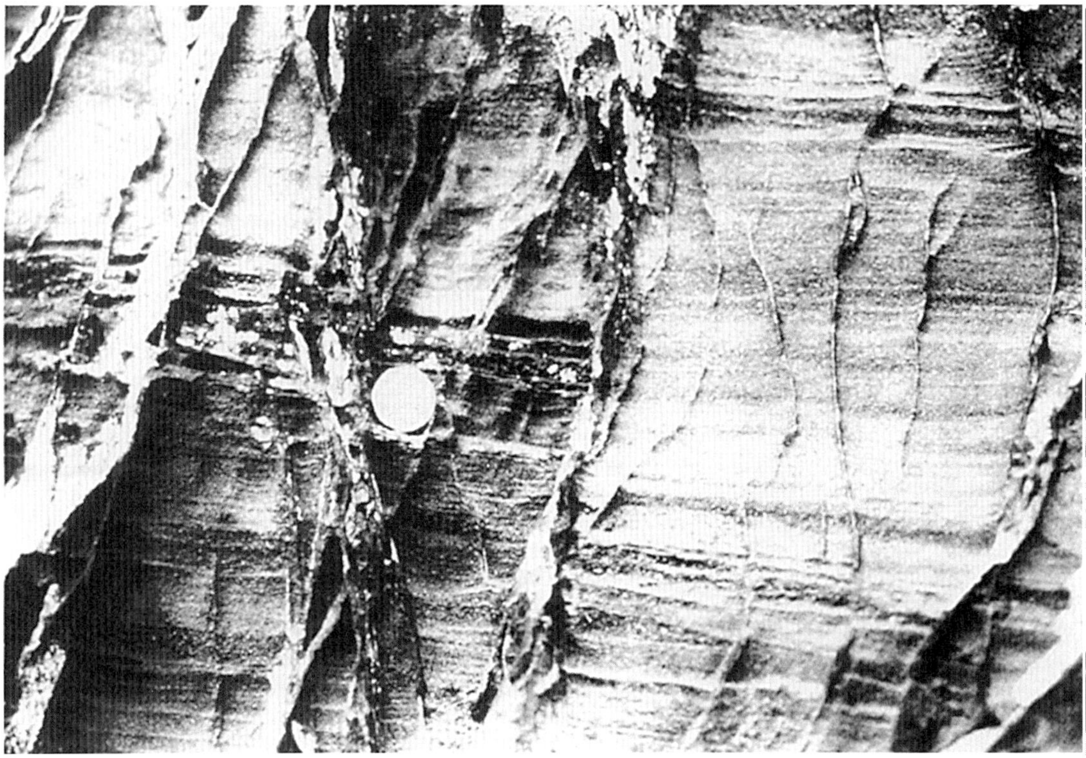

(b)

Fig. 1. Small-scale faulting in Penrith Sandstone, George Gill, Vale of Eden. (**a**) Thin anastomosing zones of bands (left) have sub-centimetre displacements. Single wavy deformation bands (right) frequently form splays and terminate as isolated fault tips. Photograph is 35 cm wide. (**b**) Fault traces from two conjugate sets of deformation bands form rhomb-shaped fault-enclosed compartments. Photograph is 8 m wide.

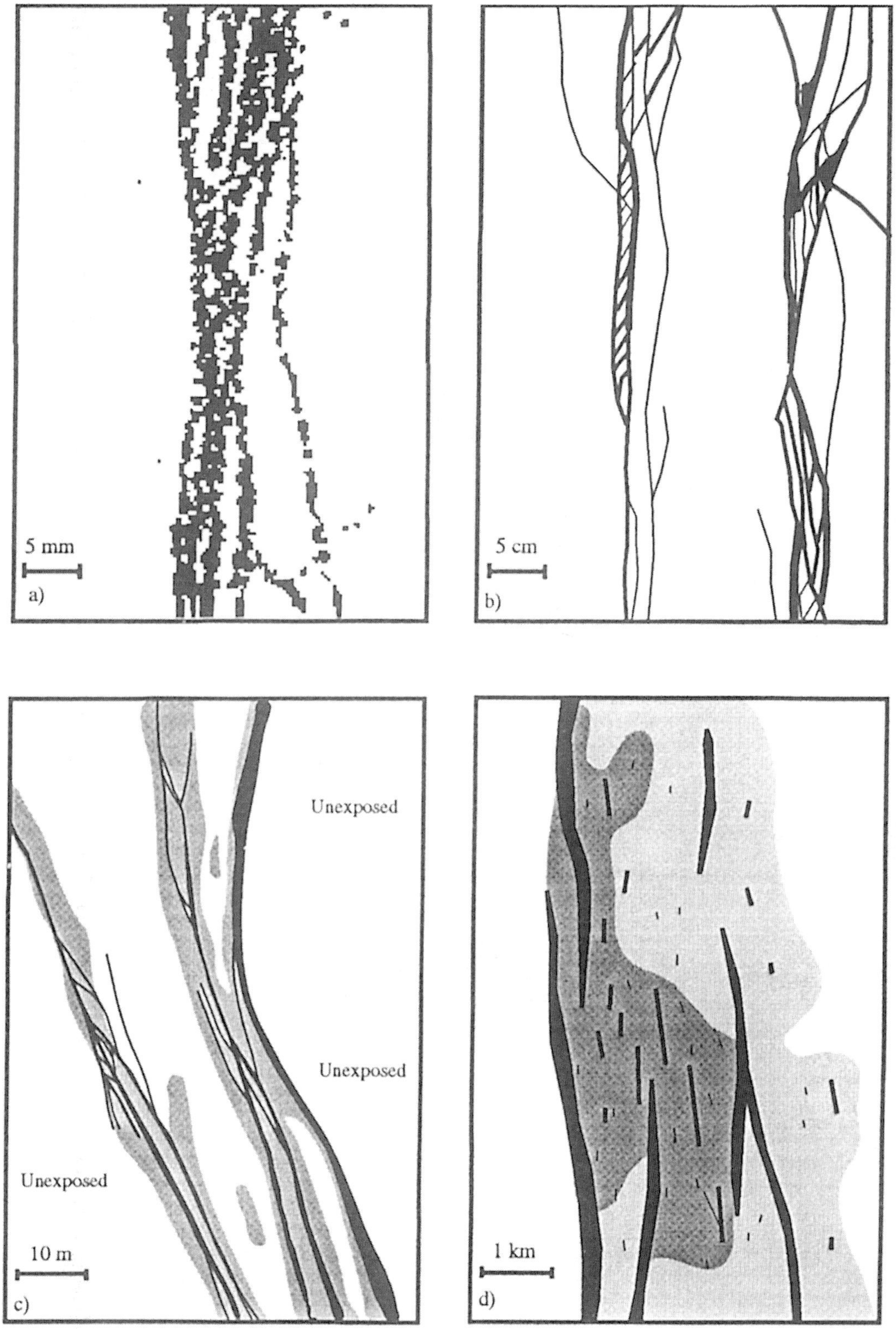

Fig. 2. Faulting at four scales. (**a**) Millimetre-scale CT image of a narrow zone of deformation bands from North Sea core. Pixel resolution is 0.2 mm × 0.2 mm. (**b**) Centimetre-scale fault map from the Permian Penrith Sandstone, Vale of Eden, N England. Resolution is that of the naked eye. (**c**) Tens of metres-scale fault map from a pavement in the Navajo Sandstone, Arches National Park, Utah. Resolution for discrete faults is 10 cm thickness. Contoured fault density for narrower faults is derived from one-dimensional scan lines. The shaded area represents fault density of more than five per metre. (**d**) Hypothetical kilometre-scale reservoir structure map. Seismic fault resolution is approximately 20 m offset.

George Gill (Fig. 2b). Two zones of deformation bands are shown, each with a variable and complex internal morphology. Again, a high frequency of splays and anastomosing strands within the zones is evident. Frequent duplex structures occur within the zones, resulting in variable gouge thicknesses along the zone traces. Duplex structures are common in faulted high-porosity sandstones (Cruikshank *et al.* 1991), and are a common way for zones of deformation bands to terminate. Because of this, the overall connectivity of deformation band systems is often very high. However, both at this and the previous scale, many of the splay faults also have isolated terminations.

Figure 2c is a fault map from the Navajo Sandstone on the Moab anticline in Arches National Park, SE Utah. The 50 m × 80 m area is two orders of magnitude larger than the previous map, and contains faults ranging in width from 1 mm to 1 m. This range is too great to allow construction of a completely scaled map, and only faults thicker than 10 cm are shown. A series of 13 scan lines perpendicular to strike have been averaged to infer the shaded regions which represent areas with over five bands per metre. This average (based on the number of faults) is only one of several possible representations and others are feasible: a fault density average is proportional to gouge thickness, and a strain average is proportional to fault displacements. Thus density and displacement information, as well as data appertaining to connectivity, is not represented by the average selected. The overall fault pattern is similar to the smaller-scale images, with tight fault clusters and frequent duplexes and splays. However, the overall proportion of fault splays to fault terminations is lower than at the previous scale. This is because large faults are frequently transferred onto structures which are below the resolution of the map. The need to average smaller faults has reduced the apparent connectivity of the fault system.

Figure 2d shows a hypothetical reservoir structure map. Only faults with offsets greater than about 20 m are resolvable by three-dimensional seismic, so this is the smallest scale of fault that can be explicitly mapped. The scale of this map is more than five orders of magnitude greater than the CT image (Figure 2a). In order to understand the effect of sub-seismic faulting on flow in this hypothetical sandstone reservoir, the nature of flow and faulting at all relevant scales, from millimetres to kilometres, must therefore be considered. Flow is a function of fault and matrix permeability, the fault system structure, and the fault density and density distribution throughout the reservoir. The object of this study is to establish which characteristics of the fault system will most affect flow.

Fault data may come from several different sources. Core may yield samples the scale of Fig. 2a. This grain and pore scale is critical as it is the scale at which permeability is determined: cataclasis in quartz arenites or clay smearing in less pure sandstones ultimately determine the petrophysical properties of the fault gouge. The centimetre–decimetre scale shown on Fig. 2b may be resolved by acoustic and electrical wellbore image logs. These may reveal information about fault density and clustering characteristics, but not about permeability which requires core. Information about fault system geometry at scales larger than this, such as that mapped on Fig. 2c, is impossible to gather from wells in anything greater than one-dimensional datasets. Faults between about 1 m and 20 m offset are generally irresolvable by either well or seismic. At these scales the use of analogue or theoretical models of faulting is vital.

Several methods exist to determine areas of likely sub-seismic faulting in reservoirs, including those based on the proximity to large structures (e.g. Gauthier & Lake 1993) or those based on curvature mapping (e.g. Lisle 1994). Field evidence in faulted sandstone systems exists for both these techniques. Jamison & Stearns (1982) have shown small faults clustering around large structures, and a relationship between curvature and fault density has been demonstrated for deformation band systems by Antonellini & Aydin (1995). Here we focus on how small-scale fault density predictions could be used to gain a quantitative assessment of hydrocarbon flow within faulted reservoirs.

An essential step in modelling sub-seismic faulting in reservoirs is to derive a fault density map based on interpretation of the relevant data: seismic, core, regional geological model, tectonic history, rheology, etc. Populating this fault density map with individual faults, each with appropriate petrophysical properties, at all relevant length scales is intractable, so statistical approaches are necessary. Figure 2c illustrates the necessity of fault system averaging, and highlights some of the assumptions and simplifications that must be made in such averaging. Even if populating a deterministic fault map were possible, calculating flow in so complex a model would be computationally exorbitant. A prediction of the flow in a faulted sandstone must therefore rely on some form of fault system averaging, incorporating the effect of flow in the faulted zones as effective properties. A fault system can be characterized by several interrelated geometrical factors. This paper considers some of the geometrical aspects of fault systems and their significance for flow. Only through an understanding of which fault system attributes are important to flow can reservoir geologists know how to characterize fault systems effectively. Similarly, only through an understanding of natural fault systems can a reservoir engineer know whether the methods or assumptions used in modelling the fault system are valid.

Previous work on predicting flow effects due to sealing or partially sealing faults has focused on the frequency of minor faults at sub-seismic scales (e.g. Gauthier and Lake 1993; Heath *et al.* 1994), with little attention placed on modelling the connectivity characteristics of the system. Conversely, many of the fault characterization and flow performance calculations for hydraulically conductive fracture systems have concentrated on the fault clustering and connectivity characteristics of the system (e.g. Odling 1992*b*). In such systems, most of the flow is concentrated in the fractures and the greater the fracture system connectivity, the greater the flow performance of the reservoir. The opposite is true for partially sealing faults. In these systems the flow is concentrated in the high-permeability matrix, and if the fault system is highly connected, flow is impaired. Fault system geometry is likely to be significant for flow through low-permeability faults, due to compartmentalization of the high-permeability matrix by low-permeability faults.

Fault system modelling

A fracture system can be statistically characterized by only a few geometrical factors (La Pointe & Hudson 1985; Dershowitz & Einstein 1988). The most fundamental factor is fault density and the other factors characterize the distribution of fault density in space. In this study, fault density refers to the total fraction of fault gouge in a volume of rock.

Individual faults (each with particular orientations, sizes, displacements, terminations and positions) occur over a wide range of scales. Fault size and displacement characteristics are governed by the scaling laws of the fault system which define the number of faults at a particular length scale that are present in the system. Fault system connectivity is controlled by the other factors: orientation, termination and position.

Fault scaling

A common technique for determining the frequency of sub-seismic faults assumes scale invariance. This method (Yielding *et al.* 1992; Gauthier & Lake 1993) relies on self-similarity of fault attribute scaling between seismic and sub-seismic scales. It is based on relationships of the form:

$$N = AL^{-m} \quad (1)$$

and

$$D = BL^{n} \quad (2)$$

where N is the cumulative number of faults of length $\geq L$, D is fault displacement and A, B, m and n are system-specific constants. Equation 1 implies that fault density is scale-invariant (Childs *et al.* 1990; Scholz & Cowie 1990), while Eqn 2 implies scale invariance of fault shape (Gillespie *et al.* 1992). From these relationships it is possible to predict the frequency of faults of any particular length scale, but implicit in the implementation of this method is the belief in a mechanical continuum of faulting at all scales covered by the extrapolation. This is not the case for fault systems in porous sandstones.

Faults in porous sandstones occur in three styles: single deformation bands, zones of deformation bands and slip surfaces (Aydin & Johnson 1978). A continuum of physical processes may exist between the first two styles as both represent strain-hardening processes, but the development of slip surfaces involves a physical transition from shear displacement accommodated across a thickening fault zone to displacement accommodated on a single plane. Slip surfaces strain-soften rather than strain-harden. It is therefore unlikely that fault systems would show a constant power-law distribution over length scales spanning this mechanical transition. In the Penrith Sandstone this transition occurs at fault displacements of about 0.3 m (Knott 1994).

Similarly, the displacement/length relationship (Eqn 2) is process-dependent. Although the relationship is broadly correct for about six orders of magnitude of faulting (Gillespie *et al.* 1992), a mechanical transition between strain-hardening deformation bands and strain-softening slip surfaces negates the possibility of complete scale invariance. High-porosity sandstones are often massively bedded with few good marker laminations and it is notoriously difficult to measure deformation band displacement gradients. Preliminary data reported by Antonellini & Aydin (1995) show that although displacement/length relationships may be scale-invariant for slip surfaces, no such relationship seems to exist for deformation bands. Cowie & Scholz (1992) have shown that an important parameter determining the displacement/length relationship is friction. In deformation bands friction is high, whereas slip surfaces have much lower friction.

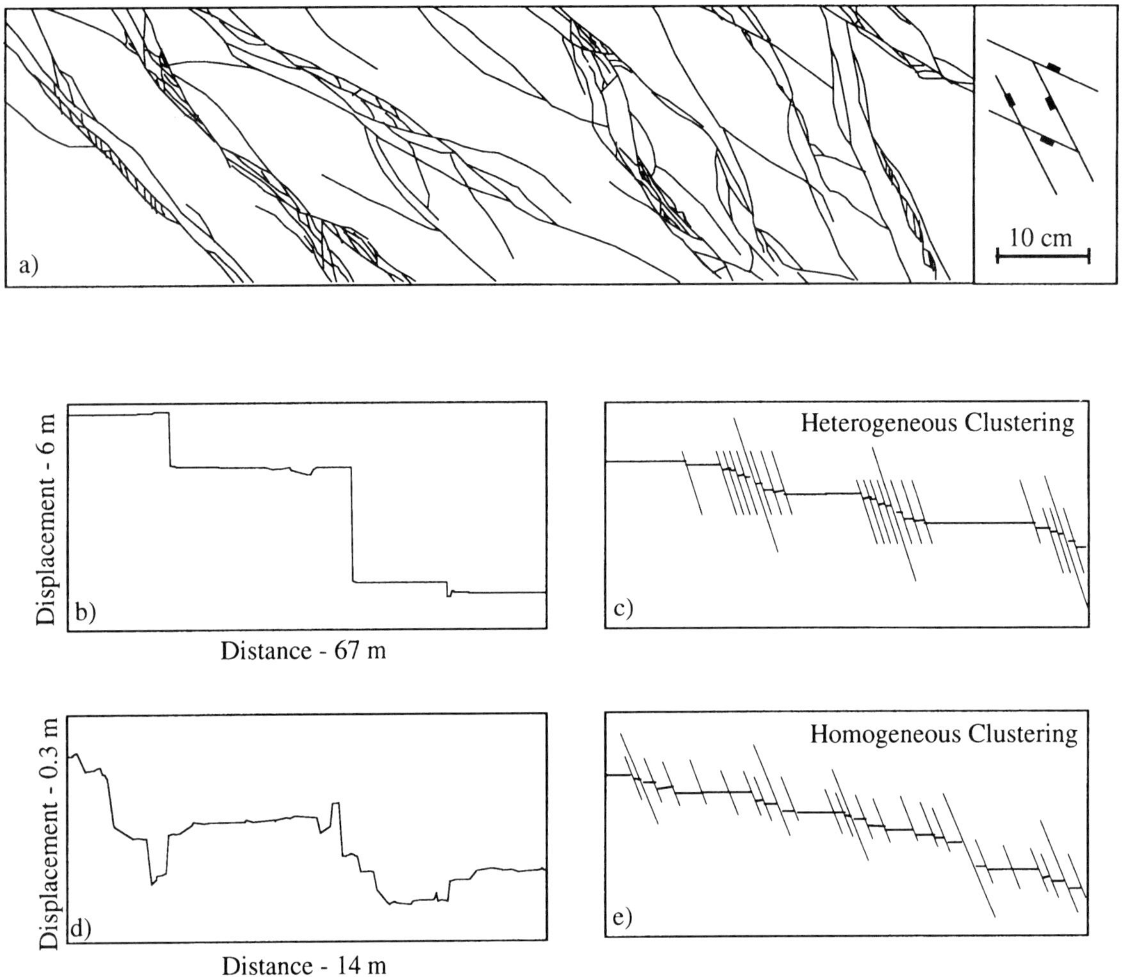

Fig. 3. (**a**) Fault trace map from the Permian Penrith Sandstone, George Gill, Vale of Eden. Fault thickness variation is not represented. (**b**) Fault displacement section for a heterogeneous cluster system, Belah Scar, Vale of Eden. (**c**) Cartoon of heterogeneous clustering – small distinct fault clusters are formed around slip surfaces. (**d**) Fault displacement section from within a cluster, George Gill, Vale of Eden. (**e**) Cartoon of a homogeneous fault system – indistinct overlapping clusters are formed.

Friction is a fundamental control on strain-hardening versus strain-softening faulting processes (Underhill & Woodcock 1987). Therefore it is reasonable to suppose that displacement gradients on deformation bands are markedly different from those on slip surfaces. The strain-hardening/strain-softening transition discussed above is peculiar to high-porosity sandstones; however, analogous mechanical and scaling transitions have been reported from other fracture styles, for example tensile fractures in basalts by Hatton *et al.* (1994).

Fault connectivity

Fault connectivity is governed by fault orientation, termination and placement characteristics. The most essential factor for connectivity is orientation: if all faults are parallel, no faults can connect. Fault systems are usually divided into orientation populations, and the stress state at the time of faulting controls the orientations of the populations present. Four orientation populations are necessary to accommodate triaxial strain (Reches 1977; Krantz 1988) and this is often observed in porous sandstones (Aydin & Reches 1982; Underhill & Woodcock 1987). Figure 3a is a small-scale fault trace map from the Penrith Sandstone, with the four principal orientation populations indicated. Usually a fault system consists of two or more orientation populations, with individual faults dispersed around the principal orientations. For a given length distribution, the fault system becomes more connected as the angle between two populations of parallel faults or the dispersion of fault orientations increases (Robinson 1983).

Fault system termination characteristics directly control fault system connectivity. Termination is characterized by the ratios of I, Y and X nodes present in a fault system. I nodes represent isolated fault tips, Y nodes are formed at fault splays or terminations against other faults, and X nodes represent fault intersections. The proportion of fault terminations in a system is likely to vary as a function of the mechanical evolution of the fault system.

Dershowitz & Einstein (1988) report two joint datasets with 42% and 60% Y nodes (it is unclear in their study whether X nodes are considered equivalent to two Y nodes, or ignored). Gillespie *et al.* (1993) show a joint map with 91% X nodes. Large-scale (seismic) fault maps often show a preponderance of I nodes (e.g. Gillespie *et al.* 1993). Fault maps of large sandstone pavements (e.g. Zhao & Johnson 1992) often show high proportions of X nodes, but this could be an effect of fault reactivation in a tensile mode. Our observations suggest that the dominant node type at the small scale in deformation band systems is Y nodes. Figure 2 shows abundant splays; Fig. 3a has 14 : 73 : 13 $I:Y:X$ node proportions.

Antonellini & Aydin (1995) have examined fault terminations in porous sandstones. Both deformation bands and zones of deformation bands form interconnected systems, while slip surfaces often terminate by transference to these smaller structures. Fault system connectivity characteristics are different above and below the mechanical scaling transition: terminations at deformation band and zone length scales are characterized by Y nodes, but slip surface/slip surface Y terminations are rare.

The spatial position of faults relates to fault clustering. In this paper, clustering refers to the relative proximity of faults (e.g. Gillespie *et al.* 1993), and not the definition used in percolation theory, in which clustering refers to the connectivity of the system (e.g. Odling 1992*a*). Fault clustering has only an indirect influence on compartmentalization: as clustering in a fault system increases, the sizes of fault-enclosed compartments become more variable.

La Pointe & Hudson (1985) relate clustering to fault system homogeneity. Highly clustered systems are heterogeneous, whereas if the faults are more evenly distributed the system is homogeneous. Clustering of deformation bands around slip surfaces is very common (e.g. Jamison & Stearns 1982; Koestler *et al.* 1994). Figure 3b is an example of such clustering, showing a displacement profile along a 67 m cliff section roughly parallel to fault dip. Two large faults with metre offsets accommodate most of the 6 m cumulative displacement, and millimetre displacement deformation bands are clustered tightly around these. A cartoon of this heterogeneous fault clustering is shown in Figure 3c.

Figure 3d shows a cross-section through a deformation band/zone of deformation band system. The faulting in this system is much more homogeneous and a simplified cartoon of one orientation population is illustrated in Fig. 3e. The faults in Fig. 3d indicate homogeneous clustering; however, what is sampled is the internal structure of a single cluster. Fault clusters are frequently observed in faulted sandstones. Many of these are not associated with a large fault, although large faults usually display clusters.

The soft-domino fault model (Walsh & Watterson 1991) has shown that linked fault systems need not be ubiquitous. In porous sandstones we observe a population of large slip surfaces which are not linked; the model predicts that the fault blocks respond by internal deformation. In high-porosity sandstones this strain is accommodated by forming deformation bands and zones, rather than by flexural slip and drag around the slip surfaces. These smaller structures form a highly connective system, and are largely responsible for fault compartmentalization.

The number of fault-enclosed compartments in a reservoir can be determined exclusively by orientation and termination characteristics of the fault system. The size distribution of these compartments is controlled by the clustering of the faulting. These influences are illustrated in Fig. 4.

Fault compartmentalization

In this section, we derive a relationship between fault connectivity and reservoir compartmentalization in two dimensions. This provides a measure of compartment density (CD) based on composite geometrical parameters of the fault system and is used in the following section for analysing the flow simulation results. We define compartment density as the number of fault-enclosed compartments present in a system, represented as a fraction of the maximum number of compartments possible for the particular number of fault terminations and intersections. This is a function of the proportion of each fault node type present (Fig. 4c), and the total number of nodes. For given $I:Y:X$ node proportions, the total number of nodes present decreases as the orientation population density ratio, x, decreases (Fig. 4b). This effect is incorporated into the CD measure by normalising

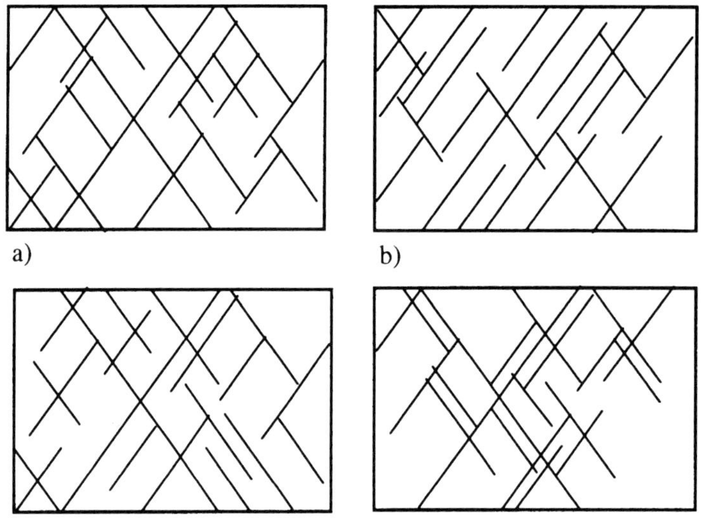

Fig. 4. Geometrical control on fault connectivity. (**a**) Datum fault system. (**b**) Effect of fault orientation: lowering the orientation population density ratio lowers the fault system connectivity. (**c**) Effect of fault termination characteristics: increasing the proportion of isolated terminations lowers the fault system connectivity. (**d**) Effect of fault clustering: increasing the heterogeneity of the system clustering has no influence on fault connectivity, but increases the range of compartment sizes.

the actual number of compartments present for a case of $x \neq 1$ against the number of compartments which would be present for the same $I:Y:X$ proportions at $x = 1$.

Figure 5 shows the fundamental compartmentalization properties of the three fault node types. Each compartment is contained by four Y or X nodes. Each X node separates four compartments (Fig. 5a), but each Y node separates only two compartments (Fig. 5b). Therefore one X node or two Y nodes are required per compartment. I nodes do not increase compartmentalization, but they can have a dispersive effect. Faults that contain an I node and one or more of any other type of node reduce the number of compartments calculated on the basis of Y and X nodes by one (Fig. 5c). However, faults that contain two I nodes and no other nodes do not affect compartmentalization (Fig. 5d). The compartment density of a fault system may therefore be determined from the proportions of each node type

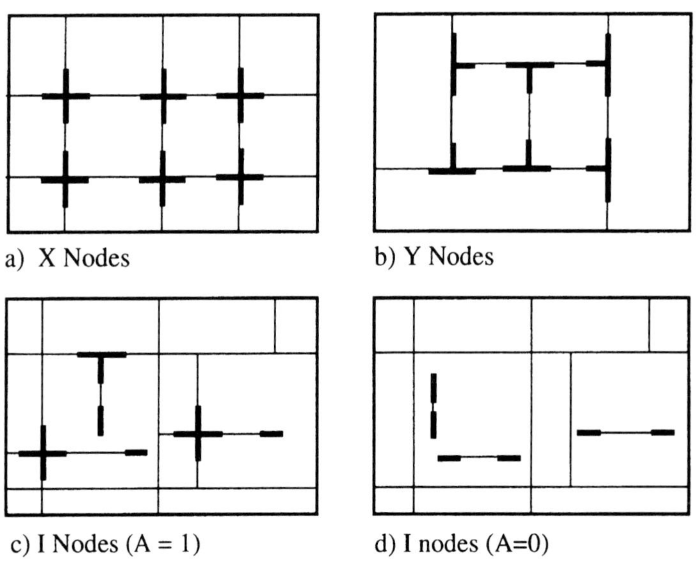

Fig. 5. Contribution of each fault node type to compartmentalization in two dimensions for two orientation populations of parallel faults. One X node (**a**) or two Y nodes (**b**) are required per compartment. I nodes may either reduce compartmentalization (**c**), or have no influence on it (**d**).

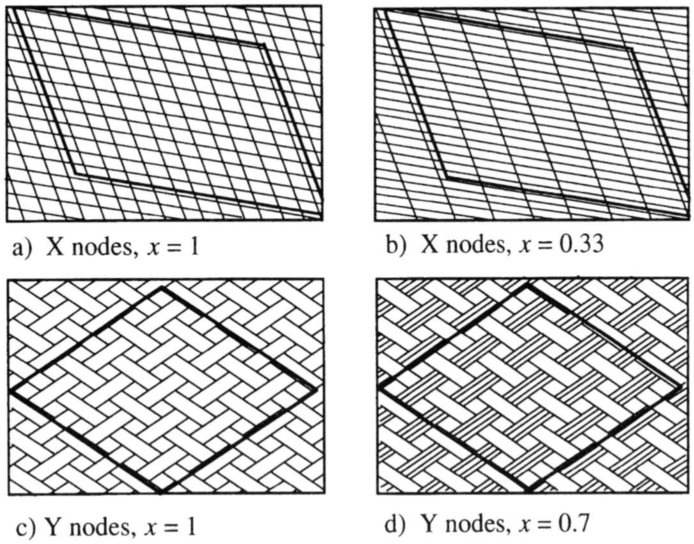

Fig. 6. Influence of the population orientation density ratio (x) on compartmentalization. Compartmentalization due to X nodes is sensitive to x. (**a**) $x = 1$, $n_a = 12$, $n_b = 12$, $n_c = n_a n_b = 144$. (**b**) $x = 0.33$, $n_a = 6$, $n_b = 18$, $n_c = n_a n_b = 108$. Compartmentalization due to Y nodes is not affected by x. (**c**) $x = 1$, $n_a = 36$, $n_b = 36$, $n_c = n_a + n_b = 72$. (**d**) $x = 0.7$, $n_a = 36$, $n_b = 51$, $n_c = n_a + n_b = 87$.

present:

$$CD = p_X + \frac{p_Y}{2} - Ap_I \quad (3)$$

Where CD is compartment density, p_X, p_Y and p_I are the proportions of each node type present, and A is a system-specific constant reflecting the proportion of I nodes that are connected to a Y or X node, and has the range $0 \leq A \leq 1$.

As well as the $I:Y:X$ node proportions, the other variable that affects the number of compartments contained by a specific number of faults is the population orientation density ratio, x (Fig. 4). The maximum number of compartments possible for a fixed number of nodes is attained when all nodes are X nodes (Eqn 3). For a population characterized entirely by X nodes, the total number of compartments (n_C) is given by (Fig. 6):

$$n_C = n_a n_b \quad (4)$$

where n_a and n_b are the total number of faults from two populations in a given area. If n_T is the total number of faults in an area, and x is the population orientation density ratio, then:

$$n_T = n_a + n_b \quad (5)$$

and

$$n_a = x n_b \quad (6)$$

Manipulating Eqns 4, 5 and 6 gives:

$$n_C = \frac{x n_T^2}{(x+1)^2} \quad (7)$$

The maximum number of compartments is achieved when $x = 1$:

$$n_{C\max} = \frac{n_T^2}{4} \quad (8)$$

For any case of $x \neq 1$ and $p_X < 1$, Eqn 3 does not give the maximum number of compartments possible. Maximum compartmentalization occurs when $x = 1$, and the number of X node-bounded compartments is determined by the product of Eqn 8 and the proportion of X nodes present. The actual compartmentalization due to X nodes is given by the product of the X node fraction and Eqn 7. Normalizing Eqn 7 by Eqn 8 provides the x-sensitive $n_C/N_{C\max}$ term for X nodes, which must be incorporated into Eqn 3. The compartmentalization due to Y nodes does not vary as a function of the x (Fig. 6), so the Y node term in Eqn 3 is correct, whatever the value of x.

CD for any case of x is therefore given by:

$$CD = \frac{4xp_X}{(x+1)^2} + \frac{p_Y}{2} - Ap_I \quad (9)$$

CD correlates linearly with n_C, is dimensionless and has the range $0 \leq CD \leq 1$. $CD = 0$ implies no compartmentalization (i.e. a case where $p_I = 1$) and $CD = 1$ represents maximum compartmentalization possible at a fixed total number of fault nodes (i.e. a case where $x = 1$ and $p_X = 1$).

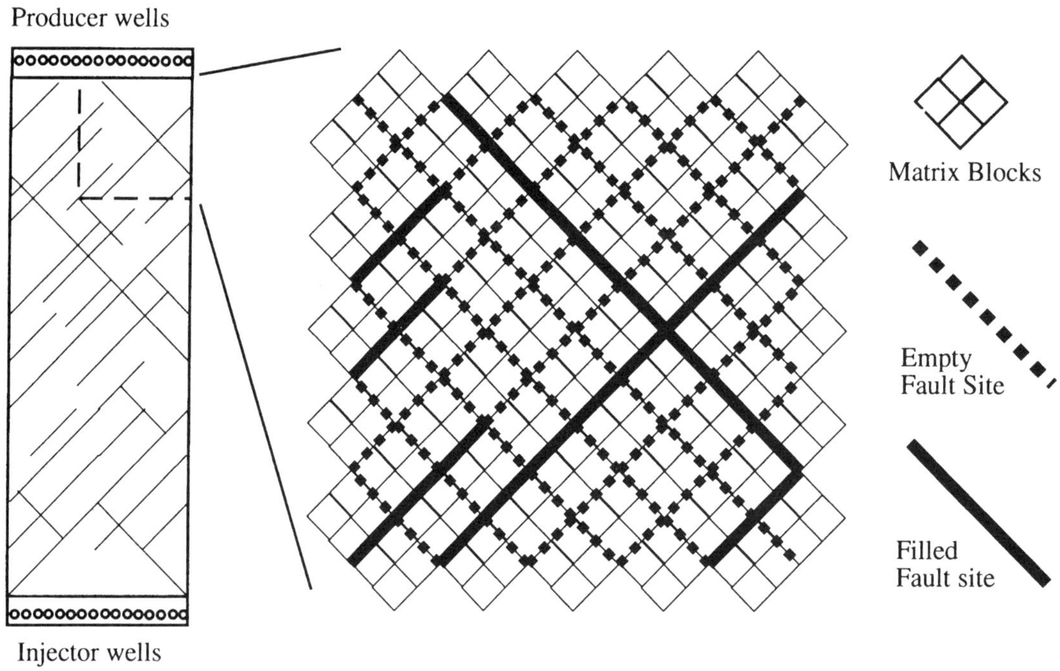

Fig. 7. Simulation grid geometry: 3553 active grid-cells are distributed at 45° to the flow direction. Each block of four matrix sites is bounded by a fault site.

Flow modelling

This section details a series of numerical simulations of single- and two-phase flow through model fault systems with varying fault connectivity characteristics. Flow simulations have been run on the ECLIPSE 100 simulator using discrete grid cells (rather than transmissibility factors) to represent low permeability faults. Fault cells are one block wide, and are separated by at least two matrix blocks (Fig. 7). All models are two-dimensional in the horizontal plane, with no-flow boundary conditions. Two orthogonal fault populations are modelled, and the mean flow direction bisects them. A 90° fault dihedral angle was used, as it allows the desired variability in fault system geometry to be modelled, whilst retaining relatively few grid-blocks in a simple configuration. Flow is modelled by fluid injection and production at controlled rates at each end of the model. The use of no-flow boundary conditions allows us to neglect the cross-flow terms in the full permeability tensor, but see Pickup *et al.* (1994) for a fuller treatment of effective permeability.

Each simulation experiment is designed to investigate the significance of a specific geometrical characteristic of the fault system. Fault density is constant in any particular experiment, and all faults have identical thickness and petrophysical properties. Four series of simulations are presented: the first investigates effective permeability as a function of compartmentalization; the second considers fault density and permeability variation; the third models oil recovery as a function of fault length scale; and the fourth, oil recovery as a function of fault system connectivity.

Compartmentalization and effective permeability

In the previous section, we showed how the orientation population ratio and termination characteristics of a fault system both contributed to compartmentalization in a faulted reservoir. The significance of these on effective permeability is discussed below.

The effect of population orientation ratio (x) is modelled with a series of 12 fault systems, each with a value of x between 0 and 1 (three are shown in Fig. 8a). A further 20 fault system models were used to calculate effective permeability variation due to fault termination characteristics, of which three are shown in Fig. 8b. The models were run with 1 D matrix and 1 mD fault permeabilities. Each fault block (Fig. 7) is 1 cm thick, and each matrix block 15 cm. The minimum spacing between faults is 30 cm and the vertical thickness of the model is 30 cm. Water was injected and produced at a constant Darcy

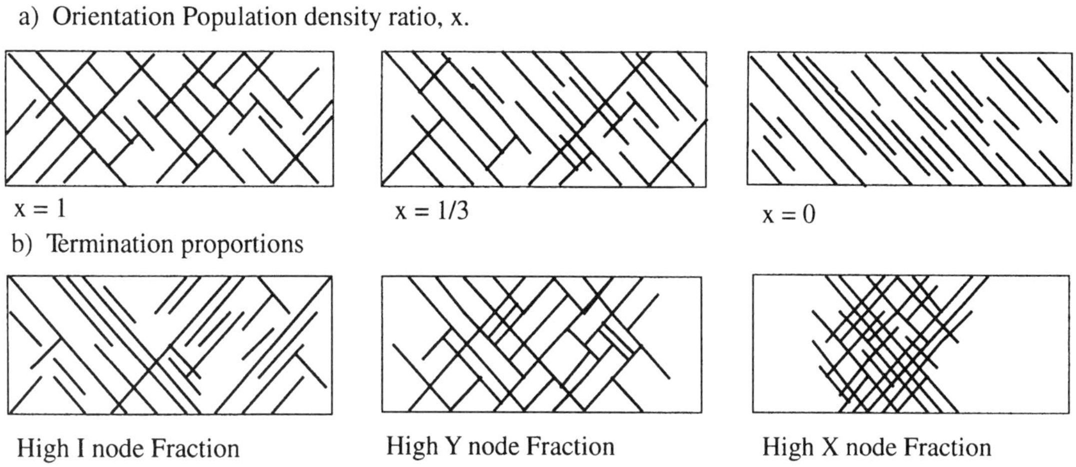

Fig. 8. Examples of the fault systems modelled to assess the control of compartmentalization on effective permeability. (**a**) Three of 12 model fault systems with variable orientation population density ratio. (**b**) Three of 20 model fault systems with variable node termination characteristics.

velocity of 0.5 m/day, and well pressures were recorded when flow had stabilized. The effective permeability of the model is calculated from the pressure drop between injector and producer wells using Darcy's law:

$$q = \frac{k_e A \Delta P}{\mu l} \tag{10}$$

where q is the volumetric flow rate, A is the cross-sectional area, $\Delta P/l$ is the pressure gradient, μ is the water viscosity and k_e is the effective permeability.

The results are shown in Fig. 9. Figure 9a is a plot of model effective permeability against orientation density ratio. Effective permeability decreases non-linearly as the fault system becomes more evenly distributed between the populations present. Figure 9b is a ternary diagram of fault nodes, with effective permeability of the 20 models indicated. Contours show an increase in effective permeability towards *I* node-dominated fault systems. In Fig. 9c effective permeability from both series of models is plotted against the compartmentalization measure *CD* (Eqn 9). These results indicate that effective permeability of a fault system decreases approximately linearly with increasing compartment density.

The results were obtained for fault systems at a constant matrix/fault permeability contrast (1000) and constant fault density. Effective permeability (k_e) can be solved analytically for end-member fault systems (Muskat 1937): if flow is perpendicular to a set of faults, effective permeability is the harmonic average permeability (k_h) of the rock-mass, and if parallel the arithmetic average (k_a). In geometrically more complex fault systems,k_e must lie between these values. The harmonic and arithmetic average permeabilities vary as a function of fault density and fault and matrix permeability, but are independent of fault system structure. The following series of flow simulations quantifies the effect of fault system structure with respect to fault thickness and permeability.

Fault density, permeability and compartmentalization

Figure 10a shows the relationship between arithmetic and harmonic average permeability and fault density for a 1 mD fault system in 1 D matrix. The separation between the two averages is equivalent to the maximum k_e range possible due to the geometrical variability of the fault system. As the matrix/fault permeability contrast (k_m/k_f) increases, the shape of the $k_a - k_h$ curve is similar but retains a higher value to lower fault densities. If we normalize $k_a - k_h$ against k_m/k_f to give $k_f(k_a - k_h)/k_m$, and plot this against fault density for three values of k_m/k_f (Fig. 10b), the distance under each curve at a particular fault density and permeability contrast is a measure of the possible effective permeability range of such a fault system. Fault system geometrical variability can be significant only if the $k_f(k_a - k_h)/k_m$ curve is high.

The term $k_f(k_a - k_h)/k_m$ reflects the heterogeneity of the system, but because it includes the $k_a - k_h$ term, a particular value does not represent a unique case of fault heterogeneity. For instance, a value of $k_f(k_a - k_h)/k_m = 0$ could imply either no faulting in the volume

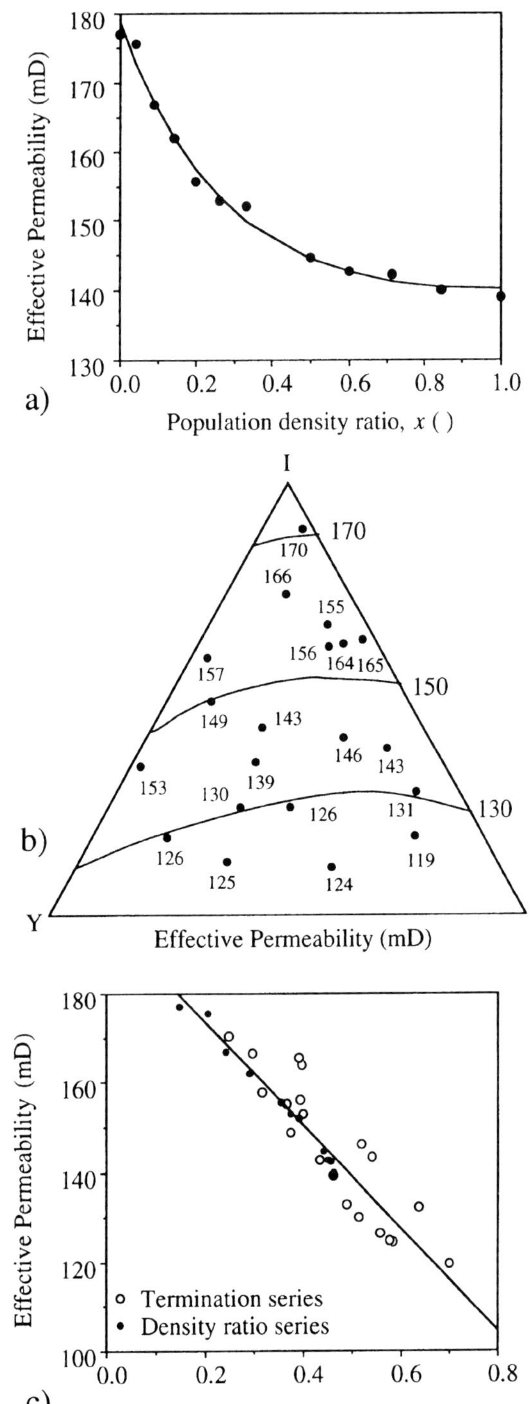

Fig. 9. Compartment density flow simulation results for 1 mD faults in 1 D matrix. Fault density = 0.015. (**a**) Effective permeability vs. orientation population density ratio. (**b**) Effective permeability vs. fault node proportions contoured on an *IYX* node ternary diagram. (**c**) Effective permeability vs. compartmentalization measure *CD* (Eqn 9) combining both geometrical variables.

under consideration, or that it is composed entirely of fault gouge. To overcome this equivocality, we use a modified form to represent fault system heterogeneity (H):

$$H = 1 - \frac{k_h}{k_m} \qquad (11)$$

This function is virtually identical to $k_f(k_a - k_h)/k_m$ for fault density (d) up to the turning points (Fig. 10b), but rather than decreasing to 0 at a fault density of 1, it increases to 1. H is similar to the Dykstra–Parsons heterogeneity measure (Dykstra & Parsons 1950; Lake & Jensen 1991), and ranges from $H = 0$ (for a case of $d = 0$ or $k_f = k_m$) to $H = 1$ (for a case of $d = 1$ or $k_f = 0$).

The nine points on Fig. 10b represent series of fault systems modelled. All models have uniform matrix permeabilities of 1 D and fault permeabilities of either 0.1 mD, 1 mD or 10 mD. Three fault densities (0.00047, 0.0047 and 0.015) were obtained by changing the relative sizes of the fault and matrix grid cells (Fig. 7). In addition, each density/permeability model is run with five geometrical templates, with varying orientation population ratio (x, as above). In this way the permeability sensitivity due to fault system structure may be compared with the permeability sensitivities due to fault thickness and fault permeability (combined in the H measure), through a total of 45 effective permeability results.

For each series, the mean of the five k_e values is shown of Fig. 10c. Unsurprisingly, the highest k_e is from the fault system with the narrowest and most permeable faults. Also shown in Fig. 10c are contours of equal H, which have been calculated from Eqn 11. The average simulated effective permeability of the fault systems varies systematically with this composite average, and one order of magnitude k_f reduction has the same effect on a particular geometrical system as one order of magnitude fault density increase.

Figure 10d shows the range in effective permeabilities for the five geometrical cases for each H case. This range is expressed as a fraction of the mean k_e of each geometrical series, and plotted against the series mean, as recorded in Fig. 10c. The results show that effective permeability is most sensitive to fault system structure when k_e is lowest. Thus k_e for the minimum case is 24 mD ± 36%, and for the maximum is 935 mD ± 0.03%.

These single-phase simulation results indicate that effective permeability is systematically influenced by fault density and fault and matrix permeability (Fig. 10c). As the heterogeneity of the fault system (H) increases, effective permeability

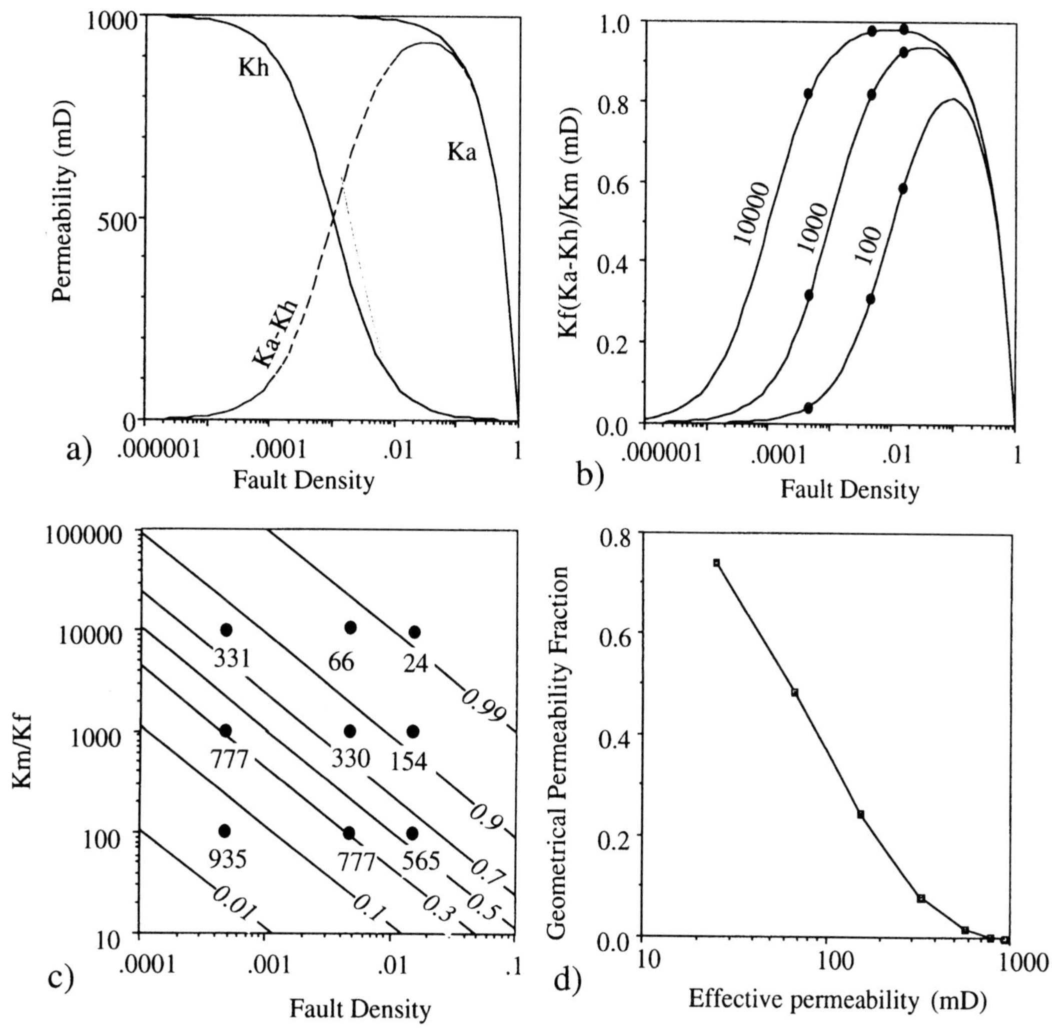

Fig. 10. Fault density, petrophysics and geometry simulation results. (**a**) Permeability vs. fault density for a system of 1 mD faults in 1 D matrix; *Ka*, arithmetic average permeability; *Kh*, harmonic average permeability. (**b**) Composite permeability average $K_f(K_a - K_h)/K_m$ vs. fault density for matrix/fault permeability contrasts of 100, 1000 and 10 000. Dots indicate positions of the nine simulation suites. (**c**) Matrix/fault permeability contrast vs. fault density. Spot values are mean effective permeability, and contours are of equal H value. (**d**) Geometrical permeability fraction vs. effective permeability.

becomes increasingly sensitive to the geometrical architecture of the fault system (Fig. 10d).

Oil recovery and length-scale

Unlike effective permeability determination, which is a function exclusively of the medium, two-phase flow processes are controlled also by the properties of the fluids involved. Saturation-dependent relative permeability and capillary pressure functions govern the phase mobilities. In this study, empirical relationships (detailed in Ringrose & Corbett 1994) have been used to generate relative permeability and capillary pressure curves for 1 D sandstone and 10 mD fault gouge.

The first two-phase simulation examines the influence of scale on oil recovery. A single fault system (Fig. 11a) was modelled at four scales. The four systems have identical H and geometry, and therefore would have identical k_e values. In the smallest model, the faults are 0.1 cm thick and the matrix grid-cells 4.95 cm. These are increased by an order of magnitude in each simulation run, and the largest has 1 m thick faults and 49.5 m matrix cells. The area of the model has been split into four regions (R1 to R4, Fig. 11a), within which the oil recovery has been recorded throughout the run. A cumulative

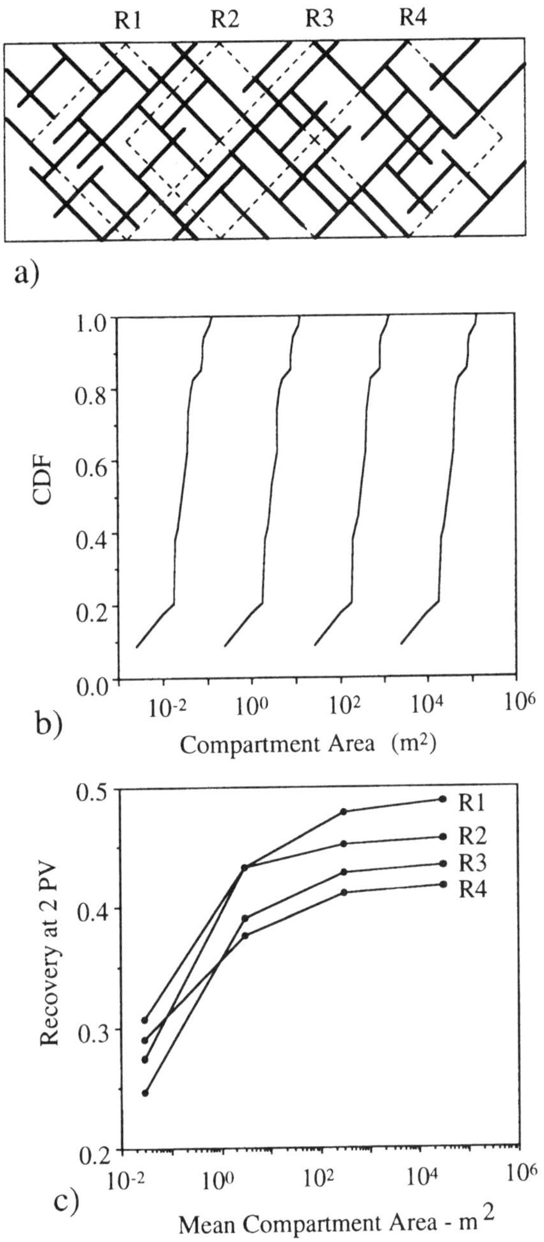

Fig. 11. Oil recovery and length scale flow simulation results for 10 mD faults in 1 D matrix. (**a**) Fault system model and reporting regions. Total flow distance is 3 m, 30 m, 300 m and 3 km. Fault thicknesses are 1 mm, 1 cm, 10 cm and 1 m. Fault density is 0.005 in every case. (**b**) Cumulative density function (CDF) of the compartment area at each length scale. (**c**) Recovery at two pore volumes (2PV) vs. mean compartment area for the four reporting regions.

density function of compartment areas for the four length scales is shown in Fig. 11b. Figure 11c shows a plot of oil recovery at two pore volumes of injected water, plotted against mean compartment area for the four models. These show that in a self-similar fault system, oil recovery is preferentially inhibited at the smaller length scales.

A set of dimensionless ratios or groups incorporates the fundamental physical processes active in two-phase flow (e.g. Shook *et al.* 1992; Ringrose *et al.* 1993). Gravitational forces can be ignored in this study, as the models are horizontal and two-dimensional, with a constant thickness. The relevant scaling group relates viscous to capillary forces. When the viscous/capillary ratio is high the system tends towards viscous-dominated flow, and recovery is enhanced. Although the Darcy velocity at each scale is modelled at a constant realistic reservoir flow rate (1 ft/day), the volumetric flow rate (q) varies as a function of model size. As the viscous/capillary ratio is directly proportional to q, the ratio increases as a function of length scale, and the flow regime for the larger length scales tends towards viscous-dominated flow with increased recovery (Fig. 11c).

Oil recovery is therefore dependent on the scale of the system, as at smaller length scales the capillary forces in the system become more significant. The fault system structure determines the distribution of high capillary pressure regions, and therefore oil recovery should be most sensitive to the structure of the fault system at capillary-dominated length scales. This sensitivity is investigated by the following simulations.

Oil recovery and fault system connectivity

Figure 12 shows four fault systems with identical fault density and different fault connectivity characteristics. System A is a highly compartmentaliszed system with a high X node fraction. Systems B and C have virtually identical node distributions, but B has continuous faulting across the model, while in C flow paths are possible entirely through the matrix. System D has 100% I nodes and therefore no fault connectivity or compartmentalization. The petrophysics are identical to the previous models, and the systems have been run at the largest and second smallest length scales used previously, which correspond to viscous- and capillary-dominated flow regimes, respectively.

Oil recovery at two pore volumes for the two reporting regions is shown in Fig. 13a and b. Oil recovery for a completely unfaulted homogeneous model should be identical at both length scales, and the discrepancy between the unfaulted recovery at each scale (labelled NF) gives an indication of the error due to numerical dispersion. Ringrose *et al.* (1993) conducted

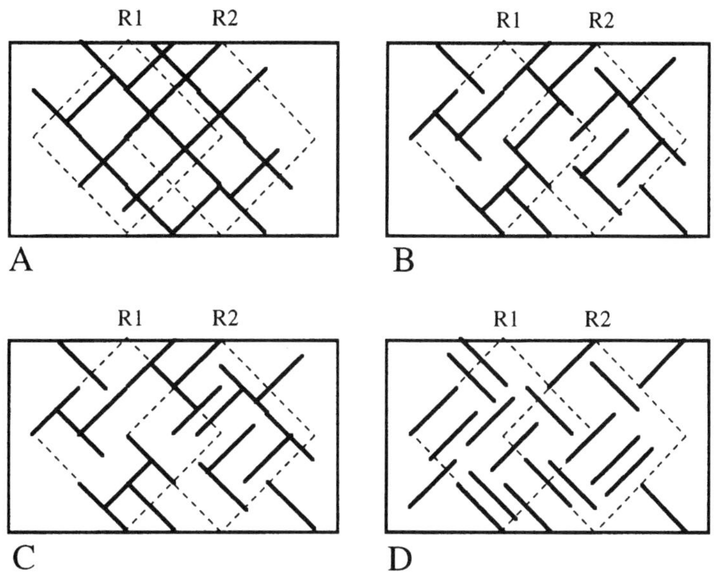

Fig. 12. Fault systems and reporting regions used to determine the influence of fault connectivity on oil recovery. System A: high *X* node fraction. Systems B and C: high *I* and *Y* node fractions. System D: high *I* node fraction.

sensitivities on numerical dispersion due to grid discretization, and concluded that dispersion is most critical at low viscous/capillary ratios. Because we are modelling different length scales rather than flow rates, the discretization in our capillary-dominated cases is much greater than in the viscous cases, which, coupled with the much shorter time-steps used in the capillary case, results in minimal numerical dispersion. These results confirm those of the previous experiment: at smaller length scales oil recovery is reduced. Numerical dispersion tends to overemphasize the effect. However, scale dependency on recovery is not merely a numerical artefact: an increase in either scale or flow rate increases the viscous forces in the system, and therefore oil recovery.

Figure 13a and b also shows that at capillary length scales the effect of fault connectivity on recovery is more influential that at viscous scales. Figure 13c and d shows recovery throughout the model at two pore volumes for the two length scales, as a fraction of the recovery for the unfaulted matrix. The more connected fault system (A) has lower recovery at both viscous and capillary length scale; however, the reduction compared to the other fault systems is much more marked at the capillary scale.

A subsidiary effect of capillary trapping is shown by fault systems B and C. Region 1 for system B has a complete downstream fault barrier into region 2, but fault system C has continuous matrix flow paths (Fig. 12). Recovery at capillary scales in region 1 is, as expected, lower for system B than for system C; however, in region 2 the relative recovery switches, and system B has a greater recovery than system C. We believe this is due to the continuous flow path present in system C. Once water has established a flow path through the model, additional injected water will preferentially follow this path, rather than displace the oil trapped in fault compartments. In system B there are no continuous matrix flow paths, and the water-front is more evenly distributed across the model. This allows a greater volume of oil to be displaced, with increased recovery. At large scales, capillary trapping is not significant, and the recovery for system C is greater than for system B (Fig. 13d). Although a completely disconnected fault system (system D) has greater recovery than connected ones, the details of the fault system connectivity, and the presence of flow paths, is significant, and recovery is not simply a function of compartmentalization and connectivity.

Summary and conclusions

This work has focused on evaluating flow behaviour in faulted high-porosity sandstone reservoirs. We have performed flow simulations through highly connected permeability-reducing fault systems. Our fault system models contain no fault scaling relationships and are all modelled with a constant flow direction relative to constant fault orientations. All faults in each model have identical thicknesses and permeabilities. Within these limitations we model fault systems containing only a few discrete faults, and invariably

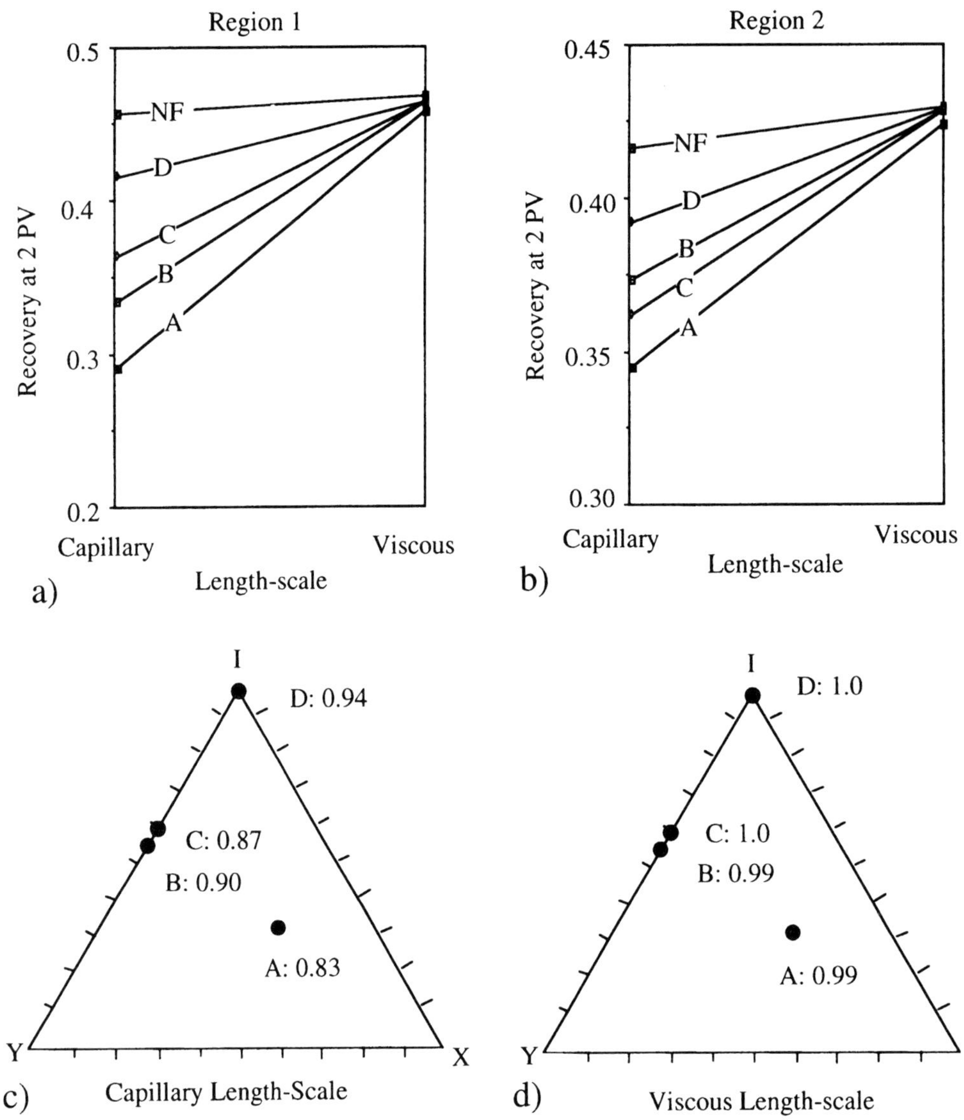

Fig. 13. Oil recovery and fault system connectivity flow simulation results for 10 mD faults in 1 D matrix. Fault density is 0.003. Oil recovery at two pore volumes (2PV) vs. length scale for systems A to D and an unfaulted model (NF) for (**a**) reporting region 1 and (**b**) reporting region 2. Total recovery at 2PV as a fraction of unfaulted recovery at (**c**), the capillary scale and (**d**) the viscous scale.

some faults pass right through the model. The fundamental flow volume we consider is the fault-enclosed compartment rather than the fault itself, and we are concerned primarily with the presence and size of compartments, and only with the thickness and petrophysical properties of the compartment-bounding faults.

Effective permeability calculations have shown that, for any fault density and fault and matrix permeability, a fault heterogeneity measure (H, Eqn 11) may be used to estimate effective permeability (Fig. 10c). Antonellini & Aydin (1994) give an example of an extremely high fault density as 100 deformation bands per metre, with each band 1.5 mm thick. This corresponds to a fault density of 0.15 and is observed only close to large fault zones. Figure 10c shows that the effective permeability of a compartmentalized system at this density is extremely low for 1 mD faults. The field map on Fig. 2c provides another example. If we take an average fault density in the shaded regions of 10 bands per metre, still assuming 1.5 mm thickness, then fault density is 0.01. Effective permeability might then be around 300 mD for a compartmentalized system of 1 mD faults in 1 D matrix.

Fault clustering appears to be extremely heterogeneous at all scales below the slip surface/zone of bands mechanical scaling transition (Figs 2 and 3), and therefore fault clusters form areas of low effective permeability, which are localized into narrow zones on a larger map. This inference suggests that the fault density necessary to severely reduce the effective permeability at any scale is likely to exist only very close to structures that are large relative to the scale under consideration.

The composite fault thickness/permeability heterogeneity measure (H) shows that a one order of magnitude change in fault density is equivalent to a one order of magnitude permeability change. The effective permeability estimates calculated above were based on realistic fault and matrix permeabilities. We infer that effective permeability is only significantly reduced at very high fault densities or low fault permeabilities. At high fault densities, increasing compartment density by enhancing fault connectivity can alter the effective permeability by about a factor of two (Figs 9c and 10d).

Antonellini & Aydin (1995) give some estimates of compartment sizes in relation to strata curvature above salt. They describe 30 m wide compartments bounded by 5 mm faults. Our two-phase experiments model a self-similar fault system and predict that compartments of this size, even at the very slow flow rates that occur away from wells during field production, do not have severe capillary trapping effects for 10 mD faults in 1 D matrix (Fig. 10c). Indeed, even in the compartments between the zones on Fig. 3a, recovery is only slightly reduced (about 38% recovery instead of about 45% for 30 m compartments). Oil recovery might be most critical around the percolation threshold of the fault system, in which case significant volumes of hydrocarbon might become trapped in compartments because the water-flood is concentrated into narrow channels exploiting tortuous flow paths through the matrix. Future work aims to carry on addressing the problem with more sensitivity studies with particular attention placed on fault clustering, two-phase flow, boundary conditions and the third dimension.

This work was carried out as part of the Reservoir Heterogeneity Project (Phase II), funded by Amerada Hess, Bow Valley, British Gas, Chevron, Conoco, The Department of Trade and Industry, Deminex, Elf, Esso, Mobil and Shell. We thank reviewers J. W. Barker and C. L. Farmer, and our colleagues J. L. Jensen and G. E. Pickup, for useful comments. The CT image has kindly been provided by L. Ball.

References

ANTONELLINI, M. & AYDIN, A. 1994. Effect of faulting on fluid flow in porous sandstones: petrophysical properties. *AAPG Bulletin*, **78**, 355–377.

—— & ——1995. Effect of faulting on fluid flow in porous sandstones: geometry and spatial distribution. *AAPG Bulletin*, **79**, 642–671.

——, —— & POLLARD, D. D. 1994. Microstructure of deformation bands in porous sandstones at Arches National Park, Utah. *Journal of Structural Geology*, **16**, 941–959.

AYDIN, A. & JOHNSON, A. M. 1978. Development of faults as zones of deformation bands and as slip surfaces in sandstone. *Pure and Applied Geophysics*, **116**, 931- 942.

—— & ——1983. Analysis of faulting in porous sandstones. *Journal of Structural Geology*, **5**, 19–31.

—— & RECHES, Z. 1982. Number and orientation of fault sets in the field and in experiments. *Geology*, **10**, 107–112.

CHILDS, C., WALSH, J. J. & WATTERSON, J. 1990. A method for estimation of the density of fault displacements below the limits of seismic resolution in reservoir formations. *In*: BULLER, A. T. *et al.* (eds) *North Sea oil and Gas Reservoirs II*. Graham and Trotman, London, 309–318.

COWIE, P. A. & SCHOLZ, C. H. 1992. Physical explanation for the displacement–length relationship of faults using a post-yield fracture mechanics model. *Journal of Structural Geology*, **14**, 1133–1148.

CRUIKSHANK, K. M., ZHAO, G. & JOHNSON, A. M. 1991. Duplex structures connecting fault segments in Entrada sandstone. *Journal of Structural Geology*, **13**, 1185–1196.

DERSHOWITZ, W. S. & EINSTEIN, H. H. 1988. Characterizing rock joint geometry with joint system models. *Rock Mechanics and Rock Engineering*, **21**, 21–51.

DUNN, D. E., LAFOUNTAIN, L. J. & JACKSON, R. E. 1973. Porosity dependence and mechanism of brittle fracture in sandstones. *Journal of Geophysical Research*, **78**, 2403–2417.

DYKSTRA, H. & PARSONS, R. L. 1950. The prediction of oil recovery by waterflood. *In: Secondary Recovery of Oil in the United States, Principles and Practice*, 2nd edition. American Petroleum Institute, 160–174.

FOWLES, J. & BURLEY, S. 1994. Textural and permeability characteristics of faulted, high porosity sandstones. *Marine and Petroleum Geology*, **11**, 608–623.

GAUTHIER, B. D. M. & LAKE, S. D. 1993. Probabilistic modeling of faults below the limit of seismic resolution in the Pelican Field, North Sea, offshore United Kingdom. *AAPG Bulletin*, **77**, 761–777.

GILLESPIE, P. A., HOWARD, C. B., WALSH, J. J. & WATTERSON, J. 1993. Measurement and characterisation of spatial distribution of fractures. *Tectonophysics*, **226**, 113–141.

——, WALSH, J. J. & WATTERSON, J. 1992. Limitations of dimension and displacement data from single faults and the consequences for data analysis and interpretation. *Journal of Structural Geology*, **14**, 1157–1172.

HATTON, C. G., MAIN, I. G. & MEREDITH, P. G. 1994. Non-universal scaling of fracture length and opening displacement. *Nature*, **367**, 160–162.

HEATH, A. E., WALSH, J. J. & WATTERSON, J. 1994. Estimation of the effect of sub-seismic sealing faults on effective permeabilities in sandstone reservoirs. *In*: AASEN, J. O. *et al.* (eds) *North Sea Oil and Gas Reservoirs III*. Kluwer, Dordrecht, 173–183.

JAMISON, W. R. & STEARNS, D. W. 1982. Tectonic deformation of Wingate Sandstone, Colorado National Monument. *AAPG Bulletin*, **66**, 2584–2608.

KNOTT, S. D. 1994. Fault zone thickness versus displacement in the Permo-Triassic sandstones of NW England. *Journal of the Geological Society, London*, **151**, 17–25.

KOESTLER, A. G., MILNES, A. G., OLSEN T. S. & BULLER A. T. 1994. A structural simulation tool for faulted sandstone reservoirs: exploratory study using data from Utah and the Gullfaks Field. *In*: AASEN, J. O. *et al.* (eds) *North Sea Oil and Gas Reservoirs III*. Kluwer, Dordrecht, 157–165.

KRANTZ, R. W. 1988. Multiple fault sets and three-dimensional strain: theory and application. *Journal of Structural Geology*, **10**, 225–237.

LA POINTE, P. R. & HUDSON, J. A. 1985. *Characterisation and Interpretation of Rock Mass Joint Patterns*. Geological Society of America, Special Paper **199**.

LAKE, L. W. & JENSEN, J. L. 1991. A review of heterogeneity measures used in reservoir characterisation. *In Situ*, **15**, 409–437.

LISLE, R. J. 1994. Detection of zones of abnormal strains in structures using Gaussian curvature analysis. *AAPG Bulletin*, **78**, 1811–1819.

MUSKAT, M. 1937. *Flow of Homogeneous Fluids*. McGraw-Hill, New York.

ODLING, N. E. 1992*a*. Network properties of a two-dimensional natural fracture pattern. *Pure and Applied Geophysics*, **138**, 95–114.

—— 1992*b*. Permeability of natural and simulated fracture patterns. *In*: LARSEN, R. M. *et al.* (eds) *Structural and Tectonic Modelling and its Application to Petroleum Geology*. Elsevier, Amsterdam, 365–380.

PICKUP, G. E., RINGROSE, P. S., JENSEN, J. L. & SORBIE, K. S. 1994. Permeability tensors for sedimentary structures. *Mathematical Geology*, **26**, 227–250.

PITTMAN, E. D. 1981. Effect of fault-related granulation on porosity and permeability of quartz sandstones, Simpson Group (Ordovician), Oklahoma. *AAPG Bulletin*, **65**, 2381–2387.

RECHES, Z. 1977. Analysis of faulting in three-dimensional strain field. *Tectonophysics*, **47**, 109–129.

RINGROSE, P. S. & CORBETT, P. W. M. 1994. Controls on two-phase flow in heterogeneous sandstones. *In*: PARNELL, J. (ed.) *Geofluids: Origin, Migration and Evolution of Fluids in Sedimentary Basins*. Geological Society, London, Special Publication, **78**, 141–150.

——, P. S., SORBIE, K. S., CORBETT, P. W. M. & JENSEN, J. L. 1993. Immiscible flow behaviour in laminated and cross-bedded sandstones. *Journal of Petroleum Science and Engineering*, **9**, 103–124.

ROBINSON, P. C. 1983. Connectivity of fracture systems – a percolation theory approach. *Journal of Physics A: Mathematical and General*, **16**, 605–614.

SCHOLZ, C. H. & COWIE, P. A. 1990. Determination of total strain from faulting using slip measurements. *Nature*, **346**, 837–838.

SHOOK, M., LI, D. & LAKE, L. W. 1992. Scaling immiscible flow through permeable media by inspectional analysis. *In Situ*, **16**, 311–349.

UNDERHILL, J. R. & WOODCOCK, N. H. 1987. Faulting mechanisms in high-porosity sandstones; New Red Sandstone, Arran, Scotland. *In*: JONES, M. E. & PRESTON, R. M. F. (eds) *Deformation of Sediments and Sedimentary Rocks*. Geological Society, London, Special Publication, **29**, 91–105.

WALSH, J. J. & WATTERSON, J. 1991. Geometric and kinematic coherence and scale effects in normal fault systems. *In*: ROBERTS, A. M. *et al.* (eds) *The Geometry of Normal Faults*. Geological Society, London, Special Publication, **56**, 193–203.

YIELDING, G., WALSH, J. J. & WATTERSON, J. 1992. The prediction of small-scale faulting in reservoirs. *First Break*, **10**, 449–460.

ZHAO, G. & JOHNSON, A. M. 1992. Sequence of deformation recorded in joints and faults, Arches National Park, Utah. *Journal of Structural Geology*, **14**, 225–236.

Polygonal faulting in the Tertiary of the central North Sea: implications for reservoir geology

LIDIA LONERGAN[1], JOE CARTWRIGHT[1], ROD LAVER[2] & JOE STAFFURTH[2]

[1] *Department of Geology, Imperial College of Science, Technology and Medicine, London SW7 2BP, UK*

[2] *Fina Exploration, Fina House, Ashley Avenue, Epsom Surrey KT18 5AD, UK*

Abstract: Interpretation of regional two-dimensional seismic surveys and three-dimensional seismic surveys in the central North Sea has demonstrated the existence of a pervasive polygonal network of normal faults affecting Tertiary shale-dominated slope and basin-floor depositional systems. The area affected by the faulting encompasses most areas of hydrocarbon production from Tertiary sandstone reservoirs. The polygonal fault networks were active during sedimentation and early burial. Throws measured on individual faults range from 8 to 100 m, with average fault plane dips of 45°. Lengths of individual fault segments range from 80 to 1400 m, and average fault spacings range from 100 to 500 m. The high density and the polygonal geometry of the faults make seismic interpretation of the Lower Tertiary interval problematic, and can lead to misinterpretation of faults as apparent seismic-stratigraphic features.

The recognition of this fault system has consequences for the development of fields in Tertiary reservoirs. The remaining hydrocarbon potential in the Lower Tertiary could be considered predominantly to be stratigraphic plays, but the presence of faults active during sedimentation and early burial implies that this play concept is too simplistic, particularly for Eocene reservoirs. The role of both syn-depositional and early-burial normal faulting on original reservoir distribution and post-depositional modification should be considered further. The presence of this fault system may also be important for secondary migration into Lower Tertiary reservoirs from Kimmeridge Clay Formation source rocks. The existence of an interconnected fault and fracture network in the low-permeability mudrocks may have provided an efficient vertical migration pathway for charging isolated lower Tertiary sandstone reservoirs. Finally, the maximum fault throws of between 50 and 100 m are large enough to represent potential barriers for lateral communication in sandstone reservoirs where individual sand bodies are commonly 25–100 m thick.

With the recent widespread application of sequence-stratigraphic concepts to the petroleum geology of the North Sea, the Lower Tertiary sandstone reservoirs (e.g. Forties-Montrose, Nelson (Paleocene) and Forth, Gryphon and Alba (Eocene)) in deep-water successions are interpreted as basin-floor or slope deposits developing in a lowstand systems tract (Stewart 1987; Milton *et al.* 1990). Within this model, basin-floor fans are predicted to be sand-rich and hence are considered to be attractive hydrocarbon prospects (Posamentier & Vail 1988). The dominant role of the sequence-stratigraphic approach has meant that subtle structural features affecting the Paleocene–Eocene reservoir intervals have been either neglected or relegated to minor significance, in comparison to stratigraphic controls on reservoir geometry and occurrence. In this paper, we describe the development of a pervasive, polygonal network of small-scale extensional faults affecting slope and basin-floor depositional systems in the central North Sea (Cartwright 1994*a*, *b*). We illustrate the detailed geometry of this polygonal fault system using as an example one of the six three-dimensional (3D) seismic datasets studied to date in the central North Sea, and discuss the implications for development and exploration strategies for Eocene reservoirs.

Stratigraphic setting of the tertiary sandstone play

The polygonal fault systems described here affect Paleocene to Miocene clastic sedimentary sequences in the central North Sea Basin (Fig. 1). The lithostratigraphy of the deformed units is dominated by mudstones and silty mudstones that accumulated in prodelta, slope, and basin-plain depositional environments in a partly enclosed marine basin (Parker 1975; Lovell 1990). Commencing in the Paleocene, a large volume of sand-rich sediment was eroded off the uplifted Scottish Highlands–Shetland Platform and transported into depocentres

From Holdsworth, R.E. & Turner, J.P. (compilers) *Extensional Tectonics: Faulting and Related Processes*. The Geological Society, Key Issues in Earth Sciences, **2 (2)**, 299–315.
1476-315X/02/$15.00
First published in: "Coward, M.P., Daltaban, T.S. & Johnson, H. (eds) *Structural Geology in Reservoir Characterization*, The Geological Society, London, Special Publications **127**, 191–207".

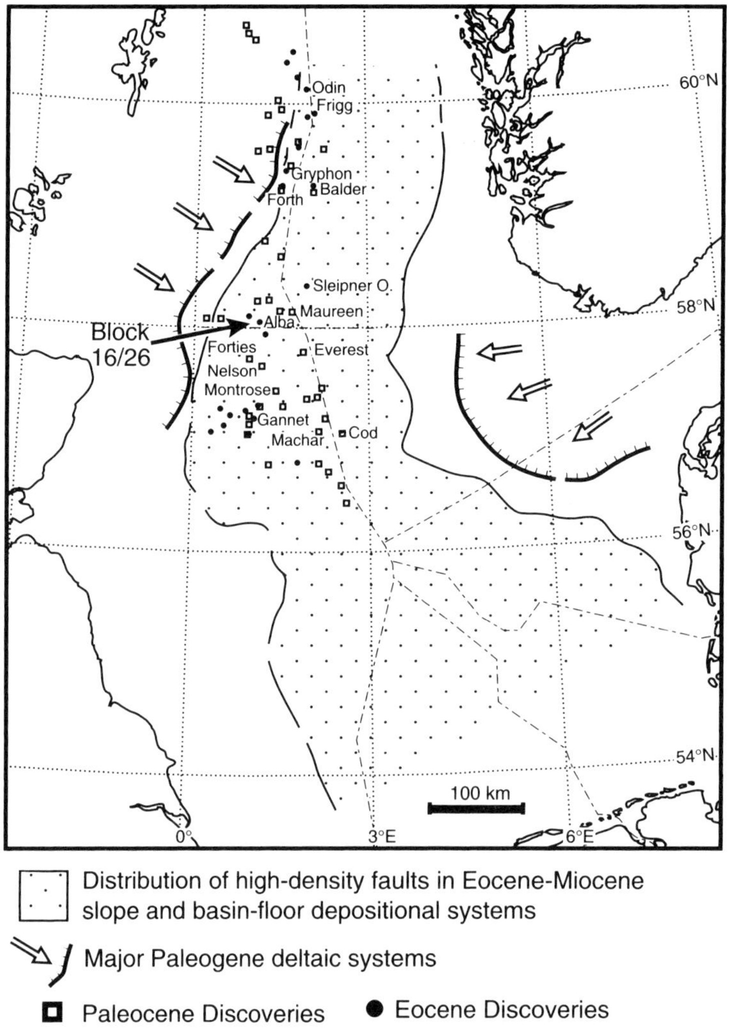

Fig. 1. Distribution of faulted Tertiary slope and basin-floor facies in the central–northern North Sea and location of main Tertiary discoveries in sandstone reservoirs.

located over the buried Mesozoic Central and Viking Grabens (Morton 1979). Large fluvial deltas built out into the Moray Firth and East Shetland Platform. At times during the Paleocene (Maureen, Andrew and Forties Formations) and during the Eocene (Balder and Frigg Formations), submarine sand-prone complexes were deposited in the deeper water environments of the Viking and Central Grabens. Channel sandstone complexes form the Paleocene (post-chalk) and Eocene reservoirs. During breaks in clastic input large areas of the shelf and coast were drowned and mudstone-dominated packages separate the major sandstone units. Channel development was much more extensive during the Paleocene than the Eocene. It is considered that during the Eocene, sands were fed to the basin during times of relative sea-level fall via canyons (e.g. the Frigg Fan (Den Hartog Jager *et al.* 1993) and the Alba sands (Harding *et al.* 1990; Newton & Flanagan 1993)), or due to slump failure of the shelf edge (Forth Field sands (Alexander *et al.* 1993)). The Paleocene sand-rich complexes tend to be more laterally extensive and have a sheet-like geometry in comparison to some of the Eocene facies, which exhibit restricted mound-like geometries with massive sandstone units (e.g. Gryphon, Forth and Balder Fields).

Three different trapping mechanisms have been described for Tertiary sandstone fields (Parsley 1990; Bain 1993).

(1) Structural dip closures have developed by differential compaction over Jurassic tilted fault blocks. Typical examples are the Paleocene Forties, Montrose and Nelson Fields.

(2) Salt-induced structural traps occur, such as the Cod, Maureen and Machar Fields.
(3) Stratigraphic traps occur where sand-rich depositional bodies are sealed and enclosed by hemipelagic shales (e.g. Alba and Frigg Fields). Later drape and differential compaction forms the trap with a mound-like morphology. Another successful stratigraphic trapping mechanism is the pinch-out of mounded sands against pre-existing topography in the basin. In Gannet East, the Eocene Tay Formation sands pinch out towards the shelf and the salt-induced highs of Gannet North and Central to the SW. The Everest field also shows a similar relationship where the sand lobes pinch out eastwards onto a topographic high in the basin (O'Connor & Walker 1993).

A stratigraphic trapping style is frequently considered more typical of the Eocene fields. Recently however, the improved high-resolution 3D seismic data acquired over fields such as Balder, have shown that trapping is achieved by a combination of structural and stratigraphic mechanisms with small faults defining the trap geometry on the sides of the sedimentary mounds (Jenssen *et al.* 1993).

Recent exploration targets in the Tertiary of the central and northern North Sea have focused on remaining subtle stratigraphic traps. If such plays are to be successful, the effect that pervasive polygonal fault systems might have on influencing the geometry of reservoir bodies must be considered. Much drilling activity aimed at Tertiary objectives is concentrating on the appraisal and development of existing discoveries. Again, success will rely on an understanding of the depositional processes and the role of faults during and after sedimentation.

Fault geometries

The structural expression of the Lower Tertiary fault system on two-dimensional (2D) seismic data has been described previously by Rundberg (1989), Higgs & McClay (1993), Clausen & Korstgård (1993), and Cartwright (1994*a*,*b*). Examples of the 2D seismic expression of these faults can be found in Fig. 2 of Cartwright (1994*b*) and Fig. 2 of Higgs & McClay (1993). The small extensional faults are generally poorly resolved on 2D seismic data. This is because the fault spacing is dense relative to the spatial resolution of the data, and because the polygonal geometry means that any 2D section will always consist of many oblique fault intersections which cannot be migrated properly. Hence 3D seismic data are required to image the faults properly. The faults are best imaged in the high-reflectivity parts of the Lower Tertiary section in mudstone-dominated lithologies. Regionally, in the central North Sea, they are most pervasively developed in slope and basin-floor depositional facies in the Eocene to Middle Miocene interval, but are recognized to a lesser extent in the more sand-prone Paleocene succession.

3D seismic case study: Block 16/26

As an example of the deformational style, two orthogonal seismic lines and two horizon maps from a 3D seismic survey in Block 16/26 (UK sector, central North Sea) are illustrated in Fig. 2. Two levels with clearly faulted horizons are visible on the seismic sections (Fig. 2A and B) at approximately 1.5 and 2.0 s two-way-travel time (TWTT) respectively. In between there is a zone of low-amplitude reflectivity where it is difficult to be confident of interpreting faults. Individual faults are thus restricted in vertical extent to only part of the entire deformed interval with only the occasional faults appearing to cross the entire interval. The deformation is layer bound and concentrated between the Middle Miocene marker and the prominent Balder horizon (base Eocene). The lower levels of the fault system affect reservoir sandstone bodies distributed within the middle Eocene interval. The faults on this dataset are best resolved in the mudrock-dominated lithologies of the uppermost Oligo-Miocene interval. Horizon data from this interval are used to illustrate the main geometrical attributes of the polygonal fault network.

We have found that structural interpretation of this highly complex fault system is best approached via a horizon-based interpretation rather than by interpreting faults on sequential vertical sections. Fault interpretation is relatively straightforward if the section intersects a fault approximately orthogonally and the faults are well imaged (e.g. centre portion of Fig. 2A). However, as a result of the polygonal organization of the fault system, most sections will have low-angle oblique or strike cuts through faults, and these cut-outs are responsible for the poor continuity of stratigraphic markers. This means that it is difficult to reproduce a unique fault interpretation on vertical sections, but it is possible to obtain accurate fault maps by interpreting clear marker horizons, on a line-by-line basis. The dense sampling of fault cuts then eliminates aliasing problems, and mis-ties are

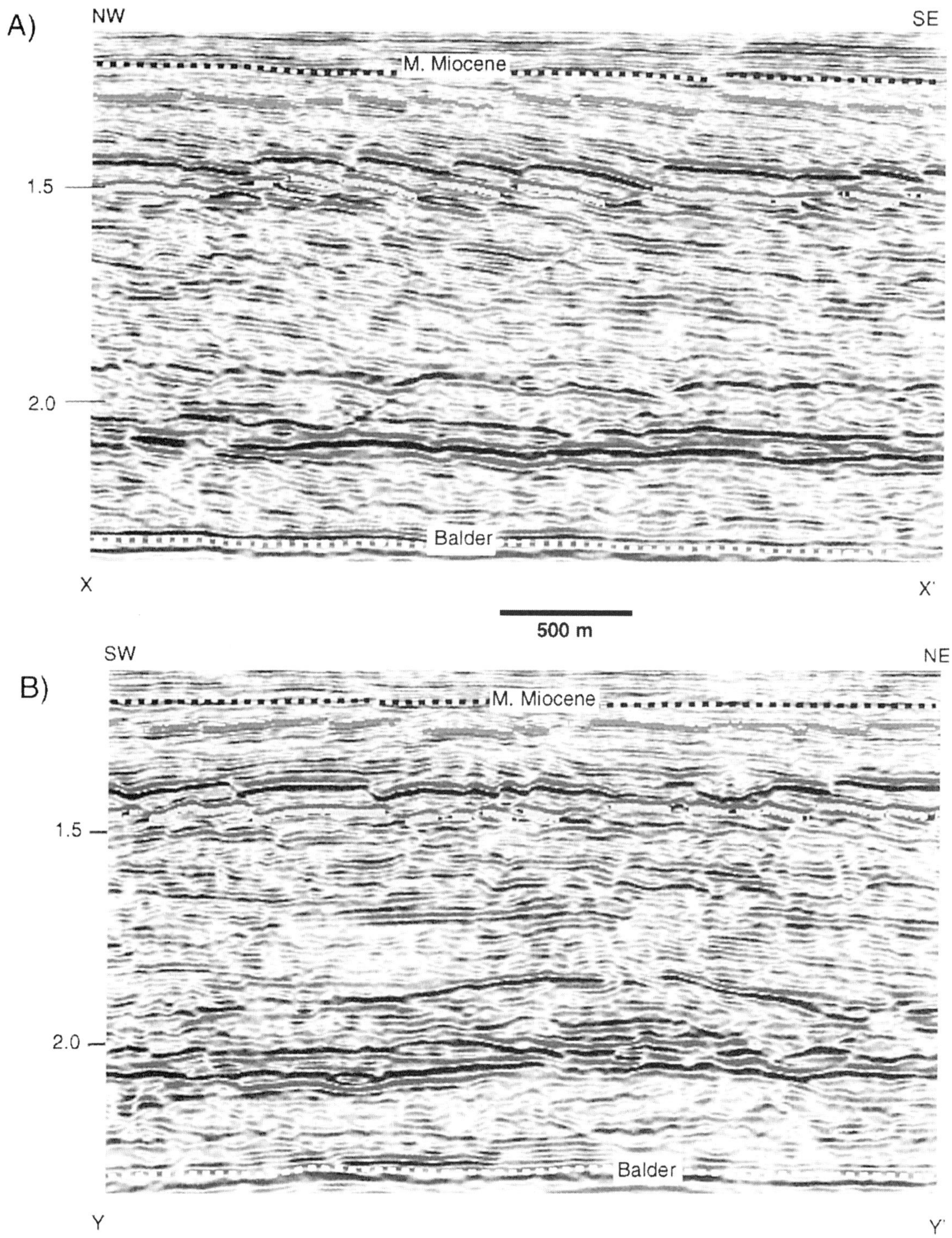

Fig. 2. (**A**) and (**B**) NW–SE (XX′) and SW–NE (YY′) seismic sections from a 3D survey of Block 16/26, in the UK Sector of the North Sea. Two interpreted horizons of Lower Miocene age (green) and Oligocene age (yellow) are illustrated on each section. The vertical scale is two-way travel time. The seismic sections (A and B) show the faulted interval bounded at the top (Middle Miocene and younger) and bottom (Balder Formation) by undeformed sequences. Faults are recognized from normal-sense offsets of stratal reflections. Seismic section XX′ (A) is perpendicular to the strike of a group of faults in the centre of the Oligocene map and hence most of the fault planes appear to dip uniformly to the NW. On the orthogonal seismic section YY′ (B) faults can be identified dipping in opposite directions, but a comparison of the fault maps and seismic sections shows that the sections intersect other faults obliquely resulting in very low angle fault offsets of horizons.

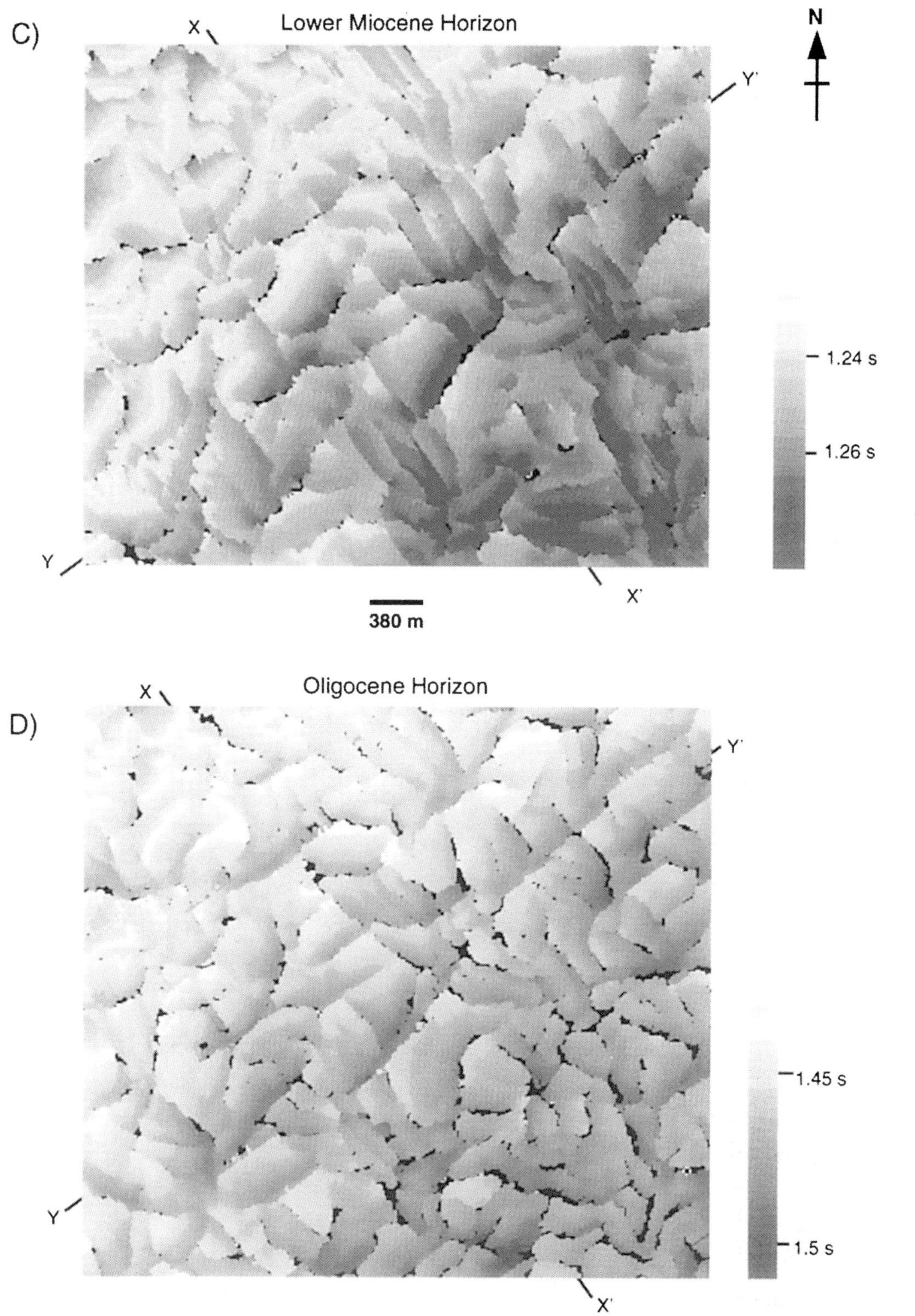

(**C**) and (**D**) Structural maps in two-way travel time of the interpreted Oligocene and Lower Miocene horizons. The horizon maps were constructed by line-by-line interpretation at 12.5 m spacing of sections oriented NW–SE and SW–NE. Each horizon is deformed by small extensional faults that are organized in a polygonal pattern.

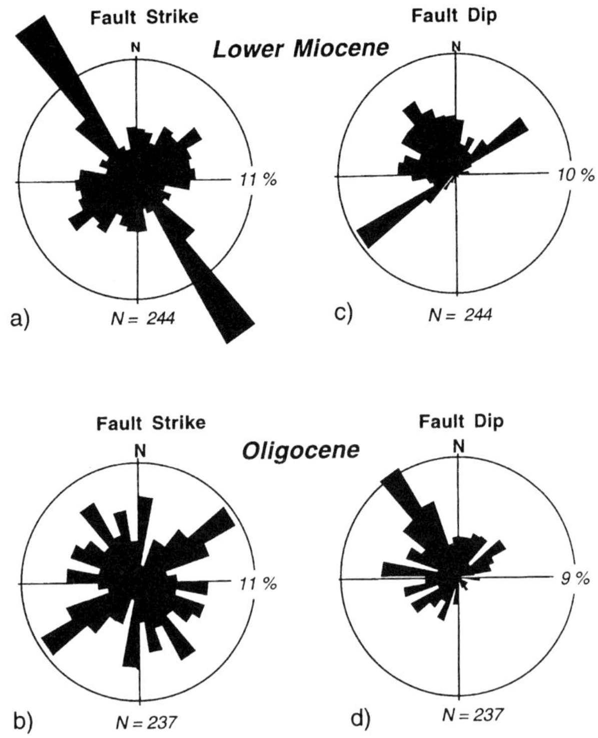

Fig. 3. Equal-area rose diagrams of fault strikes (**a**) and (**b**) and dip direction (**c**) and (**d**) for faults offsetting the Lower Miocene and Oligocene horizons. Despite the horizons being separated by only 250 m they exhibit different preferred orientations of fault strike. The Oligocene fault map has an almost equal number of faults oriented in all directions, whereas the Lower Miocene horizon has a background radial distribution of fault strikes with a superimposed strong preferred orientation of NW-striking faults. For both horizons, faults that strike between N and E, dip (and downthrow) towards the NW quadrant.

readily observable. The 3D fault structure can then be inferred by mapping several closely spaced horizons.

Figure 2C and D are fault maps constructed by interpreting a Lower Miocene and an Oligocene horizon on closely spaced lines (12.5 m intervals) throughout the 3D survey. The horizons are gently dipping at less than 1° to the SE as seen by the broad change in TWTT from grey in the NW to reds in the SE. The faults are identifiable as linear to arcuate gaps or jumps in the time contour colours. The gaps represent the fault heave and the irregularities in the fault traces reflect the resolution accuracy in interpreting fault plane positions from horizon offsets. The polygonal organization of the fault system is the most obvious characteristic of these maps, with many examples of triple and even quadruple fault intersections. The Oligocene fault map has an almost equal number of faults oriented in all directions (see rose diagram, Fig. 3b). In comparison, the Lower Miocene horizon has a background radial distribution of fault strikes, with a superimposed strong preferred orientation of NW-striking faults (Fig. 3a).

Although the two horizons are only separated by about 250 m they exhibit different fault patterns and not all of the faults connect between the two horizons. A close examination of the dataset shows that the NW-striking faults deforming the Lower Miocene tend to be short faults (in a depth direction), have small throws (<15 m), and do not extend down to the Oligocene horizon. Faults that can be traced through the two horizons tend to have accumulated the largest displacements. Closely spaced horizons with different biases in fault orientation are not unique to the Block 16/26 data and have been described from other 3D datasets (Cartwright & Lonergan 1996). An interesting feature of the deformation pattern is that faults striking between N and E preferentially dip to the NW quadrant (Fig. 3c and d), i.e. up-dip of the regional shallow tilt of the succession. Why this should be so is as yet not fully understood, but may represent a response to the palaeoslope during fault formation. Fault patterns affecting horizontal basin-floor successions do not show such a marked preferred fault dip direction. Higgs & McClay (1993) first observed the preferred dip direction to the NW in the Outer Moray Firth, which they attributed to gravitational collapse of the slope due to a tilting event in the Middle Miocene. However, their model fails to account for the polygonality of the fault system or for the occurrence of extensional faults in the horizontal region of the basin axis.

The throws on the faults shown on the two horizon maps (Figs 2C and D) range from 8 m to a maximum of 50 m, assuming a velocity of 2000 m s^{-1} for this interval (Fig. 4A). In general the throw measured on the upper portions of the faults (Lower Miocene level) is less than at the Oligocene level. Fault throws of up to 100 m have been measured on other 3D datasets from the central North Sea. In section the faults within this interval have dips ranging from 30° to 70°, with an average of 45° (Fig. 4B). Faults with arcuate map traces are often gently listric in section, and short linear fault segments tend to be planar. In plan view, lengths of individual fault segments range from 80 m to 1400 m, with the majority of the faults having segment lengths of 100–400 m on the Oligocene horizon and 200–600 m on the Lower Miocene horizon (Fig. 4C).

Timing of fault movement

The Middle Miocene 'unconformity' is a regional marker in the central–northern North Sea and other workers have assumed that the faulting affecting the Tertiary interval took place at this time (Harding *et al.* 1990; Higgs & McClay 1993; Clausen & Korstgård 1993). The fact that different levels within the deformed interval have different densities, spacings and patterns of faults (Figs 3 and 4) implies that the entire interval cannot have been faulted at the same time. Even stratigraphic horizons separated by only 250 m have different fault populations suggesting that faults must be active during early burial within the time interval separating the horizons.

Evidence for syn-sedimentary movement on the faults has been found at at least two stratigrapic levels within the 16/26 survey area. In the youngest portion of deformed succession divergent, wedge-shaped groups of reflectors observed in the hangingwall of the faults in the Oligo-Miocene interval imply that syn-sedimentary movement took place on the faults in the early Miocene. The seismic panel in Fig. 5a is a random line chosen perpendicular to the strike of a group of planar faults dipping to the NW which affect the Oligo-Miocene interval. On the centre fault block, beneath the Middle Miocene unconformity, wedge-shaped reflectors can be clearly seen.

Older syn-sedimentary fault movement is apparent from reflection configurations at deeper levels within the Eocene (Fig. 5B). The horizon map detail in Fig. 5B illustrates two faults which intersect in the SE corner of the map. The westerly fault dips and downthrows to the NE, whereas the easterly fault downthrows to the SW forming a small asymmetric depression between the two faults. On the seismic line, orthogonal to the westerly fault, a distinct thickening of strata is visible in the hangingwall of the fault above the interpreted horizon (grey marker). Other examples similar to this imply that some of the faults propagated to the seafloor generating local topography during Eocene times.

It is apparent from this description of reflection geometries associated with different sets of faults in the 16/26 survey area that the deformation occurred at different times during sediment deposition. Similar conclusions have been reached for other survey areas in the central North Sea, where up to four 'tiers' of faults have been interpreted based on fault population analysis from horizon maps (Cartwright 1996). Evidence for several episodes of fault movement places constraints on the type of process involved in the genesis of the polygonal faults. The evidence for syn-sedimentary fault activity as early as the middle Eocene also raises the possibility that faulting influenced the depositional processes at a time when reservoir-prone units were being deposited.

Fault genesis

Cartwright & Lonergan (1996) have measured the amounts of extension on Lower Tertiary faults on differently oriented sections through two 3D seismic survey areas. They found that the extensional strain in two separated areas was radially isotropic as measured on sub-horizontal marker horizons within the layer-bound deformed interval. Two explanations for this bulk strain are possible: either the extension is a true extension, and compactional flattening was accompanied by radial outward movement of the original stratigraphic horizons (Fig. 6A), or the extension is an apparent strain, and compaction was accompanied by radially uniform volumetric contraction of the beds (Fig. 6B). Since the deformation is clearly layer-bound and there is no evidence for displacement transfer to basement structures, Cartwright & Lonergan (1996) argue that the only explanation

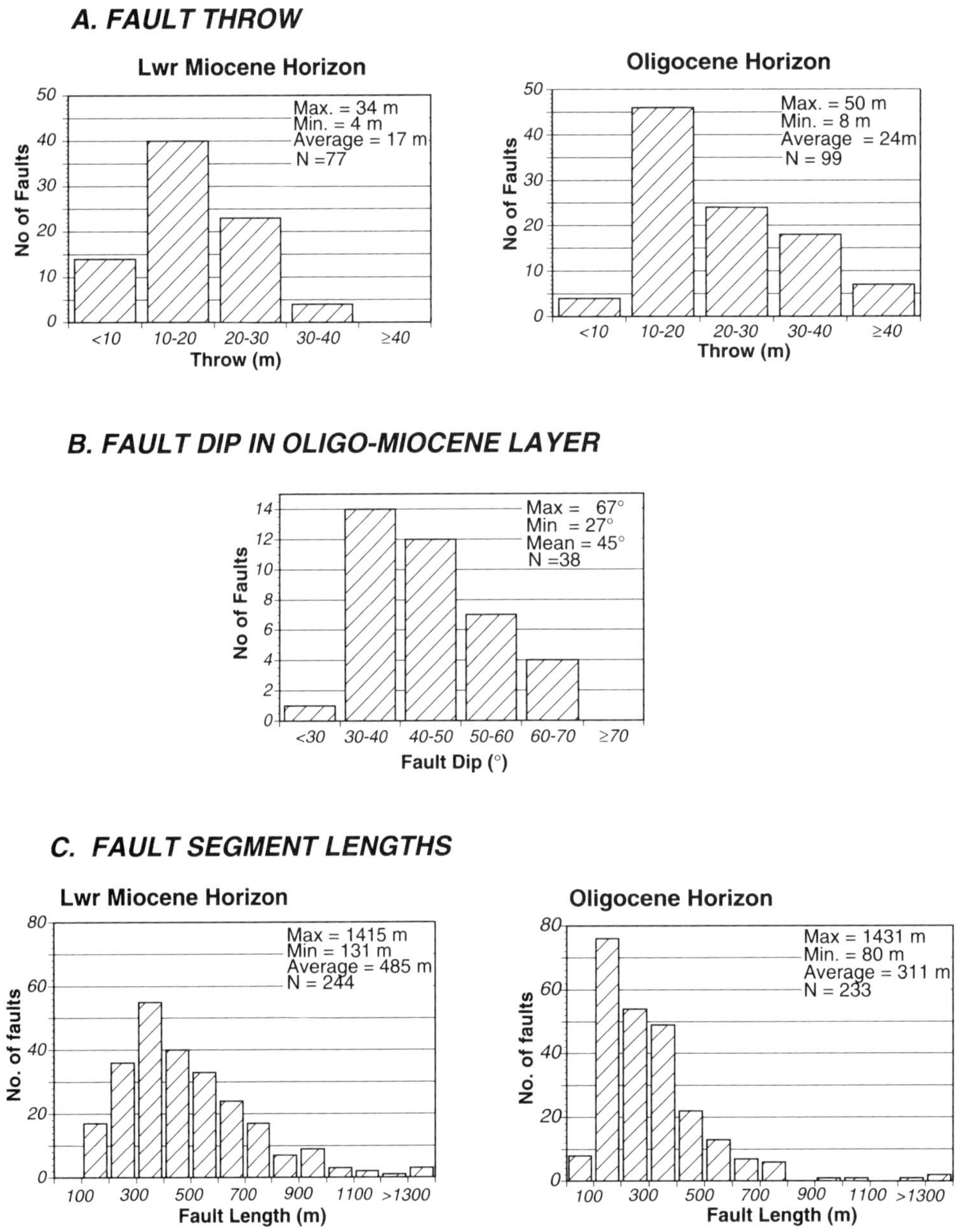

Fig. 4. Statistics for faults in the Oligo-Miocene deformed interval. (**A**) Range of throws measured on faults offsetting the mapped Oligocene and Lower Miocene horizons. Throws are maxima for each fault measured on seismic sections perpendicular to the fault strike and depth, converted assuming a seismic velocity of 2000 ms^{-1}. Dips of faults (**B**) were measured on true-scale seismic sections perpendicular to fault strike. (**C**) Histograms illustrating the range in fault segment lengths for faults intersecting the two mapped horizons.

for this apparent extension is by layer-parallel volumetric contraction. Were it to be a 'true' extension, the 10–20% measured values of apparent extension would require a totally unrealistic extrusion of 20–30 km of Lower Tertiary rocks outward at the basin margins in order to conserve volume across the 200 × 400 km dimensions of the deformed region of the basin (Fig. 1). Further evidence suggesting that the polygonal faults have not formed in response to a regional tectonic event is the observation that the fault orientation statistics and map geometries change between closely spaced horizons within one deformed interval (Figs 2 and 3; Cartwright, 1996), a feature that is atypical of tectonic fault patterns.

The mechanism of faulting by volumetric contraction can be considered partly analogous to the development of shrinkage cracks during the desiccation of muddy sediments. When mud

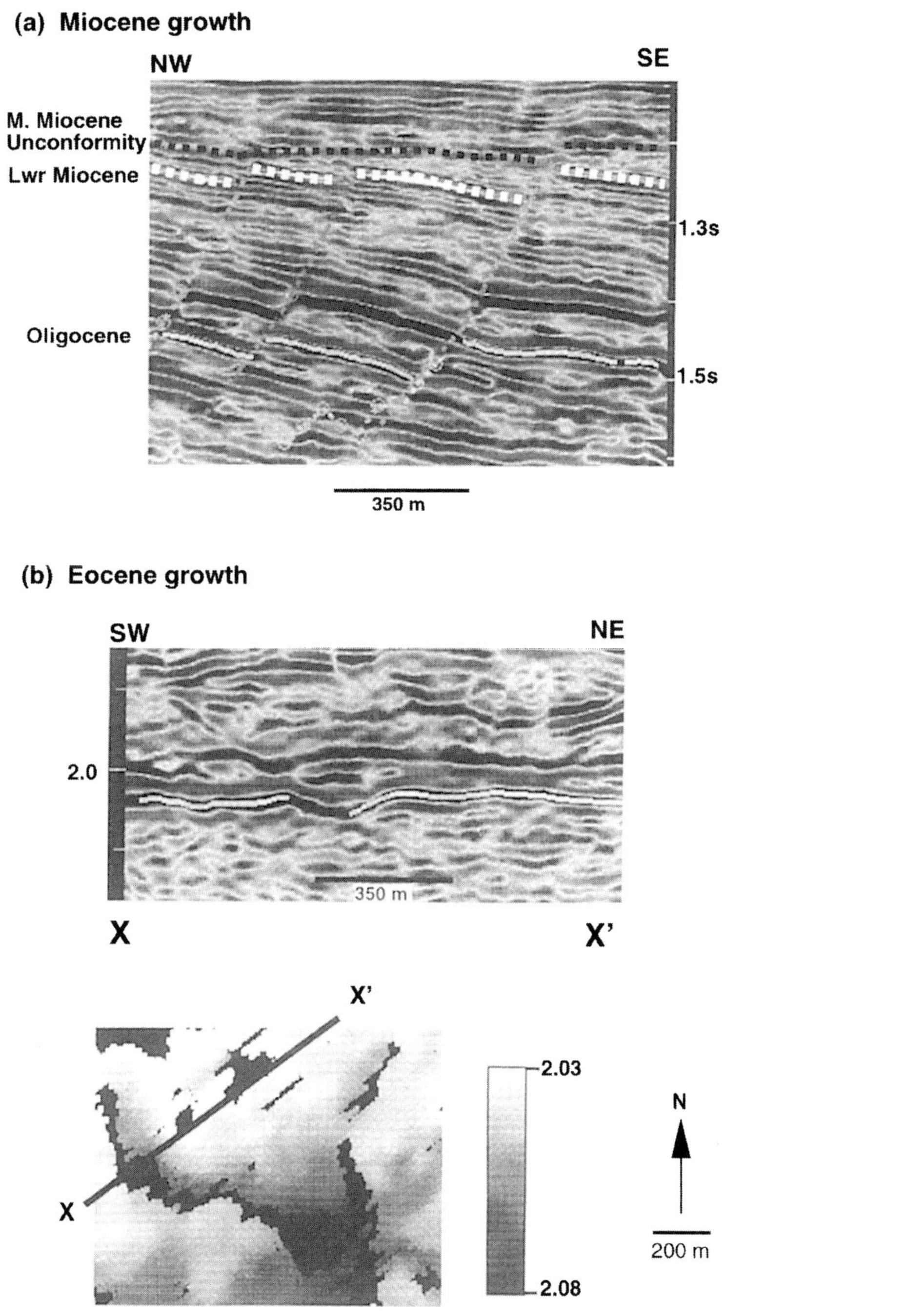

Fig. 5. Seismic sections showing syn-sedimentary fault growth in lower Miocene and Eocene times. (**A**) The seismic panel is a random line chosen perpendicular to the strike of a group of planar faults dipping to the NW in the Oligo-Miocene tier. The dashed grey lines with circles are interpreted faults. Note the low-angle strike cut of a fault parallel to the section below 1.5 s. (Not all possible faults have been interpreted on the section.) (**B**) Seismic line and map (contoured in TWTT) of an Eocene horizon (pale grey marker below 2 s on seismic line). The map shows the fault traces (black gap) and location of the seismic line. The seismic line illustrates the thickening of hangingwall strata into the fault. The grey loop above the pale marker splits into an extra back-loop close to the fault.

cracks develop there is a net loss of pore fluid resulting in the development of tension cracks creating voids, but no net extension of the shrinking mud layer. Both processes result in polygonal extensional structures, but in the case of the fault system in the Lower Tertiary, the polygonal structures are inclined normal faults, not vertical tensile fissures.

Cartwright & Lonergan (1996) propose that faults forming by volumetric contraction occur

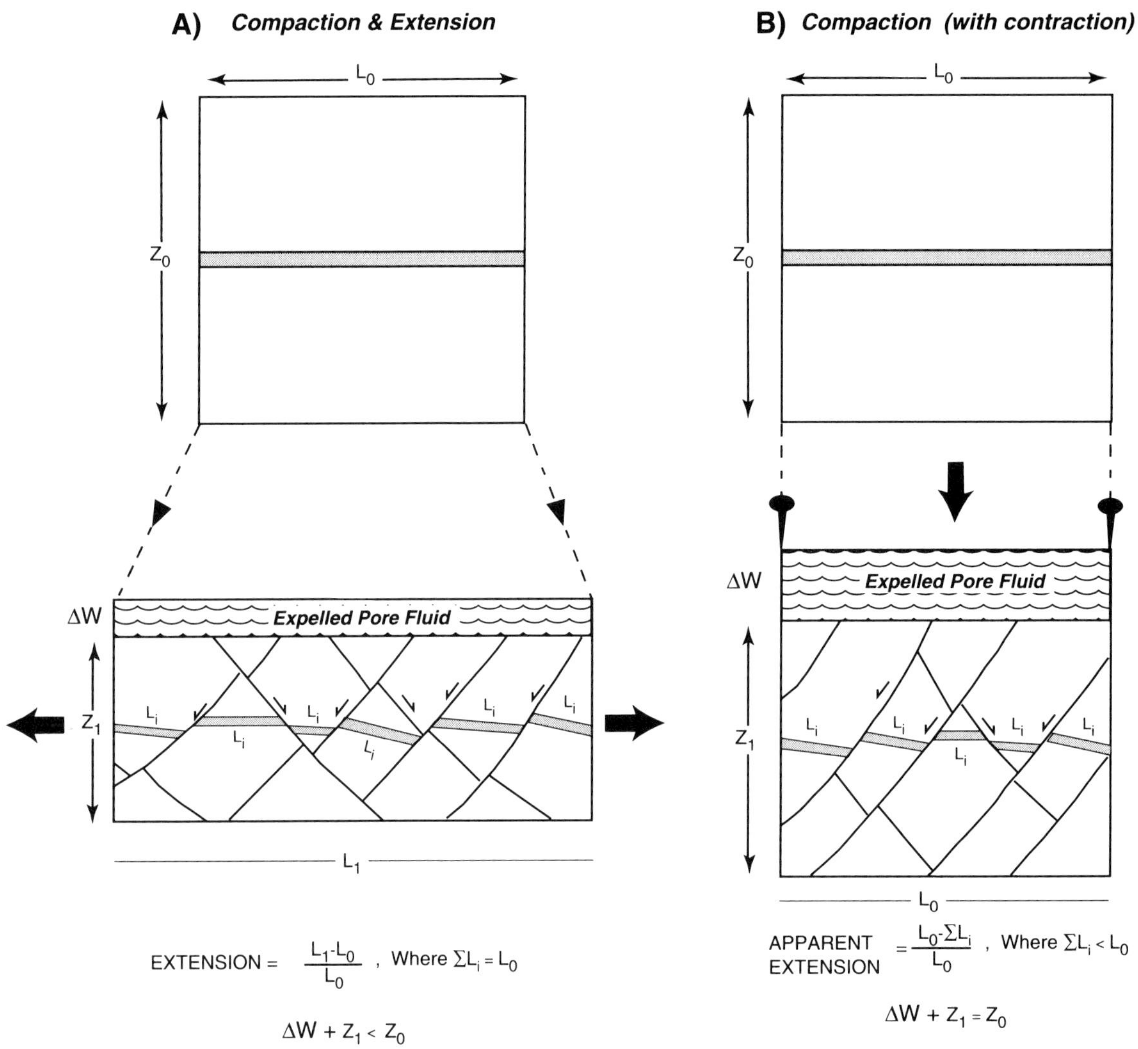

Fig. 6. Alternative mechanisms for the observed 3D deformation in the Tertiary sequence. For simplicity, the strain paths are illustrated in 2D, because the polygonal distribution of faults implies that any orientation of cross-section exhibits the same structural relationships. In both cases there is a reduction in volume due to compaction during burial. In (**A**) the compaction is accompanied by a net extension, whereas in (**B**) there is no net extension (the rectangle sides are pinned) and hence the bed lengths must contract (from Cartwright & Lonergan 1996).

in response to fluid expulsion from the mudrocks during early compaction, which is consistent with the seismic evidence for fault movement during shallow burial (Fig. 5), and that within the first two kilometres of burial mudrocks reduce their volume by up to 50% due to loss of pore fluid. The build-up of fluid overpressures may have triggered failure in the cohesive muddy sediments (e.g. Henriet *et al.* 1991; Verschuren 1992; Cartwright 1994*a*, *b*) allowing some dewatering due to enhanced permeability along and in the vicinity of faults and fractures. Alternatively, the driving force for failure may well be partly provided by physical interactions between clay minerals induced by pore fluid expulsion, rather than solely by a critical fluid overpressure (Cartwright & Lonergan 1996). The faults would continue to grow as long as the rock mass contracted volumetrically. With increasing burial, and less and less pore fluid to expel, the fault system probably ceased to be an important dewatering mechanism and became inactive beneath a threshold depth. In the meantime new sediments within the top few hundred metres beneath the sea floor began to deform in a similar way, forming a new and different set of polygonal faults.

This model explains many aspects of the fault system, including the basin-scale distribution in the low-permeability slope and basin floor lithofacies, the layer-bound distribution of the deformed intervals and the polygonal organisation of the faults. In slope settings the slope dip or wedge geometry of the sedimentary package may have had some influence on fault propagation direction, with faults dipping up-slope developing preferentially.

Implications for reservoir geology

Field development strategies

The widespread distribution of the polygonal fault network encompassing most of the areas of commercial production from Tertiary reservoirs in the North Sea (Fig. 1) means that many Lower Tertiary reservoir intervals could be intersected by normal faults with throws of up to 50–100 m. The main Lower Tertiary reservoir units have thicknesses varying between 25 m and 400 m (Fig. 7). In general the Paleocene reservoirs are thickest, but many Eocene reservoirs have thicknesses of less than 100 m over substantial parts of the productive volume, and the likelihood of the productive reservoir interval being faulted against mudrocks is high.

Given that mud-rich rocks dominate the lithostratigraphy of the Lower Tertiary, the sealing potential of faults by clay-smear along the fault plane is high. Therefore even if the throws on the faults are not large enough to offset reservoir sands, they may still be responsible for compartmentalizing the reservoirs. Sandstone injection features are commonly associated with Eocene reservoirs (Alexander *et al.* 1993; Newman *et al.* 1993; Dixon *et al.* 1995) and from both core and seismic studies are observed to be injected along fault planes. If such injected sandstones occur they could be expected to offset the effects of clay-smear along the fault plane and to improve reservoir connectivity.

When planning horizontal wells, the possibility of drilling from reservoir into shale and then back into reservoir due to <50 m offsets on faults that are seismically difficult to resolve should not be discounted (Fig. 8). Detailed horizon mapping is the best way of identifying faults. Finer resolution of subtle smaller faults, with offsets less than the cycle of seismic reflection, that only cause an inflection in the horizon can be made by making attribute (azimuth, dip, edge and amplitude) maps from interpreted horizon maps. To confirm the existence of faults that are poorly resolved in cross-section, maps of closely spaced horizons above and below the reservoir interval should be made. Faults can then be traced from one map to another and their position through the reservoir interval tracked. Although it is time-consuming, careful fault mapping will help identify small-scale faults difficult to observe on seismic sections through the reservoir body. This added information is obviously crucial when planning reservoir development and well locations.

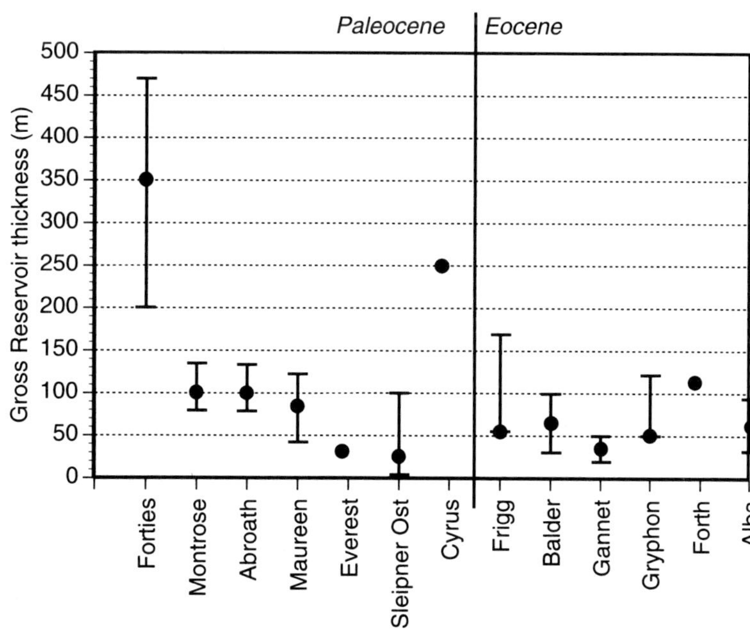

Fig. 7. Gross reservoir thicknesses for Lower Tertiary sandstone fields. Ranges in reservoir thicknesses have been plotted where available. Data compiled from numerous published sources.

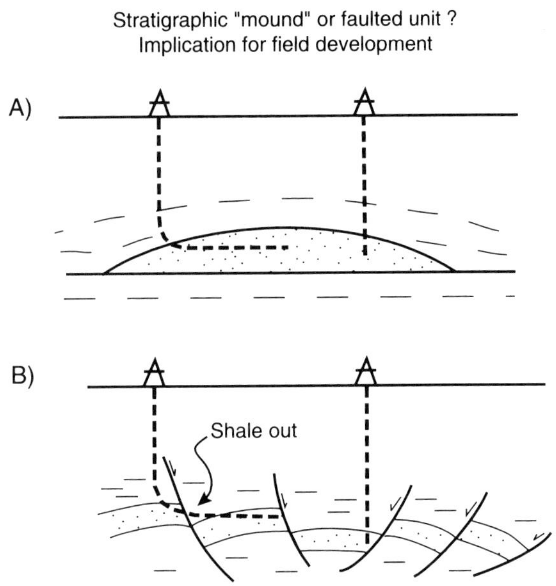

Fig. 8. Implications of polygonal faulting for reservoir geology. The misinterpretation of a complex faulted structure (**B**) for an apparently simple stratigraphic mound (**A**) may lead to surprises during field development, where horizontal wells may shale out due to small offsets on normal faults.

Seismic interpretation

Cartwright (1994*b*) has summarized the potential interpretational pitfalls of not recognizing the presence of the small Lower Tertiary faults. Changing polarity of fault dips or hangingwall roll-over on closely spaced listric faults can be mistaken for stratigraphic mound morphologies. Bain (1993, Fig. 5a) illustrates one such 'mound' from the central North Sea. When it was drilled by the operator it was found to contain no sand and the subsequent interpretation was a series of closely spaced listric normal faults. The stratal rotation within fault blocks was responsible for the apparent mound.

As previously discussed, the complex 3D fault pattern means that any seismic section contains low-angle intersections with faults parallel and oblique to the orientation of the seismic section. The resultant low-angle and subtle offsets to horizons of interest may be misinterpreted as 'stratigraphic' onlap and downlap features or even facies changes.

Poor interpretation of faults or even ignoring them can lead to substantially different and potentially erroneous reservoir correlations. Figure 9 illustrates a realistic situation where a reservoir correlation between sandstones picked in several neighbouring wells could be made in two totally contrasting ways. In the first case (Fig. 9A), the sandstones are correlated assuming a depositional model of a channelized submarine fan. This leads to a reservoir geometry of several isolated lensoid sand bodies. In the second case (Fig. 9B), the same well picks are correlated with due attention paid to fault patterns recognized on the seismic data. The resultant correlation is radically different from the stratigraphically-driven solution, and would have important consequences for reserve calculations and choice of development drilling locations.

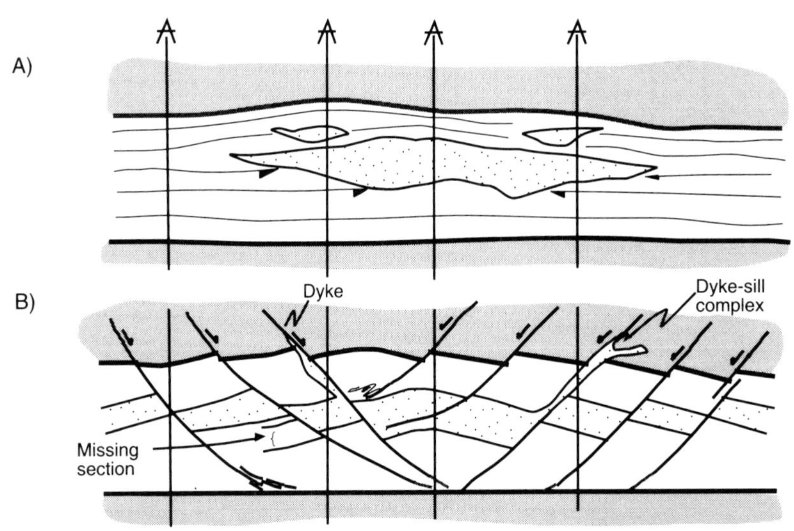

Fig. 9. Comparison of well correlations for reservoir sandstones based on (**A**) stratigraphic model, and (**B**) structural model incorporating polygonal faults. The sandstone picks are identical in both cases but in (A) the correlation depicts isolated sandstone bodies, whereas in (B) the correlation takes account of remobilization of sandstone along fault planes in the form of dykes and sills. Note also in (B) the missing section due to fault cut-out which is shown in (A) as channel incision.

Integration of the complex faulted structure into reservoir correlation schemes may in some cases reduce connectivity and downgrade the potential of an accumulation, but in other cases, syn-sedimentary faulting during reservoir deposition could lead to increased reserve potential in otherwise unexpected trapping configurations. Ignoring the complex fault system simply because it is difficult to map will lead to an incomplete assessment of potential reserves.

Depositional models

The Eocene sandstone reservoirs are distinctive in that they are predominantly massive, high-porosity sandstones with very few preserved sedimentary structures, except for dish and pillar dewatering structures (Armstrong *et al.* 1987; Brewster 1991; Mattingly & Bretthauer 1992; Newton & Flanagan 1993; Alexander *et al.* 1993; Newman *et al.* 1993; Jenssen *et al.* 1993; Shanmugam *et al.* 1995; Dixon *et al.* 1995). Sandstone bodies can be up to 100 m thick but exhibit rapid lateral thickness variation and often pinch out over distances of less than 500 m. Post-depositional modification of the massive sandstone bodies is a common feature of many Eocene reservoirs. Soft-sediment deformation structures and sand and mud injection features have been described in cored intervals through the Alba, Forth, Gryphon and Balder fields (Mattingly & Bretthauer 1992; Jaffri 1993; Newton & Flanagan 1993; Alexander *et al.* 1993; Newman *et al.* 1993; Jenssen, *et al.* 1993; Dixon *et al.* 1995). In the Forth and Alba Fields, extensive sandstone dykes and sills have been intruded into the Eocene muds above the main massive sandstone reservoirs (Alexander *et al.* 1993; Newton & Flanagan 1993). In some cases, these sandstone dykes are found to be oil-bearing. More recently Dixon *et al.* (1995) have recognized discordant sandstone intrusions on seismic data that have a lateral extent of 200–300 m.

Recent sedimentary interpretation of the massive sandstone bodies with mounded topography suggests that they are mass-flow deposits (slumps or debris flows) as opposed to sand-rich turbidites (Newman *et al.* 1993; Shanmugam *et al.* 1995). A puzzling feature is the steep-sided slopes (up to 15°) on the sides of the mounds in both the Gryphon and Balder Fields. Such dips are considered too steep to be depositional in origin, even after accounting for compaction (Jenssen *et al.* 1993). Recent work on both the Gryphon and Balder Fields (Newman *et al.* 1993; Jenssen *et al.* 1993; Rye-Larsen 1994), which benefits from new 3D seismic surveys acquired over the fields, recognizes the importance of small-scale faulting in controlling the geometry of the sandstone mounds. Essentially, the workers on these fields concluded that the faults are the result of slumping. Jenssen *et al.* (1993) and Rye-Larsen (1994) consider that slump faults alone cannot account for the mounding observed in the Balder Field and suggest that some of the mound-like geometries can be attributed to remobilization and injection of sandstones from underlying low-relief turbidite sheet sands. More recently, Dixon *et al.* (1995) have concluded that post-depositional sandstone diapirism and intrusions of clastic dykes and sills associated with major gravity slide faulting are responsible for the reservoir geometries observed in the Forth and Gryphon Fields.

Our studies of the polygonal fault systems imply that minor extensional faults form as part of the compactional dewatering process during early burial and we suggest that the faults recognized locally in the Eocene fields and attributed to gravity sliding or differential compaction are part of this wider polygonal system. The injection and soft-sediment deformation structures widespread in the Eocene sandstone bodies are all consistent with syn-sedimentary and early burial dewatering deformation. The patterns of sandstone dyke injection described by Alexander *et al.* (1993) for the Forth Field in particular are highly reminiscent of the complex polyhedral geometries of the faults described in this paper. Dyke injection requires elevated pore fluid pressures in order to fluidize the sand in the original reservoir unit, and to provide the pressure differential for upward flow (Jaffri 1993). Upward injection would exploit the weaknesses due to the presence of already formed extensional faults in the overlying mudstones. More direct evidence of the interrelationship between the polygonal faults and final reservoir configurations is provided by a horizon slice through the Alba Field published by Newton & Flanagan (1993, Fig. 8). This illustrates the close spatial relationships between the sandstone reservoirs and the polygonal fault pattern in the surrounding hemipelagic shales. In a number of places the edges of the Alba 'channel' change strike and follow a polygon side, which implies that either the distribution of sandstones in the deep-water Eocene successions was influenced by contemporaneous, near-surface faulting or that faults controlled the post-depositional modifications to the original sedimentary geometry.

Soft-sediment deformation structures are often interpreted as having formed due to slope instability, but we consider that dewatering and polygonal fault development could lead to a similar range of structures with identical characteristics in core samples. However, the exact role of the polygonal fault deformation system in controlling sand-body geometries requires further study. Syn-sedimentary faulting generating an irregular sea-floor topography (Fig. 5B) may have exerted a primary control on the location of sand-rich mass-flow deposits. Alternatively the effects of faulting may have been more pronounced during early burial by facilitating remobilization and injection of existing sands, thus contributing to the isolated mound-like geometries typical of some Eocene fields. It is possible that in near-surface conditions of generally elevated pore fluid pressures, considerable volumes of sand could have been fluidized, remobilized and injected as large sandstone sills (Jaffri 1993; Dixon *et al.* 1995). Clearly, this possibility raises some interesting questions about final reservoir configurations. Whilst this is speculative at present, this type of process would help to explain the lack of intra-formational mudstones in the Eocene reservoir units, their lack of internal sedimentary structures, and the abundance of dewatering textures. It may also help to explain some of the more enigmatic aspects of the external form of the reservoirs, such as the steep flank dips, the limited erosion at the base of features described as channels and the isolated nature of many sand bodies with no intervening transitional facies. By recognizing the importance of dewatering phenomena, it may be that a reappraisal of traditional depositional models is now required.

Migration of hydrocarbons and sealing potential of shales

There is general consensus that the hydrocarbons in the Tertiary reservoirs are sourced from the Jurassic Kimmeridge Clay Formation (Barnard & Cooper 1981; Cornford 1990) which began to generate oil in the latest Cretaceous and earliest Tertiary in the centres of the Central and Viking Grabens (Cornford 1990). Oil sourced from the Upper Jurassic in isolated lensoid sandstone bodies, sealed above and below by shales, has long posed a problem for the definition of a plausible migration route. Even more enigmatic is the presence of biodegraded oils in some apparently closed Tertiary reservoirs (e.g. Alba), which also requires groundwater carrying bacteria to have entered the reservoirs (Mason *et al.* 1995).

For some fields, such as Gryphon, located above Mesozoic tilted fault blocks, a vertical migration pathway through Mesozoic faults is generally assumed. The fracture system is then assumed to extend into the Tertiary rocks above the structural high in order to facilitate charging of the Tertiary reservoirs (e.g. Newman *et al.* 1993). For other fields, such as Forties, long-distance migration pathways have been proposed. First the oil migrates vertically via salt features in the Central Graben, and then, once it enters the distal facies of the Paleogene reservoir fan sands,

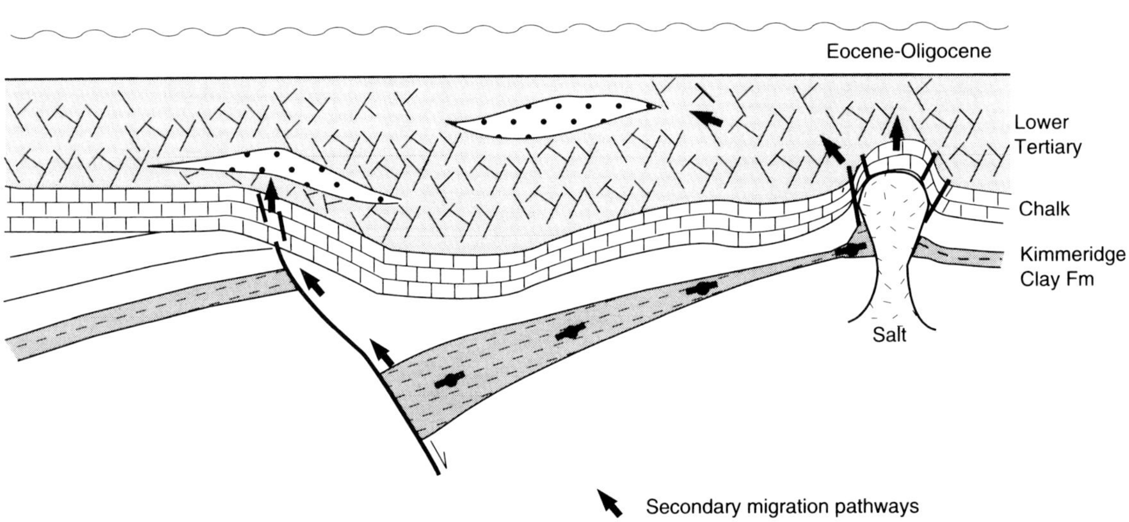

Fig. 10. Hydrocarbons generated from the Kimmeridge Clay Formation in deep kitchens can migrate upwards via basin-bounding faults and salt structures. Once the hydrocarbons encounter the mudrock-dominated lower Tertiary successions, the pervasive system of polygonal faults and fractures facilitates migration into isolated reservoir sandstones.

it migrates laterally up-dip into traps (Cornford *et al.* 1986; Cayley 1987).

The existence of an extensive network of syn-depositional and early-burial faults within the Lower Tertiary shale successions provides an alternative secondary migration pathway for charging the isolated Tertiary reservoirs. Once the oil migrated from the Jurassic kitchens through the Cretaceous by mainly sub-vertical pathways (via basin margin faults and salt structures) it could have encountered the pervasive polygonal fault and fracture system in the mudrocks, allowing migration into isolated sandstones (Fig. 10). Biodegraded oils may be the result of migration into leaky shallow traps where seawater was still present, or, as favoured by Mason *et al.* (1995), aquifers may have been recharged by meteoric waters during the Early Miocene when large areas of the Paleogene shelf were subaerially exposed.

The fact that oil is found in Eocene sandstones means that at some point the fault and fracture network became impermeable and the shale sequences became mainly sealing. The model of fault genesis put forward by Cartwright & Lonergan (1996) proposes that faults were active during early burial when the shales were expelling large volumes of pore water. Once the shales compacted beyond a critical value, the faults ceased to be active and the vertical permeability of the shales was much reduced. Hence the lower Tertiary mudstone units have become better, tighter seals through time.

Conclusions

The chronostratigraphic studies of the Tertiary in the North Sea have clearly aided our understanding of how and when sand was supplied and transported into the slope and basin environments during the lower Tertiary, but until recently subtle structural controls on the Tertiary plays have been considered to be of minor significance. In the Eocene (and possibly in the Paleocene), small-scale extensional faults organized in remarkable polygonal patterns were active during the deposition of the sediments. The role that such faulting played in influencing the locus of sandstone bodies within the depositional environment, and the subsequent modification and remobilization of original reservoir configurations during early burial, has not been adequately considered in reservoir models published to date.

The existence of a dense 3D fault network with extensional throws ranging from 10 to 100 m may explain the compartmentalization of some Tertiary reservoirs discovered during field development. Given the density and pattern of faults observed on the seismic datasets studied so far, there is a high probability that any elongate reservoir sandstone bodies occurring within the mud-dominated slope and deep-water succession of the Eocene in the North Sea Basin will be faulted.

The polygonal faulting is now seen to be an intrinsic product of water loss and compaction in the fine-grained mud-dominated sediments. The faults formed during the earliest phases of burial and acted as active dewatering conduits. They may also have been the migration pathways for the hydrocarbons that charged the isolated Tertiary sandstone reservoirs.

We wish to thank Fina Exploration Ltd for financial support and permission to publish this paper, and an anonymous referee for helpful comments. Chevron UK Ltd, as operator, and their partners in Alba are acknowledged for permission to publish seismic data from Block 16/26. The views expressed in this paper are those of the authors and are not necessarily those of the above-named companies.

References

ALEXANDER, R. W. S., SCOFIELD, K. & WILLIAMS, M. C. 1993. Understanding the Eocene Reservoirs of the Forth Field, UKCS Block 9/23b. *In*: SPENCER, A. M. (ed.) *Generation, Accumulation and Production of Europe's Hydrocarbons III*. Special Publication, European Association of Petroleum Geoscientists, 3–15.

ARMSTRONG, L. A., TEN HAVE, A. & JOHNSON, H. D. 1987. The geology of the Gannet fields, Central North Sea, UK sector. *In*: BROOKS, J. & GLENNIE, K. (eds) *Petroleum Geology of North west Europe*. Graham & Trotman, London, 533–548.

BAIN, J. S. 1993. Historical overview of exploration of Tertiary plays in the UK North Sea. *In*: PARKER, J. R. (ed.) *Petroleum Geology of Northwest Europe: Proceedings of the 4th Conference*. Geological Society, London, 5–13.

BARNARD, P. C. & COOPER, B. S. 1981. Oils and source rocks of the North Sea area. *In*: ILLING, L. V. & HOBSON, G. D. (eds) *Petroleum Geology of the Continental Shelf of North-West Europe*. Heyden, London, 169–175.

BREWSTER, J. 1991. The Frigg Field, Block 10/1 UK North Sea and 25/1, Norwegian North Sea. *In*: ABBOTTS, I. L. (ed.) *United Kingdom Oil and Gas Fields, 25 years Commemorative Volume*. Geological Society, London, Memoir **14**, 117–126.

CARTWRIGHT, J. A. 1994*a*. Episodic basin-wide fluid expulsion from geopressured shale sequence in the North Sea basin. *Geology* **22**, 447–450.

—— 1994*b*. Episodic basin-wide hydrofracturing of overpressured Early Cenozoic mudrock sequences in the North Sea Basin. *Marine and Petroleum Geology* **11**, 587–607.

——1996. Polygonal fault systems: A new type of fault structure revealed by 3-D seismic data from the North Sea Basin. *In*: WEIMAR, P. & DAVIS, T. L. (eds) *Application of 3D Seismic Data to Exploration & Production.* AAPG Studies in Geology No **42** and SEG Geophysical Developments Series No **3**, 225–230.

—— & LONERGAN, L. 1996. Volumetric contraction during the compaction of mudrocks: a mechanism for the development of regional-scale polygonal fault systems. *Basin Research* **8**, 183–193.

CAYLEY, G. T. 1987. Hydrocarbon migration in the central North Sea. *In*: BROOKS, J. & GLENNIE, K. (eds) *Petroleum Geology of North West Europe.* Graham & Trotman, London, 549–555.

CLAUSEN, O. R. & KORSTGÅRD, J. A. 1993. Small scale faulting as an indicator of deformation mechanism in the Tertiary sediments of the northern Danish Central Trough. *Journal of Structural Geology*, **15**, 1343–1358.

CORNFORD, C. 1990. Source rocks and hydrocarbons of the North Sea. *In*: GLENNIE, K. W. (ed.) *Introduction to the Petroleum Geology of the North Sea*, 3rd ed., Blackwell, Oxford, 294–361.

——, NEEDHAM, C. E. J. & DE WALQUE, L. 1986. The geochemical habitat of North Sea oils and gases. *In*: SPENCER, A. M. (ed.) *Habitat of Hydrocarbons on the Norwegian Continental Shelf.* Graham & Trotman, London, 39–54.

DIXON, R. J., SCHOFIELD, K., ANDERTON, R., REYNOLDS, A. D., ALEXANDER, R. W. S., WILLIAMS, M. C. & DAVIES, K. G. 1995. Sandstone diapirism and clastic intrusion in the Tertiary sub-marine fans of the Bruce-Beryl Embayment, Quadrant 9, UKCS. *In*: HARTLEY, A. J. & PROSSER, D. J. (eds) *Characterisation of Deep Marine Clastic Systems.* Geological Society, London, Special Publication **94**, 77–94.

DEN HARTOG JAGER, D., GILES, M. R. & GRIFFITHS, G.˜R. 1993. Evolution of Paleogene submarine fans of the North Sea in space and time. *In*: PARKER, J. R. (ed.) *Petroleum Geology of Northwest Europe: Proceedings of the 4th Conference.* Geological Society, London, 59–71.

HARDING, A. W., HUMPHREY, T. J., LATHAM, A., LUNSFORD, M. K. & STRIDER, M. H. 1990. Controls on Eocene submarine fan deposition in the Witch Ground Graben. *In*: HARDMAN, R. F. P. & BROOKS, J. (eds) *Tectonic Events Responsible for Britain's Oil and Gas Reserves.* Geological Society, London, Special Publication **55**, 353–367.

HENRIET, J. P., DE BATIST, M. & VERSCHUREN, M. 1991. Early fracturing of Palaeogene clays, southernmost North Sea: Relevance to mechanisms of primary hydrocarbon migration. *In*: SPENCER, A. M. (ed.) *Generation, Accumulation and Production of Europe's Hydrocarbons.* European Association of Petroleum Geoscientists, Special Publication **1**, 217–227.

HIGGS, W. G. & MCCLAY, K. R. 1993. Analogue sandbox modelling of Miocene extensional faulting in the Outer Moray Firth. *In*: WILLIAMS, G. D. & DOBB, A. (eds) *Tectonics and Seismic Sequence Stratigraphy.* Geological Society, London, Special Publication **71**, 141–162.

JAFFRI, F. 1993. *Cross-cutting sand bodies of the Tertiary. Beryl Embayment, North Sea.* PhD Thesis, University of London.

JENSSEN, A. I., BERGSLIEN, D., RYE-LARSEN, M. & LINDHOLM, R. M. 1993. Origin of complex mound geometry of Paleocene submarine-fan sandstone reservoirs, Balder Field, Norway. *In*: PARKER, J. R. (ed.) *Petroleum Geology of Northwest Europe: Proceedings of the 4th Conference.* Geological Society, London, 135–143.

LOVELL, J. P. B. 1990. Cenozoic. *In*: GLENNIE, K. W. (ed.) *Introduction to the Petroleum Geology of the North Sea*, 3rd edition, Blackwell, Oxford, 273–293.

MASON, P. C., BURWOOD, R. & MYCKE, B. 1995. The reservoir geochemistry and petroleum charging histories of Palaeogene-reservoired fields in the Outer Witch Ground Graben. *In*: CUBITT, J. M. & ENGLAND, W. A. (eds) *The Geochemistry of Reservoirs.* Geological Society, London, Special Publication, **86**, 281–301.

MATTINGLY, G. A. & BRETTHAUER, H. H. 1992. The Alba Field. A Middle Eocene deep water channel system in the UK North Sea. *In*: HALBOUTY, M. T. (ed.) *Giant Oil and Gas Fields of the Decade, 1978–1988.* AAPG, Memoir **54**, 78–88.

MILTON, N. J., BERTRAM, G. T. & VANN, I. R. 1990. Early Palaeogene tectonics and sedimentation in the Central North Sea. *In*: HARDMAN, R. F. P. BROOKS, J. (eds) *Tectonic Events Responsible for Britain's Oil and Gas Reserves.* Geological Society, London, Special Publication, **55**, 339–351.

MORTON, A. C. 1979. The provenance and distribution of the Paleocene sands of the Central North Sea. *Journal of Petroleum Geology*, **2**, 11–21.

NEWMAN, M. ST. J., REEDER, M. L., WOODRUFF, A. H. W. & HATTON, I. R. 1993. The geology of the Gryphon oil field. *In*: PARKER, J. R. (ed.) *Petroleum Geology of Northwest Europe: Proceedings of the 4th Conference.* Geological Society, London, 123–133.

NEWTON, S. K. & FLANAGAN, K. P. 1993. The Alba Field: evolution of the depositional model. *In*: PARKER, J. R. (ed.) *Petroleum Geology of Northwest Europe: Proceedings of the 4th Conference.* Geological Society, London, 161–171.

O'CONNOR, S. J. & WALKER, D. 1993. Paleocene reservoirs of the Everest trend. *In*: PARKER, J. R. (ed.) *Petroleum Geology of Northwest Europe: Proceedings of the 4th Conference.* Geological Society, London, 145–160.

PARKER, J. R. 1975. Lower Tertiary sand development in the central North Sea. *In*: WOODLAND, A. W. (ed.) *Petroleum and the Continental Shelf of North-west Europe 1. Geology.* Elsevier, Barking, 447–453.

PARSLEY, A. J. 1990. North Sea hydrocarbon plays. *In*: GLENNIE, K. W. (ed.) *Introduction to the Petroleum Geology of the North Sea*, 3rd edition, Blackwell, Oxford, 362–388.

POSAMENTIER, H. W. & VAIL, P. R. 1988. Eustatic controls on clastic deposition II – sequence and systems tract models. *In*: WILGUS, C. K. *et al.* (eds) *Sea-level Changes: An Integrated Approach.* Society of Economic Palaeontologists & Mineralogists, Special Publication **42**, 125–154.

RUNDBERG, Y. 1989. *Tertiary sedimentary history and basin evolution of the Northern North Sea*. Dr. Ing. Thesis, Trondheim, Norway.

RYE-LARSEN, M. 1994. The Balder Field: refined reservoir interpretation with the aid of high resolution seismic data and seismic attribute mapping. *In*: AASEN, J. O. *et al.* (eds) *North Sea Oil and Gas Reservoirs III*. Kluwer, Dordrecht, 115–124.

SHANMUGAM, G., BLOCH, R. B., MITCHELL, S. M. *et al.* 1995. Basin-floor fans in the North Sea: Sequence stratigraphic models vs. sedimentary facies. *AAPG Bulletin*, **79**, 477–512.

STEWART, I. J. 1987. A revised stratigraphic interpretation of the Early Palaeogene of the central North Sea. *In*: BROOKS, J. & GLENNIE, K. W. (eds) *Petroleum Geology of North-West Europe*. Graham & Trotman, London, 557–576.

VERSCHUREN, M. 1992. *An integrated 3D approach to clay tectonic deformation*. PhD thesis, Ghent University.

Index